T0207105

Lecture Notes in Computer Science   14509

The series Lecture Notes in Computer Science (LNCS), including its subseries Lecture Notes in Artificial Intelligence (LNAI) and Lecture Notes in Bioinformatics (LNBI), has established itself as a medium for the publication of new developments in computer science and information technology research, teaching, and education.

LNCS enjoys close cooperation with the computer science R & D community, the series counts many renowned academics among its volume editors and paper authors, and collaborates with prestigious societies. Its mission is to serve this international community by providing an invaluable service, mainly focused on the publication of conference and workshop proceedings and postproceedings. LNCS commenced publication in 1973.

Jaideep Vaidya · Moncef Gabbouj · Jin Li
Editors

Artificial Intelligence Security and Privacy

First International Conference
on Artificial Intelligence Security and Privacy, AIS&P 2023
Guangzhou, China, December 3–5, 2023
Proceedings, Part I

 Springer

Editors
Jaideep Vaidya
Rutgers University
Newark, NJ, USA

Moncef Gabbouj
Tampere University
Tampere, Finland

Jin Li
Guangzhou University
Guangzhou, China

ISSN 0302-9743   ISSN 1611-3349 (electronic)
Lecture Notes in Computer Science
ISBN 978-981-99-9784-8   ISBN 978-981-99-9785-5 (eBook)
https://doi.org/10.1007/978-981-99-9785-5

This Springer imprint is published by the registered company Springer Nature Singapore Pte Ltd.
The registered company address is: 152 Beach Road, #21-01/04 Gateway East, Singapore 189721, Singapore

Paper in this product is recyclable.

Preface

The first International Conference on Artificial Intelligence Security and Privacy (AIS&P 2023) was held in Guangzhou, China during December 3–5, 2023. AIS&P serves as an international conferences for researchers to exchange the latest research progress in all areas such as artificial intelligence, security and privacy, and their applications. This volume contains papers presented at AIS&P 2023.

The conference received 115 submissions. The committee accepted 40 regular papers and 23 workshop papers to be included in the conference program. Every paper received 2 or 3 Single-blind reviews. These proceedings contain revised versions of the accepted papers. While revisions were expected to take the referees' comments into account, this was not enforced and the authors bear full responsibility for the content of their papers.

AIS&P 2023 was organized by Huangpu Research School of Guangzhou University. The conference would not have been such a success without the support of these organizations, and we sincerely thank them for their continued assistance and support.

We would also like to thank the authors who submitted their papers to AIS&P 2023, and the conference attendees for their interest and support. We thank the Organizing Committee for their time and effort dedicated to arranging the conference. This allowed us to focus on the paper selection and deal with the scientific program. We thank the Program Committee members and the external reviewers for their hard work in reviewing the submissions; the conference would not have been possible without their expert reviews. Finally, we thank the EasyChair system and its operators, for making the entire process of managing the conference convenient.

November 2023

Chunsheng Yang
Haibo Hu
Changyu Dong

Organization

General Chairs

Chunsheng Yang   National Research Council, Canada
Haibo Hu   Hong Kong Polytechnic University, China
Changyu Dong   Guangzhou University, China

Program Chairs

Jaideep Vaidya   Rutgers University, USA
Moncef Gabbouj   Tampere University, Finland
Jin Li   Guangzhou University, China

Track Chairs

Muhammad Khurram Khan   King Saud University, Saudi Arabia
Yun Peng   Guangzhou University, China
Kwangjo Kim   KAIST, South Korea
Shaowei Wang   Guangzhou University, China

Publication Chairs

Weizhi Meng   Technical University of Denmark, Denmark
Francesco Palmieri   University of Salerno, Italy

Publicity Chairs

Hongyang Yan   Guangzhou University, China
Yu Wang   Guangzhou University, China

Steering Committee

Albert Zomaya   University of Sydney, Australia
Jaideep Vaidya   Rutgers University, USA
Moncef Gabbouj   Tampere University, Finland
Jin Li   Guangzhou University, China

Contents – Part I

Contents – Part II

Fine-Grained Searchable Encryption Scheme Against Keyword Brute-Force Attacks

Yawen Feng[ID], Shengke Zeng[✉][ID], Jixiang Xiao, Shuai Cheng,
and Fengchun Zhang

School of Computer Science and Technology, Xihua University, Chengdu, China
{feisongan,zengsk,xiaojixiang,chengshuai}@stu.xhu.edu.cn

Abstract. The inherent security threat of public key encryption with keyword search (PEKS) is the inside guessing attack since the ciphertext of keyword is generated publicly. Sever-aided schemes and keyword search with authenticated encryption schemes are proposed to resist inside keyword guessing attack in PEKS. Unfortunately, these solutions have limitations due to the security and privacy. To overcome these weakness, we propose an encrypted keyword search with fine-grained access control. The access policy of our scheme is semi-hidden in order to prevent the behavior of online keyword guessing attacks. At the same time, our scheme achieves offline keyword guessing attacks resistant by the private generation of ciphertext (without using the private key). Security proofs and experiment results show that our solution is feasible in terms of security and performance.

Keywords: privacy protection · fine-grained access control · semi-hidden access policy · public encryption with keyword search · keyword guessing attack

1 Introduction

Data sharing and outsourcing are popular with the development of cloud computing. However, data leakage threatens the user security and privacy. Traditional encryption is a direct approach to ensure the data confidentiality. However, it limits the data usage. Song *et al.* [1] proposed searchable encryption to make the encrypted data searchable without decryption. Song's searchable encryption is based on symmetric encryption thus it cannot be applied to the scenario that the data receiver wants to share confidential data with owner. Boneh *et al.* [2] proposed a public key encryption with keyword search (PEKS) scheme to handle this problem. Nevertheless, it incurs insider guessing attacks due to the limited keyword space and public encryption.

On the other hand, attribute-based encryption (ABE) [3] brings fine-grained access control to encrypted data. Combining ABE with PEKS allows to search encrypted data with fine-grained access control. The attribute-based keyword

J. Vaidya et al. (Eds.): AIS&P 2023, LNCS 14509, pp. 1–15, 2024.
https://doi.org/10.1007/978-981-99-9785-5_1

search (ABKS) [4–12] access policies based on the attributes of the data user, The data owner uploads the access policy and keyword index ciphertext to the server. Because the attribute matrix of the access policy is exposed to the server, the server can choose its own secret values for key distribution. This results in generating a valid keyword index ciphertext with the received trapdoor for keyword testing. The above operation will cause keyword leakage.

It is found that the reason for the existence of keyword guessing attacks in the ABKS scheme is the leakage of access policy, so the privacy of access policy in the scheme must be protected. We think about the following two ways: the first method is full-hidden access policy, but most of the full-hidden access policy schemes are based on the "AND" gate to realize the resistance to keyword guessing attacks, which has the problem of restricted attribute expression and large computation; the second method is semi-hidden access policy to realize the resistance to keyword guessing attacks, which separates the attribute value and the attribute name. The data owner sends the ciphertext to the server after hiding the attribute value, and any third party such as the server cannot forge the keyword ciphertext index and ciphertext components based on the exposed access policy, in which the semi-hidden access policy has the advantage of richer attribute expression and higher computation rate than the fully hidden access policy. The scheme is based on the idea of a semi-hidden access policy, which uses the data owner's secret value to protect the attribute value. Meanwhile, the authorization center assists the keyword trap to hide the attribute values to achieve the secure delivery of the trap over the public channel.

1.1   Related Works

Since PEKS schemes are faced with guessing attacks, then some scholars proposed dual-server [13] and authenticated encryption [14] to resist keyword online guessing attacks. However, the dual-server-based solution may meet two servers colluding, which can expose data privacy. The authentication-based solution uses the data owner's own private key in constructing the keyword index ciphertext, which can lead to the exposure of the privacy of the data owner's identity. This is a serious violation of user privacy in special application scenarios such as e-healthcare.

In order to adapt to a wider range of application scenarios, ABKS schemes such as distributed, malicious user traceability, and multi-data owners are proposed. Scheme [4] considered multiple data owner authentication of files as the research background, which not only realized hidden access policy, but also allowed tracking of malicious users. Scheme [5] proposed attribute-based keyword search encryption for secure multi-authority, which prevented single points of failure and protected data privacy. Scheme [7] achieved verifiable query results while protecting the privacy of access policies. However, the above schemes are also suffer from keyword guessing attacks as its limited keyword space. In order to solve such attacks, the indistinguishability of ciphertexts and the indistinguishability of trapdoors are proposed. However, scheme [8, 11] can only ensure the privacy of the keyword index ciphertext while exposing the privacy of the keywords in the trapdoor. An attacker can execute a bilinear pair operation by exhausting keywords based on the system public key and the trapdoor component generated by the data user itself, and this operation leads to keyword leakage. Some schemes [5,6,12] can only guarantee the privacy of the trapdoor but expose the keywords in the keyword-indexed ciphertext. The attacker can also get the keyword information by running a bilinear pair operation based on the system public key and the ciphertext components generated by the data owner itself to extract the keywords. The threat of any of the above attacks remains unsafe for the entire scheme. The scheme [9] proves the privacy of the ciphertext and the privacy of the trapdoor. The scheme is guaranteed to be secure from guessing attacks by offline attackers. However, the scheme suffers from online keyword-guessing attacks. Since M is visible to the server, which can generate its own random secret value and distribute the key based on M, then generate a valid keyword indexed ciphertext to match with the received keyword trapdoor. This operation results in keyword leakage.

After research, it is found that the following two methods are realized to resist keyword attack. First, using the private key of the data owner, the authentication method is used to resist the keyword guessing attack. Second, correlate keywords with attributes and full hidden access policy.

In 2021, Miao et al. [4] achieved to resist online keyword guessing attacks. The scheme is based on an authentication method that requires authorization from the data owner for data access interactions. And if the data user sends authorization information to other people, those people will know the identity of the data owner (knowing only the authorization information is not enough to match the keyword index and decrypt the ciphertext). Chaudhari et al. [15] proposed an access policy full hiding approach to prevent the server from forging a valid keyword-indexed ciphertext based on the privacy of the ciphertext and the privacy of the trapdoor achieved. However, the access policy of this scheme is based on the "AND" calculation of multiple attributes at the same time, and the secret value obtained from the polynomial calculation of multiple attribute values for each attribute, and the attribute expression is limited [16,17]. In addition, the data owner needs to interact with the authorized authority when constructing the keyword index ciphertext to secure the keywords based on the master key. In 2022, chaudhar et al. [18] proposed an attribute-based keyword search scheme for hiding attribute values. Although the scheme hides the access structure and effectively improves the computational efficiency compared to the scheme [15], the keywords and attributes are separated. It is not possible to protect the privacy of keywords in trapdoors. Liu et al. [19] proposed a multi-valued attribute structure and multi-keyword access control scheme based on "AND" gate, which achieves the privacy of trapdoors, but the existing ciphertexts lead to the leakage of keywords in the keyword-indexed ciphertexts. The scheme [7] proposed by Niu is based on an attribute tree construction, which enriches the expression of attributes in the access policy. Although the scheme states hidden access structure, it exists attribute value guessing attack and also suffers from the same problem of keyword leakage due to existing ciphertexts.

1.2   Contributions

To overcome these limitations, we propose a fine-grained searchable encryption scheme with a semi-hidden access policy, which not only supports fine-grained access control in PEKS but also prevents the keyword guessing by the semi-hidden access policy. Specifically, the access policy in the ciphertext component is divided into an attribute name part and an attribute value part. To prevent keyword online guessing attacks, the attribute value components are unknown to third parties such as servers. It is impossible for a server to forge a valid keyword ciphertext index without knowing the full access policy. Thus, it is guaranteed that the authenticity of the keyword index ciphertext origin is not forged by any other attacker. Our contributions are shown as follows:

- Prevent keyword guessing attacks. In this paper, we construct a semi-hidden access policy to solve the keyword online guessing attack in fine-grained searchable encryption. Moreover, keyword offline guessing attack is prevented by using the secret value of the access policy.
- Allow the trapdoors passed publicky. The keyword trapdoor is generated based on the key returned to the data user by the authorized authority, ensuring the privacy of the keyword. It enables trapdoors to can be passed over a public channel without exposing any bottom plaintext information.
- A security model of fine-grained searchable encryption with semi-hidden access policies is proposed. It not only proves the privacy of the ciphertext and the privacy of the trapdoor, but also proves that the server cannot forge a valid keyword-indexed ciphertext for matching.

1.3   Organization of the Paper

The organization of this paper is as follows: Sect. 2 is the basic techniques; Sect. 3 proposes the definition of scheme and security model; Sect. 4 describes our scheme and security proof; Sect. 5 presents the performance analysis; Sect. 6 is conclusion.

2   Preliminaries

2.1   Bilinear Mapping

Let $\mathcal{G}, \mathcal{G}_T$ be multiplicative cyclic groups of prime order q; g is a generator of $\mathcal{G}$, and $g_1, g_2 \in \mathcal{G}$. Let $\hat{e} : \mathcal{G} \times \mathcal{G} \rightarrow \mathcal{G}_T$ be defined as follow [15]:

- Bilinear: $\forall a, b \in Z_q, \hat{e}(g_1^a, g_2^b) = e(g_1, g_2)^{ab}$.
- Non degeneracy: $\hat{e}(g_1, g_2) \neq 1$.
- Computability: $\hat{e}(g_1, g_2)$ can be computed efficiently.

2.2 Access Structure

Let $A = \{A_1, A_2, \cdots A_n\}$ be a set of attributes [20]. $\mathcal{A} \subseteq 2^{\{A_1, A_2, \cdots A_n\}}$ is monotone for any subset B, C: if $B \in \mathcal{A}$ and $B \subseteq C$, then $C \in \mathcal{A}$, and $\mathcal{A}$ is called access structure. If the set $D \in \mathcal{A}$, D is called authorized set; otherwise it is called a non-authorized set.

2.3 Complexity Assumption

Definition 1. (Bilinear Decisional Diffie-Hellman Problem): Given $a, b, c, d \in Z_q^*$ and (g, g^a, g^b, g^c, g^d), there is no probabilistic polynomial time (PPT) algorithm $\mathcal{B}$ can distinguish the tuple $(g, g^a, g^b, g^c, g^{abc})$ and the tuple (g, g^a, g^b, g^c, g^d) by a non-negligible advantage ε.

$$\left| Pr\left[g, g^a, g^b, g^c, g^{abc}\right] - Pr\left[g, g^a, g^b, g^c, g^d\right] \right| < \varepsilon$$

Definition 2. (Decision Bilinear Diffie-Hellman Problem) [21]: Given $a, b, c \in Z_q^*$ and $(g, \hat{e}, g^a, g^b, g^c, Z)$, where $Z \in \mathcal{G}_T$. There is no PPT algorithm $\mathcal{B}$ can distinguish the tuple $\left(g, g^a, g^b, g^c, \hat{e}(g, g)^{abc}\right)$ and the tuple (g, y^a, g^b, g^c, Z) by a non-negligible advantage ε.

$$\left| Pr\left[g, g^a, g^b, g^c, \hat{e}(g, g)^{abc}\right] - Pr\left[g, g^a, g^b, g^c, Z\right] \right| < \varepsilon$$

3 System Solutions

3.1 System Model

The system architecture of this paper is shown in Fig. 1, in which the solid line represents the secure channel and the dotted line represents the open channel. In our scenario, even if the trapdoor is passed over the open channel, no keyword information is given away to the adversary. The solution consists of 4 main entities: the Authorised Centre (AC), the Cloud Server (CS), the Data Owner (DO), and the Data Users (DU). The main responsibilities of each entity are as follows:

- AC is considered to be a trusted entity, it is mainly responsible for initializing the system machinery;
- DO encrypts the file and keywords using the designated access policy to generate the cipher component and uploads the cipher to the CS;
- DU interacts with the AC to obtain the authorization key, when the DU wants to access the file, the trapdoor generated by the authorization key is presented to the CS;
- The main task of CS is to store the cipher of DO and match the keyword-indexed ciphertext with the trapdoor.

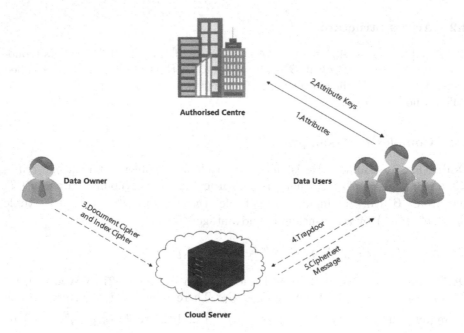

Fig. 1. System Model

3.2 Definition of Scheme

We describe the notations used in scheme construction in Table 1. The scheme consists of 6 algorithms named *Setup, KeyGen, Enc, Trap, Search* and *Dec*, which are defined as the following.

(1) **Setup**$(\kappa) \rightarrow (PP, MSK)$: AC executes the algorithm. Given a security parameter κ, this will generate the public parameters PP and the master secret key MSK, where MSK is owned by AC.

(2) **KeyGen**$(PP, MSK, \mathcal{U}) \rightarrow (SK_u)$: DU sends a request to the AC, and AC executes the algorithm. Take the public parameters PP, the master secret key MSK and DU's attribute set $\mathcal{U}$ as input, AC computes a decryption key SK_u and sends it to the corresponding DU over a secret channel.

(3) **Enc**$(PP, F, \mathcal{L}, w) \rightarrow (C, I_w)$: DO executes the algorithm. The public parameters PP, the data file F, the keyword w, and the access structure $\mathcal{L}$ as input, the ciphertext components C and index components I_w are output.

(4) **Trap**$(PP, w', SK_u) \rightarrow (T)$: DU executes the algorithm. Output the search token T by taking the private key SK_u of DU and the search keyword set w' as input, then send it to CS.

(5) **Search**$(PP, T, C, I_w) \rightarrow (C')$: CS executes the algorithm. After obtaining the token T, CS first matches it with the index I_w. If the token T is valid, then the relevant search results C' is returned to DU, otherwise, "*Invalid*" is returned.

Table 1. Notation Definitions

Notions	Descriptions
PP	System param
MSK	Master key
$\mathcal{L}$	Access structure
$\mathcal{U} = \{I_U, U\}$	Attribute set of DU
$U = \{Att_1, Att_2, \ldots Att_n\}$	Attribute values of DU
SK_u	Decryption key of DU
F	File
C	Ciphertext Components
w	Keyword
I_w	Index for keyword w
T	Trapdoor for queried keyword w'

(6) **Dec**$(PP, C', SK_u) \rightarrow (F)$: DU executes the algorithm. The search results C' and the decryption key SK_u as input, the plaintext file F is output.

3.3   Security Model

The security of this scheme is based on the privacy of the ciphertext and the privacy of the trapdoor to resist keyword guessing attacks.

Trapdoor Privacy. The privacy of trapdoors in the scheme is described by a game between an attacker $\mathbb{A}$ and a challenger $\mathbb{C}$. Privacy of the trapdoor is proved if the algorithm without PPT has a non-negligible advantage in winning the following game. It is described as follows.

- **Setup:** The attacker $\mathbb{A}$ chooses an access policy $\mathcal{L}$ and the security parameter κ itself and sends them to the challenge $\mathbb{C}$. The $\mathbb{C}$ for system initialization, then generates the system public key PP and the system master key MSK, where PP is sent to $\mathbb{A}$ and $\mathbb{C}$ stores MSK itself.
- **Phase1:** Allow the $\mathbb{A}$ to send queries to the following oracles for polynomially multiple times adaptively.
  - Trapdoor Oracle $\mathcal{O}_T$: The $\mathbb{C}$ creates an empty keyword list L_{tw}. $\mathbb{A}$ sends keyword w and the set of attribute values Att to $\mathbb{C}$ for query. The $\mathbb{C}$ computes the keyword trapdoor T_w based on **Trap**(PP, w, SK_u), where SK_u is generated by **KeyGen**$(PP, MSK, \mathcal{U})$. Then $\mathbb{C}$ records it in the L_{tw} and returns T_w to $\mathbb{A}$.
- **Challenge:** The $\mathbb{A}$ sends two keywords w_0^* and w_1^* with equal length to the $\mathbb{C}$, where they are not queried before. After the $\mathbb{C}$ randomly chooses a keyword w_b^* to compute, which $b \in \{0, 1\}$. And return the encrypted keyword trapdoor $T_{w_b^*}$ to $\mathbb{A}$.

- **Phase2:** This phase is a repeat of the Phase1. Among (w_0^*, w_1^*) cannot be asked.
- **Guess:** The $\mathbb{A}$ outputs a guess b'. If $b' = b$, the $\mathbb{A}$ wins the game. Otherwise the $\mathbb{A}$ fails. The $\mathbb{A}$ advantage of the winning game is defined as follows:

$$Adv\,(1^\kappa) = \left| Pr\left[b' = b\right] - \frac{1}{2} \right|$$

Ciphertext Privacy. The privacy of the ciphertext in the scheme is described by a game between an attacker $\mathbb{A}$ and a challenger $\mathbb{C}$. The privacy of the ciphertext is proved if the algorithm without PPT has a non-negligible advantage in winning the following game. It is described as follows.

- **Setup:** The operation executed in this phase is the same as **Setup** in Trapdoor Privacy.
- **Phase1:** Allow the $\mathbb{A}$ to send queries to the following oracles for polynomially multiple times adaptively.
    - Ciphertext Oracle $\mathcal{O}_C$: The $\mathbb{C}$ creates an empty keyword list L_{cw}. $\mathbb{A}$ sends keyword w to $\mathbb{C}$ for query. The $\mathbb{C}$ computes the keyword indexing ciphertext I_w based on **Enc**$(PP, \mathcal{L}, w)$, then records it in the L_{cw} and returns I_w to $\mathbb{A}$.
- **Challenge:** The $\mathbb{A}$ sends two keywords w_0^* and w_1^* with equal length to the $\mathbb{C}$, where they are not queried before. After the $\mathbb{C}$ randomly chooses a keyword w_b^* to compute, which $b \in \{0, 1\}$. And return the encrypted keyword indexing ciphertext I_w to $\mathbb{A}$.
- **Phase2:** This phase is a repeat of the Phase1. Among (w_0^*, w_1^*) cannot be asked.
- **Guess:** The $\mathbb{A}$ outputs a guess b'. If $b' = b$, the $\mathbb{A}$ wins the game. Otherwise the $\mathbb{A}$ fails. The $\mathbb{A}$ advantage of the winning game is defined as follows:

$$Adv\,(1^\kappa) = \left| Pr\left[b' = b\right] - \frac{1}{2} \right|$$

4   Detailed Construction of the Scheme

4.1   Algorithm Description

This section describes the steps of the scheme implementation in detail. We introduce the partial hidden access policy of LSSS in generating the ciphertext component to prevent keyword online guessing attacks and implement ciphertext privacy and trapdoor privacy to prevent keyword offline guessing attacks. Due to the privacy of the trapdoor and attribute value hiding effects, the trapdoor cannot reveal any underlying plaintext information even when passed over a public channel.

Setup (κ): Authorised Centre chooses two multiplicative cyclic groups $\mathcal{G}$ and $\mathcal{G}_T$ separately. Let $\mathcal{G}, \mathcal{G}_T$ be a bilinear group of order prime q, where the length of q is determined by the security parameter κ. g is the generator of group $\mathcal{G}$.

- Define one bilinear mapping $\hat{e} : \mathcal{G} \times \mathcal{G} \to \mathcal{G}_T$.
- Randomly pick $\alpha, \beta, \delta \in \mathcal{Z}_q$ and $h, u \in \mathcal{G}$, then calculate relevant parameters $\hat{e}(g,g)$, $\hat{e}(h,g)^\alpha$, $\hat{e}(u,g)^\beta$, h^δ, u^δ.
- Select a cryptographic hash function $H : \{0,1\}^* \to \mathcal{G}$.

Authorised Centre returns the master key $MSK = (\alpha, \beta, \delta)$ and the system param $PP = (q, \mathcal{G}, \mathcal{G}_T, \hat{e}, g, h, u, H, \hat{e}(g,g), \hat{e}(h,g)^\alpha, \hat{e}(u,g)^\beta, h^\delta, u^\delta)$.

KeyGen ($PP, MSK, \mathcal{U}$): The data user's attribute is $\mathcal{U} = \{I_U, U\}$, where I_U denotes the attribute name and U denotes the attribute value.

- The data user randomly selects an $\gamma \in Z_q^*$, then sends $u^{\frac{1}{\gamma}}$ and attribute $\mathcal{U}$ to the authorisation centre over a secure channel.
- The authorisation centre randomly selects $r \in Z_q^*$, generates a corresponding attribute key $D = u^{\frac{\beta}{\gamma}} h^\alpha, D_1 = g^{\frac{\alpha-r}{\delta}}, D_{2i} = g^{rAtt_i} u^{r-\alpha}, D_3 = h^r$ for each attribute value $Att_i \in U (i \in [1, 2, ..., n])$ of the data user, and returns $SK_u = (D, D_1, D_2, D_3)$ to the data user over the secure channel.

Enc ($PP, F, \mathcal{L}, w$): The data owner selects a symmetric key $k \in Z_q^*$ to encrypt the file F with the symmetric encryption algorithm (Enc, Dec) to generate the ciphertext $E = Enc(k, F)$.

- The data owner designates the access policy $\mathcal{L} = (M, \rho, \mathcal{T})$, where M is a matrix of $l \times n$, ρ maps each row of the matrix to an attribute name and $\mathcal{T} = \{t_{\rho_i}\}_{i \in 1,2...,l}$ to be the attribute value related with (M, ρ). A secret value s is randomly selected as the first value of vector $\boldsymbol{\nu}$, and then $n - 1$ values $\nu_2, ..., \nu_n$ are randomly selected to be added to vector $\boldsymbol{\nu}$, where $s, \nu_2, ..., \nu_n \in Z_{q,}$. Then $\lambda_i = M_i \cdot \boldsymbol{\nu}$ is calculated and the secret value s is shared to the attribute name M_i.
- The symmetric key k and the keyword w extracted from the F are encrypted using $\mathcal{L}$ to generate the ciphertext component $C = (C_1, C_2, C_{i,1}, C_{i,2}, C_{i,3})$ and the keyword indexed ciphertext component I_w, where the attribute name M_i and the attribute value t_{ρ_i} in the access policy generate separate ciphertext components, then the attribute value is hidden using the key-blinding technique.
- For each attribute $i \in [1, 2, ..., n]$ in the access policy $\mathcal{L}$, the data owner randomly selects the parameter $t_i \in Z_q^*$ and computes the ciphertext component $C_1 = k \cdot \hat{e}(g,g)^{s\alpha}, C_2 = g^s, C_{i,1} = h^{\delta\lambda_i} u^{\delta t_i}, C_{i,2} = g^{-t_i t_{\rho_i} + \lambda_i}, C_{i,3} = g^{t_i}$ and the keyword-indexed ciphertext component $I_w = (\hat{e}(g,g)^{H(w)} \hat{e}(u,g)^\beta)^s$.

Finally, the data owner uploads the ciphertext $(E, C, I_w, \bar{\mathcal{L}})$ to the server, where $\bar{\mathcal{L}} = (M, \rho)$ denotes an access policy without attribute values.

Trap (PP, w', SK_u): The data user enters the random number γ, the keyword w' and their own attribute key SK_u to calculate the trapdoor T, where

$$T = \left(T_w = g^{H(w')}D^\gamma, T_1 = D_1^\gamma, \{T_{2i} = D_{2i}^\gamma\}_{i \in [1,2,...,n]}, T_3 = D_3^\gamma\right).$$ Then T is sent to the cloud server.

   Search (PP, T, C, I_w): After CS receives the search trapdoor T from DU. Let $I = \{i : \rho(i) \in U\}$ $(i \subseteq 1, 2, ..., l)$, where $U \in \mathcal{A}$. There are coefficients $\{c_i \mid i \in I\}$ so that $\sum_{i \in I} c_i M_i = (1, 0, 0, ..., 0)$. Then we have $\sum_{i \in I} c_i \lambda_i = s$.

- CS checks if the attributes in the trapdoor T satisfy the access policy in the ciphertext uploaded by DO by computing $R_i = \hat{e}(T_1, C_{i,1})\hat{e}(T_3, C_{i,2})\hat{e}(T_{2i}, C_{i,3}) = \hat{e}(g, h)^{\alpha \gamma \lambda_i}$ and $R = \sum_{i \in I} R_i^{c_i} = \hat{e}(g, h)^{\alpha \gamma s}$.
- If the attributes do not satisfy the access policy, the search is stopped; otherwise, the following equation $I_w = \frac{\hat{e}(T_w, C_2)}{R}$ is executed.

   The server checks the keywords in the indexed cipher I_w for consistency with the keywords in the trapdoor T_w, if the match is successful, the server returns 1 and returns $C' = (E, C_1, R)$ to the data user, otherwise 0.

   Dec (PP, C', SK_u): The DU based on the results returned by the CS, computes $k = \frac{C_1}{R^{\frac{1}{\gamma}}}$, thus decrypting the file $F = Dec(k, E)$.

4.2   Correctness

In this section, we will show the correctness of the formulas above. We can verify that the keywords are the same and that the attributes satisfy the access policy by using the following formula.

$$\begin{aligned}
R_i &= \hat{e}(T_1, C_{i,1})\hat{e}(T_3, C_{i,2})\hat{e}(T_{2i}, C_{i,3}) \\
&= \hat{e}(D_1^\gamma, h^{\delta \lambda_i}u^{\delta t_i})\hat{e}(D_3^\gamma, g^{-t_i t_{\rho_i} + \lambda_i})\hat{e}(D_{2i}^\gamma, g^{t_i}) \\
&= \hat{e}(g, h)^{\alpha \gamma \lambda_i}
\end{aligned}$$

Next, the algorithm computes:

$$\begin{aligned}
\frac{\hat{e}(T_w, C_2)}{R} &= \frac{\hat{e}(g^{H(w')}D^\gamma, g^s)}{\hat{e}(g, h)^{\alpha \gamma s}} \\
&= \frac{\hat{e}(g^{H(w')}u^\beta h^{\alpha \gamma}, g^s)}{\hat{e}(g, h)^{\alpha \gamma s}} \\
&= \hat{e}(g^{H(w')}u^\beta, g^s) \\
&= I_w
\end{aligned}$$

4.3   Security Proof

Due to the space limitation, we give the sketch of the security proofs only here. The readers can refer to our full version for the concrete proof steps.

Theorem 1. *The trapdoor privacy is preserved under BDDH assumption.*

Proof. Assume $\mathbb{A}$ is an attacker to break the keyword privacy with non-negligible advantage. We construct a challenger $\mathbb{C}$ to break the $BDDH$ assumption. Technically, $\mathbb{C}$ is given a challenge (g, g^a, g^b, g^c, g^d) where $a, b, c, d \in Z_q^*$, its goal is to distinguish $g^{abc} = g^d$ or g^d is random in $\mathcal{G}$. In our simulation, $\mathbb{C}$ should return the corresponding keyword trapdoor about the queries on some keyword kw (issued by $\mathbb{A}$). Then, $\mathbb{A}$ sends two keywords kw_0, kw_1 to $\mathbb{C}$ as challenge. $\mathbb{C}$ randomly chooses $\tau \in \{0,1\}$ and returns the target keyword trapdoor $T_{kw_\tau} = g^{H(w)}(u^\beta g^d), T_1 = g^{\gamma \frac{\alpha - r}{\delta}}, T_{2i} = g^{\gamma r Att_i} g^{a(r-\alpha)}, T_3 = h^{\gamma r}$, where $h^{\frac{1}{\gamma}} = g^{\frac{1}{a}}$, $g^\alpha = g^b$ and $u^\beta = g^c$. Certainly, $\mathbb{A}$ can continue quering the encryption of kw' except $kw' = kw_\tau$. Finally, $\mathbb{A}$ guesses the value of τ. We can see that $\mathbb{A}$'s output helps $\mathbb{C}$ to break $BDDH$ assumption indeed. If g^d is a random value in $\mathcal{G}$, T_{kw_τ} is random regardless of kw_τ. In this case, $\mathbb{A}$ has no advantage to guess bit τ in kw_τ. If $g^d = g^{abc}$, it is the real environment of our scheme. Therefore, if $\mathbb{A}$ has non-negligible advantage to guess the right τ, it implies that $g^d = g^{abc}$ which violates the $BDDH$ assumption.

Theorem 2. *The ciphertext privacy is preserved under DBDH assumption.*

Proof. Let us assume $\mathbb{A}$ is an attacker who can break the ciphertext privacy with non-negligible advantage. We construct a challenger $\mathbb{C}$ to break the $DBDH$ assumption. Technically, $\mathbb{C}$ is given a challenge (g, g^a, g^b, g^c, Z) where $Z \in \mathcal{G}_T$, its goal is to distinguish $Z = \hat{e}(g^{ac}, g^b)$ or Z is random in $\mathcal{G}_T$. In our simulation, $\mathbb{C}$ should return the corresponding keyword ciphertext about the queries on some keyword kw (issued by $\mathbb{A}$). $\mathbb{A}$ sends the keyword kw as a challenge to $\mathbb{C}$. $\mathbb{C}$ randomly chooses $\tau \in \{0,1\}$ and returns the keyword ciphertext $I_{kw_\tau} = \hat{e}(g,g)^{cH(kw_\tau)}\hat{e}(g^{ac}, g^b)$, where $u^\beta = g^{ab}$ and $g^s = g^c$. Certainly, $\mathbb{A}$ can continue quering the encryption of kw' except the challenge. Finally, $\mathbb{A}$ guesses the value of τ. $\mathbb{C}$ can solve the $DBDH$ problem with the help of $\mathbb{A}$ output. If Z is a random value in $\mathcal{G}_T$, I_{kw_τ} is random regardless of kw_τ. Therefore $\mathbb{A}$ has no advantage in guessing the bit τ in I_{kw_τ} in this instance. If $Z = \hat{e}(g^{ac}, g^b)$, it means that $\mathbb{A}$ can guess the value of τ with non-negligible probability. This is contrary to the $DBDH$ assumption.

Table 2. Computation Cost Comparison

Schemes	IndexGen	Trapdoor	Search
Miao et al. [4]	$(2n+2)t_E + t_M$	$(2U+1)t_E$	$(2U+1)t_P + t_E$
Chaudhari et al. [15]	$(n*m+m+1)t_E + t_M$	$(n*m+2)t_E$	$(n*m+1)(t_P+t_M)$
Chaudhari et al. [18]	$2t_P + (4+n)t_E + nt_M$	$4t_E$	$4t_P + 2t_M$
Liu et al. [19]	$t_P + (n+2)t_E + nt_M$	$(n+2)t_E + t_M$	$(n+1)(t_P+t_M)$
Ours	$(4n+3)t_E + (n+1)t_M$	$(3+U)t_E$	$(1+3U)t_P + Ut_E + (1+2U)t_M$

[1] Note.t_P, t_M, t_E denotes the time of pairing operation; multiplication and exponentiation operation in group $\mathcal{G}$; U: Number of possible values of DUs; n: Number of attributes in system; m $= max(|V_i|)_{1 \leq i \leq n}$, where $V_i =$ valueset for attribute i; $|\mathcal{G}|, |\mathcal{G}_T|$ represent the lengths of the elements in $\mathcal{G}$ and $\mathcal{G}_T$ respectively.

Table 3. Communication Cost Comparison

Schemes	Encrypted Index size	Trapdoor size
Miao et al. [4]	$(2n+1)\lvert\mathcal{G}\rvert + \lvert\mathcal{G}_T\rvert$	$(2n+1)\lvert\mathcal{G}\rvert$
Chaudhari et al. [15]	$(n*m+1)\lvert\mathcal{G}\rvert$	$(n*m+2)\lvert\mathcal{G}\rvert$
Chaudhari et al. [18]	$3\lvert\mathcal{G}\rvert + \lvert\mathcal{G}_T\rvert$	$4\lvert\mathcal{G}\rvert$
Liu et al. [19]	$2\lvert\mathcal{G}\rvert + \lvert\mathcal{G}_T\rvert$	$(n+2)\lvert\mathcal{G}\rvert$
Ours	$(3n+1)\lvert\mathcal{G}\rvert + \lvert\mathcal{G}_T\rvert$	$(3+n)\lvert\mathcal{G}\rvert$

[2] Note. $\lvert\mathcal{G}\rvert$, $\lvert\mathcal{G}_T\rvert$ represent the lengths of the elements in $\mathcal{G}$ and $\mathcal{G}_T$ respectively.

5   Performance Analysis

In this section, we compare the performance of our scheme with existing schemes (including [4,15,18,19]) according to the computational cost and size of the different phases. The performance evaluation of the scheme thought experiment.

5.1   Computation and Communication Analysis

The data for the computational cost analysis are shown in Table 2. In the keyword index ciphertext generation phase, our scheme uses $4n+3$ exponential calculations and $n+1$ multiplication operations. Compare the scheme [18] with $2t_P + (4+n)t_E + nt_M$ and the scheme [19] with $t_P + (n+2)t_E + nt_M$. Our scheme is more efficient than scheme [18] and [19]. And it slightly higher than scheme [15] with $(n*m+m+1)t_E + t_M$ while satisfying the privacy of the index and resisting keyword online attacks. In the trapdoor generation phase, our scheme uses exponential operations at the data user attribute level to compare scheme [4] with $(2U+1)t_E$, scheme [15] with $(n*m+2)t_E$ and scheme [19] with $(n+2)t_E + t_M$. There are advantages in this paper. The scheme [18] is efficient, but does not ensure the privacy of the trapdoor and there is a keyword guessing attack. In the keyword search phase, the computation of our scheme is $(1+3U)t_P + Ut_E + (1+2U)t_M$, which is a bit more efficient than the scheme [15] that satisfies both keyword offline guessing attack and keyword online guessing attack.

The data for the communication cost analysis are shown in Table 3. In storage consumption, our index cipher size is $(3n+1)\lvert\mathcal{G}\rvert + \lvert\mathcal{G}_T\rvert$ and the trapdoor size is $(3+n)\lvert\mathcal{G}\rvert$. In general, our scheme has obvious advantages compared to scheme [4,15]. The scheme [19] takes up less storage space, there is a keyword guessing attack of the scheme [19] on keyword indexed ciphertexts. The scheme [18] also takes up less storage space, but the scheme uses multi-valued "AND" gate expressions with a restricted access structure. In addition, there are offline keyword guessing attacks. In our scheme, the access policy includes "AND" gate and "OR" gate, and the attributes are more richly expressed, and the key is to realize offline keyword guessing attack and online keyword guessing attack.

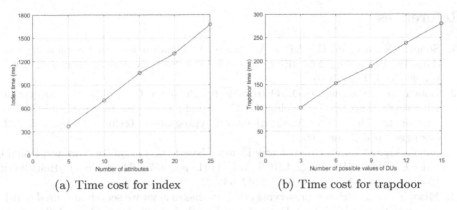

(a) Time cost for index (b) Time cost for trapdoor

Fig. 2. Actual performance

5.2 Experimental Analysis

To prove the effectiveness of our proposed scheme, we performed a data evaluation by using a real data set. We implemented the experiment using JAVA language on a Lenovo AMD A8-6410 APU using AMD Radeon R5 Graphics 2.00 GHz and 8 GB RAM with Windows 10. The JPBC-based cryptographic library was simulated and tested in terms of index encryption time, trapdoor time, search time, decryption time. The results of the experiments are the average of five operations. We set the number of attributes in the access policy to be in the range of 5 to 25. As shown in Fig. 2(a), generation time of keyword index ciphertext tends to increase linearly with the attribute values. We set the number of attribute values for the user in the range of 3 to 15. As shown in Fig. 2(b), the generation time of the trapdoor tends to increase linearly with the increase of the attribute values. Experimental results show that our scheme achieves resistance to keyword guessing attacks while The computational efficiency and storage space are slightly better than other schemes. And the computation results are in the acceptable range.

6 Conclusion

In this paper, we propose a fine-grained searchable encryption scheme with semi-hidden access policy to resist keyword guessing attacks around the security of data. The data owner of this paper can generate the ciphertext component using the parameters generated by himself and the system public key. Moreover, the access policy includes "AND" gate and "OR" gate, and the attribute expression is richer. It better achieves the requirement of resisting online keyword guessing attack and offline keyword guessing attack. Performance and experimental analysis show that there are computational and storage advantages with the keyword security achieved.

References

1. Song, D.X., Wagner, D., Perrig, A.: Practical techniques for searches on encrypted data. In: Proceeding 2000 IEEE Symposium on Security and Privacy. S&P 2000, pp. 44–55. IEEE (2000)
2. Boneh, D., Crescenzo, G.D., Ostrovsky, R., Persiano, G.: Public key encryption with keyword search. In: Cachin, C., Camenisch, J.L. (eds.) International Conference on the Theory and Applications of Cryptographic Techniques, pp. 506–522. Springer, Heidelberg (2004)
3. Sahai, Amit, Waters, Brent: Fuzzy identity-based encryption. In: Cramer, Ronald (ed.) EUROCRYPT 2005. LNCS, vol. 3494, pp. 457–473. Springer, Heidelberg (2005). https://doi.org/10.1007/11426639_27
4. Miao, Y., et al.: Privacy-preserving attribute-based keyword search in shared multi-owner setting. IEEE Trans. Dependable Secure Comput. **18**(3), 1080–1094 (2021)
5. Miao, Y., Deng, R.H., Liu, X., Choo, K.-K.R., Wu, H., Li, H.: Multi-authority attribute-based keyword search over encrypted cloud data. IEEE Trans. Dependable Secure Comput. **18**(4), 1667–1680 (2021)
6. Sun, J., Xiong, H., Nie, X., Zhang, Y., Wu, P.: On the security of privacy-preserving attribute-based keyword search in shared multi-owner setting. IEEE Trans. Dependable Secure Comput. **18**(5), 2518–2519 (2021)
7. Niu, S., Song, M., Fang, L., Yu, F., Han, S., Wang, C.: Keyword search over encrypted cloud data based on blockchain in smart medical applications. Comput. Commun. **192**, 33–47 (2022)
8. Liu, Z., Liu, Y., Fan, Y.: Searchable attribute-based signcryption scheme for electronic personal health record. IEEE Access 6, 76 381–76 394 (2018)
9. Li, H., Jing, T.: A lightweight fine-grained searchable encryption scheme in fog-based healthcare IoT networks. Wireless Commun. Mob. Comput. **2019**, 15 (2019)
10. Zhang, K., Jiang, Z., Ning, J., Huang, X.: Subversion-resistant and consistent attribute-based keyword search for secure cloud storage. IEEE Trans. Inf. Forensics Secur. **17**, 1771–1784 (2022)
11. Yu, J., Liu, S., Xu, M., Guo, H., Zhong, F., Cheng, W.: An efficient revocable and searchable MA-ABE scheme with blockchain assistance for C-IoT. IEEE Internet Things J. **10**(3), 2754–2766 (2023)
12. Huang, Q., Yan, G., Wei, Q.: Attribute-based expressive and ranked keyword search over encrypted documents in cloud computing. IEEE Trans. Serv. Comput. **16**(2), 957–968 (2023)
13. Chen, R., Mu, Y., Yang, G., Guo, F., Wang, X.: Dual-server public-key encryption with keyword search for secure cloud storage. IEEE Trans. Inf. Forensics Secur. **11**(4), 789–798 (2015)
14. Huang, Q., Li, H.: An efficient public-key searchable encryption scheme secure against inside keyword guessing attacks. Inf. Sci. **403**, 1–14 (2017)
15. Chaudhari, P., Das, M.L.: Privacy preserving searchable encryption with fine-grained access control. IEEE Trans. Cloud Comput. **9**(2), 753–762 (2021)
16. Yang, Y., et al.: Dual traceable distributed attribute-based searchable encryption and ownership transfer. IEEE Trans. Cloud Comput. **11**(1), 247–262 (2023)
17. Niu, S., Hu, Y., Su, Y., Yan, S., Zhou, S.: Attribute-based searchable encrypted scheme with edge computing for industrial Internet of Things. J. Syst. Architect. **139**, 102889 (2023)
18. Chaudhari, P.: KeySea: keyword-based search with receiver anonymity in attribute-based searchable encryption. IEEE Trans. Serv. Comput. **15**(2), 1036–1044 (2022)

19. Liu, X., Yang, X., Luo, Y., Zhang, Q.: Verifiable multikeyword search encryption scheme with anonymous key generation for medical internet of things. IEEE Internet of Things J. **9**(22), 22 315–22 326 (2022)
20. Miao, Y., Deng, R.H., Choo, K.-K.R., Liu, X., Ning, J., Li, H.: Optimized verifiable fine-grained keyword search in dynamic multi-owner settings. IEEE Trans. Dependable Secure Comput. **18**(4), 1804–1820 (2021)
21. Miao, Y., Ma, J., Liu, X., Li, X., Jiang, Q., Zhang, J.: Attribute-based keyword search over hierarchical data in cloud computing. IEEE Trans. Serv. Comput. **13**(6), 985–998 (2020)

Fine-Grained Authorized Secure Deduplication with Dynamic Policy

Jixiang Xiao, Shengke Zeng[✉], Yawen Feng, and Shuai Cheng

School of Computer Science and Technology, Xihua University, Chengdu 610000, Sichuan, China
{xiaojixiang,zengsk,feisongan,chengshuai}@stu.xhu.edu.cn

Abstract. Attribute-based encryption (ABE) is a promising technology to provide fine-grained access control for the encrypted data and hence is widely used in the outsourced storage. However, it can not support the secure deduplication for its encryption feature. In order to handle this problem that users with different access policy to the ciphertext of the same plaintext should retrieve this ciphertext normally, we propose a novel ABE scheme supporting fine-grained authorized secure deduplication. Compared with the related works, we consider the dynamic policy update which adapts to the real-world environment more.

Keywords: Attribute-based Encryption · Policy Updating · Deduplication · Hidden Policy

1 Introduction

Cloud computing provides the feasibility of sharing data for users. Considering the sensitive data, encryption is the indispensable measure to support secure cloud storage. However, it raises challenges such as access control and duplicates check for encrypted data.

Attribute-based encryption (ABE) [18] is a special encryption to support fine-grained access control for the encrypted data according to key-policy (KP-ABE) [8] or ciphertext-policy (CP-ABE) [3]. If and only if certain identities of receivers meet the access policy, the plaintext can be successfully decrypted. Since ABE implements more flexible and fine-grained access control policies, many ABE schemes [9,16,20] have been proposed for more robust security, more functionality, and higher efficiency.

Secure deduplication technology [14] can check duplicates for the encrypted data existing in cross users. It hence improves the storage space greatly even for the encrypted data. However, when extending the deduplication functionality of ABE fine-grained access control, it shows some difficulties for data consumers to retrieve data after deduplication. For example, we consider the following situation: Alice and Bob have the same message M that needs to be uploaded to the cloud, however Alice encrypts message M under an access policy $\mathbb{A}$. Bob encrypts message M under access policy $\mathbb{A}'$. Assuming that the cloud removes

J. Vaidya et al. (Eds.): AIS&P 2023, LNCS 14509, pp. 16–32, 2024.
https://doi.org/10.1007/978-981-99-9785-5_2

Bob's duplicate message M, then legitimate users under access policy A' lose the right to access message M. It is also based on considering this problem that many existing deduplication schemes [1,2,5,12] are based on something other than ABE.

1.1 Related Work

In order to provide a more flexible access control mechanism and meet fine-grained data sharing requirements, Sahai and Waters first proposed a concept of attribute-based encryption (ABE) [18]. Then, they implemented the ABE scheme [8] with access control policies support, which is also the first KP-ABE scheme. Compared with KP-ABE, the CP-ABE [3] scheme in which the user's authority depends on its own attributes is more flexible and applicable in practical applications. Given that a clear form of access policy may expose some sensitive information, Song et al. [19] proposed an ABE scheme that hides access policies and verifies user attributes instead. However, this scheme has security risks since attackers may obtain decrypted content directly. Lai et al. [10] proposed a novel ABE model that partially hides the access strategy through attribute key-value pairs to achieve user privacy protection.

The data provider may require changing the authorized user in the ciphertext access policy in the application environment. Goyal et al. [8] and Sahai et al. [17] first discussed the policy update structure. Subsequently, Yang et al. [20] gave specific policy update algorithms under three ABE structures: Boolean expression, access tree and access matrix. The schemes proposed by Liu [13] and Fugkeaw [7] et al. combined multi-authorization agencies and access policy update, and they respectively adopted the structure of LSSS matrix and access tree to implement policy update. However, these solutions are all applied to the scenario of a single data provider.

The main difficulty encountered in the application of ABE in data deduplication technology is how to securely deduplicate the same data with different access policies under the premise of retaining permissions. Lai et al. [11] first proposed the adaptive CP-ABE as a cryptographic primitive. It brings a solution to the problem of applying ABE in data deduplication. On the premise of authorized data deduplication, Cui et al. [6] used the idea of Lai scheme [11] to implement a fine-grained access control scheme that supports deduplication. However, their scheme does not support policy updates. The deduplication scheme that supports fine-grained access control proposed by Premkamal et al. [15] also addresses the problem of different access policies formulated by different data providers. Unfortunately, it still does not deal with the possible follow-up policy update requirements of the data provider.

1.2 Our Contribution

In this paper, we introduce a CP-ABE scheme for fine-grained data deduplication with enhanced user privacy. Our scheme allows dynamic access policy updates, making it a practical choice for data deduplication. Key contributions include:

1. Partially Hidden Policy: Inspired by [10] scheme, we implemented a partially hidden access policy to prevent disclosure of sensitive information. Moreover, our scheme adopts a more flexible (t, n) threshold gate access control structure instead of an AND-OR gate.
2. Dynamic Policy Update: Our scheme achieves correct and efficient dynamic policy updates in the context of deduplication without re-encryption. In Sect. 6, we verify the efficiency of our dynamic policy update in various scenarios through experimental simulations.
3. Data Deduplication: For eliminating duplicate copies of confidential data, we implement an attribute-based secure deduplication. Our scheme supports that the legitimate users covered by different access policies on the same data can retrieve the deleted data.

1.3 Paper Organization

The remaining of the paper is organized as follows. Some preliminaries are introduced in the Sect. 2. In Sect. 3, we provide definitions for our scheme and its security models. We propose a concrete construction of FASD-DP-CPABE in Sect. 4. Section 5 and Sect. 6 respectively provides the security analysis and the experimental results of the proposed scheme. Finally, we summarize our result in Sect. 7.

2 Preliminaries

2.1 Notions

Let G and G_T be two multiplicative cyclic groups of prime order p, and g be a generator of G. A bilinear map $\hat{e}$ is a map $\hat{e} : G \times G \longrightarrow G_T$ with the following properties:

- Bilinearity: for $\forall g_1, g_2 \in G$, and $a, b \in Z_p$, we have $\hat{e}\left(g_1^a, g_2^b\right) = \hat{e}\left(g_1, g_2\right)^{ab}$.
- Non-degeneracy: $\hat{e}\left(g, g\right) \neq 1$, where $g \in G$.
- Computability: For $\forall g_1, g_2 \in G$, there is an efficient algorithm to calculate $\hat{e}\left(g_1, g_2\right)$.

Definition 1. *(Discrete Logarithm(DL) Problem) : The Discrete Logarithm Hardness Assumption is described as follows: For $g \in G, a \in Z_p$, Given a tuple (g, g^a), there is no PPT algorithm $\mathcal{B}$ with a non-negligible advantage ε that can calculate a.*

$$Pr\left[\mathcal{B}\left(g, g^a\right) = a\right] \leq \varepsilon$$

Definition 2. *(Decisional Bilinear Diffie-Hellman (DBDH) Assumption) : $a, b, c, z \in Z_p$ are randomly chosen and g is a generator of G. The decision BDH assumes that there is no polynomial-time algorithm $\mathcal{B}$ capable of distinguishing the given tuples $\left\langle g^a, g^b, g^c, \hat{e}(g, g)^{abc}\right\rangle$ from $\left\langle g^a, g^b, g^c, \hat{e}(g, g)^z\right\rangle$ with non-negligible advantage ε.*

$$\left|Pr\left[g^a, g^b, g^c, e(g, g)^{abc}\right] - Pr\left[g^a, g^b, g^c, \hat{e}(g, g)^z\right]\right| \leq \varepsilon$$

2.2   Definitions

Access Structure. Let $P = \{P_1, P_2, \ldots, P_n\}$ be a set of parties, a collection $\mathbb{A} \subseteq 2^P$ is monotone for $\forall B, C$: if $B \in \mathbb{A}$ and $B \subseteq C$ then $C \in \mathbb{A}$. A monotone access structure [38] is a monotone collection $\mathbb{A}$ of non-empty subsets of P, i.e., $\mathbb{A} \subseteq 2^{\{P_1, P_2, \ldots, P_n\}} \setminus \{\emptyset\}$. The sets in $\mathbb{A}$ are called authorized sets; otherwise, the sets are called unauthorized sets.

Access trees $\mathcal{T}$. Let $\mathcal{T}$ be a tree representing an access structure. Each non-leaf node in the tree represents a threshold gate, which is used to describe its child nodes and a threshold, and the root node also includes a *uid* to identify the user. For convenience, we define *parent* (x) to represent the parent node of node x, *att* (x) to represent the attribute name of leaf node x, and *index* (x) to represent the unique index of node x.

3   System and Security Models

In this section, we introduce the formal definition of authorized secure deduplication with dynamic policy.

3.1   System Model

System model such as Fig. 1 mainly includes 4 entities: Attribute Center (AC), Cloud Service Provider (CSP), Data Owner (DO) and Data User (DU).

- *Attribute Center* (AC). AC is also the authorization center which is fully trusted third party and is responsible for generating system public parameters PP and master private key MSK, and generating decryption keys for users.
- *Cloud Service Provider* (CSP). CSP provides storage and computing services for entities in the system. It is mainly responsible for storing encrypted data uploaded by users, updating access policies, and updating ciphertext operations after receiving the user's updated key. CSP are considered semi-trusted.
- *Data Owner* (DO). DO specifies the access policy, encrypts the message according to the access policy, and then sends the ciphertext to the cloud storage. When the access policy needs to be updated, the DO generates an update key and sends it to the CSP for ciphertext update.
- *Data User* (DU). DU can access the ciphertext on the cloud, and correctly recover the plaintext through the key if and only if its properties satisfy the access policy.

3.2   Formal Definition of Scheme

In order to implement attribute-based encryption with access policy updates in data deduplication, our scheme includes the following 6 algorithms, which are formally defined as follows:

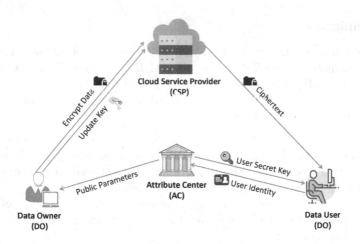

Fig. 1. System Model

- $Setup\,(\lambda) \longrightarrow (PP, MSK)$: AC runs this algorithm and takes the security parameter λ as input to initialize the system. It outputs the public parameter PP and the master private key MSK kept by AC.
- $Encrypt\,(PP, \mathcal{T}, m) \longrightarrow CT$: DO runs this algorithm, with taking public parameter PP, access policy $\mathcal{T}$ and message m as input. It generates ciphertext CT encrypted under access policy $\mathcal{T}$.
- $KeyGen\,(MSK, \mathcal{S}) \longrightarrow sk$: AC runs this algorithm, taking the master private key MSK and the user's attribute key-value pair set $\mathcal{S} = (A_{uid}, V_{uid})$ as input. It outputs the decryption key sk and returns it to the user.
- $UKeyGen\left(uid, Y', Y, UType, \mathcal{T}_{uid}\right) \longrightarrow UK$: DO runs the update key generation algorithm, with taking user identity uid, new leaf node Y', currently leaf node Y, update type $UType$ and access tree $\mathcal{T}_{uid}$ as input. It outputs the update key UK and sends it to the CSP.
- $CTUpdate\,(uid, UK, CT) \longrightarrow CT'$: CSP runs the ciphertext update algorithm, with taking user identity uid, updated key UK and ciphertext CT as input. It outputs the updated ciphertext CT'.
- $Decrypt\,(sk, \mathcal{S}, CT) \longrightarrow m$: DU runs the decryption algorithm, with taking the decryption key sk, the user's attribute key-value pair $\mathcal{S}$, and ciphertext CT as input. If and only if the user attributes satisfy the access policy $\mathcal{T}$, the algorithm outputs the message plaintext m.

3.3   IND-CPA Security Model

Our scheme should be secure against the indistinguishability of the ciphertext policy through a security game between a challenger $\mathcal{C}$ and an adversary $\mathcal{A}$. The challenge-adversary model is as follows:

- **Init.** $\mathcal{A}$ chooses a challenge access strategy $W^* = (\mathcal{T}, V_{\mathcal{A}})$, where $\mathcal{T}$ is the access tree under the attribute name set γ, and $V_{\mathcal{A}}$ is the specific attribute value associated with the attribute name.
- **Setup.** Give the security parameter λ. $\mathcal{C}$ runs the $Setup(\lambda)$ algorithm to output the public parameter PP and returns it to adversary $\mathcal{A}$.
- **Phase 1.** $\mathcal{A}$ requests multiple decryption keys to access policy W_j from $\mathcal{C}$, where $\gamma \notin W_j$ for all j.
- **Challenge.** $\mathcal{A}$ sends two messages m_0 and m_1 of equal length to $\mathcal{C}$. $\mathcal{C}$ randomly selects a message m_b $(b \in \{0,1\})$ and encrypts it under the access policy W^*. The ciphertext CT_b is returned to adversary A.
- **Phase 2.** Phase2 is a repeat of Phase1.
- **Guess.** $\mathcal{A}$ outputs a guess b' of b. If $b = b'$, the adversary $\mathcal{A}$ wins the game. The advantage of A in this game is defined as $\varepsilon = Pr\left[b = b'\right] - \frac{1}{2}$.

Definition 3. *The FASD-DP-CPABE scheme is secure in IND-CPA if no polynomial-time adversary has a non-negligible advantage in the game.*

4   Fine-Grained Authorized Secure Deduplication with Dynamic Policy

4.1   Setup

This algorithm is initialized by AC call, and generates public parameters and the system master key. The AC selects a security parameter λ, and generates the public parameter and the master secret key by performing the following steps:

- Choose two multiplicative cyclic groups G and G_T of prime order p, where p is determined by a security parameter λ.
- Choose g as the generator of the group G, and define a bilinear pairing $\hat{e} : G \times G \longrightarrow G_T$.
- Define the Lagrangian coefficients $\triangle_{i,S}$ for $i \in Z_p$ and a set,S , of elements in $Z_p : \triangle_{i,S(x)} = \prod_{j \in S, j \neq i} \frac{x-j}{i-j}$.
- Let there be n attributes in the universe and the attribute names are denoted using the notation $A = \{A_1, A_2, \dots, A_n\}$. The attribute value corresponding to each attribute is represented by the notation $V_i = \{v_{i,1}, v_{i,2}, \dots, v_{i,m_i}\}$, where i represents the attribute A_i.
- Randomly pick $a, \alpha \in Z_p, h, u \in G$.
- Choose two collision-resistant cryptographic hash functions $H_1 : \{0,1\}^* \longrightarrow Z_p$ and $H_2 : \{0,1\}^* \longrightarrow Z_p$.
- Let UID represent the set of identities of all users in the system.

Publish public parameters as $PP = \langle p, G, G_T, e, g, h, u, \hat{e}(g,g)^\alpha, g^a, UID, H_1, H_2 \rangle$, and the master key as $MSK = \langle a, \alpha \rangle$.

4.2   Data Upload

In this section, we describe the details of the data encryption and deduplication part of our scheme. Data upload includes two stages: encrypted data upload by DO and duplicates check by CSP. The details are as follows.

Encrypt. The data owner executes the encryption algorithm, and the output message is the ciphertext under the attribute set $\gamma \subseteq A$. Input public parameters PP, message m, and access tree $\mathcal{T}$ described in Sect. 2.4, each leaf node in $\mathcal{T}$ corresponds to an attribute name in set A. DO does the following.

1. The data owner randomly selects $k \in Z_p$ and calculates the $g_{uid}^{\frac{a}{k}}$ to identify the root node of the access tree and keeps the access policy update key k.
2. From the file hash, calculate the secret value $s = H_1(m)$. The DO choose a polynomial q_x for each node x in the tree $\mathcal{T}$. These polynomials q_x are chosen from top to bottom, starting from the root node r. For each node, x in the tree, set the degree d_x of the polynomial q_x to be one less than the threshold value k_x of that node, that is $d_x = k_x - 1$. For the root node r, set $q_r(0) = s$ and d_r other points of the polynomial q_r randomly to define it completely. For any other node x, set $q_x(0) = q_{parent(x)}(index(x))$ and choose d_x other points randomly to completely define q_x. Then computes some of the ciphertext components associated with the access policy:

$$C = m\hat{e}(g,g)^{\alpha s}, \ C_0 = g^{as}$$
$$\left\{ C_{i,uid,1} = h^{q_x(0)} u^{H_1(i)}, C_{i,3} = g^{H_1(i)} \right\}_{i \in \gamma, uid \in UID} \tag{1}$$

3. The data owner realizes the association between the attribute name and the attribute value by calculating the encryption component

$$\left\{ C_{i,uid,2} = g^{-H_1(i)H_2(v_{i,j})} \right\}_{i \in \gamma, 1 \leq j \leq m_i}$$

4. The final ciphertext CT will be sent to the CSP. The attribute value is transparent to both CSP and DO, thus a partially hidden access strategy is realized.

$$CT = \left\langle C, C_0, \{C_{i,uid,1}, C_{i,uid,2}\}_{i \in \gamma}, \{C_{i,3}\}_{i \in \gamma}, g_{uid}^{\frac{a}{k}}, \mathcal{T} \right\rangle$$

Deduplication. After the CSP receives the ciphertext uploaded by the user, the scheme is similar to Cheng et al. [5], which judges whether it is duplicate data through the label. There are two cases:

Case 1: If there is no same label, it means that the data is uploaded for the first time, and the CSP directly stores the ciphertext without any operation.

Case 2: If the same label exists, the CSP performs access policy fusion and ciphertext update to ensure that legitimate users satisfying two different access policies can decrypt data normally. Supposed CT_1 is the ciphertext already stored in the CSP, CT_2 is the ciphertext of the same data uploaded by the user, then the cloud executes as follows.

1. For trees $\mathcal{T}_1$ and $\mathcal{T}_2$, CSP adds a new OR node as the root node, $\mathcal{T}_1$ and $\mathcal{T}_2$ as the left subtree and right subtree of this newly added node, to combine the access trees in the two ciphertexts to generate a new tree $\mathcal{T}_{new}$. The new tree generated is the access policy as the ciphertext.

2. In the two ciphertexts of CT_1 and CT_2, C, C_0 and $\{C_{i,3}\}_{i \in \gamma_1 \cap \gamma_2}$ are common encryption components, thus CSP combines $C_{i,uid,1}$, $C_{i,uid,2}$ and $C_{i,3}$ in CT_2 to generate the ciphertext CT_{new} under the encryption of the access tree $\mathcal{T}_{new}$.

$$CT_{new} = \left\langle \begin{array}{c} C, C_0, \{C_{i,u_1,1}, C_{i,u_1,2}\}_{i \in \gamma_1}, \\ \{C_{i,u_2,1}, C_{i,u_2,2}\}_{i \in \gamma_2}, \{C_{i,3}\}_{i \in \gamma_2 \cup \gamma_2}, \mathcal{T}_{new} \end{array} \right\rangle$$

4.3 KeyGen

The AC invokes the algorithm to generate a decryption key for it through the attributes related to the data user. Specific steps are as follows:

1. The algorithm takes master private key MSK and user attribute $\mathcal{S} = (I_S, S)$ as input, where $I_S \subseteq A$ is the set of user attribute names, and $S = \{s_\tau\}_{\tau \in I_S}$ is the set of attribute values. It picks $r \in Z_p$ at random and for $\forall \tau \in I_S$ computes

$$\left\langle K_1 = g^{\frac{\alpha}{a}} h^r, K_2 = g^r, K_3 = g^{ar}, \{K_\tau = g^{s_\tau r} u^{-ar}\}_{\tau \in I_S} \right\rangle$$

2. It outputs an attribute-based decryption key sk with respect to a set $\mathcal{S}$ of user attributes and returns it to the user.

$$sk = \left\langle K_1, K_2, K_3, \{K_\tau\}_{\tau \in I_S} \right\rangle$$

4.4 Access Policy Update

In our scheme, the access policy in the ciphertext can be updated dynamically. Each data owner can add, delete and modify the attributes in its encrypted access policy without affecting the normal access of legitimate users of other data owners' access policies. For this purpose, it contains two algorithms: UKeyGen and CTUpdate. First, the data owner runs the algorithm UKeyGen to generate an update key and sends it to the CSP. Then the CSP runs the algorithm CTUpdate and uses the received update key to update the ciphertext.

1. The UKeyGen algorithm takes user identity uid, new leaf node Y', currently leaf node Y, update type $UType$ and access tree $\mathcal{T}_{uid}$ as input. Let $i = att(Y)$ and $i' = att\left(Y'\right)$, there are three types of policy updates:

   Type1 means the update type is modification. Since the (t, n) threshold value has not been changed, as well as the access tree structure. Then it calculates the update key as

$$UK_{Y,Y',uid} = \left\langle \begin{array}{c} g_{uid}^k, UK_{Y,Y',uid}^1 = u^{H_1\left(i'\right) - H_1(i)}, g^{H_1\left(i'\right)} \\ , UK_{Y,Y',uid}^2 = g^{H_1(i) H_2(v_{i,j}) - H_1\left(i'\right) H_2\left(v_{i',j'}\right)} \end{array} \right\rangle$$

where $1 \leq j \leq m_i$ and $1 \leq j' \leq m_{i'}$.

Type2 means the update type is add. When new leaf node Y' is added, the threshold gate of Y' parent node changes from (t, n) to $(t, n + 1)$. Similar to Type1, the user obtains the encrypted component of the attribute value associated with new leaf node Y from the CSP and then calculates the update key

$$UK_{Y,Y',uid} = \left\langle \begin{array}{c} g_{uid}^k, UK_{Y,Y',uid}^1 = h^{q_{Y'}(0)} u^{H_1\left(i'\right)}, \\ g^{H_1\left(i'\right)}, UK_{Y,Y',uid}^2 = g^{-H_1\left(i'\right) H_2\left(v_{i',j'}\right)} \end{array} \right\rangle$$

where $1 \leq j' \leq m_{i'}$.

Type3 means the update type is deleted. For the threshold gate (t, n) of the parent node of currently leaf node Y, three cases are respectively handled when deleting node Y:

(a) When $t < n$ and $n > 2$, the currently leaf node Y is directly deleted, and the threshold gate of its parent node is changed from (t, n) to $(t, n - 1)$. Then DO generates update key

$$UK_{Y,uid} = \left\langle g_{uid}^k \right\rangle$$

(b) When $t = n$ and $n > 2$, After deleting currently leaf node Y, the threshold gate of its parent node is changed from (t, n) to $(t - 1, n - 1)$, then DO selects the polynomial q' of $parent(Y)$ to redistribute the secret values of all affected leaf nodes. Let S_x be the set of all affected leaf nodes of node $parent(Y)$ except node Y, then computes the update key

$$UK_{Y,uid} = \left\langle \left\{ UK_{x,uid}^1 = h^{q'_x(0) - q_x(0)} \right\}_{x \in S_x}, g_{uid}^k \right\rangle$$

(c) When $n = 2$, the difference from case (ii) is that the sibling node Y_{bro} of currently leaf node Y replaces $parent(Y)$ and redistributes the secret values of all affected leaf nodes under the subtree rooted at node Y_{bro}. The update key is

$$UK_{Y,uid} = \left\langle \left\{ UK_{x,uid}^1 = h^{q'_x(0) - q_x(0)} \right\}_{x \in S_x}, g_{uid}^k \right\rangle$$

In particular, when the node $parent(Y)$ is an OR gate, that is, the threshold gate is $(1, 2)$, Y_{bro} inherits the secret value of $parent(Y)$ without redistribution.

Finally, the DO sends $UK = \left\langle UK_{Y,uid}, Y, Y', UType \right\rangle$ to CSP, where Y' is the sibling node of Y when the update type is Type3.

2. The CTUpdate algorithm takes uid, UK and CT as input, and the CSP update ciphertext CT' is as follows.

First, the CSP verifies the user identity by calculating $e\left(g^{\frac{a}{k}}, g^k\right) = e\left(g, g^a\right) \neq 1$, if the verification fails, the algorithm terminates, otherwise proceed to the next step.

(a) For Type1, the ciphertext component C_Y' is computed as:

$$C_{Y,uid,1}' = C_{Y,uid,1} \cdot UK_{Y,Y',uid}^1 = h^{q_{Y'}(0)} u^{H_1(i)}$$

$$C_{Y,uid,2}' = C_{Y,uid,2} \cdot UK_{Y,Y',uid}^2 = g^{-H_1(i)H_2(v_{i,j})}$$

$$C_{Y,3} = g^{H_1(i')}$$

(b) For Type2, the ciphertext component C_Y' is computed as:

$$C_{Y,uid,1}' = UK_{Y,Y',uid}^1 = h^{q_{Y'}(0)} u^{H_1(i)}$$

$$C_{Y,uid,2}' = UK_{Y,Y',uid}^2 = g^{-H_1(i)H_2(v_{i,j})}$$

$$C_{Y,3} = g^{H_1(i')}$$

(c) For Type3, CSP delete $C_{Y,uid,1}$ and $C_{Y,uid,2}$, then the ciphertext component C_Y' is computed as:

  i. When $t < n$ and $n > 2$, CSP only needs to remove the encryption component $C_{Y,uid,1} = h^{q_Y(0)} u^{H_1(i)}$ corresponding to attribute Y and the leaf nodes in the access tree.

  ii. When $t = n$ and $n > 2$,

$$C_{x,uid,1}' = C_{x,uid,1} \cdot UK_{x,uid}^1 = h^{q_x'(0)} u^{H_1(i_x)}$$

  where $x \in S_x$.

  iii. When $n = 2$ and $t = 2$,

$$C_{x,uid,1}' = C_{x,uid,1} \cdot UK_{x,uid}^1 = h^{q_x'(0)} u^{H_1(i_x)}.$$

  where $x \in S_x$.

In updating the ciphertext, CSP only re-encrypts the modified part of the ciphertext. The update key UK only expresses the change caused by the access policy update and cannot reveal any information about the encrypted message. Therefore, CSP cannot further snoop on encrypted data.

4.5   Decrypt

The user runs the decryption algorithm $DecryptNode$ with sk, S and CT as input, which performs the following two steps:

1. First, the algorithm traverses each subtree $\mathcal{T}_{uid}$ with uid as the root node of the access tree $\mathcal{T}$ from top to bottom and judges whether the user attributes satisfy the access policy formulated by a certain DO.

   - When $A_{DO} \in S \models \mathcal{T}_{uid}$, and the attribute value matches the detection:

$$\hat{e}\left(g^{H_1(i_{DO})H_2(v_{i,j,DO})}, C_{i,uid,2}\right) = 1,$$

   the algorithm performs decryption calculation.

– Otherwise, the algorithm continues to traverse the next subtree $\mathcal{T}_{uid'}$ until the access policy is met for decryption calculation or the algorithm terminates and outputs $\perp$.

2. The algorithm performs decryption calculation if and only when the attribute satisfies the access policy formulated by a data owner. The plaintext can be decrypted correctly and recovered.

Let $i = att(x)$, if x is a leaf node, then do the following calculation:

$$
\begin{aligned}
E &= \hat{e}\left(K_3, C_{i,uid,1}\right) \hat{e}\left(K_2, C_{i,uid,2}\right) \hat{e}\left(K_{V_i}, C_{i,3}\right) \\
&= \hat{e}\left(g^{ar}, h^{q_x(0)} u^{H_1(i)}\right) \cdot \hat{e}\left(g^r, g^{-H_1(i) H_2(v_{i,j})}\right) \cdot \hat{e}\left(g^{s_\tau r} u^{-ar}, g^{H_1(i)}\right) \\
&= \hat{e}\left(g, h\right)^{ar q_x(0)}
\end{aligned}
$$

When x is not a leaf node, we consider the case of recursion. The $DecryptNode\,(CT, sk, x)$ algorithm works as follows: For all nodes z of node x, it calls $DecryptNode$ and stores the output as F_z. Let S_x be an arbitrary set of size k_x of children z such that $F_z \neq \perp$. If no such set exists, then this node does not satisfy the access policy, and the algorithm returns $\perp$. Otherwise, we compute:

$$
\begin{aligned}
F_x &= \prod_{z \in S_x} F_z^{\triangle_{i, S_x'}(0)}, where \quad \begin{array}{l} i = index\,(z), \\ S_x'\,(0) = \{index\,(z) : z \in S_x\} \end{array} \\
&= \prod_{z \in S_x} \left(\hat{e}\left(g, h\right)^{ar q_z(0)}\right)^{\triangle_{i, S_x'}(0)} \\
&= \prod_{z \in S_x} \left(\hat{e}\left(g, h\right)^{ar q_{parent(z)}(index(z))}\right)^{\triangle_{i, S_x'}(0)} \\
&= \prod_{z \in S_x} \left(\hat{e}\left(g, h\right)^{ar q_x(i)}\right)^{\triangle_{i, S_x'}(0)} \\
&= \hat{e}\left(g, h\right)^{ar q_x(0)}
\end{aligned}
$$

and return the result.

The decryption algorithm starts from the root node of the access tree and is called from top to bottom. We note that $DecryptNode\,(CT, sk, x) = \hat{e}\,(g, h)^{ars}$ when user attributes satisfy the access tree $\mathcal{T}$. The algorithm recovers the plaintext of the message by calculating as follows:

$$
\begin{aligned}
D &= \hat{e}\left(K_1, C_0\right) = \hat{e}\left(g^{\frac{\alpha}{a}} h^r, g^{as}\right) \\
&= \hat{e}\left(g, g\right)^{\alpha s} \cdot \hat{e}\left(g, h\right)^{ars}
\end{aligned}
$$

Finally, the plain text of the message is $m = \frac{C \cdot E}{D}$.

5   IND-CPA Security Analysis

Theorem 1. *In the following, we prove that our scheme is IND-CP-CKA secure in the random oracle model under the DBDH assumption.*

We follow the scheme of Sahai and Waters [18] as our basic framework. The security model in Sect. 3.4 is also similar to theirs. Compared to their security model, our model adds the ability to allow an adversary to issue queries to UKeyGen.

Proof. Suppose that there exists a PPT adversary $\mathcal{A}$ with a non-negligible advantage ε capable of breaking our scheme. Then, we build a simulator $\mathcal{B}$ that can break the DBDH assumption with advantage $\frac{\varepsilon}{2}$. The execution process of $\mathcal{B}$ is as follows:

Let G and G_T be two multiplicative cyclic groups of prime order p, g be a generator of G, and map $\hat{e} : G \times G \longrightarrow G_T$ be a bilinear map. The challenger $\mathcal{C}$ flips a fair binary coin $\mu = \{0, 1\}$. For random $a, b, c, z \in Z_p$, if $\mu = 0$, $\mathcal{C}$ sets the tuple $\left\langle g^a, g^b, g^c, \hat{e}(g, g)^{abc} \right\rangle$; if $\mu = 1$, it sets the tuple to $\left\langle g^a, g^b, g^c, \hat{e}(g, g)^z \right\rangle$.

Init. $\mathcal{A}$ chooses a challenge access strategy $W^* = (\mathcal{T}, V_{\mathcal{A}})$, where $\mathcal{T}$ is the access tree under the attribute name set γ, and $V_{\mathcal{A}}$ is the specific attribute value associated with the attribute name.

Setup. The simulator sets the parameter $\hat{e}(g, g)^{\alpha} = \hat{e}(g, g)^{ab}$. The challenger $\mathcal{C}$ generates the public parameter PP under the security parameter λ and returns it to the adversary $\mathcal{A}$. Two random oracles $\mathcal{O}_{H1} : \{0, 1\}^* \to Z_p$ and $\mathcal{O}_{H2} : \{0, 1\}^* \to Z_p$ are defined to simulate hash functions.

Phase 1. Adversary $\mathcal{A}$ requests any user attribute set $\mathcal{S} = (I_S, S)$ that does not satisfy W^* to obtain the relevant decryption key, where $I_S \subseteq A$ is the set of attribute names, and $S = \{s_\tau\}_{\tau \in I_S}$ is the set of corresponding attribute values. $\mathcal{C}$ runs the KeyGen algorithm. There are two situations:

1. For $s_\tau \in W^*$, $\mathcal{C}$ queries $\mathcal{O}_{H2}$ for $H_2(s_\tau)$. Let $s_\tau = H_2(s_\tau)$, calculate

$$\{K_\tau = g^{s_\tau r} u^{-ar}\}_{\tau \in I_S, s_\tau \in W^*}$$

2. For the remaining $s_\tau \notin W^*$, $\mathcal{C}$ randomly chooses $s^{'} \in Z_p$. Let $s_\tau = s^{'}$, calculate

$$\{K_\tau = g^{s_\tau r} u^{-ar}\}_{\tau \in I_S, s_\tau \notin W^*}$$

where $r \in Z_p$ is chosen randomly. Output the private key sk and return it to $\mathcal{A}$.

$$sk = \left\langle K_1 = g^{\frac{\alpha}{a}} h^r, K_2 = g^r, K_3 = g^{ar}, K_\tau \right\rangle$$

Challenge. $\mathcal{A}$ submits two messages, m_0, and m_1, of the same length to $\mathcal{C}$, and $\mathcal{A}$ does not obtain any key that satisfies the access policy W^* in Phase 1. Let $s \in Zp$ be the secret value of the encrypted message. Using the results of oracles

$\mathcal{O}_{H1}$ and $\mathcal{O}_{H2}$, $\mathcal{C}$ flips a fair binary coin $b \in \{0,1\}$ and outputs the challenge ciphertext as:

$$CT = \left\langle C = m_b Z, C_0, \{C_{i,uid,1}, C_{i,uid,2}, C_{i,3}\}_{i \in \gamma}, \mathcal{T} \right\rangle$$

If $\mu = 0$, then $Z = \hat{e}(g,g)^{abc}$. We let $s = c$, then we can get $Z = \hat{e}(g,g)^{abs} = \hat{e}(g,g)^{abc}$ and $C_0 = g^{ac}$. Therefore it can represent that the ciphertext is a valid DBDH tuple.

Otherwise $\mu = 1$, then $Z = \hat{e}(g,g)^z$. Since z is a random element, it can represent that the adversary $\mathcal{A}$ cannot obtain any information about b from the ciphertext m_b.

Phase 2. Phase2 is a repeat of Phase1.

Guess. Adversary $\mathcal{A}$ outputs a guess b' of b. When $\mu = 1$, since the adversary cannot obtain any information about the message m_b, we can get $Pr\left[b \neq b' | \mu = 1\right] = \frac{1}{2}$. When $b \neq b'$, the simulator guesses $\mu' = 1$ and we have $Pr\left[\mu' = \mu | \mu = 1\right] = \frac{1}{2}$. When $\mu = 0$, the adversary $\mathcal{A}$ has the advantage ε of being able to see the ciphertext m_b. Similarly, $Pr\left[b = b' | \mu = 0\right] = \frac{1}{2} + \epsilon$. When $b = b'$, the simulator guesses that $\mu' = 0$ and we have $Pr\left[\mu' = \mu | \mu = 0\right] = \frac{1}{2} + \epsilon$. Then the overall advantage in DBDH game simulation is $\frac{1}{2}Pr\left[\mu' = \mu | \mu = 0\right] + \frac{1}{2}Pr\left[\mu' = \mu | \mu = 1\right] - \frac{1}{2} = \frac{1}{2}\left(\frac{1}{2} + \varepsilon\right) + \frac{1}{2}\frac{1}{2} - \frac{1}{2} = \frac{1}{2}\epsilon$. That is, if the adversary has an advantage of ϵ, we can construct a simulator B that can break the DBDH assumption with only a negligible advantage of $\frac{1}{2}\epsilon$.

Table 1. Comparison of characteristics with previous works.

Scheme	Access Structure	Hidden Policy	Policy Update	Deduplication
Bethencourt et al. [3]	Tree	✗	✗	✗
Liu et al. [13]	LSSS	✗	✔	✗
Fugkeaw et al. [7]	Tree	✗	✔	✗
Cui et al. [6]	LSSS	✗	✗	✔
Premkamal et al. [15]	LSSS	✗	✗	✔
Ours	Tree	✔	✔	✔

Table 2. The summary of some notations.

Notation	Description
$|p|$	Size of elements in the Z_p group
$|g|$	Size of elements in the G group
$|g_T|$	Size of elements in the G_T group
$|T|$	Size of access policy
$|uk|$	Update key size
N	The number of users in the merged access policy
n_u	The number of attributes that identify the user
n_c	Number of attributes in access policy when encrypted
n_d	The number of attributes involved in decryption
n_s	The number of nodes involved in the policy update
E	One exponentiation operation
P	One paring operation

6 Performance Evaluations

In this section, we theoretically analyze the storage overhead, computation overhead, and performance of the FD-MU-CPABE scheme. Finally, we conduct experiments to evaluate the practicality and show simulation results.

6.1 Theoretical Analysis

As shown in Table 1, we compare the functionality of our scheme with other schemes. In addition, we summarize some notations in terms of storage and computation overhead in Table 2. Table 3 describes the storage overhead of each entity in our scheme. We can see that the primary storage overhead of AC comes from the master private key MSK. The size of the ciphertext components and the access tree is the primary storage overhead of CSP. The storage cost of the data owner comes from the private key sk related to its attributes and the access strategy formulated by it. The user's storage cost mainly comes from the public parameter PP and update key UK. In addition, we also describe the computational overhead in the ABE encryption and decryption process of our scheme. The time complexity of the algorithm Setup run by AC is constant. The calculation overhead of the user's private key sk is related to its associated attributes, and the relationship is linear. Encryption and decryption algorithms scale linearly with the size of the access tree.

6.2 Experimental Evaluation

To evaluate the expressive performance of our scheme, we implemented it using the Stanford pairing-based cryptographic library [4]. A bilinear pairing is constructed on the curve $y^2 = x^3 + x$ on the field F_q of the prime number $q = 3mod4$,

Table 3. Storage overhead and ABE scheme calculation overhead

Storage	AC	CSP	DO	DU
	$2\,|p|$	$\left(2\,|n_c|\,N + 2 + n_c'\right)|g| + |g_T| + |\mathcal{T}|$	$|p| + |\mathcal{T}| + 5\,|g| + |g_T| + |uk|$	$(|n_u| + 3)\,|g|$
Computation	**Setup**	**Encryption**	**KeyGen**	**Decryption**
	$O\,(1)\,(P + E)$	$O\,(n_c)\,E$	$O\,(n_u)\,E$	$(O\,(n_d) + O\,(N))\,P$

where the order of the groups G and G_T is a prime number of 160 bits, and the length of q is 512 bits. We performed experiments on a laptop running 64-bit Windows equipped with an Intel(R) Core(TM) i7-12700H 2.70 GHz and 40.00 GB of RAM.

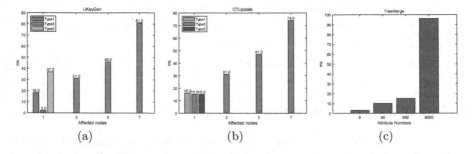

(a)　　　　　　　　(b)　　　　　　　　(c)

Fig. 2. Policy Update

Limited by the length of the paper, we focus on the policy update process. Fig. 2(a) and Fig. 2(b) show the time overhead of the update key generation algorithm and the ciphertext update algorithm under the influence of the number of affected nodes in the access tree during an update. For Type1 and Type2, since the affected nodes are only update nodes, the time overhead is constant. We took the average of twenty experimental results to get the time overhead of Type 1 and Type 2. For the three threshold gate situations of Type 3, the time overheads show linear growth with the number of affected nodes.

The time cost of access tree merging is affected by the experimental code language and the data structure of the access tree in the code. Our implementation uses the Java coding language and the list structure to express the access tree. Its time overhead is shown in Fig. 2(c), and the number of attributes included in the access strategy has little impact on merging the access trees.

7　Conclusion

In this paper, we introduce a CP-ABE scheme for enhancing access control in cloud storage through data deduplication. Our scheme is proven secure under the IND-CPA model with a random oracle. Simulation experiments validate its effectiveness, especially for multiple data owners. Our future work aims to further explore extended attribute-based encryption's functionality and security.

References

1. Bellare, M., Keelveedhi, S.: Interactive message-locked encryption and secure deduplication. Cryptology ePrint Archive (2015)
2. Bellare, M., Keelveedhi, S., Ristenpart, T.: Message-locked encryption and secure deduplication. In: Johansson, T., Nguyen, P.Q. (eds.) EUROCRYPT 2013. LNCS, vol. 7881, pp. 296–312. Springer, Heidelberg (2013). https://doi.org/10.1007/978-3-642-38348-9_18
3. Bethencourt, J., Sahai, A., Waters, B.: Ciphertext-policy attribute-based encryption. In: 2007 IEEE Symposium on Security and Privacy (SP 2007), pp. 321–334. IEEE (2007)
4. Boneh, D., Lynn, B., Shacham, H.: The stanford paired encryption library. http://crypto.stanford.edu/pbc/ (2007)
5. Cheng, S., Zeng, S., Zeng, H., Feng, Y., Xiao, J.: Secure single-server fuzzy deduplication without interactive proof-of-ownership in cloud. Cryptology ePrint Archive, Paper 2023/005 (2023). https://eprint.iacr.org/2023/005. https://eprint.iacr.org/2023/005
6. Cui, H., Deng, R.H., Li, Y., Wu, G.: Attribute-based storage supporting secure deduplication of encrypted data in cloud. IEEE Trans. Big Data **5**(3), 330–342 (2017)
7. Fugkeaw, S., Sato, H.: Scalable and secure access control policy update for outsourced big data. Futur. Gener. Comput. Syst. **79**, 364–373 (2018)
8. Goyal, V., Pandey, O., Sahai, A., Waters, B.: Attribute-based encryption for fine-grained access control of encrypted data. In: Proceedings of the 13th ACM Conference on Computer and Communications Security, pp. 89–98 (2006)
9. Joshi, M., Joshi, K., Finin, T.: Attribute based encryption for secure access to cloud based EHR systems. In: 2018 IEEE 11th International Conference on Cloud Computing (CLOUD), pp. 932–935. IEEE (2018)
10. Lai, J., Deng, R.H., Li, Y.: Expressive CP-ABE with partially hidden access structures. In: Proceedings of the 7th ACM Symposium on Information, Computer and Communications Security, pp. 18–19 (2012)
11. Lai, J., Deng, R.H., Yang, Y., Weng, J.: Adaptable ciphertext-policy attribute-based encryption. In: Cao, Z., Zhang, F. (eds.) Pairing 2013. LNCS, vol. 8365, pp. 199–214. Springer, Cham (2014). https://doi.org/10.1007/978-3-319-04873-4_12
12. Liu, J., Asokan, N., Pinkas, B.: Secure deduplication of encrypted data without additional independent servers. In: Proceedings of the 22nd ACM SIGSAC Conference on Computer and Communications Security, pp. 874–885 (2015)
13. Liu, Z., Jiang, Z.L., Wang, X., Yiu, S.M.: Practical attribute-based encryption: outsourcing decryption, attribute revocation and policy updating. J. Netw. Comput. Appl. **108**, 112–123 (2018)
14. Meyer, D.T., Bolosky, W.J.: A study of practical deduplication. ACM Trans. Storage (ToS) **7**(4), 1–20 (2012)
15. Premkamal, P.K., Pasupuleti, S.K., Singh, A.K., Alphonse, P.: Enhanced attribute based access control with secure deduplication for big data storage in cloud. Peer-to-Peer Networking Appl. **14**, 102–120 (2021)
16. Rouselakis, Y., Waters, B.: Efficient statically-secure large-universe multi-authority attribute-based encryption. In: Böhme, R., Okamoto, T. (eds.) FC 2015. LNCS, vol. 8975, pp. 315–332. Springer, Heidelberg (2015). https://doi.org/10.1007/978-3-662-47854-7_19

17. Sahai, A., Seyalioglu, H., Waters, B.: Dynamic credentials and ciphertext delegation for attribute-based encryption. In: Safavi-Naini, R., Canetti, R. (eds.) CRYPTO 2012. LNCS, vol. 7417, pp. 199–217. Springer, Heidelberg (2012). https://doi.org/10.1007/978-3-642-32009-5_13
18. Sahai, A., Waters, B.: Fuzzy identity-based encryption. In: Cramer, R. (ed.) EUROCRYPT 2005. LNCS, vol. 3494, pp. 457–473. Springer, Heidelberg (2005). https://doi.org/10.1007/11426639_27
19. Song, Y., Han, Z., Liu, F., Liu, L.: Attribute-based encryption with hidden policies in the access tree. J. Commun. 36(9), 119–126 (2015)
20. Yang, K., Jia, X., Ren, K.: Secure and verifiable policy update outsourcing for big data access control in the cloud. IEEE Trans. Parallel Distrib. Syst. 26(12), 3461–3470 (2014)

Deep Multi-image Hiding with Random Key

Wei Zhang[1], Weixuan Tang[1(✉)], Yuan Rao[1], Bin Li[2], and Jiwu Huang[2]

[1] Institute of Artificial Intelligence, Guangzhou University, Guangzhou, China
`tweix@gzhu.edu.cn`
[2] Guangdong Key Laboratory of Intelligent Information Processing and Shenzhen
Key Laboratory of Media Security, Shenzhen University, Shenzhen, China

Abstract. Multi-image hiding is the technique of hiding multiple secret images within one cover image. In most existing methods, it is possible for one receiver to reveal other receivers' secret images. To improve the privacy and secrecy among different receivers, one possible solution is to introduce the key mechanism, wherein only the receiver with private key has the permission to reveal the corresponding secret image. In this paper, a multiple image hiding method called DEMIHAK (Deep Multiple Image Hiding with Random Key) is proposed, which utilizes deep neural networks to implement a secure key verification. From the side of the sender, each secret image is assigned with a random key, according to which can sample a key map. Then, BindNet is utilized to incorporate a secret image and its key map into a processed secret image, and HideNet is adopted to conceal multiple processed secret images within cover image and generate a stego image. From the side of the receiver, according to a transmitted private key, RevealNet can be applied to reveal the corresponding secret image from the stego image. Experimental results show that DEMIHAK outperforms existing method from the perspective of visual quality, security, and secrecy.

Keywords: Image hiding · Steganography · Multi-channel communication · Private key

1 Introduction

Information hiding techniques, including steganography and watermarking, is the study of secretly hiding secret information into multimedia carriers. Conventional steganographic methods were mainly carried under minimal-distortion framework. Their distortion functions were calculated according to heuristic principles [1,2], statistical models [3,4], or deep learning techniques [5,6]. With near-optimal steganographic codes, message embedding and extraction can be fulfilled according to pre-defined distortion functions. However, due to the restriction of the form of distortion, their embedding capacities were still to be improved. Therefore, to achieve higher capacities, recent methods [7,8] abandoned the settings of minimal-distortion framework, and directly employed neural networks to implement message embedding and extraction.

© The Author(s), under exclusive license to Springer Nature Singapore Pte Ltd. 2024
J. Vaidya et al. (Eds.): AIS&P 2023, LNCS 14509, pp. 33–41, 2024.
https://doi.org/10.1007/978-981-99-9785-5_3

Besides hiding message bits, information hiding technique can also be applied to hide secret image. Baluja [9] proposed to hide secret image within cover image with encoder-decoder structure. Rehman *et al.* [10] introduced loss function that ensured joint end-to-end training of encoder-decoder networks. Yu [11] utilized the attention mechanism to find spotlights and inconspicuous areas of cover images. Zhang *et al.* [12] proposed UDH to disentangle the encoding of secret image from cover image, and implement image hiding in a cover-agnostic manner. Guan *et al.* [13] and Lu *et al.* [14] respectively applied INN to implement image embedding and extraction.

Note that multi-channel steganography [15] can transmit different secret data to multiple receivers via one cover image, which can avoid unnecessary data communication on public channel and enhance the transmission secrecy. Therefore, it is desired to apply multi-channel steganography in deep image hiding, and implement multi-image hiding [12,13]. However, these methods have not fully considered the privacy and secrecy among different receivers, *i.e.*, there is no mechanism to guarantee that one particular receiver cannot reveal other receiver's secret images. Kweon made an early attempt to solve the issue [16], which utilized the feature maps of secret image as its key. However, it has two limitations. Firstly, it has severe security loophole. Such method utilizes high-level feature of secret image as its corresponding key. Therefore, it is possible to directly reveal the secret image from the key. Secondly, the length of its key is rather long. In the case of hiding secret image with size of 256×256, the length of key becomes $4 \times 4 \times 512$. As a result, it is inconvenient to privately transmit such key.

In this paper, to overcome the above issue, a secure multiple image hiding method called DEMIHAK (Deep Multiple Image Hiding with Random Key) is proposed. On the side of hiding, the sender assigns each secret image with a specific key, according to which can sample a key map. Then, BindNet is utilized to tie the secret image with key map and generate a processed secret image. Afterwards, HideNet is adopted to conceal multiple processed secret images within cover image as stego image. Such stego image and key can be transmitted to the receiver through a public and private channel. On the side of revealing, the receiver utilizes RevealNet to extract the secret image from the stego image according to the given key, wherein such secret image and key are previously tied by BindNet. By this means, the randomly generated private key can be utilized to control the access right to secret image. Results show that our proposed DEMIHAK outperforms existing method in terms of visual quality, security, and secrecy.

2   Proposed Method

2.1   Overview

As shown in Fig. 1, the proposed DEMIHAK is composed of a binding network (*BindNet*), a hiding network (*HideNet*), and a revealing network (*RevealNet*). In order to protect the privacy and secrecy of different receivers in multi-channel

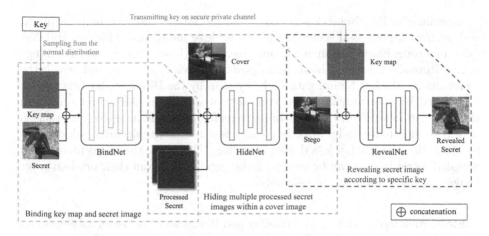

Fig. 1. Overall framework of our proposed DEMIHAK.

communication of deep image hiding, the sender and receiver utilize BindNet and RevealNet to implement image hiding and revealing with shared key, respectively.

The working flow of DEMIHAK is as follows. The sender executes three steps. Firstly, each secret image is assigned with a key, which is represented as an integer. Such key is further applied as a triggered seed to generate a key map with the same size of cover and secret image, wherein each element is a floating-point value sampled from the normal distribution. Note that the same key corresponds to the same key map. Secondly, the BindNet is utilized to tie each secret image with a specific key map. It takes in a pair of secret image and key map, and binding them into a processed secret image. Thirdly, to perform information hiding, HideNet is adopted to conceal multiple processed secret images within a cover image, and generates a stego image. Afterwards, the stego image can be transmitted on public channel, and the keys are transmitted on a secure private channel. On another side, the receiver receives the stego image and a specific key. Such key is served as a triggered seed to generate a key map, as it did by the sender. Finally, the RevealNet is applied to reveal secret image which is bound with such key by BindNet.

Note that in DEMIHAK, the BindNet, HideNet, and RevealNet are trained in an end-to-end manner. Therefore, a specific key can be bound with a secret image, and only the receiver with the shared key has the access right to reveal the corresponding secret image. By this means, privacy can be protected in multi-image hiding within different receivers.

2.2 Network Architecture

In our proposed DEMIHAK, BindNet, HideNet, and RevealNet share the same architecture of UNet, which is based on the encoder-decoder structure with skip connections between mirrored layers, as shown in Table 1.

Specifically, BindNet takes a pair of secret image and key map as input, and outputs a processed secret image. To achieve multi-image hiding, HideNet takes in a cover image and multiple processed secret image, and generates a stego image. RevealNet receives the stego image and a specific key map, and reveals the secret image which is bound with such key map by the BindNet. By this means, the shared key can provide access control to a specific hidden secret image. In HideNet and RevealNet, the skip connections facilitate the image reconstruction of fine-grained textures by directly passing features to the decoder, which compensates the information loss during encoding. Therefore, the stego images and revealed secret images can be visually indistinguishable from their original cover images and secret images, respectively.

Table 1. Architecture of BindNet, HideNet and RevealNet. The kernel configurations of each layer are given in the following format: (kernel size, stride, padding)

Index	Type	Kernel	Input	Out	Concat
1	Conv2d+BN+ReLU	4, 2, 1	In	64	N/A
2	Conv2d+BN+ReLU	4, 2, 1	64	128	N/A
3	Conv2d+BN+ReLU	4, 2, 1	128	256	N/A
4	Conv2d+BN+ReLU	4, 2, 1	256	512	N/A
5	Conv2d+BN+ReLU	4, 2, 1	512	512	N/A
6	DeConv2d+BN+ReLU	4, 2, 1	512	512	N/A
7	DeConv2d+BN+ReLU	4, 2, 1	1024	256	#4
8	DeConv2d+BN+ReLU	4, 2, 1	512	128	#3
9	DeConv2d+BN+ReLU	4, 2, 1	256	64	#2
10	DeConv2d+Sigmoid	4, 2, 1	128	out	#1

2.3   Loss Function

In this paper, the cover and stego image are denoted as C and C', respectively. The n-th secret and revealed secret image are denoted as S_n and S'_n, respectively. And N is the number of hidden secret images. The DEMIHAK is trained by minimizing the multi-loss function as

$$\mathcal{L} = \mathcal{L}_{\text{MIX}}\left(C', C\right) + \alpha \cdot \sum_{n=1}^{N} \mathcal{L}_{\text{MIX}}\left(S'_n, S_n\right), \tag{1}$$

$$\mathcal{L}_{\text{MIX}}(x, y) = \beta \cdot \mathcal{L}_{\text{MS-SSIM}}(x, y) + (1 - \beta) \cdot G \cdot \mathcal{L}_1(x, y), \tag{2}$$

$$\mathcal{L}_{\text{MS-SSIM}}(x, y) = 1 - \text{MS-SSIM}(x, y), \tag{3}$$

where $\mathcal{L}_{\mathrm{MIX}}$ is the weighted sum of multi-scale structural similarity (MS-SSIM) loss and $\mathcal{L}_1$ loss. Specifically, $\mathcal{L}_{\mathrm{MS\text{-}SSIM}}$ is employed to preserve the contrast in high-frequency textural regions, while $\mathcal{L}_1$ loss is utilized to preserve the color and brightness. α and β control the balance between different losses, and G are the values for computing multi-scale SSIM.

3 Experimental Results

3.1 Experiment Setups

To the best of our knowledge, [16] is the only method which applies private keys for multi-image hiding. Therefore, it was applied as the comparative method. The MS-COCO dataset with 164,000 images was applied, which were divided into a training set, validation set, and testing set. All the images were resized into 256×256. As for DEMIHAK, the model was trained for 264 epochs, and the batch size was set to 32. AdamW optimization and cosine annealing learning rate strategy was applied, wherein the maximum learning rate was set to 0.001. Hyper-parameter α and β were set to 1 and 0.025 respectively, and G was [0.5, 1.0, 2.0, 4.0, 8.0]. As for [16], its settings followed the released official code. The experiments were conducted on PyTorch with an NVIDIA Tesla V100S GPU.

Table 2. APE, PSNR, and SSIM of different methods.

N	Pair	Method	APE ↓	PSNR ↑	SSIM ↑
2	(C, C')	[16]	3.608	34.83	0.932
		DEMIHAK	**1.991**	**39.00**	**0.972**
	(S, S')	[16]	4.363	33.01	0.939
		DEMIHAK	**3.516**	**33.98**	**0.950**
3	(C, C')	[16]	5.086	31.82	0.886
		DEMIHAK	**2.245**	**38.07**	**0.968**
	(S, S')	[16]	6.947	28.69	0.861
		DEMIHAK	**5.298**	**30.32**	**0.895**
4	(C, C')	[16]	5.449	31.41	0.879
		DEMIHAK	**2.541**	**37.08**	**0.961**
	(S, S')	[16]	6.898	28.49	0.841
		DEMIHAK	**6.108**	**29.01**	**0.866**

3.2 Visual Quality

In this part, the visual quality of stego and revealed secret image are investigated. From the aspect of objective evaluation, the visual quality is evaluated by

APE (average pixel error), PSNR (peak-signal-to-noise-ratio), and SSIM (structural similarity). These metrics are calculated by a pair of C and C', and a pair of S and S'. The results are shown in Table 2. It can be observed that our proposed DEMIHAK significantly outperforms [16] on all metrics. For example, in the case of hiding 3 images and evaluating for a pair of C and C', the performance gap is 2.841, 6.25 and 0.082 for APE, PSNR and SSIM, respectively. Similar phenomenon can be observed when hiding different numbers of images and evaluating for a pair of S and S'.

Fig. 2. Visualization of cover image (C), stego image (C'), secret image (S), revealed secret image (S'), and their differences $|C - C'|$ and $|S - S'|$.

From the aspect of subjective evaluation, the texture details of a pair of C and C', and a pair of S and S' are compared. Local regions are zoomed in for more clear presentation. The results are shown in Fig. 2. It can be seen that the differences between C and C', and the differences between S and S' in DEMIHAK are much smaller than those in [16], indicating that DEMIHAK achieves better visual quality. Besides, it can be observed that the images generated by [16] have severe visible defects. Firstly, its stego image has stripe artifacts. Secondly, its certain revealed secret image may carry the information of other secret image, which discloses the privacy of different receivers. For example, the upper part of the red circle signpost in the third secret image can be seen in the first revealed secret image, as shown in the zoomed in area. These visual artifacts do not appear in our DEMIHAK.

3.3 Security Performance Against Steganalyzer

In this part, the security performance of different image hiding methods are evaluated by StegExpose [17], which is a traditional steganalyzer combined with several steganalytic methods including Chi-squared attack, RS analysis, and sample pair analysis.

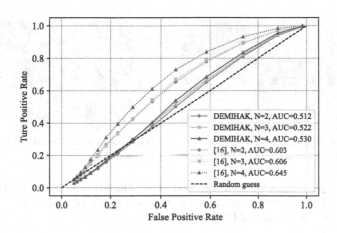

Fig. 3. The ROC curves of StegExpose for detecting DEMIHAK and [16].

8000 pairs of cover and stego images from the testing set are utilized to plot the receiver operating characteristic (ROC) curves, wherein stegos are regarded as positive samples and covers are regarded as negative samples. The results are given in Fig. 3. It can be observed that the ROC curves of DEMIHAK are closer to the random guessing line compared with those of [16]. Moreover, the area under the curve (AUC) values of DEMIHAK are much smaller than those of [16]. Such results demonstrate that our DEMIHAK is harder to detect and achieves better security performance.

3.4 Effectiveness of Key Mechanism

The goal of applying key mechanism in multi-image hiding is to protect the privacy among different receivers from two aspects. On one hand, the receiver with the transmitted correct key can reveal the corresponding secret image. On the other hand, the malicious receiver with the mismatched key cannot disclose the image content. In this part, to verify the effectiveness of key mechanism, we make further study of the mismatched key scenario under two settings.

In the first setting, the mismatched keys are randomly sampled from the normal distribution, and the results are given in Fig. 4(a). It can be observed that for [16], the local semantics of secret image are exposed within specific image blocks. In the second setting, all values of mismatched keys are fixed to -1, and the results are given in Fig. 4(b). It can be seen that for [16], specific regions of the revealed secret images are highly similar with the original secret images, leading to more severe information leakage issues. By contrast, DEMIHAK generates rather blurred revealed secret images in both settings, and thus the content of secret images can be kept hidden. As unauthorized individuals without the correct keys cannot obtain any portion of secret images, the key mechanism in DEMIHAK can well protect the privacy and secrecy.

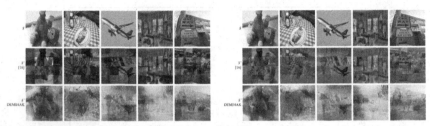

(a) The elements of mismatched keys are sampled from normal distribution.   (b) The elements of mismatched keys are fixed to -1.

Fig. 4. Revealing images with mismatched keys. The first row indicates the original secret image, the second row indicates the revealed image of [16], and the third row indicates the revealed image of DEMIHAK.

4   Conclusion

In this paper, a secure multiple image hiding method called DEMIHAK is proposed, which combines the learning ability of deep neural networks and the security of key mechanism. Specifically, a private key is bound with a secret image by BindNet in multi-image hiding, and is regarded as a verification code by RevealNet in secret image revealing. Extensive experiments have been conducted to verify its effectiveness from the aspect of image visual quality and security against steganalyzer. Further studies have been given to analyze the effectiveness of key mechanism. In the future, we would consider the robustness of DEMIHAK against different noisy channels, and try to extend it for hiding secret images within JPEG image.

Acknowledgements. This work was supported by NSFC (Grant 62002075), Guangdong Basic and Applied Basic Research Foundation (Grant 2023A1515011428), the Science and Technology Foundation of Guangzhou (Grant 2023A04J1723).

References

1. Pevný, T., Filler, T., Bas, P.: Using high-dimensional image models to perform highly undetectable steganography. In: Proceedings of the 12th Information Hiding Workshop, pp. 161–177 (2010)
2. Li, B., Wang, M., Huang, J., Li, X.: A new cost function for spatial image steganography. In: Proceedings of the IEEE International Conference on Image Processing, pp. 4026–4210 (2014)
3. Sedighi, V., Cogranne, R., Fridrich, J.: Content-adaptive steganography by minimizing statistical detectability. IEEE Trans. Inf. Forensics Secur. **11**(2), 221–234 (2016)
4. Guo, L., Ni, J., Su, W., Tang, C., Shi, Y.Q.: Using statistical image model for jpeg steganography: uniform embedding revisited. IEEE Trans. Inf. Forensics Secur. **10**(12), 2669–2680 (2015)
5. Yang, J., Ruan, D., Huang, J., Kang, X., Shi, Y.Q.: An embedding cost learning framework using GAN. IEEE Trans. Inf. Forensics Secur. **15**, 839–851 (2019)
6. Tang, W., Li, B., Li, W., Wang, Y., Huang, J.: Reinforcement learning of non-additive joint steganographic embedding costs with attention mechanism. Sci. China Inf. Sci. **66**(3), 132305:1-132305:14 (2023)
7. Zhu, J., Kaplan, R., Johnson, J., Fei-Fei, L.: HiDDeN: hiding data with deep networks. In: Ferrari, V., Hebert, M., Sminchisescu, C., Weiss, Y. (eds.) ECCV 2018. LNCS, vol. 11219, pp. 682–697. Springer, Cham (2018). https://doi.org/10.1007/978-3-030-01267-0_40
8. Zhang, K.A., Cuesta-Infante, A., Veeramachaneni, K.: Steganogan: High capacity image steganography with GANs. arXiv preprint arXiv:1901.03892 (2019)
9. Baluja, S.: Hiding images in plain sight: deep steganography. In: Proceedings of the 31th International Conference on Neural Information Processing Systems, pp. 2069–2079 (2017)
10. ur Rehman, A., Rahim, R., Nadeem, S., ul Hussain, S.: End-to-end trained CNN encoder-decoder networks for image steganography. In: Leal-Taixé, L., Roth, S. (eds.) ECCV 2018. LNCS, vol. 11132, pp. 723–729. Springer, Cham (2019). https://doi.org/10.1007/978-3-030-11018-5_64
11. Yu, C.: Attention based data hiding with generative adversarial networks. In: Proceedings the AAAI Conference on Artificial Intelligence, pp. 1120–1128 (2020)
12. Zhang, C., Benz, P., Karjauv, A., Sun, G., Kweon, I.S.: UDH: Universal deep hiding for steganography, watermarking, and light field messaging. In: Proceedings of the 33th Conference on Neural Information Processing Systems, pp. 10223–10234 (2020)
13. Guan, Z., et al.: Deepmih: deep invertible network for multiple image hiding. IEEE Trans. Pattern Anal. Mach. Intell. **45**(1), 372–390 (2023)
14. Lu, S.P., Wang, R., Zhong, T., Rosin, P.L.: Large-capacity image steganography based on invertible neural networks. In: Proceedings of the IEEE Conference on Computer Vision and Pattern Recognition, pp. 10816–10825 (2021)
15. Wang, Z., Feng, G., Ren, Y., Zhang, X.: Multichannel steganography in digital images for multiple receivers. IEEE Multimedia **28**(1), 65–73 (2021)
16. Kweon, H., Park, J., Woo, S., Cho, D.: Deep multi-image steganography with private keys. Electronics **10**(16), 1906:1–1906:10 (2021)
17. Boehm, B.: Stegexpose-a tool for detecting LSB steganography. arXiv preprint arXiv:1410.6656 (2014)

Member Inference Attacks in Federated Contrastive Learning

Zixin Wang[1], Bing Mi[2,3], and Kongyang Chen[1,3(✉)]

[1] Institute of Artificial Intelligence and Blockchain, Guangzhou University,
Guangzhou, China
[2] Guangdong University of Finance and Economics, Guangzhou, China
[3] Pazhou Lab, Guangzhou, China
kychen@gzhu.edu.cn

Abstract. In the past, the research community has studied privacy issues in federated learning, self-supervised learning, and deep models. However, privacy investigations into the domain of federated contrast learning are rarely exploited. Consequently, our research endeavours to unveil the potential privacy risks intrinsic to federated contrast learning. In this paper, we introduce four types of membership inference attacks to probe into and analyse the privacy protection performance of federated contrast learning models. To gain a more holistic understanding of the privacy concerns in federated contrast learning, we systematically assess the efficacy of various membership inference attacks within this realm. Simultaneously, we scrutinise the potential risks posed by these attack methods from multiple perspectives and examine their applicability in real-world settings. Through these evaluations, our objective is to furnish the academic community with a more lucid viewpoint, thereby fostering a comprehensive appreciation of the privacy safeguarding capabilities of federated contrast learning models.

Keywords: Contrastive Learning · Federated Learning · Membership Inference Attack

1 Introduction

Investigations into the privacy of deep neural networks commenced early and have significantly matured, with the membership inference attack as a classic methodology for scrutinising the privacy implications of such network models. The exploration into the privacy risks associated with federated learning was thoroughly examined. However, with the recent burgeoning of contrastive learning methods, and the incremental adoption of contrast learning, the exploration into contrast learning approaches based on contrastive learning remains relatively sparse. Recent work utilized cosine similarity within contrastive learning to implement membership inference attacks, thereby unmasking the inherent privacy risks in contrast learning based on contrastive learning. This suggests

J. Vaidya et al. (Eds.): AIS&P 2023, LNCS 14509, pp. 42–52, 2024.
https://doi.org/10.1007/978-981-99-9785-5_4

that federated contrast learning architectures may harbour similar privacy risks, an area which has hitherto remained largely unexplored. Therefore, our research confirms these privacy risks and addresses this knowledge gap by employing distinct membership inference attack methodologies and perspectives.

Our contributions include employing membership inference attacks to evaluate the privacy risks in federated contrast learning. First, we utilise the similarity value from cosine similarity for membership inference attacks. Second, we apply gradient ascent to overfitted models, observing that member data exhibit a slower rise in data loss following gradient ascent compared to non-member data, enabling member/non-member discrimination. Third, we concatenate the cosine similarity and the one value with the highest confidence in the data loss and model encoder predictions into a three-dimensional array and subsequently train a linear classifier for membership/non-membership prediction. Lastly, we expose the internal privacy risks inherent in contrastive learning models. Given that the internal encoder of such models is typically a deep neural network such as ResNet, we extract the encoder and subject it to traditional membership inference attacks.

The remainder of this paper is organized as follows. Section 2 discusses the related work. Section 3 presents our method. Section 4 shows the experiment results. Finally, Sect. 5 concludes this paper.

2   Related Work

Researchers have explored distributed contrastive learning methods and proposed privacy-preserving federated contrastive learning (FCL) strategies [4]. These strategies keep raw data local and train a global encoder by integrating local encoders. Recent studies primarily focus on the non-independent identically distributed (Non-IID) issue in FCL. For instance, Zhang et al. [9] designed a dictionary module and an alignment module for acquiring superior feature representation in Non-IID data. Zhuang et al. proposed a divergence-aware module to mitigate the weight divergence issue in Non-IID data. Zhuang et al. utilized local knowledge to alleviate the Non-IID problem. Other contemporary solutions have explored the application of FCL in various application domains.

Recently, to integrate distributed unlabeled data while keeping the raw data local, researchers have been focusing on applying contrastive learning in the context of federated learning. One of the primary challenges that FCL faces is the Non-IID data across distributed clients. Zhang et al. [9] introduced a method called FedCA, which includes a dictionary module to maintain the consistency of the representation space and an alignment module for aligning feature representations. FedCA manages to extract superior feature representations from Non-IID data. Zhuang et al. [11] revealed that Non-IID data could cause weight divergence and introduced a divergence-aware module to tackle this issue. Zhuang et al. discovered that retaining local knowledge is beneficial for Non-IID data. Based on this finding, they proposed a FedEMA method to tackle the Non-IID data issue. Other concurrent studies mainly focus on applying FCL to specific domains. Also, there are similar solutions to infer the security [3,8] and privacy [10] issues. For example, Pan et al. [5] presented the membership inference attacks and backdoor attacks for model architectures. Chen et al. exploited the membership inference attacks for model unlearning [1,2].

As far as we are aware, we are the first to study member inference attacks in the context of federated contrastive learning. Our work builds on existing research into member inference attacks in federated learning and contrastive learning. For instance, Shokri et al. [7] provided a comprehensive evaluation of active and passive member inference attacks on federated supervised learning models. Liu et al. [6] uncovered that self-supervised learning models, like contrastive learning models, also pose privacy risks, by leveraging the distinct cosine similarity features in contrastive learning for member inference attacks. Their research thus indicates that federated contrastive learning could also be susceptible to privacy leaks, laying the groundwork for member inference attacks in a distributed context.

3   Our Method

In this section, we first review the paradigm of federated contrastive learning and briefly analyze the privacy risks in the federated contrastive model.

3.1   General Framework of Federated Contrastive Learning

Server Parameter Aggregation: Federated Contrastive Learning (FCL) is a label-free distributed learning system. Traditional supervised learning requires manual annotation of a large number of labels, which can lead to high human resource costs. Compared to labeled data, unlabeled data are easier to collect and can be obtained in greater quantities. In the process of federated contrastive learning, each client first receives the model structure and parameters issued by the server, and then trains with their own unlabeled data Di locally. Assuming the number of clients is N, these N clients collaboratively train an encoder under the coordination of the server. The goal of federated contrastive learning is to minimize the average loss of each client's model. The loss function is as follows:

$$\arg \min_{w \in R^d} f(w) = \frac{1}{N} \sum_{i=1}^{N} f_i(w) \tag{1}$$

where fi is the loss function of the ith client. Specifically, in the t-th round, the server sends the current server's encoder Gi to n and designated clients. Each designated client independently trains the server's Gi encoder on its local unlabeled data Di, resulting in the locally trained encoder Li(t+1). Then, the client sends the encoder update, Li(t+1)-Gt, to the server. The server averages the collected model parameters to obtain a new server-side encoder Gi+1, as shown below:

$$G_{t+1} = G_t + \frac{\eta}{n} \sum_{i=1}^{n} (L_{t+1}^i - G_t) \tag{2}$$

The training process will iterate until the server's encoder converges. Obviously, Federated Contrastive Learning (FCL) is significantly different from Federated Learning (FL). Firstly, FL aims to learn a model for a specific task, while FCL aims to learn an encoder that can be used for multiple downstream tasks. Secondly, FL's local training algorithm is supervised, while FCL's local training algorithm is unsupervised.

Local Contrastive Learning: The goal of contrastive learning is to make the similarity of the same data's augmented data as high as possible, and the similarity of different data outputs as low as possible. Most existing contrastive learning models use InfoNCE as the training loss function. We use MoCo in our experiments for unsupervised learning. The objective of MoCo is to learn features in the representation space that clearly distinguish between positive and negative examples. The main idea of MoCo is to introduce a momentum encoder and a larger memory queue to construct the contrastive loss.

The MoCo architecture includes two encoders, a query encoder and a key encoder with momentum. The query encoder is used to extract feature representations h_q from the original data. The key encoder is an encoder with a momentum update strategy, it extracts feature representations h_k from the original data. The parameters of the query encoder are updated at each training step, while the parameters of the key encoder are updated based on the parameters of the query encoder with a certain momentum coefficient.

MoCo uses data augmentation techniques to generate two views (i.e., two samples). Views from the same sample form a positive pair, while views from different samples form a negative pair. The memory queue stores past keys, allowing for consideration of more negative samples when calculating the contrastive loss. Each element in the queue is a feature vector that has been extracted and encoded from the original data.

For the positive pair (i, j), MoCo's loss function is defined as follows:

$$L(i,j) = -\log\left(\frac{\exp\left(f(q_i)\cdot f(k_j)/\tau\right)}{\sum_{n=1}^{N}\exp\left(f(q_i)\cdot f(k_n)/\tau\right)}\right) \tag{3}$$

In which, $f(q_i)$ and $f(k_j)$ represent the feature representations of the query and key respectively, N is the number of negative samples, and τ is the temperature parameter, which acts as a smoothing factor in the loss function. The loss function aims to maximize the similarity between positive samples while minimizing the similarity between negative samples. Compared with SimCLR, MoCo introduces more negative samples into the loss function through the momentum encoder and memory queue. This method shows good performance in unsupervised learning tasks and can effectively extract meaningful feature representations.

3.2   Security Issues in Federated Contrastive Learning

Due to its distributed nature, Federated Contrastive Learning (FCL) may also encounter similar security issues as those in Federated Learning (FL). In FCL, membership inference attacks are also a potential security risk. The aim of a membership inference attack is to determine whether a specific data sample has participated in the model's training. In a membership inference attack, the attacker tries to analyze the global model to identify which data samples are part of the training data. This attack can lead to data leaks and privacy violations, especially when the data involved in the training contains sensitive information. However, membership inference attacks against FCL are significantly different from those against FL. Firstly, in FL, the training of the model usually relies on labeled data. Attackers can leverage this label information to launch membership inference attacks. In contrast, in FCL, each participant learns an encoder in an unsupervised way, meaning all local data are unlabeled. This makes executing membership inference attacks in FCL more difficult. Secondly, the distributed nature and learning objectives of FCL are different from those of FL. While FL aims to learn a model for a specific task, FCL aims to learn a general encoder that can be applied to multiple downstream tasks. Since the data distribution may differ from the downstream tasks, this makes strategies for membership inference attacks based on data distribution less feasible.

3.3   Membership Inference Attacks Against Federated Contrastive Learning

Due to the distributed nature of federated contrastive learning, an attacker can potentially come from a client. The goal of the attacker is to infer whether the data is training member data by obtaining the output of model inference. Assuming the attacker comes from the client, they have two ways to carry out membership inference attacks, which we refer to as passive and active membership inference attacks. Passive attackers do not participate in training or interfere with it, but only obtain the model parameters after training is completed, and then launch membership inference attacks. Active attackers will perform gradient ascent on their local model, execute gradient ascent on the data to be inferred, upload it to the server and other aggregated model parameters, and then compute the loss for the data that just underwent gradient ascent.

4   Experiments

4.1   Experimental Settings

Datasets: We conducted experiments on the SVHN, CIFAR10, CIFAR100 data sets, as illustrated in Table 1.

Table 1. Datasets description

Datasets	Shape	Classes	Number of training data	Number of testing data
CIFAR-10	$32 \times 32 \times 3$	10	50,000	10,000
CIFAR-100	$32 \times 32 \times 3$	100	50,000	10,000
SVHN	$32 \times 32 \times 3$	10	73,257	26,032

Experimental Details: We implemented a series of attacks mentioned above using PyTorch in Python 3.7. Our computational resources included 4 NVIDIA V100 GPUs. For each experiment, although it was successfully implemented on each dataset, some attack diagrams were not significantly different, so some attack diagrams are represented by CIFAR10. During the experiment, we set the number of clients to 4, which allowed federated learning of the model and made the model more likely to overfit, thus reducing the consumption of computational resources. All of our experiments were conducted using Non-IID distributions, because IID distribution is a better way to defend against overfitting, but in real life, most federated learning scenario data distributions are Non-IID. In the passive membership inference attack, we record the model parameters every hundred rounds of aggregation and then perform a membership inference attack on it. To allow the model to overfit faster, we pre-trained the model parameters initially issued by the server for 500 rounds. To control variables, we used a combination of 10,000 training data and 10,000 test data on all datasets as the experiment, and in general, member data and non-member data were used for the experiment operation during membership inference. In the active membership inference attack, we recorded the loss of member and non-member data at each aggregation round, and saved the parameters of the aggregated model every hundred rounds for static active membership inference attack. The number of local training rounds was 10. The learning rate parameters of SGD during gradient ascent and training were both 0.1, but the learning rate of SGD during gradient ascent could be appropriately reduced to prevent large fluctuations in the loss record and make the whole curve smoother.

To evaluate the performance of the encoder, common methods include linear evaluation and weighted KNN evaluation. Linear evaluation measures the feature representation extracted by the encoder by training a linear model. Weighted KNN evaluation classifies by comparing the cosine similarity of feature representations and using a weighted voting k-nearest neighbor method. During training, we used weighted KNN evaluation for monitoring, and used linear evaluation to test the final performance.

Evaluation Metrics: The membership inference attacks mainly evaluate the classifier, so our evaluation metrics are accuracy, precision, and recall. Accuracy is the ratio of the number of correctly classified samples to the total number of samples.

Precision represents the proportion of samples that actually belong to the positive class in the samples predicted as the positive class by the classifier.

Precision is concerned with the accuracy of the samples predicted as positive by the classifier. Recall represents the proportion of samples that actually belong to the positive class and are correctly predicted as the positive class by the classifier. Recall is concerned with the classifier's ability to identify positive samples.

4.2 Experimental Results

During the membership inference attack, we first used the method based on cosine similarity for the experiment. This method is mainly used to detect the overfitting characteristics of federated self-supervised learning. The experimental results prove that the federated self-supervised learning framework indeed has privacy risks caused by overfitting, as shown in Fig. 1. Although the federated learning framework has stronger overfitting resistance compared to single training learning, there is still a risk of overfitting. The following Table 2 shows the performance data of membership inference attacks based on cosine similarity performed on the model obtained after multiple rounds of federated aggregation.

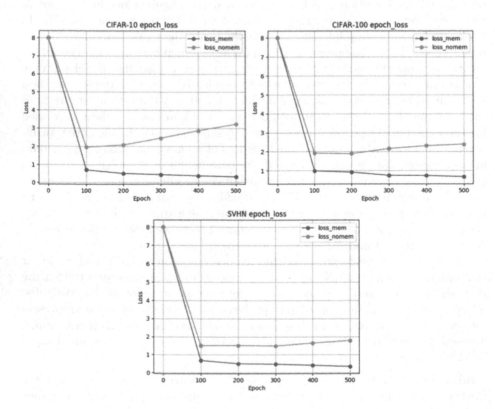

Fig. 1. The overfitting characteristics for different datasets.

Table 2. Datasets description of membership inference attacks.

Datasets	Accuracy	Precision	Recall
CIFAR-10	0.914	0.897	0.943
CIFAR-100	0.932	0.911	0.964
SVHN	0.905	0.901	0.924

We collected the model parameters after every hundred rounds of aggregation during model training, and used the model to calculate the cosine similarity and loss for member and non-member data as shown in Fig. 2. It can be found that the model's generalization ability declines after overfitting. Although the contrast learning model itself has a good ability to resist overfitting, the generalization of the model is slowly declining as the number of training epochs increases. After the model is overfitted, the loss of non-member data is slowly rising, indicating that the generalization ability of the model is gradually declining. Similarly, as the overfitting of the model becomes more and more serious, the cosine similarity of non-member data is also slowly decreasing.

These results provide valuable insights into the privacy risks associated with federated self-supervised learning due to overfitting. They highlight the need for further research into strategies for mitigating these risks while maintaining the effectiveness of the learning process.

Next, we focus on the generic membership inference attack based on the internals of the model. In our experiments, we found that if the contrast learning model is overfitted, then the encoder extracted from it is even more severely overfitted. And because in most cases this encoder is a deep neural network model, like ResNet, VGG, we can directly conduct the traditional membership inference attack on it. This entails having it predict the probabilities of classes, and using the highest probabilities as input training data for the classifier. For this, we only need to inform the model how many classes there are, without providing definite labels, and the model will give out prediction probabilities. There will be a high degree of confidence in these prediction probabilities, meaning the values of the top few predictions will be very high. Therefore, we believe that this method can be used to further investigate the privacy risks of the model.

We extracted the output of each layer within the encoder for the membership inference attack and found that the output of the last few layers within the model can yield a fairly high accuracy when used as a binary classifier. Figure 3 shows the success rate of the membership inference attack obtained for each layer of the neural network within the encoder of the extracted model for each dataset compared to the accuracy of the first membership inference attack as illustrated in Table 3.

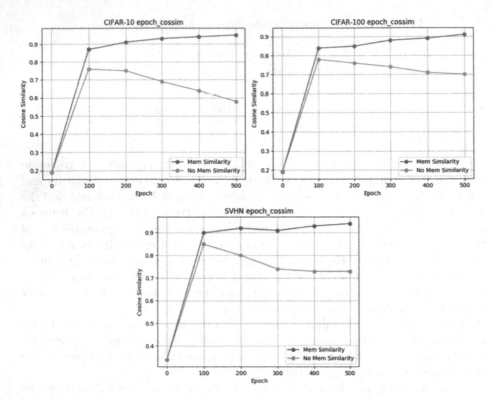

Fig. 2. The overfitting characteristics for different datasets.

Table 3. The Top3 confidence scores for member and non-member data across different layers in CIFAR10.

	Member	Non-member
Encoder	[68.1, 14.8, 5.9]	[18.1, 7.7, 5.1]
Avgpool	[1.5, 1.0, 0.7]	[0.3, 0.3, 0.2]
Layer4	[36.7, 11.1, 5.4]	[4.4, 1.2, 0.5]

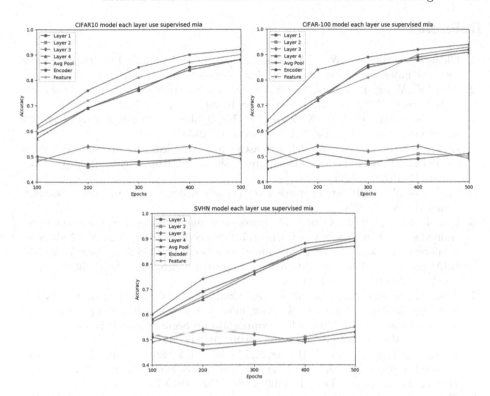

Fig. 3. The cosine similarity comparison of each layer inside the model's encoder and the membership inference attack.

5 Conclusion

Our study confirms the risks of federated self-supervised learning. When the overall model is overfitted due to improper parameter selection or extremely uneven data distribution across clients, the model will have more serious privacy risks. The privacy issue of federal self-supervised learning also needs to be taken seriously. However, when the number of clients increases, the amount of data increases, and the data distribution of clients is close to iid are better able to resist overfitting, and then adding some defenses such as differential privacy, early stop, etc., can make the federal self-supervised learning model obtain better generalization ability to resist membership inference attacks and privacy leaks.

Acknowledgments. This work is supported by National Natural Science Foundation of China (No. 61802383), Research Project of Pazhou Lab for Excellent Young Scholars (No. PZL2021KF0024), Guangzhou Basic and Applied Basic Research Foundation (No. 202201010330, No. 202201020162), Guangdong Regional Joint Fund Project (No. 2022A1515110157), and Research on the Supporting Technologies of the Metaverse in Cultural Media (No. PT252022039).

References

1. Chen, K., Huang, Y., Wang, Y., Zhang, X., Mi, B., Wang, Y.: Privacy preserving machine unlearning for smart cities. Ann. Telecommun. (2023)
2. Chen, K., Wang, Y., Huang, Y.: Lightweight machine unlearning in neural network. CoRR abs/2111.05528 (2021). https://arxiv.org/abs/2111.05528
3. Chen, K., Zhang, H., Feng, X., Zhang, X., Mi, B., Jin, Z.: Backdoor attacks against distributed swarm learning. ISA Transactions (2023)
4. Li, Q., He, B., Song, D.: Model-contrastive federated learning. In: Proceedings of the IEEE/CVF Conference on Computer Vision and Pattern Recognition (CVPR), pp. 10713–10722 (2021)
5. Li, Y., et al.: Model architecture level privacy leakage in neural networks. Sci. China Inf. Sci. (2022)
6. Liu, H., Jia, J., Qu, W., Gong, N.Z.: Encodermi: membership inference against pre-trained encoders in contrastive learning. In: Proceedings of the 2021 ACM SIGSAC Conference on Computer and Communications Security. CCS 2021, New York, NY, USA, pp. 2081–2095. Association for Computing Machinery (2021). https://doi.org/10.1145/3460120.3484749
7. Nasr, M., Shokri, R., Houmansadr, A.: Comprehensive privacy analysis of deep learning: Passive and active white-box inference attacks against centralized and federated learning. In: 2019 IEEE Symposium on Security and Privacy (SP), pp. 739–753 (2019). https://doi.org/10.1109/SP.2019.00065
8. Wang, Y., Chen, K., Tan, Y., Huang, S., Ma, W., Li, Y.: Stealthy and flexible trojan in deep learning framework. IEEE Trans. Dependable Secur. Comput. **20**(3), 1789–1798 (2023). https://doi.org/10.1109/TDSC.2022.3164073
9. Zhang, F., et al.: Federated unsupervised representation learning. CoRR abs/2010.08982 (2020). https://arxiv.org/abs/2010.08982
10. Zhou, D., Chen, W., Chen, K., Mi, B.: Fast and accurate SNN model strengthening for industrial applications. Electronics **12**(18), 1–12 (2023)
11. Zhuang, W., Gan, X., Wen, Y., Zhang, S., Yi, S.: Collaborative unsupervised visual representation learning from decentralized data. In: Proceedings of the IEEE/CVF International Conference on Computer Vision (ICCV), pp. 4912–4921 (2021)

A Network Traffic Anomaly Detection Method Based on Shapelet and KNN

Si Yu[✉][iD], Xin Xie[iD], Zhao Li[iD], Wenbing Zhen[iD], and Tijian Cai[iD]

East China Jiaotong University, Nanchang 330013, China
niliyuer@163.com

Abstract. Network traffic anomaly detection is the foundation for discovering malicious attacks and securing network security. With the emergence of new technologies such as port masquerading and traffic encryption, traditional traffic anomaly detection methods face many difficulties in dealing with large-scale, high-dimensional, and diverse network traffic data, such as traffic features needing to be more abstract and the model being uninterpretable. In this paper, we construct a network traffic anomaly detection model based on shapelet and KNN (K-Nearest Neighbor). First, the backpropagation and k-shape algorithm are used to learn the set of shapelet instances; second, the DTW of the shapelet and the original sequence is calculated as attribute values to generate the transformed dataset of test set and shapelet; finally, combine with KNN classifier for network traffic anomaly detection. In this paper, multi-classification experiments are conducted on one available dataset, NSL-KDD with 99.18% accuracy, and the experimental results are analyzed for model solvability.

Keywords: shapelet · interpretability · traffic anomaly detection

1 Introduction

The explosive growth of network traffic makes it increasingly tricky to rationally allocate bandwidth resources, prevent network congestion, and ensure the regular operation of critical services [1]. Therefore, making fast and accurate identification of abnormal traffic can provide knowledge support for network security posture assessment and immunization decisions and improve the overall response capability of network security emergency organizations [2]. Advances in network technology have driven improvements in network traffic anomaly detection methods to adapt to Internet changes.

The traditional network traffic anomaly detection methods are mainly port-based and Deep Packet Inspection (DPI). The former is suitable for the case of a few types of network traffic, and network applications generally correspond to fixed port numbers [3]. But the classification accuracy obtained by relying only on the port number for traffic classification is low. The DPI-based approach, on

J. Vaidya et al. (Eds.): AIS&P 2023, LNCS 14509, pp. 53–64, 2024.
https://doi.org/10.1007/978-981-99-9785-5_5

the other hand, achieves a more accurate classification of network traffic by analyzing the content of the packets. Sen et al. [4] accurately identified point-to-point (P2P) traffic by parsing the load content of the traffic packets and determining the application signature embedded inside them. With the widespread use of technologies such as port masquerading and traffic encryption, these methods have many limitations [5].

To address the limitations of traditional methods, machine learning methods have been introduced into the traffic classification field. The researchers extracted network traffic features, including statistical descriptive features and self-learning high-dimensional features. Then they trained machine learning models to classify different traffic feature patterns [6]. Among them, KNN [7] and k-means [8] are clustering-based detection methods that achieve anomaly detection by finding dense regions in the data and defining anomalies as values far from these dense regions. The above methods depend highly on feature selection, so selecting features significantly impacts classification accuracy.

Unlike machine learning methods, deep learning methods do not rely on artificially designed features but learn the high-dimensional features of the input data autonomously through multilayer networks. For example, the anomaly detection method based on a recurrent neural network [9] which uses Long Short-term Memory (LSTM) model to preserve time series information, has a robust temporal feature learning capability. The above methods often take raw traffic data as input for building end-to-end deep learning models and learn high-dimensional features from the raw traffic autonomously through deep learning methods, i.e., raw input data and output anomaly detection results. The whole anomaly detection process then becomes a black box and completely agnostic, leading to many safety accidents [10]. Consequently, interpretability will become an inevitable choice for the development of AI.

In recent years in the field of time series classification, shapelet-based time series classification techniques have been widely used with high interpretability and high accuracy and can be used for high-dimensional data. Hence, this paper constructs a network traffic anomaly detection model based on shapelet and KNN that can both automatically learn to generate network traffic features and have interpretability. The main work is as follows:

(1) The backpropagation and k-shape algorithm are used to learn the set of shapelet instances;
(2) Construct a transformed dataset of the test set and shapelet instances to convert the original time series set into a feature vector set that can be input to most machine learning algorithms;
(3) And combine transformed dataset with the KNN classifier to realize the anomaly judgment of network traffic data, and the experimental results are analyzed interpretably in conjunction with professional knowledge.

2   Related Work

2.1   Shapelet Basic Knowledge

A shapelet exists as a subseries of a time series, which can appear at any position of the time series data and can characterize a category of data to the maximum extent. The concept of shapelet was first proposed in the literature [11], and it was used to classify nettle leaves and verbena, and the classification effect is shown in Fig. 1, where verbena is on the left and nettle leaves on the right. From Fig. 1, it can be seen that the two leaves show different pinch angles at the corresponding positions of the shapelet. In the subsequent classification, it is only necessary to compare whether the corresponding shapelet is present in the one-dimensional time series transformed from the leaf contour to classify the two leaves.

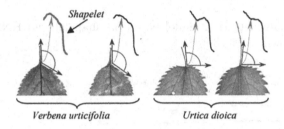

Fig. 1. Classification of nettle and verbena based on shapelet. [11]

2.2   K-Nearest Neighbor Algorithm

The KNN algorithm [12] is a simple and intuitive supervised learning algorithm for classification and regression proposed by Cover and Hart, i.e., making decisions concerning only these K nearest neighbor samples. The principle is shown in Fig. 2:

The three most essential elements of the KNN algorithm are calculating the distance, determining the k-value, and the classification decision. The distance measure measures the correlation between the test data and the known data set. The choice of k-value is critical and will directly influence the prediction results. When the K value is small, the model will depend on nearby neighboring samples and has better sensitivity. The stability will be weaker, quickly leading to overfitting. When the K value is small, the model will depend on nearby neighboring samples and has better sensitivity, but the stability will be weaker, which will quickly lead to overfitting. When the K value is larger, the stability increases, but the sensitivity will weaken, soon leading to underfitting. The decision is divided into voting and weighted voting. Voting is minority-majority, and the category with the highest number of points among each neighbor of the input test data K is classified as that point. Weighted voting weighs each of K's immediate neighbors according to the distance; the closer the distance, the greater the weight.

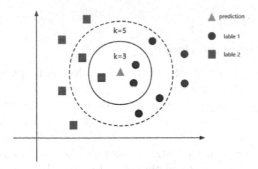

Fig. 2. KNN algorithm principle diagram.

3 Propose Methods

The traffic anomaly detection model based on shapelet and KNN proposed in this article is shown in Fig. 3.

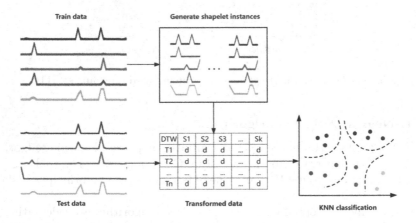

Fig. 3. Traffic anomaly detection model based on shapelet and KNN.

3.1 Generating Shapelet Instances

First, define the set of traffic data time series:

$$T = \{T_1, \ldots, T_n\}. \tag{1}$$

Assume that the length of all traffic data time series in T is Q, the ith traffic data time series T_i consists of Q elements:

$$T_i = \{T_{i,1}, \ldots, T_{i,Q}\} \tag{2}$$

where $T_{i,Q}$ denotes the Qth element of the ith traffic data time series. $T_{i,j:L}$ denotes the jth segment of length L in the ith traffic data time series:

$$T_{i,j:L} = \{T_{i,j}, \ldots, T_{i,j+L-1}\} \tag{3}$$

Next, define a subset of the shapelet sequence:

$$S = \{S_1, \ldots, S_k\} \tag{4}$$

Assume that the length of all shapelets in S is L, the kth shapelet sequence consists of L values:

$$S_k = \{S_{k,1}, \ldots, S_{k,L}\} \tag{5}$$

$S_{k,L}$ is the Lth element of the kth shapelet sequence. The Euclidean distance between the kth shaplet sequence Sk and $T_{i,j:L}$ is represented as:

$$D_{i,k,j} = \frac{1}{L} \sum_{l=1}^{L} (T_{i,j+l-1} - S_{k,l})^2 \tag{6}$$

where $T_{i,j+L-1}$ denotes the (j+L-1)th element of the ith traffic data time series. $S_{k,l}$ is the first element of the kth shaplet sequence, $l = 1, 2, \ldots L$, the shapelet transformation between the kth shapelet sequence S_k and the ith traffic data time series T_i is represented as:

$$C_{i,k} = \min_{j \in \{1, \ldots, Q-L+1\}} D_{i,k,j} \tag{7}$$

Given a subset of shapelet sequences S, C_i is the shapelet transform of the ith traffic data time series T_i, and the component of C_i is $\{C_{i,k}\}_{1 \leq k \leq K}$.

Let T_i and T_{i+1} be two adjacent time series in the traffic data time series set T, the shapelet transformations of T_i and T_{i+1} are denoted as C_i and C_{i+1}, and record the DTW between T_i and T_{i+1} as: $DTW(T_i, T_{i+1})$, the loss $\mathcal{L}(T_i, T_{i+1})$ is defined as:

$$\mathcal{L}(T_i, T_{i+1}) = \frac{1}{2}(DTW(T_i, T_{i+1}) - \beta \parallel C_i - C_{i+1} \parallel_2)^2 \tag{8}$$

where β is the scale parameter and the total loss of the N traffic data time series T is:

$$\mathcal{L}(T) = \frac{2}{N(N-1)} \sum_{i_1=1}^{N-1} \sum_{i_2=i_1+1}^{N} \mathcal{L}(T_i, T_{i+1}) \tag{9}$$

The backpropagation algorithm is used to learn the subset S of shapelet sequences corresponding to the traffic data time series T and the scale parameter β to minimize the overall loss.

3.2   K-Shape Algorithm

The center of mass of the set of shapelet instances is calculated using k-shape clustering [13] and used as the final shapelet instance. First, the similitude between the group of traffic data time series T and the subset of shapelet sequences S is expressed utilizing the intercorrelation a, defined as:

$$CC_w(S,T) = (c_1, \ldots, c_w) \tag{10}$$

where CC_w is a sequence of correlations of length n+k-1, c_w denoting the wth mutuality value, $w \in \{1,2,...,n+k-1\}$. The coefficient normalization of CC_w is performed to remove the amplitude and phase effects. Coefficient normalization is the geometric mean of the intercorrelation series divided by the autocorrelation of the individual series, and here in this paper, a biased estimator NCC_b is used, defined as follows:

$$NCC_b = \frac{CC_w(S,T)}{k} \tag{11}$$

The center of mass is the data point with the smallest sum of square distances from all other data points. Assume that a shapelet sequence exists with n observation points and define it as:

$$S_k = \{x_1, x_2, ..., x_i, ..., x_n\} \tag{12}$$

Clustering is the partitioning of a sequence of k shapelets into M pairwise disjoint clusters, defining the set of clusters as:

$$P = \{p_1, ..., p_m, ..., p_M\} \tag{13}$$

where p_m represents the mth cluster, $M < k$, $m \in \{1,2,...M\}$, then $x_i \in p_m$. To avoid the problem that the center of mass does not make effective use of category features, the k-shape converts the center of mass computation to find the minimum value of the sum of the squares of the distances of the center of mass from all other time series u_i:

$$u_i = \arg\min_{x_i \in P_m} \sum \text{dist}(w, x_i)^2 \tag{14}$$

where w is a representative symbol for a virtual center of mass, which can be interpreted as many points of mass to be selected. Since the interrelationship number measures the similarity of the time series rather than the dissimilarity, converting u_i to ascertain the maximum value of the observation point and the center of mass NCC_b.

$$u_i = \arg\max_{x_i \in p_m} \sum NCC_b(x_i, w)^2 \tag{15}$$

$$u_i = \arg\max_{x_i \in p_m} \sum \left(\frac{\max_w CC_w(x_i, w)}{k} \right)^2 \tag{16}$$

The final k-shape clustering method is implemented by recalculating the center of mass, reassigning each sequence to a different cluster established on their distance from the new center of mass, and iterating in a loop until it no longer changes.

4 Experiment

4.1 Experimental Setup

Experimental Environment. All experiments were run on a CentOS machine with Intel(R) Xeon(R) Gold 6126 CPU @2.60 GHz configuration. The Pytorch framework was used for generating shapelet instances, and the knn classification was done using the Keras framework. The programming language and framework versions used for the experiments were: python3.8.8, pytorch1.10.0, TensorFlow-gpu2.10.0, and keras2.10.0.

Evaluation Metrics. In this experiment, Accuracy, Precision, Recall, and F1 values are used to evaluate the performance of the model, and the relevant definitions are as follows:

$$Accuracy = \frac{TP + TN}{TP + TN + FP + FN} \tag{17}$$

$$Precision = \frac{TP}{TP + FP} \tag{18}$$

$$Recall = \frac{TP}{TP + FN} \tag{19}$$

$$F1 = \frac{2 \times Pr\ ecision \times Recall}{Pr\ ecision + Recall} \tag{20}$$

In the above equation (17)(18)(19), TP indicates that the true positive category is judged as positive; FN indicates that the true positive category is misclassified as negative; FP indicates that the true negative category is misclassified as positive; and TN indicates that the true negative category is judged as negative at the same time. In the multiclassification model, the experiment used the weighted average method to assign weights based on the proportion of different class. [13].

4.2 Dataset and Preprocessing

NSL-KDD. Each record in the NSL-KDD dataset contains 43 dimensions of features. The types of attacks in the training and test datasets of NSL-KDD are classified into DoS, U2R, and Probe. The training and test data category features were classified into five classes according to the characteristics of network attacks. Then the category labels of string type were converted into numerical labels, and the correspondence of the converted labels is shown in Table 1:

Dataset Preprocessing. Data pre-processing is separated into two parts; the first part is data cleaning: processing character type features and converting character type features in the original dataset into numerical features; the second part is data transformation: processing numerical type features, normalizing and normalizing the data input to the model according to the model settings.

Table 1. NSL-KDD dataset labels

Original category labels	Category labels after tagging	Number
Normal	0	77054
DoS	1	53363
Probe	2	14230
R2L	3	3418
U2R	4	252

4.3 Experimental Results and Analysis

Experiment 1: Selection of KNN Hyperparameters. KNN algorithm determines the category of new samples according to the nearest k training samples by the classification decision rule, so this paper conducted experiments on the selection of k value, and the results are shown in Table 2:

Table 2. Results for different values of k

k neighbors	NSL-KDD Acc
1	**0.9918**
2	0.9901
3	0.9898
4	0.9881
5	0.9871

Obviously when k=1, both datasets have the highest accuracy, which is 0.9918.

Experiment 2: Multi-classification Anomaly Detection Results. The value of k was set to 1, and traffic multiclassification experiments were conducted. The results of the NSL-KDD dataset are shown in Table 3:

Table 3. NSL-KDD multi-classification results

k neighbors	Accuracy	Classification	precision	recall	F1
1	0.9918	normal	0.99	1.00	1.00
		DoS	0.99	1.00	0.99
		probe	0.98	0.97	0.98
		r2l	0.96	0.96	0.96
		u2r	0.82	0.62	0.71

Experiment 3: Comparison with Existing Methods. For the datasets, five methods with better classification effects in the past five years were selected for comparison to verify the anomaly detection effect of this model. The comparison results for the NSL-KDD are shown in Table 4:

Table 4. Compare accuracy with other methods.

NSL-KDD Methods	Accuracy
Shapelets-Based [14]	95
DFEL-KNN [15]	98.82
IDS-KNN [16]	98.64
CFS-RF [17]	98.60
BWO-CONV-LSTM [18]	98.67
Ours	**99.18**

Experiment 4: Ablation Experiment. To verify the usefulness of the Generate shapelet instances module, this experiment uses the KNN algorithm to detect anomalies in the raw data of dataset, and the results obtained are compared with the best values in Table 1. The results are shown in Table 5:

Table 5. Ablation experiment results.

Dataset	Generate shapelet	KNN	Accuracy
	-	✓	74.13
NSL-KDD	✓	✓	**99.18**

From the data analysis in the table, there is a substantial improvement in the multiclassification accuracy of the NSL-KDD dataset. It can be proved that the Generate shapelet instance module can enhance the accuracy of traffic anomaly detection.

Experiment 5: Interpretability Experiments. Compared with other methods, the shapelet-based classification method has some interpretability. However, these interpretations usually require some expertise. Take the normal and DoS abnormal traffic of NSL-KDD as an example: there are significant differences in src_bytes, dst_bytes, count, and dst_host_counts values between the two. For this reason, we design experiments to visualize the results of these two types

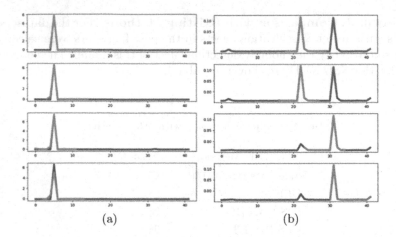

Fig. 4. Visualization of the generated shapelet instances for normal and DoS traffic.

of traffic Generate shapelet instance, as shown in Fig. 4, Figure (a) is normal traffic, and Figure (b) is DoS abnormal traffic.

The specific meanings of these four columns are described below.

"src_bytes": number of data bytes transferred from the source to the target in a single connection; "dst_bytes": number of data bytes transferred from the target to the original in a single bond; "count": number of connections to the same target host as the current connection in the last two seconds; "dst_host_count": number of contacts with the same target host IP address.

By comparing figures (a) and (b), we can find that the values of "src_bytes" and "dst_bytes" in normal traffic are significantly higher than those of DoS abnormal traffic. In comparison, the values of count and "dst_host_counts" in normal traffic are significantly lower than those of DoS abnormal traffic. This is because DOS attacks are denial-of-service attacks that exploit program vulnerabilities to execute resource exhaustion on a one-to-one basis [19]. Creating large amounts of useless data causes congestion in the network leading to the attacked host, preventing the host from communicating with the outside world properly. From this analysis, this paper's proposed shaplet-based anomaly detection method has a certain extent of interpretability.

5   Conclusion

We propose a traffic anomaly detection model based on shapelet and KNN, which uses a backpropagation and k-shape algorithm to learn shapelets, constructing its transformed dataset with the original time series, changing the initial set of time series into a set of feature vectors that can be input to most machine learning algorithms, and finally combining with KNN classifier for anomaly detection of traffic.

Experiments on NSL-KDD datasets demonstrate that the model proposed in this paper can achieve better accuracy than existing deep learning-based shapelet methods and combine expertise to analyze the solvability of the shapelet visualization part.

Acknowledgments. This paper is supported by the National Natural Science Foundation of China, under Grant No. 62162026, Science and Technology Project supported by the education department of Jiangxi Province, under Grant No. GJJ210611, and the Science and Technology Key Research and Development Program of Jiangxi Province, under Grant No. 20203BBE53029.

References

1. Wang, W., Wang, C., Guo, Y.: Industrial control malicious traffic anomaly detection system based on deep autoencoder. Front. Energy Res. **8**, 555145 (2021)
2. Xie, X., Ning, W., Huang, Y.: Graph-based Bayesian network conditional normalizing flows for multiple time series anomaly detection. Int. J. Intell. Syst. **37**, 10924–10939 (2022)
3. Moore, A.W., Papagiannaki, K.: Toward the accurate identification of network applications. In: Dovrolis, C. (ed.) PAM 2005. LNCS, vol. 3431, pp. 41–54. Springer, Heidelberg (2005). https://doi.org/10.1007/978-3-540-31966-5_4
4. Sen S., Spatscheck O., Wang D.: Accurate, scalable in-network identification of p2p traffic using application signatures. In: The Web Conference (2004)
5. Dainotti A., Pescapé A., Claffy K.: Issues and future directions in traffic classification (2012)
6. Cai L., Janowicz K., Mai G.: Traffic transformer: capturing the continuity and periodicity of time series for traffic forecasting. In: Transactions in GIS vol. 24, pp. 736–755 (2020)
7. Tian, J., Azarian, M., Pecht, M.: Anomaly detection using self-organizing maps-based k-nearest neighbor algorithm. In: PHM Society European Conference (2014)
8. Moisés F., Bruno B., Lucas D.: Anomaly detection using baseline and K-means clustering. In: SoftCOM 2010, 18th International Conference on Software, Telecommunications and Computer Networks, pp. 305–309(2020)
9. Radford B., Apolonio L., Trias A.: Network traffic anomaly detection using recurrent neural networks. arXiv:1803.10769 (2018)
10. Hong W., Wang Y.: Prediction method of lane changing frequency based on neural network and Markov chain. In: Journal of East China Jiaotong University (2019)
11. Ye L., Keogh E.: Time series shapelets: a new primitive for data mining. In: Knowledge Discovery and Data Mining (2009)
12. Zhu H., Basir O.: An adaptive fuzzy evidential nearest neighbor formulation for classifying remote sensing images. In: IEEE Transactions on Geoscience and Remote Sensing, vol. 43, pp. 1874–1889 (2005)
13. Qu, Y., Bao, T., Li, L.: Do we need to pay technical debt in blockchain software systems. Connect. Sci. **34**, 2026–2047 (2022)
14. Kim Y., Sa J., Kim S.: Shapelets-based intrusion detection for protection traffic flooding attacks. In: DASFAA Workshops (2018)
15. Zhou Y., Han M., Liu L.: Deep learning approach for cyberattack detection. In: IEEE INFOCOM 2018 - IEEE Conference on Computer Communications Workshops (INFOCOM WKSHPS), pp. 262–267 (2018)

16. Abrar I., Ayub Z., Masoodi F.: A machine learning approach for intrusion detection system on NSL-KDD dataset. In: International Conference on Smart Electronics and Communication (ICOSEC), pp. 919–924 (2020)
17. Abrar I., Ayub Z., Masoodi F.: A machine learning approach for intrusion detection system on NSL-KDD Dataset. In: Sensors (Basel, Switzerland) 20 (2020)
18. Kanna, P., Santhi, P.: Hybrid intrusion detection using MapReduce based black widow optimized convolutional long short-term memory neural networks. Expert Syst. Appl. **194**, 116545 (2022)
19. Xie, X., Li, X., Xu L.: HaarAE: an unsupervised anomaly detection model for IOT devices based on Haar wavelet transform. Appl. Intell. 1–13 (2023). https://doi.org/10.1007/s10489-023-04449-z

Multi-channel Deep Q-network Carrier Sense Multiple Access

Xunxun Pi[1] and Junhang Qiu[2]

[1] School of Information Engineering, Jiangxi University of Science and Technology,
No. 115, Yaohu University Park, Nanchang, Jiangxi, China
[2] School of Information Engineering, East China Jiao Tong University,
Nanchang 33000, China
2020031010000221@ecjtu.edu.cn

Abstract. With the continuous development of network communication and the application of many specific scenarios, the dynamics of network traffic continues to increase, making the optimization of routing problems an NP-hard problem. When using traditional routing algorithms, accuracy and efficiency cannot be balanced. Recently, Deep Q-Network (DQN) has shown great potential for solving dynamic network problems. However, existing DQN-based routing solutions often overlook network environment issues related to packet level, packet size, expected transmission time, and do not generalize well when the network changes. In this paper, we present a new carrier sense multiple access (CSMA) protocol called MC-DQN CSMA, which employs Deep Q-Network to improve the performance of the network. First, we propose a distance constraint under the signal-to-interference-to-noise ratio (SINR) model, which effectively avoids interference and improves the probability of success. Based on the dynamic and unpredictable needs of Ad Hoc networks, we try to use DQN strategy to train the network's agents without expert knowledge. Furthermore, we demonstrate the performance of the proposed algorithm by comparing it with other methods and describing it graphically, which focus on transmitting packets in multi-channel Ad hoc networks.

Keywords: SINR model · Multi-channel · Ad Hoc networks · Reinforcement learning · Deep Q-network

1 Introduction

In our paper, we study the communication in Ad Hoc networks, and propose a improved Carrier Sense Multiple Access (CSMA) mechanism, i.e. MC-DQN CSMA. Consistent with the traditional CSMA, in MC-DQN CSMA, a node has the ability to sense signals on the channel and has to sense the channel before transmitting a message. We use Request To Send/Clear To Send (RTS/CTS)

Supported by Jiangxi Province 03 Special Project (No. 20203ABC03W07).

J. Vaidya et al. (Eds.): AIS&P 2023, LNCS 14509, pp. 65–80, 2024.
https://doi.org/10.1007/978-981-99-9785-5_6

message to achieve the handshake between senders and receivers, thereby delivering the information about the channel, sub-slot and so on.

The main contributions of the paper can be summarized as follows.

- In order to solve the interference in Ad Hoc networks, we propose the distance constraint for the algorithm under Signal to Interference plus Noise Ratio (SINR) model.
- We divide the slot into control slot and the data slot which consists of multiple sub-slots. Besides, multiple channels are adopted for transmitting the packet.
- We use the Deep Q-network algorithm, treat each sender as an Agent, and train the Agent by building a simulated Ad hoc environment. Finally, the Agent will interact with and analyze the network environment, and then choose the best action.

2   Related Works

The Ad Hoc network is one of the mainstream networks and the hop topic of research. Aiming to improve the performance of the network, there exists a lot of research that has made the contribution for achieving high throughout in Ad Hoc networks [1–4]. In fact, many factors can make a influence to the throughput, such as the property of hardware, the traffic load. It has been discussed that the transmission probability can also affect the throughput. [5] uses energy harvesting to power a mobile Ad Hoc network by a stochastic-geometry model, in which each sender has a transmission probability. And the authors prove that the maximum network throughput increases with the optimal transmission probability. Moreover, some studies investigate other methods to achieve high throughput in Ad Hoc networks.

A noticeable issue is how to solve the interference in Ad Hoc networks. In recent years, researchers tend to use SINR model to deal with the inherent problem. Compared with other interference models, SINR model is more realistic and has been applied for various scenarios, such as link scheduling [6–9], spanner [10,11], dominating set [12], broadcasting [13]. In SINR model, a transmission is considered successful if and only if the SINR value of the link is greater than a given threshold. Thus, a link needs to take the interference caused by its neighbor nodes into account. However, the interference will be decreased along with the improvement of distance between two nodes due to the pass loss.

Although multiple-channel situation is more complicated than single-channel situation, it can greatly increase the throughput of the network [14], and many works have been devoted to utilize the multiple channels in communication protocols [15–17]. In multiple-channel situation, the most important issue is the assignment of the channels. At the early stage, [18] proposes a new multi-channel MAC protocol for wireless mobile Ad Hoc networks, which follows an "on-demand" style to assign channels to the nodes. In this method, the number of required channel is independent of the size or topology of the network. [19] proposes a channel-switching scheme to deal with channel congestion, thereby achieving the throughput improvement. In our work, channel congestion is solved under SINR model, and nodes select the channel randomly, which can ensure the uniformly distribution. But the RTS message is transmitted with a probability.

Labeled datasets are rare in the field of communication networks, as a result, people have tried reinforcement learning (RL) that can directly generate an action forwarding behavior, which does not require other auxiliary algorithms for routing strategies. For example, Hu et al. [20] have used Q-learning to manage the routing based on the energy in the nodes of a wireless sensor network (WSN). Such Q-learning-based methods can only calculate simple input information, as the Q values are stored using value tables. Sun et al. [21,22] have proposed the use of DRL to adjust the link weights in a communication network, based on which the global routing of the network is adjusted. Xu [23]has used Multi-agents deep deterministic policy gradient (MADDPG) to solve routing problems in a distributed manner. In our work, we treat the sender as an Agent and train it using the DQN algorithm. At the same time, we fully consider the state of the dynamic network environment, and choose to incorporate the distance constraint with the receiver in the current channel, the current packet level, the current packet size, and the expected transmission time into the state space. This allows the Agent to accurately grasp environmental information and make the best decision.

3 Models and Definitions

In this section, we will introduce the SINR model and the multi-channel model in the following subsections.

3.1 SINR Model

As an inherent problem in wireless networks, nodes transmitting simultaneously on the same channel can interfere with each other, which can lead to collisions and retransmission costs. To describe the interference in Ad Hoc networks, we use the SINR model, which is more practical than other interference models and more in line with practical applications.

Assume that there exist n nodes in the Ad Hoc network. At the initialization stage, arbitrary node i should send a probe message to its neighbors, aiming to calculate the distance between itself and neighbors. The distance is recorded as a sequence, such as d_{iu}, d_{iv}, d_{iw}, which represents the distance to node u, v, w and so on. As for the interference in the network, we use SINR model to evaluate the tolerant interference of links, which is more practical and more realistic than other interference protocols. Now consider a link l_{uv}, for which the sender is node u and the receiver is node v. According to SINR model, link l_{uv} can transmit successfully if and only if it satisfies the function as follows:

$$SINR = \frac{Pd_{uv}^{-\alpha}}{I_{sum} + N} = \frac{Pd_{uv}^{-\alpha}}{\sum_{w \in \Psi_{u,v}} Pd_{wv}^{-\alpha} + N} \geq \beta. \qquad (1)$$

where $\mathcal{P}$ is the transmission power; α is the path-loss exponent, whose value is normally between 2 and 6; N indicates the noise of the background; and I_{sum} represents the interference experienced by the receiver v caused by all the simultaneously transmitting nodes, which are involved in set Ψ, on the same channel with node u and v. In particularly, link l_{uv} can transmit successfully if and only if its $SINR$ value is above or equal to the value of the threshold β.

It can be known that the noise of the background exists in the normal situation, and the interference will be generated once the neighbor nodes start to transmit a packet. Then $N > 0$ and $I_{sum} \geq 0$ can be assured. Therefore, the tolerant sum of the distance (i.e. Φ) can be induced by the inequality in formula 1 as follows:

$$\frac{P d_{uv}^{-\alpha}}{\sum_{w \in \Psi_{u,v}} P d_{w\tilde{v}}^{-\alpha} + N} \geq \beta \Rightarrow D_{sum} = \sum_{w \in \Psi\{u,v\}} d_{wv}^{-\alpha} \leq \frac{\mathcal{P} d_{uv}^{-\alpha} - \beta N}{\mathcal{P} \cdot \beta} = \Phi.$$

$$(2)$$

A transmission can be successful if and only if the sum of the distance is lower than the tolerant sum of the distance, i.e. $D_{sum} \leq \Phi$, and we define D_{sum} as the distance constraint. Since the distance constraint can be traced back to $SINR$ model definition, a transmission can be considered to meet the $SINR$ requirement when it satisfies the distance constraint. As mentioned in the basic model, nodes calculate the distance to neighbors at the initialization stage. Thus Φ can be determined at once when d_{uv} has been known or, in other words, when receiver v has confirmed that node u is its sender.

3.2   Multi-channel Model

As for the channel mode, we study the transmission in a multi-channel communication with $\Gamma \geq 2$ available channels. And slots can be divided into control time slots and data time slots. Figure 1 shows the illustration of the channels and slots. In particularly,there exists a special channel, i.e. the public channel, which aims for transmitting RTS/CTS messages and is marked as c_0. Other $\Gamma - 1$ channels $c \in \{c_1, c_2, \cdots, c_j, \cdots, c_{\Gamma-1}\}$ are selected by nodes for transmitting packets. As shown in Fig. 1, the public channel is only used in the control slot, while other channels are used in the data slot.

We assume that the number of nodes, which tend to transmit a packet in the data slot, is greater than $\Gamma \log n$, thus multiple nodes will operate on different channels with high probability when nodes uniformly select the channels.

Since each node may start to transmit its packet at various time points, we further divide the data slot into multiple sub-slots $t \in \{t_1, t_2, \cdots, t_k, \cdots, t_\sigma\}$, in which σ is the scale of the data slot and satisfies that $\sigma > \frac{\log n}{\Gamma}$. Therefore, the sender and the receiver can confirm the specific time point by transmitting RTS/CTS messages with each other.

As a kind of handshake protocol, RTS/CTS messages are the classic method in CSMA mechanism, which is adopted to confirm the related information of the transmission, thereby reducing the potential problems, such as hidden terminal.

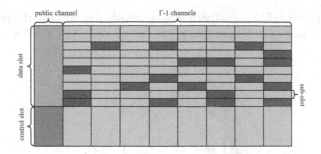

Fig. 1. The illustration of channels and slots

In our algorithms, RTS/CTS messages are transmitted on channel c_0 in the control slot. The process of receiving RTS message is illustrated in Fig. 2.

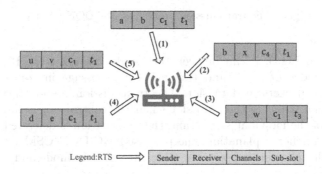

Fig. 2. RTS messages received by node

As shown in Fig. 2, the number sequence represents the sequence that node v receives the RTS messages. Receiver v has received the RTS messages respectively from node a, b,c and d before receiving the RTS message from its sender u. By decoding the RTS message from sender u, receiver v confirms that it is the destination of the RTS message, and obtains the information of predetermined channel and sub-slot, which are channel c_1 and t_1, respectively. Then receiver v can calculate its $D_{sum} = d_{av} + d_{dv}$. Now, the value of D_{sum} can be verified whether it satisfies the distance constraint: if $D_{sum} \leq \Phi$, receiver v will send a CTS message back to sender u with the sub-slot unchanged; if $D_{sum} > \Phi$, receiver v will adjust the sub-slot, and send a CTS message back to sender u with the information of the adjusted sub-slot.

4 Algorithm Description

Based on the models in Sect. 3, the algorithm of MC-DQN CSMA is described in this section. As mentioned above, the slot consists of the control slot (i.e. $\mathcal{CT}$)

and the data slot (i.e. $\mathcal{DT}$). Furthermore, we divide the slot into the multiple sub-slots. Figure 3 shows the transmission process of MC-DQN CSMA mechanism.

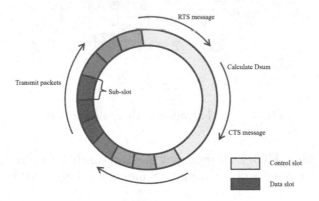

Fig. 3. The transmission process of MC-DQN CSMA

As shown in Fig. 3, it can be seen that nodes transmit the RTS message, calculate the value of D_{sum} and transmit CTS message in control slot, and nodes transmit packets in data slot. The transmission round starts from the control slot, and launches the data slot when the control slot is finished. Then a new round of the loop starts over, and the information is delivered along with the loop. For further explanation, the proposed MC-DQN CSMA algorithm is described in the following subsections, which are receiver mode and sender mode respectively.

4.1   Receiver Mode

If a node has no packet to be transmitted in a round, it is a potential receiver. It is required to listen on the public channel in a CT.

The potential receiver may receive one or multiple RTS messages. We define the "target RTS" to indicate the RTS message whose destination is the potential receiver. Then potential receiver decodes the message and confirm whether the RTS message is its target RTS. If the message is not the target RTS, the information of the message will only be recorded in the RTS list of the receiver. Otherwise, according to the distance constraint, the receiver calculates the value of D_{sum} based on the information in the target RTS. When the distance constraint is satisfied, the receiver can send a CTS message back to its sender, and adjusts the selected sub-slot if necessary.

The implementation details for adjusting sub-slots are as follows: If the sub-slot reserved by the sending end conflicts with sub-slots reserved by other nodes in previous control time slots, the receiving end will adjust the proposed sub-slot from the sending end. It will find the nearest sub-slot in the data time slot of this transmission round that avoids conflicts and include the adjusted sub-slot in the CTS message.

The receiver, which has not received the target RTS, keeps listening on the public channel. However, the receiver will turn to the selected channel in the DT after it sends a CTS message successfully.

Algorithm 1 below describes the relevant actions of receiver in MC-DQN CSMA.

Algorithm 1. MC-DQN CSMA(Receiver).

Require: $flag = 1 \rightarrow$ listening on public channel c_0;
$\qquad\qquad flag = 0 \rightarrow$ otherwise
Ensure: $\mathcal{CT}$ - control slot; $\mathcal{DT}$ - data slot
1: **while** $slot = \mathcal{CT}$ && $flag$ **do**
2: $\quad$ listen on the channel;
3: $\quad$ **if** receive a RTS message **then**
4: $\qquad$ decode the message and confirm the address
5: $\qquad$ **if** the address is not correct **then**
6: $\qquad\quad$ record the information about sender s_i, channel c_j, sub-slot t_k, etc.;
7: $\qquad$ **else**
8: $\qquad\quad$ confirm the information about channel c_j, sub-slot t_k, etc.;
9: $\qquad\quad$ calculate D_{sum} and check the distance constraint;
10: $\qquad\quad$ adjust sub-slot to t'_k;
11: $\qquad\quad$ **Let** $t'_k =$ current scheduled sub-slot;
12: $\qquad\quad$ **if** t'_k has caused a conflict **then**
13: $\qquad\qquad$ **Let** $M =$ all of the sub-slots;
14: $\qquad\qquad$ **for** *iteration* $i = t'_k$ **to** M:
15: $\qquad\qquad\quad$ **if** i is not occupied **then**
16: $\qquad\qquad\qquad$ $t'_k = i$;
17: $\qquad\qquad\qquad$ **Break**;
18: $\qquad\qquad\quad$ **end if**
19: $\qquad\qquad$ **end for**
20: $\qquad\quad$ **end if**
21: $\qquad\quad$ send a CTS message back to the sender;
22: $\qquad\quad$ turn to channel c_j;
23: $\qquad\quad$ $flag = 0$
24: $\qquad$ **end if**
25: $\quad$ **end if**
26: **end while**
27: **while** $slot = \mathcal{DT}$ **do**
28: $\quad$ **if** $flag == 0$ **then**
29: $\qquad$ wait for the packet on channel c_j in sub-slot t'_k;
30: $\qquad$ turn to public channel c_0;
31: $\qquad$ $flag = 1$;
32: $\quad$ **end if**
33: **end while**

4.2   Sender Mode

The CT can be regarded as the period for preparing a transmission. A sender, which has a packet to be transmitted, should pick up a channel c_j from $\gamma - 1$ channels and randomly select a sub-slot t_k of data slot in the period. After determining the information about the channel, sub-slot, etc., the sender sends a RTS message with probability P_i ($0 < P_i < 1$) to its receiver.

After the corresponding receiver calculates the value of D_{sum} and check the distance constraint, the selected sub-slot may be adjusted to avoid the collision with other transmission. Therefore, the determined sub-slot can be known by decoding the CTS message, which is sent from the receiver. If the sender has not received the CTS message from its corresponding receiver when the control slot is finished, the transmission between itself and the receiver cannot be operated in the data slot.

Algorithm 2 below describes the relevant actions of sender in MC-DQN CSMA.

Algorithm 2. MC-DQN CSMA(Sender).

Require: $flag = 1 \rightarrow$ listening on public channel c_0;
$\qquad\qquad flag = 0 \rightarrow$ otherwise
Ensure: $\quad CT$ - control slot; DT - data slot
1: **while** $slot = CT$ && $flag$ **do**
2: $\quad$ listen on the channel;
3: $\quad$ **if** have a message to transmit **then**
4: $\qquad$ randomly select channel c_j for transmitting the packet;
5: $\qquad$ randomly select a sub-slot t_k;
6: $\qquad$ **while** the current channel is occupied **do**
7: $\qquad\quad$ randomly select a time period to wait;
8: $\qquad$ **end while**
9: $\qquad$ send a RTS message to receiver with probability P_i;
10: $\qquad$ listen on the channel;
11: $\qquad$ **if** receive a CTS message **then**
12: $\qquad\quad$ adjust the sub-slot to t'_k;
13: $\qquad\quad$ turn to channel c_j;
14: $\qquad$ **else**
15: $\qquad\quad$ try to send the RTS again;
16: $\qquad$ **end if**
17: $\quad$ **end if**
18: **end while**
19: **while** $slot = DT$ **do**
20: $\quad$ **if** $flag == 0$ **then**
21: $\qquad$ transmit the packet on channel c_j in sub-slot t'_k;
22: $\qquad$ turn to public channel c_0;
23: $\qquad$ $flag = 1$;
24: $\quad$ **end if**
25: **end while**

Transmitting a message with a certain probability is a classic method in CSMA mechanism, which can improve the success probability and reduce the collision in a transmission. Regarding the probability of transmitting the RTS message, since all senders desire to reserve the initial time slot for delivering messages promptly, conflicts are likely to arise. Therefore, it is essential to minimize the potential collision that can occur when multiple RTS/CTS messages are sent concurrently.

5   CSMA Based on DQN

In this section, we will introduce the key elements of CSMA in DQN algorithm, including state, action, and reward function. The following is a summary of these elements, combined with the nature of CSMA algorithm.

5.1   State Space

There are four network parameters in the state space, which are distance constraints, priority of data packets to be transmitted, current data packet size and expected transmission time.

Distance constraint: According to the definition in the SINR model described above, when the link L meets the formula 2, the current link is considered to be able to transmit successfully, otherwise, it cannot be successful, and the model does not consider the transmission in the link L, which means that the link does not participate in the game.

Priority of Packets to be Transmitted: Each sender plays a game without knowing the probability selection of other nodes, and sets the gain value of the packet according to the importance of the information. Due to the different contributions of data packets to the network, some data packets have higher priority. This article defines five message levels for distinguishing the importance of the data packet currently being sent, so the message level is a discrete value from 1 to 5.

Current Data Packet Size: The size of data packets affects the transmission time, the success rate during transmission, and the degree of network congestion. This is crucial for determining when the sender should send RTS messages and how to allocate communication resources. When there are many large data packets in the network, communication resources may become scarce, potentially resulting in a higher probability of collisions and delays. In this situation, intelligently adjusting the probability of sending RTS messages helps improve network performance.

Expected Transmission Time: Expected transmission time is an important indicator for measuring link transmission performance. A shorter expected transmission time usually implies better link quality, while a longer expected transmission time may indicate poorer link quality or higher network congestion.

The expected transmission time will be calculated in real-time based on the following introduction: To calculate the expected transmission time, the data packet size (DPS), the channel transmission rate (CTR), and the channel reliability (CR) are required. Specifically, formula 3 can be used to calculate the expected transmission time:

$$ETT = DPS/(CTR \times CR). \tag{3}$$

In this case, the DPS represents the size of the data packet, and the unit can be bytes or bits. The CTR represents the time required for the data packet to travel from the sender to the receiver, and its unit is usually bits per second (bit/s). CR refers to the probability of errors occurring in the network during the data transmission process.

5.2   Action Definition

The output of the DQN neural network simulates the Q-values of the Q-function. In this case, the dimension of the action space should be discrete, representing different probability levels for sending RTS messages. For example, we can divide the probabilities into 10 levels(0%,10%,20%,...,100%). After the output layer, the Q-value corresponding to each probability level is calculated separately, and the agent will ultimately choose the action with the highest Q-value.

5.3   Reward Function

Since our primary goal is to improve network throughput while minimizing the conflicts caused by simultaneous message transmissions, ensuring the optimal benefit of the entire network. To achieve this, we choose the link benefit value between the agent (i.e., the sender) and the receiver as the reward function's measurement metric. The formula for calculating the link benefit value is as follows:

$$\lambda_u(T) = S_u g_u P_u - \frac{c}{\zeta^{T_u}} + lnP_u - P_u. \tag{4}$$

Specifically, $G_u(T) = g_u P_u$ is the gain value of the data packet, where g_u is the unit gain when the data packet is successfully transmitted; $C_u(T) = \frac{c}{\zeta^{T_u}} + lnP_u - P_u$ is the cost required to transmit the data packet, where is the unit cost required for data packet transmission, and ζ is a factor to ensure that the cost is reduced when the transmission time ends.

In this case, the calculation of g_u is related to the expected transmission time and the level of the data packet. Specifically, first divide the data packet size (DPS) by the expected transmission time (ETT) to calculate the amount of data that can be transmitted per unit of time. Multiply this by the corresponding weight (W), where different levels of data packets have different weights. The final result is the unit gain brought by the successful transmission of the current data packet. The calculation formula is as follows:

$$g_u = W \times (DPS / ETT). \tag{5}$$

The unit cost of data packet transmission (c) can be calculated by comprehensively evaluating three key factors: energy consumption, processing time, and channel quality. For example, let the average distance a data packet can be transmitted be L, the average energy consumption required for a node to transmit a single data packet be E, the average time for processing, sending, and transmitting a data packet be T, the current channel quality be Q, and ε represents a very small number to prevent the denominator from being zero. The calculation formula for the unit cost of data packet transmission is as follows:

$$c = \frac{E \times L + T \times Q}{1 - Q + \varepsilon}. \tag{6}$$

The value of the reward function is related to the change of linked benefit value in formula 6. When the benefit value of the link increases, reward Agent, $r_{t+1} = 1$. When the benefit value of the link decreases, the agent will be punished, $r_{t+1} = 0$.

$$r(s, a) = \begin{cases} 0 & \lambda_u \leqslant \lambda_u' \\ 1 & \lambda_u \leqslant \lambda_u' \end{cases} \tag{7}$$

5.4 Loss Function

In the DQN algorithm, the loss function is mainly used to measure the difference between the Q-values predicted by the neural network and the actual target Q-values. Typically, DQN uses the Mean Squared Error (MSE) as the loss function, as shown in formula 8.

$$Loss = \frac{1}{n} \sum_{i=0}^{n} \left(Q^i(s, a; \theta) - Q^i_{target}(s, a) \right)^2 \tag{8}$$

In formula 8, n represents the number of data samples extracted from the experience replay buffer in this training round, $Q^i(s, a; \theta)$ is the Q-value output by the main neural network for the i_{th} data sample, and $Q^i_{target}(s, a)$ is the Q-value output by the target neural network for the i_{th} data sample.

6 Simulation

In this section, we prove the performance of proposed algorithm by comparing with other methods, which focus on transmitting packet in multi-channel Ad Hoc networks.

In this section, a simulation environment was implemented using the Python programming language. In the simulation, the experimental environment is stable, and the background noise remains constant. Nodes are randomly distributed in the network and can use multiple channels for data packet transmission. The range of the number of nodes considered in this section is from 0 to 100, and the average energy consumption required for a single data packet transmission by a node is 50J. The performance measured in the experiment includes unit cost, gain value, and benefit value, which will vary with the number of nodes participating in the transmission.

Next, we present the training curve of our model. We trained the agent for 40,000 iterations. As shown in Fig. 4, the value of the loss function in each episode decreased significantly as the number of iterations increased. Moreover, similar to most existing DQN schemes, the agent learned more rapidly in the beginning, and the increase in performance tended to slow down during the training process. For this figure, the unit of measurement on the x-axis is in thousands, revealing that the loss remained relatively stable after approximately 19,000 iterations.

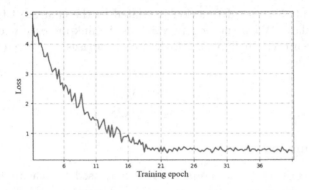

Fig. 4. The training curve.

The methods compared with the algorithm in this chapter are the random (Random) algorithm and the first-in-first-out (FIFO) algorithm. Both of these algorithms are classic transmission strategies in wireless networks. In the random algorithm, nodes are randomly selected to transmit data, while other nodes need to maintain a listening state and wait for the next opportunity to transmit. The FIFO algorithm follows the principle that nodes that collect data first should be prioritized for transmission. Although this method can reduce the probability of conflicts, it may also cause unnecessary delays.

As mentioned earlier, the performance studied in the experiment includes unit cost, gain value, and benefit. The average values of these performance metrics are shown in Figs. 5, 6, and 7 respectively.

Figure 5 shows the changes in unit cost values with the increase in the number of nodes for three methods. The calculation formula for unit cost is as per formula 6. As can be seen, the unit cost of all three methods rises with the increase in the number of nodes, but the MC-DQN CSMA mechanism can achieve a higher unit cost than the other two methods.

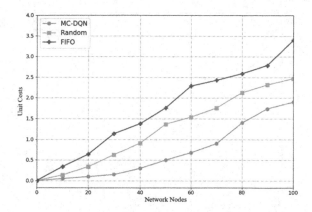

Fig. 5. The unit costs increase with the total number of the network nodes.

Figure 6 shows the changes in gain values with the increase in the number of nodes for three methods. The calculation formula for gain values is as per formula 5. In Fig. 6, the gain values of all three methods decrease rapidly at first and then gradually stabilize with the growth in the number of nodes. However, the gain value of the MC-DQN CSMA mechanism is always higher than that of the other methods.

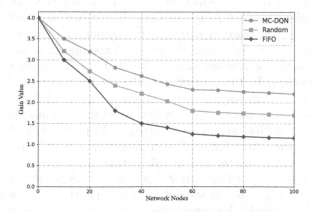

Fig. 6. The overall gain decreases with the total number of the network nodes.

The results in Fig. 7 show that the benefits of the three methods decrease with the increase in the number of nodes. Also, the benefit value is calculated from the unit cost and gain values, as shown in formula 4. According to the simulation results of unit cost and gain values, as well as the theoretical analysis, it can be inferred that the MC-DQN CSMA algorithm can achieve higher benefit values. This is also evidenced by the experimental results, as shown in Fig. 7.

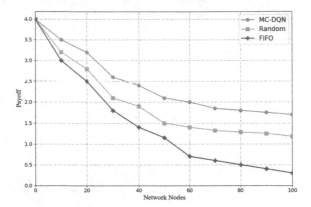

Fig. 7. The overall payoff decreases with the total number of the network nodes.

In summary, according to the performance analysis of unit cost, gain value, and benefit value, the MC-DQN CSMA algorithm proposed in this chapter can achieve better performance in multi-channel Ad hoc networks than other algorithms.

7   Conclusions

This paper proposes the MC-DQN CSMA algorithm, which controls the probability of the sender sending RTS messages to the receiver in an Ad hoc network, thereby driving the increase in network throughput and achieving maximum network benefits. First, we regard the entire transmission process as a transmission round, which is divided into a control slot and a data slot composed of multiple sub-slots. In the control slot, the sender selects a channel and a sub-slot to transmit data packets, and then communicates with the receiver via RTS/CTS messages to reserve the sending time; in the data slot, the sender sends data packets with a certain set of settings. We proposed a SINR model for judging whether the transmission of data packets under the currently selected link can be successful, and for ease of calculation, we derived the SINR model and finally judged whether the current link can be successfully transmitted by calculating the $D_s um$ value. In the Ad hoc network, the network environment status changes frequently. The sender needs to adjust the sending probability of the RTS message in real time according to the environment status. For this, we proposed a Deep Q-learning strategy, and proposed the state space, action space, and reward function of the sender as an Agent. The sender interacts with the environment to choose the best RTS message sending probability, to increase the overall network throughput and maximize network benefits. Finally, we compared the proposed algorithm with other algorithms that focus on transmitting data packets in multi-channel Ad hoc networks, and graphically described the gap between them.

Acknowledgements. This work was supported by Jiangxi artificial intelligence production-education integration innovation center.

References

1. Wang, Y.-L., Mei, S., Wei, Y.-F., Wang, Y.-H., Wang, X.-J.: Improved ant colony-based multi-constrained QoS energy-saving routing and throughput optimization in wireless Ad-hoc networks. J. Chin. Univ. Posts Telecommun. **21**, 43–59 (2014)
2. Luo, J., Zhang, J., Yu, L., Wang, X.: Impact of location popularity on throughput and delay in mobile Ad hoc networks. IEEE Trans. Mobile Comput. **14**, 1004–1017 (2015)
3. Mahdian, M., Yeh, E.M.: Throughput and delay scaling of content-centric Ad hoc and heterogeneous wireless networks. IEEE/ACM Trans. Netw. **25**, 3030–3043 (2017)
4. Zafar, S., Tariq, H., Manzoor, K.: Throughput and delay analysis of AODV, DSDV and DSR routing protocols in mobile Ad hoc networks. Int. J. Comput. Netw. Appl. (IJCNA), **3**, 1–7 (2016)
5. Huang, K.: Spatial throughput of mobile Ad hoc networks powered by energy harvesting. IEEE Trans. Inf. Theory **59**, 7597–7612 (2013)
6. Wang, C., Yu, J., Yu, D., Huang, B., Yu, S.: An improved approximation algorithm for the shortest link scheduling in wireless networks under SINR and hypergraph models. J. Combinatorial Optimiz. **32**, 1052–1067 (2014)
7. Huang, B., Yu, J., Cheng, X., Chen, H., Liu, H.: SINR based shortest link scheduling with oblivious power control in wireless networks. J. Netw. Comput. Appl. **77**, 64–72 (2017)
8. Yu, J., Huang, B., Cheng, X., Atiquzzaman, M.: Shortest link scheduling algorithms in wireless networks under the SINR model. IEEE Trans. Vehicular Technol. **66**, 2643–2657 (2016)
9. Yu, D., Wang, Y., Hua, Q., Yu, J., Lau, F.: Distributed wireless link scheduling in the SINR model. J. Combinat. Optimiz. **32**, 278–292 (2016). https://doi.org/10.1007/s10878-015-9876-8
10. Yu, D., Ning, L., Zou, Y., Yu, J., Cheng, X., Lau, F.C.: Distributed spanner construction with physical interference: constant stretch and linear sparseness. IEEE/ACM Trans. Netw. **25**, 2138–2151 (2017)
11. Zhang, X., Yu, J., Li, W., Cheng, X., Yu, D., Zhao, F.: Localized algorithms for Yao graph-based spanner construction in wireless networks under SINR. IEEE/ACM Trans. Netw. **25**, 2459–2472 (2017)
12. Yu, J., Jia, L., Li, W., Cheng, X., Wang, S., Bie, R., Yu, D.: A self-stabilizing algorithm for CDS construction with constant approximation in wireless networks under SINR model. 2015 IEEE 35th International Conference on Distributed Computing Systems, pp. 792–793. IEEE (2015)
13. Tian, X., Yu, J., Ma, L., Li, G., Cheng, X.: Distributed deterministic broadcasting algorithms under the SINR model. In: IEEE INFOCOM 2016-The 35th Annual IEEE International Conference on Computer Communications, pp. 1–9. IEEE (2016)
14. Ning, L., Yu, D., Zhang, Y., Wang, Y., Lau, F., Feng, S.: Uniform information exchange in multi-channel wireless Ad hoc networks. arXiv preprint arXiv:1503.08570 (2015)

15. Daum, S., Ghaffari, M., Gilbert, S., Kuhn, F., Newport, C.: Maximal independent sets in multichannel radio networks. In: Proceedings of the 2013 ACM Symposium on Principles of Distributed Computing, pp. 335–344 (2013)

16. Shi, W., Hua, Q.-S., Yu, D., Wang, Y., Lau, F.C.M.: Efficient information exchange in single-hop multi-channel radio networks. In: Wang, X., Zheng, R., Jing, T., Xing, K. (eds.) WASA 2012. LNCS, vol. 7405, pp. 438–449. Springer, Heidelberg (2012). https://doi.org/10.1007/978-3-642-31869-6_38

17. Wang, Y., Wang, Y., Yu, D., Yu, J., Lau, F.: Information exchange with collision detection on multiple channels. J. Combinat. Optimiz. **31**, 118–135 (2016). https://doi.org/10.1007/s10878-014-9713-5

18. Wu, S.-L., Lin, C.-Y., Tseng, Y.-C., Sheu, J.-L.: A new multi-channel mac protocol with on-demand channel assignment for multi-hop mobile Ad hoc networks. Proceedings International Symposium on Parallel Architectures, Algorithms and Networks. I-SPAN 2000, pp. 232–237. IEEE (2000)

19. Inokuchi, M., Ueda, H., Motoyoshi, G.: Throughput improvement by disruption-suppressed channel switching in multi-channel Ad-hoc networks. In: 2017 International Conference on Computing, Networking and Communications (ICNC), pp. 615–619. IEEE (2017)

20. Hu, T., Fei, Y.: QELAR: a machine-learning based adaptive routing protocol for energy-efficient and lifetime-extended underwater sensor networks. IEEE Trans. Mob. Comput. **9**, 796–809 (2010)

21. Sun, P., Li, J., Guo, Z., Xu, Y., Lan, J., Hu, Y.: SINET: enabling scalable network routing with deep reinforcement learning on partial nodes. In: Proceedings of the ACM SIGCOMM 2019 Conference Posters and Demos, pp. 88–89 (2019)

22. Sun, P., Hu, Y., Lan, J., Tian, L., Chen, M.: TIDE: time-relevant deep reinforcement learning for routing optimization. Future Gener. Comput. Syst. **99**, 401–409 (2019)

23. Xu, Q., Zhang, Y., Wu, K., Wang, J., Lu, K.: Evaluating and boosting reinforcement learning for intra-domain routing. In: 2019 IEEE 16th International Conference on Mobile Ad Hoc and Sensor Systems (MASS), pp. 265–273. IEEE (2019)

DFaP: Data Filtering and Purification Against Backdoor Attacks

Haochen Wang[1(✉)], Tianshi Mu[2], Guocong Feng[3], ShangBo Wu[1], and Yuanzhang Li[1]

[1] Beijing Institute of Technology, Beijing, China
wanghc@bit.edu.cn
[2] China Southern Power Grid Digital Grid Group Co., Ltd., Guangzhou, China
muts_csg@yeah.net
[3] China Southern Power Grid Co., Ltd., Guangzhou, China
fenggc_csg@yeah.net

Abstract. The rapid development of deep learning has led to a dramatic increase in user demand for training data. As a result, users are often compelled to acquire data from unsecured external sources through automated methods or outsourcing. Therefore, severe backdoor attacks occur during the training data collection phase of the DNNs pipeline, where adversaries can stealthily control DNNs to make expected or unintended outputs by contaminating the training data. In this paper, we propose a novel backdoor defense framework called DFaP (Data Filter and Purify). DFaP can make backdoor samples with local-patch or full-image triggers added harmless without needing additional clean samples. With DFaP, users can safely train clean DNN models with unsecured data. We have conducted experiments on two networks (AlexNet, ResNet-34) and two datasets (CIFAR10, GTSRB). The experimental results show that DFaP can defend against six state-of-the-art backdoor attacks. In comparison to the other four defense methods, DFaP demonstrates superior performance with an average reduction in attack success rate of 98.01%.

Keywords: artificial intelligence · deep learning · AI security · backdoor defense · data filtering · data purification

1 Introduction

Deep Neural Networks (DNNs) have found extensive applications in various fields, including self-driving [1–3], object detection [4,5], and natural language processing [6,7]. However, as a data-starved model, DNNs require a large amount of training data. To achieve optimal performance, users have to obtain training data from unverified external data sources through automated methods or outsourcing.

As shown in Fig. 1(A), it is a common strategy to obtain training data from the public. However, it is hard to guarantee whether an unverified external data

© The Author(s), under exclusive license to Springer Nature Singapore Pte Ltd. 2024
J. Vaidya et al. (Eds.): AIS&P 2023, LNCS 14509, pp. 81–97, 2024.
https://doi.org/10.1007/978-981-99-9785-5_7

source is trustworthy. This makes backdoor attacks a severe threat to the data collection phase of the DNNs pipeline [8], where adversaries can implant backdoors into the model by contaminating the training data, causing severe damage in various situations.

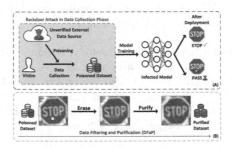

Fig. 1. (A) The DNNs pipeline suffers severe backdoor attacks during the data collection phase. **(B)** To ensure the dataset is non-threatening before the training phase, we propose DFaP implement unsupervised filtering and purification of the training data.

Specifically, the adversary stealthily implants backdoors into DNNs; in the inference phase, the infected model can behave like a clean model for clean samples, but predict malicious, adversaries expected target labels for backdoor samples with triggers. For example, backdoor attacks can easily fool DNNs to recognize a "stop" sign with a small, even invisible trigger as a "pass" sign, thus causing a catastrophic traffic accident. Therefore, various backdoor defense methods are proposed to meet the expectation that the dataset is clean before the training phase.

However, existing countermeasures based on data filtering are facing the following challenges.

Specifically, the adversary stealthily implants backdoors into DNNs. During the inference phase, the infected model can mimic a clean model for regular samples but will predict malicious, adversary-expected target labels for backdoor samples with triggers. For instance, a backdoor attack could deceive a DNN into perceiving a "stop" sign as a "pass" sign with a small, even invisible trigger, potentially leading to a catastrophic traffic accident. Consequently, various backdoor defense methods have been proposed to ensure the cleanliness of the dataset prior to the training phase.

Nevertheless, existing countermeasures based on data filtering encounter the following challenges.

(1) Require additional clean samples. Most backdoor defense methods require a certain percentage of clean samples. Realistic scenarios where the victim collects extra clean samples and ensures the samples are i.i.d with the dataset to be detected are difficult.

(2) Challenging to suppress multiple types of attacks. The triggers of backdoor samples may be a tiny local-patch block [9,10] or hidden full-image perturbation [11,12]. It is challenging to filter multiple types of backdoor samples with high performance for existing defense methods.

(3) False positive results. Current filtering-based methods are difficult to avoid false positive results (i.e., filtering out clean samples), which reduces the model performance [13].

In this paper, we propose a backdoor defense framework referred to as DFaP (Data Filter and Purify) to address the above challenges and ensure that the dataset is clean. To address challenges (1) and (2), DFaP employs CAM Filter for data filtering. This enables DFaP to identify and filter local-patch backdoor samples without requiring additional clean samples. Expressly, the decision features of the backdoor samples are limited to the trigger features [14], while the decision features of the clean samples contain multiple semantic features of the corresponding classes. Therefore, CAM Filter performs erasure repair on the critical decision features of the samples and compares the predicted labels before and after the erasure. The prediction labels change before and after erasure for local-patch backdoor samples, and the rest of the samples remain consistent. To address challenges (2) and (3), DFaP performs data purification via ED Purifier. ED Purifier enables DFaP to purify backdoor samples from local-patch or full-image triggers. The purified samples are still used to train the model, avoiding information loss due to false positive results. Specifically, this component exploits the data-dependent feature of the Encoder-Decoder structure (i.e., it can only compress and recover features similar to the training data) and the self-supervised technique to corrupt triggers while preserving the main semantic features. This makes the backdoor samples non-threatening, while the purified samples can still contribute to the model training phase.

As a result, DFaP enables unsupervised filtering and purification of unsecured DNNs training data.

We summarize our contributions below:

- We propose an unsupervised sample filtering technique called CAM Filter. DFaP can filter backdoor samples without additional clean samples.
- We propose a sample purification technique called ED Purifier. DFaP can purify backdoor samples from local-patch or full-image triggers rather than discard them, thus avoiding information loss due to filtering out clean samples.
- We illustrate the effectiveness and versatility of DFaP through comprehensive experiments involving various backdoor attacks across different datasets.

2   Related Work

2.1   Backdoor Attack

BadNets [9] is the first proposed backdoor attack method against DNNs. The adversary adds backdoor samples with attached triggers to the dataset, thus implanting the backdoor into DNNs. By this method, the prediction results of the model can be manipulated, such as identifying the "stop" sign as a "pass" sign. In order to increase the concealment of triggers, [11,15] proposed to construct invisible triggers based on perturbation. A more attractive attack is the Clean-Label Attack [16,17]. The Clean-Label Attack preserves the label of the backdoor samples and makes the backdoor samples look like benign samples. In the Clean-Label Attack, the source image of the backdoor data may come from different classes. [18,19] proved that adding triggers only to the target class data is also feasible. [10] adds a perturbation to the image based on GAN [20] or FGSM [21], then attaches triggers to build a poisoned dataset. [22,23] propose attacks against federated learning. [24] propose attacks based on image scaling. Trojan Attack [25] demonstrates the generation of triggers to implant backdoors by reverse engineering. [11,26] proposed to construct triggers based on norm constraints. [11] also used Steganography to embed full-image triggers stealthily. [27] proposes frequency domain attack can be realized. FTrojan [12] adds triggers in the frequency domain that behave as a full-image perturbation in RGB space.

2.2 Backdoor Defense

Backdoor defense consists of model inspection and data inspection. Neural Cleanse [28] is a typical model inspection method. It calculates the perturbation that changes the data label, and the slightest perturbation is called a trigger. Fine-Pruning [29] provides defense by pruning model neurons of low relevance for clean samples. NAD [30] proposes a distillation method to reverse the elimination of the trigger. I-BAU [31] propose a minimax formulation and a solver for backdoor removal. The following methods are data inspection methods. [32] proposed to extract the spectral features of the potential representation and filter the backdoor samples based on the outliers. SPECTRE [33] further proposed to amplify the spectral features by whitening the data and using QUantum Entropy as the outlier score. STRIP [34] proposed filtering backdoor data by overlaying multiple samples and computing entropy. [35] proposes a federated learning defense based on secure aggregation strategy. Februus [36] proposed a data purification method based on an interpretable method and the GAN model. SCAn [37] decomposes latent features into a class-specificity identity and a variation component. Thus, the backdoor data is filtered by statistical analysis of the decomposition component. The above defense methods are enlightening. However, addressing the following challenges remains a difficulty, including requiring additional clean samples, not being able to suppress multiple types of attack, and false positive results. This inspires us to build a backdoor defense framework that addresses the above challenges.

3   Methodology

3.1   Threat Model

Adversaries' Abilities: We assume that the attack occurs during the data collection phase, and the victim needs to train the classifier $f(\Theta, x)$ using an unverified external dataset. The adversaries have access to the victim's data, network, and training algorithm. Instead of training the model, the adversaries construct the backdoor sample x_b with the trigger Δ. Then, x_b is added to the victim's dataset.

Adversaries' Goal: The adversaries expect the victim to train with the poisoned dataset D_b and thus the victim unknowingly acquires the infected model $f^b(\Theta, x)$. Specifically, the adversaries have two goals. First, $f^b(\Theta, x)$ still classifies correctly for x without Δ. Second, when x_b with Δ is fed into $f^b(\Theta, x)$, the prediction will always be the target class y_t set by the adversaries.

3.2   DFaP Framework Overview

We consider a realistic threat scenario in the DNNs pipeline as in Fig. 1(A). To ensure the dataset is clean before the training phase, we propose the DFaP framework to achieve unsupervised filtering and purification of the DNNs training data. An overview of DFaP is shown in Fig. 2.

Data Filtering. DFaP includes the data filtering component: CAM Filter, which filters out local-patch backdoor samples without additional clean samples. CAM Filter first extracts key decision features for sample classification and performs erasure repair. For backdoor samples with tiny local-patch triggers added, the remaining semantic features after erasing the backdoor patches cause their predicted labels to change. Therefore, the local-patch backdoor samples are erasure-prone samples. For the samples without local-patch triggers, the remaining semantic features ensure the consistency of the prediction labels before and after erasure. In the second step, CAM Filter predicts the samples before and after erasure, and the erasure-prone samples are judged as the local-patch backdoor samples.

Data Purification. DFaP includes the data purification component: ED Purifier. With the data-dependent feature of the Encoder-Decoder structure and the self-supervised technique, ED Purifier can discard the non-existent backdoor features (i.e., local-patch or full-image trigger) and retain the main semantic features of the samples. Therefore, ED Purifier achieves data purification. As shown in Fig. 2(B-1), the ED Purifier is trained with the filtered dataset. As shown in Fig. 2(B-2), the samples are then purified to continue contributing to model training instead of being discarded.

3.3   Data Filtering

We propose CAM Filter to implement erasure repair of decision features and filter local-patch backdoor samples.

Erase Phase. This stage is to extract the critical decision features of the sample for erasure and repair.

Grad-CAM [38] is a method designed to interpret DNNs decisions, and we utilize it to locate pixels with high weight for model decisions. First, the highest convolutional layer feature map A of the model is extracted. Second, the weight of the kth feature map on class c is defined as α_k^c, which is represented as Eq. (1). Here, Z is the number of feature map pixels, y^c is the score of class c (i.e., the predicted score of class c in logits), and A_{ij}^k is the pixel value at position (i, j) in the kth feature map.

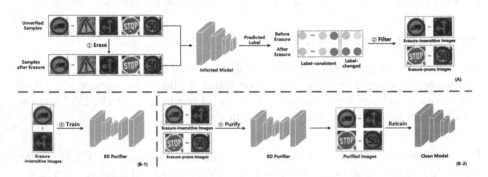

Fig. 2. The overview of DFaP. **(A) CAM Filter:** This component performs data filtering. **(B) ED Purifier:** (B-1) shows the **Train Phase**. (B-2) shows the **Purify Phase**. Eventually, the purified dataset can be used safely train DNNs models.

$$a_k^c = \frac{1}{Z} \sum_i \sum_j \frac{\partial y^c}{\partial A_{ij}^k} \tag{1}$$

Then, the class activation mapping can be obtained by using the weighted sum of weight α_k^c and feature graph A with an additional $ReLU$ operation. The significance of $ReLU$ is that we only care about pixels that positively impact class c. The activation diagram L^c for class c is shown in Eq. (2).

$$L^c = ReLU(\sum_k \alpha_k^c A^k) \tag{2}$$

The value range of L^c is standardized to $[0, 1]$, the size of L^c is changed to the same as the input sample by interpolation, and thus we can get the activation value of any pixel in the input sample. The pixels with activation values higher than the threshold T will be erased.

Since the erased samples need to be reclassified, they should more closely resemble the original samples. So we use a fast image restoration method (i.e., TELEA [39]) for restoration. Unlike the GAN used by Februus, we do not need additional clean samples for training.

Filter Phase. This stage is predicted for the samples before and after erasure separately. As shown in Fig. 3, the samples with changed labels (i.e., erasure-prone samples) are identified as local-patch backdoor samples, thus achieving data filtering.

3.4 Data Purification

The ED Purifier implements purification to render the backdoor samples harmless, while the clean samples retain the semantic features.

Train Phase. In this stage, we trained ED Purifier as shown in Algorithm 1.

In terms of training data, we consider that preparing an additional clean dataset is a demanding requirement. In the DFaP framework, the ED Purifier is trained with the filtered dataset (i.e., erasure-insensitive samples) without the need of additional clean samples.

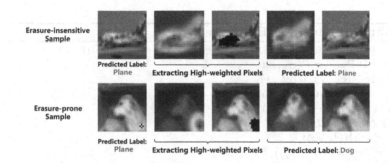

Fig. 3. The samples before and after erasure repair are predicted, respectively. The label of the erasure-insensitive sample in the first row remains unchanged, and the label of the erasure-prone sample in the second row is changed.

The ED Purifier consists of two independent Encoder-Decoder structures, including a traditional AutoEncoder $A(\Theta_a, x)$ and UNet $U(\Theta_u, x)$.

$A(\Theta_a, x)$ aims to remove local-patch trigger features and retain the main semantic features for erasure-prone samples. DFaP selects the AutoEncoder structure as $A(\Theta_a, x)$. Besides, the characteristic of ℓ_{mse} (denoted as Eq. (3)) is that it only considers the average size of all pixel errors, which means it tends to produce an output closer to the overall average and is more likely to destroy trigger features. Therefore, ℓ_{mse} is chosen as the loss function for $A(\Theta_a, x)$.

$$\ell_{mse} = \frac{1}{n} \sum_{i=1}^{n} (x_i' - x_i)^2 \tag{3}$$

$U(\Theta_u, x)$ aims to recover the original semantic features of erasure-prone samples while purifying the possible hidden full-image triggers in erasure-insensitive samples. Therefore, we adopted a self-supervised training based on the $\mathcal{J}$-invariant function [40].

The $\mathcal{J}$-invariant function $f(.)$ is defined as: when $\mathcal{J} = \{J1, J2, ...\}$ is a partition of the sample (where J is a disjoint subset of the sample and the concatenation of all j is the sample x), $f(.)$ is $\mathcal{J}$-invariant if for any $J \in \mathcal{J}$ the function value $f(x_J)$ is independent of the value of x_J (where x_J is an element of x restricted to the coordinates in J).

The full-image backdoor sample x is constructed as in Eq. (4), where y is the original sample and t is the full-image trigger perturbation value.

$$x = y + t \tag{4}$$

The self-supervised optimization is to minimize the ℓ_f as in Eq. (5).

$$\ell_f = \sum_{J \in \mathcal{J}} \mathbb{E}_x \left\| f(x_{Jc})_J - x_J \right\|^2 \tag{5}$$

Algorithm 1. Training the ED Purifier

Input: Filtered dataset D; AE model $A(\Theta_a, x)$; UNet model $U(\Theta_u, x)$;
Output: ED Purifier $P(\Theta_p, x)$;
 1: Initialize $A(\Theta_a, x)$, $U(\Theta_u, x)$
 2: **for** number of training iterations **do**
 3:     $B \leftarrow$ GetImagesBatch(D)
 4:     **for** (x_i, y_i) in B **do**
 5:         $x_i' = A(\Theta_a, x_i))$
 6:         x_i'', $x_i''' = U(\Theta_u, x_{iJ}'), U(\Theta_{u'}, x_{iJ})$
 7:         $\mathcal{L}_{mse} \mathrel{+}= \ell_{mse}(x_i, x_i')$
 8:         $\mathcal{L}_s \mathrel{+}= \ell_f(x_{iJ}, x_{iJ}'') + \lambda\ell_{ms}(x_i, x_i'')$
 9:         $\mathcal{L}_{s'} \mathrel{+}= \ell_f(x_{iJ}, x_{iJ}''') + \lambda\ell_{ms}(x_i, x_i''')$
10:     **end for**
11:     Update $\Theta_a \leftarrow$ AdamOptimizer($\Theta_a, \mathcal{L}_{mse}$)
12:     Update $\Theta_u \leftarrow$ AdamOptimizer($\Theta_u, \mathcal{L}_s$)
13:     Update $\Theta_{u'} \leftarrow$ AdamOptimizer($\Theta_{u'}, \mathcal{L}_{s'}$)
14: **end for**
15: $P(\Theta_p, x) \leftarrow$ Concatenate(A, U)
16: **return** $P(\Theta_p, x)$

Here, f is a $\mathcal{J}$-invariant function(i.e. the $U(\Theta_u, x)$), in which $\|\cdot\|$ denotes the l_2 norm, the partition J of x is a set of disjoint subsets of x whose union is all x, and the Jc is the complement of J.

We use the masking procedure (denoted as Eq. (6)) to modify the x to x_J.

$$x_J = 1_J \cdot s(x) + 1_{Jc} \cdot x \tag{6}$$

Here Jc is the complement of J, 1_J is the indicator function, the value of elements whose coordinates belong to J is 1, and the value of elements whose coordinates belong to Jc is 0; and $s(\cdot)$ is a convolution operation.

In addition, to preserve the semantic information of the erase-insensitive samples and to further recover the semantic information of the erase-prone samples, we add the MS-SSIM constraint term to the self-supervised loss.

MS-SSIM is a multi-scale structural similarity that can measure the repair quality of samples, referred to as MS.

$$\ell_{ms} = 1 - \prod_{m=1}^{M} (\frac{2\mu_x\mu_{x''} + c_1}{\mu_x^2 + \mu_{x''}^2 + c_1})^{\beta_m} (\frac{2\sigma_{xx''} + c_2}{\sigma_x^2 + \sigma_{x''}^2 + c_2})^{\gamma_m} \tag{7}$$

Here, M denotes different scales, μ_x and $\mu_{x''}$ represent the mean of the original and purified samples, σ_x and $\sigma_{x''}$ represent the variance of the original and purified samples, $\sigma_{xx''}$ represents the covariance, β_m and γ_m represent the relative importance, c_1 and c_2 represent the constant terms.

The self-supervised loss ℓs is shown in Eq. (8). Therefore, $U(\Theta_u, x)$ achieves a balance between purifying full-image triggers and retaining semantic features.

$$\ell_s = \ell_f + \lambda\ell_{ms} \tag{8}$$

Then, the ED Purifier $P(\Theta_a, x)$ is obtained by training and concatenating according to Algorithm 1.

Purify Phase. In this stage, the ED Purifier performs two rounds of data purification. For the filtered erasure-prone sample x, after feeding into $P(\Theta_p, x)$, x' that remove the local-patch trigger features but retain the main semantic features can be obtained by $A(\Theta_a, x)$ part. Then, the x'' that further restores the semantic features can be obtained by the $U(\Theta_u, x')$ part. For the remaining erasure-sensitive sample x, feeding $U(\Theta_{u'}, x)$ eliminates the potential full-image trigger while retaining almost complete semantic features by constraining MS-SSIM. The purification effect for the triggers is shown in Fig. 4.

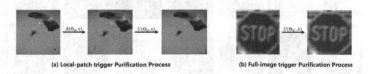

(a) Local-patch trigger Purification Process   (b) Full-image trigger Purification Process

Fig. 4. The purification process of the backdoor sample with local-patch trigger is shown in (a). The purification process of the backdoor sample with full-image trigger is shown in (b).

4 Experiments

4.1 Experimental Setting

Datasets and Models. To evaluate the performance of DFaP, we conducted experiments on the GTSRB [41], and CIFAR-10 [42] datasets. GTSRB (German Traffic Sign Benchmark dataset) is a widely used traffic sign dataset. GTSRB contains 43 classes of traffic sign images of different sizes that can be used to simulate autonomous driving scenarios. For model training, we resize the data size to 32 × 32. CIFAR-10 is a widely used visual object classification dataset. CIFAR-10 consists of 32 × 32 size images with 10 classes. Besides, we choose AlexNet [44] as the base model for GTSRB-based experiments and ResNet-34 [45] as the base model for CIFAR-10 and ImageNet-based experiments.

Backdoor Attacks Configurations. In this paper, we have selected six state-of-the-art attacks, including four local-patch backdoor attacks and two full-image backdoor attacks, respectively: (1) BadNet Attack [9], (2) CL Attack [10], (3) Trojan Attack (Trojan SQ) [46] (4) ℓ_0 norm constraint invisible trigger Attack (ℓ_0 inv) [11] (5) Steganography Attack [11], (6) FTrojan Attack [12]. The effect of adding triggers is shown in Fig. 5. More details about the attack configuration are summarized in Appendix A.

Defense Configurations. We compare DFaP with four backdoor defense methods, respectively:(1) Fine-Pruning [29], (2) STRIP [34], (3) Februus [36], and (4) I-BAU [31]. Fine-Pruning achieves backdoor defense by pruning neurons. STRIP filters backdoor samples by overlaying multiple samples and computing entropy. Februus proposes a backdoor erasure method based on GAN. I-BAU proposes a minimax formulation for removing backdoors from the infected model. For a fair comparison, we follow the defense configuration in the original paper.

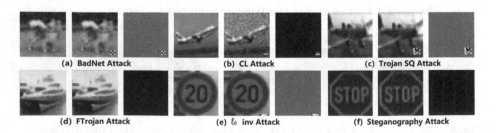

Fig. 5. Comparison between clean and backdoor samples.

Table 1. The performance of the five defense methods is evaluated based on six attacks. Deviation indicates the change in BA/ASR of the defense method compared to before the defense. BA/ASR pairs with poor defense (i.e., ASR above 10%) are marked by a red background. The best performance is highlighted. For Steganography, the experiments are based on GTSRB, and others are based on CIFAR-10.

Backdoor Attack	Before		Fine-Pruning		STRIP		Februus		I-BAU		DFaP(ours)	
	BA	ASR	BA	ASR	BA	ASR	BA	ASR	BA	ASR	BA	ASR
BadNets	88.40%	100%	67.46%	36.11%	87.27%	0.79%	84.14%	6.54%	87.28%	15.79%	**87.43%**	0.40%
CL	84.77%	100%	76.56%	88.63%	86.17%	1.50%	81.64%	6.82%	81.82%	2.27%	86.11%	2.34%
Trojan SQ	86.36%	100%	71.80%	1.58%	80.23%	2.97%	79.72%	36.70%	86.56%	14.14%	**85.47%**	1.05%
FTrojan	87.52%	100%	74.89%	**1.17%**	86.72%	100%	84.77%	99.44%	86.52%	97.87%	82.83%	7.73%
ℓ_0 inv	96.76%	100%	78.15%	0.80%	96.10%	1.19%	94.35%	5.50%	97.76%	0.60%	**96.30%**	0.40%
Steganography	96.46%	100%	69.16%	0.20%	96.23%	98%	93.31%	87.00%	96.20%	0.20%	**96.26%**	0%
Deviation	-	-	17.04% ↓	78.59% ↓	1.26% ↓	65.93% ↓	3.72% ↓	59.67% ↓	**0.69%** ↓	78.19% ↓	0.98% ↓	**98.01%** ↓

For DFaP, to ensure the practicality of the framework under different classification tasks, we set the erasure repair threshold for multiple classification tasks uniformly to 0.5 and the label consistency rate threshold for the reselection mechanism to 0.03.

Evaluation Metrics. To evaluate the performance of defense methods on the poisoned dataset, we consider two aspects: whether the defense method avoids the activation of backdoor behavior by triggers and whether it maintains a high accuracy rate of benign sample classification. Therefore, we evaluate the attack success rate(ASR) and benign sample accuracy(BA). ASR represents the percentage of backdoor samples that successfully activate backdoor behavior. BA represents the classification accuracy of benign test samples. The more the ASR drops and the less the BA drops, the stronger the defense method is.

4.2   Defense Performance

To evaluate the defense performance of our proposed DFaP framework, we use two metrics (i.e., ASR and BA) to evaluate its performance against six backdoor attacks. We also compare the performance of DFaP with the other three defense methods as shown in Table 1. The more the ASR drops and the less the BA drops, the stronger the defense method is. In addition, we consider cases where the ASR is still above 10% to represent the defense method's poor effectiveness, and such cases are marked with a red background.

STRIP achieves good defense results on four local-patch backdoor attacks (i.e., BadNets and CL), comparable to DFaP. However, STRIP needs to prepare additional clean samples, making it less practical than DFaP. Also, STRIP cannot defend against full-image backdoor attacks (i.e., Steganography and FTrojan). Our analysis that STRIP achieves defense only if the triggers in the overlay samples are still valid. However, the full-image triggers tend to behave as minor perturbations that cannot be triggered in the overlay samples, making STRIP unable to filter.

Februus takes a similar idea to CAM Filter but with image restoration via GAN, and requires additional clean samples. Compared to DFaP, Februus fails to defend against 6.14% of attacks at BadNets, 4.48% at CL, 5.1% at ℓ_0 inv, and 35.65% at Trojan SQ. Also, since Februus focus on local-patch backdoor attacks, it has almost no defense effect for Steganography and FTrojan.

Fine-Pruning effectively reduces the ASR of ℓ_0 inv, Trojan SQ, Steganography, and FTrojan. However, Fine-Pruning achieves poor effectiveness for BadNets and CL. The reason for this in our analysis is that under attacks such as BadNets, backdoor neurons associated with triggers overlap with benign neurons, resulting in Fine-Pruning unable to achieve backdoor defense by pruning. Meanwhile, Fine-Pruning has a significant decrease in BA.

I-BAU achieved significant defensive results against all five attacks except FTrojan, and had the least impact on BA, with an average drop of 0.69%. However, I-BAU required additional clean samples to retrain the infected model. The ASR of BadNets and Trojan SQ remained above 10% after the defense, and FTrojan escaped the defense.

DFaP achieved the highest ASR reduction (i.e., 98.01%) for the six attacks, while the reduction in BA was insignificant (i.e., 0.98%) and only higher than I-BAU. In summary, DFaP achieved effective backdoor defense against the six attacks and the best defense effect compared to the other four defense methods. Furthermore, we compare our approach with recently four state-of-the-art defense methods as summarized in Table 2.

Table 2. Comparison between DFaP and other backdoor defense methods.

Work	Domain	Run-time	No additional clean samples required	Against all employed local-patch backdoor attacks	Against all employed full-image backdoor attacks
Fine-Pruning	Network	×	✓	×	✓
STRIP	Input	✓	×	✓	×
Februus	Input	✓	×	×	×
I-BAU	Network	×	×	×	✓
DFaP(ours)	Input	✓	✓	✓	✓

5   Conclusion

To eliminate backdoor threats in the data collection phase, we propose a backdoor defense framework called DFaP. DFaP enables unsupervised filtering and purification of DNNs training data from unverified external data sources. Experiments show that DFaP can effectively defend against six state-of-the-art local-patch and full-image backdoor attacks. DFaP achieves superior performance compared to four other defense methods and is robust under more advanced attack configurations. In summary, our work explores an approach based on data filtering and purification to implement backdoor defense. This provides a feasible solution for training a clean DNN model with unsecured data.

Appendix

A. Backdoor Attacks Configurations

The six state-of-the-art backdoor attacks detailed in Table 3 employ various methodologies. BadNet achieves backdoor implantation by introducing a local-patch trigger into the model. In contrast, CL employs the FGSM [21] technique to add perturbations to samples, effectively implanting the backdoor without altering the labels. Trojan WM takes a reverse engineering approach to generate triggers. On the other hand, ℓ_0 inv formulates trigger generation as a regularization optimization problem. Utilizing steganography, a covert full-image trigger is incorporated by modifying the least significant bits of images. Lastly, FTrojan introduces a frequency domain-based backdoor attack, discreetly embedding a full-image trigger in the RGB space.

Table 3. A configuration summary for the backdoor attacks.

Backdoor Attack	BadNets	CL	Trojan SQ	FTrojan	ℓ_0 inv	Steganography
Dataset	CIFAR-10	CIFAR-10	CIFAR-10	CIFAR-10	GTSRB	GTSRB
Model	ResNet-34	ResNet-34	ResNet-34	ResNet-34	AlexNet	AlexNet
Injection Rate	0.1	0.2	0.1	0.2	0.1	0.2
Target Size	3 × 3	3 × 3	5 × 5	Full Image	1 × 7	Full Image
Target Label	1	1	1	1	40	40
ASR	100%	100%	100%	100%	100%	100%
BA	88.40%	84.77%	86.36%	87.52%	96.76%	96.46%

B. Defend Against Different Injection Rate Attacks

Adversaries may increase the backdoor sample injection rate to increase the difficulty of data filtering. Therefore, we test DFaP by implementing BadNets and Steganography attacks with different injection rates (i.e., 10% to 50%). We compare the best defense methods under each attack as a reference. Specifically, STRIP, Frbruus, and DFaP were tested under BadNets attack with different injection rates, and Fine-Pruning, I-BAU, and DFaP were tested under Steganography attack with different injection rates. The ASRs for different injection rates attacks under various defense methods are shown in Fig. 6(a). Fine-Pruning and Februus failed to defend after the injection rates reached 30% and 40%. The rest of the defense methods have consistent defense performance under different injection rates. The BAs (based on CIFAR-10) for different injection rates of BadNets under various defense methods are shown in Fig. 6(b). DFaP and STRIP achieved comparable results, higher than Februus. The BAs (based on GTSRB) for different injection rates of Steganography under various defense methods are shown in Fig. 6(c). DFaP and I-BAU achieved comparable results above Fine-Pruning. The experimental results show that the ASR of the retrained model after DFaP dropped from 100% to nearly 0%, and there was no gap between the BA of the retrained model and the BA of the infected model. Therefore, the performance of DFaP is robust to backdoor attacks with different injection rates, despite higher injection rates being more challenging to defend.

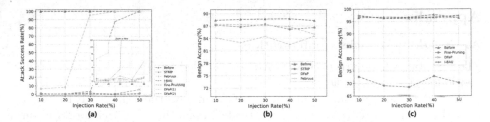

Fig. 6. The performance of DFaP is evaluated based on different injection rates. We show the ASRs for different injection rates attacks under various defense methods in (a), the BAs (based on CIFAR-10) for different injection rates of BadNets under various defense methods in (b), and the BAs(based on GTSRB) for different injection rates of Steganography under various defense methods in (c).

C. Sensitivity Study

In this section, we evaluate the sensitivity of the CAM Filter to the erasure repair threshold through a sensitivity study. We calculate the True Acceptance Rate (TAR) and the False Acceptance Rate (FAR) to measure the data filtering capability of DFaP. TAR and FAR represent the percentage of local-patch and clean samples judged as erasure-prone samples.

In Fig. 7, we demonstrate the data filtering effect of DFaP based on different erasure repair thresholds (from 0.3 to 0.6) for three datasets based on a 10% injection rate of BadNets. The experimental results reveal that when the threshold is dropped from 0.6 to 0.3, the TAR still reaches about 80%, indicating that DFaP at lower threshold can still ensure the dataset's usefulness. Meanwhile, even with a higher threshold, such as CIFAR-10 under the erasure repair threshold of 0.6, the FAR reaches 6.74% (i.e., an injection rate of 0.006), which is still far below the injection rate required for a successful backdoor attack. Therefore, DFaP does not necessitate complex hyperparameter selection.

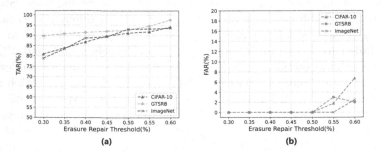

Fig. 7. The performance of DFaP with different erasure repair thresholds.

References

1. Chen, C., Seff, A., Kornhauser, A., et al.: DeepDriving: learning affordance for direct perception in autonomous driving. In: Proceedings of the IEEE International Conference on Computer Vision, pp. 2722–2730 (2015)
2. Tian, Y., Pei, K., Jana, S., et al.: DeepTest: automated testing of deep-neural-network-driven autonomous cars. In: Proceedings of the 40th International Conference on Software Engineering, pp. 303–314 (2018)
3. Jung, C., Shim, D.H.: Incorporating multi-context into the traversability map for urban autonomous driving using deep inverse reinforcement learning. IEEE Robot. Autom. Lett. 6(2), 1662–1669 (2021)
4. Redmon, J., Farhadi, A.: Yolov3: an incremental improvement. arXiv preprint arXiv:1804.02767 (2018)
5. Guo, J., Han, K., Wang, Y., et al.: Distilling object detectors via decoupled features. In: Proceedings of the IEEE/CVF Conference on Computer Vision and Pattern Recognition, pp. 2154–2164 (2021)
6. Devlin, J., Chang, M.W., Lee, K., et al.: Bert: pre-training of deep bidirectional transformers for language understanding. arXiv preprint arXiv:1810.04805 (2018)
7. Xie, W., Feng, Y., Gu, S., et al.: Importance-based neuron allocation for multilingual neural machine translation. arXiv preprint arXiv:2107.06569 (2021)
8. Gao, Y., Doan, B.G., Zhang, Z., et al.: Backdoor attacks and countermeasures on deep learning: a comprehensive review. arXiv preprint arXiv:2007.10760 (2020)
9. Gu, T., Dolan-Gavitt, B., Garg, S.: BadNets: identifying vulnerabilities in the machine learning model supply chain. arXiv preprint arXiv:1708.06733 (2017)
10. Turner, A., Tsipras, D., Madry, A.: Label-Consistent Backdoor Attacks. stat 1050, 6 (2019)
11. Li, S., Xue, M., Zhao, B.Z.H., et al.: Invisible backdoor attacks on deep neural networks via steganography and regularization. IEEE Trans. Dependable Secure Comput. 18(5), 2088–2105 (2020)
12. Wang, T., Yao, Y., Xu, F., et al.: Backdoor attack through frequency domain. arXiv preprint arXiv:2111.10991 (2021)
13. Pang, R., Zhang, Z., Gao, X., et al.: TROJANZOO: towards unified, holistic, and practical evaluation of neural backdoors. In:2022 IEEE 7th European Symposium on Security and Privacy (EuroS&P), pp. 684–702. IEEE (2022)
14. Chou, E., Tramer, F., Pellegrino, G.: SentiNet: detecting localized universal attacks against deep learning systems. In: 2020 IEEE Security and Privacy Workshops (SPW), pp. 48–54. IEEE (2020)
15. Zhong, H., Liao, C., Squicciarini, A.C., et al.: Backdoor embedding in convolutional neural network models via invisible perturbation. In: Proceedings of the Tenth ACM Conference on Data and Application Security and Privacy, pp. 97–108 (2020)
16. Shafahi, A., Huang, W.R., Najibi, M., et al.: Poison frogs! targeted clean-label poisoning attacks on neural networks. In: Advances in Neural Information Processing Systems, 31 (2018)
17. Zhu, C., Huang, W.R., Li, H., et al.: Transferable clean-label poisoning attacks on deep neural nets. In: International Conference on Machine Learning. PMLR, pp. 7614–7623 (2019)
18. Barni, M., Kallas, K., Tondi, B.: A new backdoor attack in CNNs by training set corruption without label poisoning. In: 2019 IEEE International Conference on Image Processing (ICIP), pp. 101–105. IEEE (2019)

19. Quanxin, Z., Wencong, M.A., Yajie, W., et al.: Backdoor attacks on image classification models in deep neural networks. Chin. J. Electron. (2022). https://doi.org/10.1049/cje.2021.00.126
20. Goodfellow, I., Pouget-Abadie, J., Mirza, M., et al.: Generative adversarial networks. Commun. ACM **63**(11), 139–144 (2020)
21. Goodfellow, I.J., Shlens, J., Szegedy, C.: Explaining and harnessing adversarial examples. arXiv preprint arXiv:1412.6572 (2014)
22. Li, Y., Sha, T., Baker, T., et al.: Adaptive vertical federated learning via feature map transferring in mobile edge computing. Computing, 1–17 (2022). https://doi.org/10.1007/s00607-022-01117-x
23. Yang, J., Baker, T., Gill, S.S., et al.: A federated learning attack method based on edge collaboration via cloud. Softw. Pract. Exp. (2022)
24. Zheng, J., Zhang, Y., Li, Y., et al.: Towards evaluating the robustness of adversarial attacks against image scaling transformation. Chin. J. Electron. **32**(1), 151–158 (2023)
25. Liu, Y., Ma, S., Aafer, Y., et al.: Trojaning attack on neural networks. In: 25th Annual Network and Distributed System Security Symposium (NDSS 2018). Internet Soc (2018)
26. Zhang, Y., Tan, Y., Sun, H., et al.: Improving the invisibility of adversarial examples with perceptually adaptive perturbation. Inf. Sci. **635**, 126–137 (2023)
27. Wang, Y., Tan, Y., Lyu, H., et al.: Toward feature space adversarial attack in the frequency domain. Int. J. Intell. Syst. **37**(12), 11019–11036 (2022)
28. Wang, B., Yao, Y., Shan, S., et al.: Neural cleanse: identifying and mitigating backdoor attacks in neural networks. In: 2019 IEEE Symposium on Security and Privacy (SP), pp. 707–723. IEEE (2019)
29. Liu, K., Dolan-Gavitt, B., Garg, S.: Fine-Pruning: defending against backdooring attacks on deep neural networks. In: Bailey, M., Holz, T., Stamatogiannakis, M., Ioannidis, S. (eds.) RAID 2018. LNCS, vol. 11050, pp. 273–294. Springer, Cham (2018). https://doi.org/10.1007/978-3-030-00470-5_13
30. Li, Y., Lyu, X., Koren, N., et al.: Neural attention distillation: erasing backdoor triggers from deep neural networks. arXiv preprint arXiv:2101.05930 (2021)
31. Zeng, Y., Chen, S., Park, W., et al.: Adversarial unlearning of backdoors via implicit hypergradient. In: International Conference on Learning Representations
32. Tran, B., Li, J., Madry, A.: Spectral signatures in backdoor attacks. In: Advances in Neural Information Processing Systems, 31 (2018)
33. Hayase, J., Kong, W., Somani, R., et al.: SPECTRE: defending against backdoor attacks using robust statistics. In: International Conference on Machine Learning, pp. 4129–4139. PMLR (2021)
34. Gao, Y., Xu, C., Wang, D., et al.: STRIP: a defence against trojan attacks on deep neural networks. In: Proceedings of the 35th Annual Computer Security Applications Conference, pp. 113–125 (2019)
35. Yang, J., Zheng, J., Zhang, Z., et al.: Security of federated learning for cloud-edge intelligence collaborative computing. Int. J. Intell. Syst. **37**(11), 9290–9308 (2022)
36. Doan, B.G., Abbasnejad, E., Ranasinghe, D.C.. Februus: input purification defense against trojan attacks on deep neural network systems. In: Annual Computer Security Applications Conference, pp. 897–912 (2020)
37. Tang, D., Wang, X.F., Tang, H., et al.: Demon in the variant: statistical analysis of DNNs for robust backdoor contamination detection. In: USENIX Security Symposium, pp. 1541–1558 (2021)

38. Selvaraju, R.R., Cogswell, M., Das, A., et al.: Grad-cam: visual explanations from deep networks via gradient-based localization. In: Proceedings of the IEEE International Conference on Computer Vision, pp. 618–626 (2017)
39. Telea, A.: An image inpainting technique based on the fast marching method. J. Graph. Tools **9**(1), 23–34 (2004)
40. Batson, J., Royer, L.. Noise2self: blind denoising by self-supervision. In: International Conference on Machine Learning. PMLR, pp. 524–533 (2019)
41. Stallkamp, J., Schlipsing, M., Salmen, J., et al.: Man vs. computer: benchmarking machine learning algorithms for traffic sign recognition. Neural Netw. **32**, 323–332 (2012)
42. Krizhevsky, A., Hinton, G.: Learning multiple layers of features from tiny images (2009)
43. Deng, J., Dong, W., Socher, R., et al.: ImageNet: a large-scale hierarchical image database. In: 2009 IEEE Conference on Computer Vision and Pattern Recognition, pp. 248–255. IEEE (2009)
44. Krizhevsky, A., Hinton, G.: Learning multiple layers of features from tiny images (2009)
45. He, K., Zhang, X., Ren, S., et al.: Deep residual learning for image recognition. In: Proceedings of the IEEE Conference on Computer Vision and Pattern Recognition, pp. 770–778 (2016)
46. Guo, W., Wang, L., Xing, X., et al.: TABOR: a highly accurate approach to inspecting and restoring trojan backdoors in AI systems. arXiv e-prints (2019). arXiv: 1908.01763
47. Subramanya, A., Pillai, V., Pirsiavash, H.: Fooling network interpretation in image classification. In: Proceedings of the IEEE/CVF International Conference on Computer Vision, pp. 2020–2029 (2019)

A Survey of Privacy Preserving Subgraph Matching Methods

Xingjiang Cheng[1], Fuxing Zhang[1], Yun Peng[1], Xianmin Wang[1(✉)],
Teng Huang[1(✉)], Ziye Zhou[2], Duncan S. Wong[1], and Changyu Dong[1]

[1] Artificial Intelligence Research Institute, Guangzhou University,
Guangzhou 510006, China
xingjiangc@e.gzhu.edu.cn, {xianmin,huangteng1220}@gzhu.edu.cn
[2] Guangdong Mogulinker Technology Co. Ltd., Guangzhou 518110, China
zhouzy@mogulinker.com

Abstract. Due to the widespread of large-scale graph data and the increasing popularity of cloud computation, more and more graph processing tasks are outsourced to the cloud. Since graph data has rich information such as node information and edge information, a fundamental challenge is to minimize the overhead of subgraph matching without leakage of the sensitive information of graphs. This paper presents a survey of recent methods for privacy-preserving subgraph matching. Finally, this paper provides valuable insights and possible future directions.

Keywords: Privacy computing · Subgraph matching · Outsourced graph data

1 Introduction

Since graphs contain rich semantic and structural information, graphs are widely used in fields such as social networks, biological networks, and transportation networks. In many popular applications, such as Google's knowledge graph and Facebook's graph search, graphs are used to meet query needs, so graph data management has become a research topic of great concern. This paper focuses on subgraph matching queries [12,18,39], which is a key component and a fundamental query in many applications.

At the same time, more and more enterprises are choosing to migrate their IT infrastructure to cloud platforms. Some graph database systems (such as TuGraph [8] and Neo4j [1]) also provide cloud-based software as a service (SaaS), allowing users to upload graph data to the cloud platform and obtain cloud computing services by outsourcing graph data. This kind of outsourcing service is suitable for subgraph query services. However, this also brings up an important issue, namely the risk of privacy leakage, because SPs are not always trustworthy.

One of the significant privacy leakage problems is the "identity leakage" [13,20], where an adversary can locate the target entity t in a social network graph G with a high probability. This is caused by structural attacks such as

J. Vaidya et al. (Eds.): AIS&P 2023, LNCS 14509, pp. 98–113, 2024.
https://doi.org/10.1007/978-981-99-9785-5_8

Degree Attack, Sub-graph Attack, 1-Neighbor-Graph Attack, and Hub Fingerprint Attack [10,13,24]. Various privacy-preserving graph publishing techniques [28,29] have been proposed to address these threats, with the k-automorphism model being a typical solution that will be discussed in detail later. Another concern is "content disclosure" [9,10], which compromises sensitive label information like a user's name or salary. To prevent content leakage, three classic privacy protection techniques have been proposed: k-anonymity, ℓ-diversity, and t-closeness. These techniques aim to generalize multiple labels into equivalence classes, hiding the sensitive information of individual labels and only revealing the generalized label information to attackers.

In order to address the problems of subgraph matching efficient and secure on the cloud, many researchers have proposed various methods. This survey focuses on privacy-preserving subgraph matching in cloud, aiming to minimize the computational overhead of subgraph matching both on the cloud and the client side, while ensuring the confidentiality of users' sensitive information. In this paper, we present the development trajectory and important techniques in the field, discussing the key challenges and advancements in privacy-preserving subgraph matching. By exploring the evolving landscape of privacy-preserving techniques in the context of cloud-based graph processing, this survey provides valuable insights into state-of-the-art solutions. To summarize, our main contributions are as follows:

1. This paper comprehensively summarizes the research progress of privacy-preserving subgraph matching in cloud, including the development status, key issues, and technical progress, and provides readers with a comprehensive overview.
2. This paper focuses on the key technologies and frameworks of privacy-preserving subgraph matching in cloud, such as GPG algorithm, star decomposition algorithm, and provides valuable insights and possible future directions.

The rest of this paper is organized as follows. Section 2 presents the background and problem statement. Section 3 overviews privacy-preserving graph publishing and anonymization technologies. Section 4 provides an overview of various frameworks for graph outsourcing and subgraph matching. Section 5 concludes this paper.

2   Background and Problem Statement

2.1   Graph Query

A graph is denoted as $G = (V(G), E(G), L)$, where $V(G)$ is a set of vertices; $E(G)$ is a set of undirected edges; L is a set of vertex labels. This graph modeling has only one label per vertex, but this model is not suitable for linking complex graph data, such as social networks. The following is the definition of attributed graph modeling:

Definition 1. *Attributed Graph [16]. An attributed graph is defined as $G = (V(G), E(G), T, \Gamma, L)$, T is a set of vertex types, where each vertex has and only has one vertex type; Γ is a set of vertex attributes.*

Compared with ordinary graphs, attributed graphs have multiple vertex types, and each type has one or more attribute values. The vertex type, vertex attributes, and vertex labels of vertex v are denoted as $T(v), \Gamma(v), L(v)$, respectively.

Definition 2. *Subgraph Matching [16]. Given a graph $G = V(G), E(G), T, \Gamma, L$ and a query graph $Q = V(Q), E(Q), T, \Gamma, L$, Q is subgraph isomorphic to G, if and only if there exists at least one injective function $g : V(Q) \to V(G)$ such that*

i) $\forall q_i \in V(Q), g(q_i) \in V(G) \Rightarrow L(q_i) \subseteq L(g(q_i))$; and
ii) $\forall q_i, q_j \in V(Q)$, edge $\overline{q_i q_j} \in E(Q) \Rightarrow$ edge $g(q_i)g(q_j) \in E(G)$.

Note that the definition of subgraph matching here is for a data graph, and changing it to a database (many graphs) is also universally established. The set of subgraph matches of Q on G is denoted as $R(Q, G)$. Table 1 lists the symbols that will be used later.

Table 1. Notations [16]

G	The original data graph
G^k	The data graph after k-automorphism algorithm
G^o	The outsourced data graph
Q	The original query graph
Q^o	The outsourced query graph
$R(Q, G)$	The set of subgraph matches of Q on G

2.2   Problem Statement

In the system model, the data owner (DO) outsources the encrypted data graph to the cloud server (provided by the service provider (SP)), the user (client) submits a query request (encrypted query) to the SP, and the SP returns the final matching result. We assume that the server is "semi-honest", which is widely used in many papers [32–34]. The server will execute the query honestly but will try to infer private information about the data graph and the query graph. However the cloud server cannot learn any label and edge structure information about the data graph and query graph, and the user does not know the relevant information about the data graph. Our problem can now be formalized as follows:

Problem 1. Our problem is finding all subgraph matches of a query graph Q on graph G in cloud, denoted as $R(Q, G)$, while preserving the privacy of both the data graph G and the query graph Q.

2.3 Secure Multi-party Computation

Secure Multi-party Computation (MPC) [36] is a calculation between multiple participants to ensure that the input data of each participant remains private, and the calculation results can be obtained correctly without revealing any private information. Under the system architecture of this technology, the power of the cloud server is divided into multiple parties, such as three parties, which called CS_1, CS_2, and CS_3 respectively. The following two definitions were proposed by [19].

Replicated Secret Sharing. Given a secret bit $x \in \mathbb{Z}_2$, replicated secret sharing (RSS) [37] splits it into three shares $\langle x \rangle_1, \langle x \rangle_2$ and $\langle x \rangle_3 \in [Z]_2$, where $x = \langle x \rangle_1 \oplus \langle x \rangle_2 \oplus \langle x \rangle_3$. Three pairs of shares $(\langle x \rangle_1, \langle x \rangle_2), (\langle x \rangle_2, \langle x \rangle_3)$ and $(\langle x \rangle_3, \langle x \rangle_1)$ are held respectively by three parties P_1, P_2 and P_3, where P_i holds the i-th pair. With this, we use $(\langle x \rangle_i, \langle x \rangle_{i+1})$ to represent the shares held by $P_i(i \in \{1, 2, 3\})$ and denote such a sharing of x as $[\![x]\!]$.

Function Secret Sharing. Function Secret Sharing (FSS) [38] allows partitioning of a private function f into compact function keys. These keys are designed in such a way that each individual key does not disclose any private information about the original function f. Each key can be evaluated at a given point x, producing an output. By combining the evaluation results of these function keys, the original function $f(x)$ can be computed without revealing sensitive details of f.

3 Privacy-Preserving Graph Publishing

Many previous works focus on privacy-preserving graph publishing [2,28,29,45], these works focus on protecting graph privacy from structural attacks, but assume that the attacker only launches one form of structural attack. Actually, the attacker has strong background knowledge to launch multiple types of structural attacks to identify targets. We mainly introduce the k-automorphism method proposed by zou [2], which claims to be able to resist all structural attacks. When considering the protection of label information, we mainly introduce generalization techniques.

3.1 K-Automorphism

k-automorphism [2] is a privacy-preserving model that claims to defend all existing structural attacks, including degree attack, 1 neighbor-graph attack, subgraph attack, and hub-fingerprint attack [10,13,24]. It is a method proposed based on the symmetry of the graph. The basic idea is as follows: given a graph G, by introducing noise edges, transform G into a k-automorphic graph G^k (Definition 3), where each vertex has at least k - 1 symmetric vertices identical to it. This means that there is no structural difference between v and the other k - 1 symmetric vertices, so the probability of an attacker identifying v is no more than $\frac{1}{k}$. Note that this approach assumes that the graph is unlabeled. Figure 1 shows an example of the algorithm in [2].

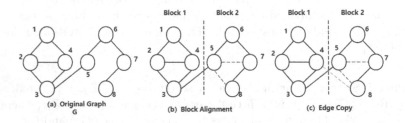

Fig. 1. An example of the k-automorphism algorithm [16]

Definition 3. *K-Automorphic Graph [2]. A k-automorphic graph G^k is defined as $G^k = \left(V(G^k), E(G^k)\right)$, where $V\left(G^k\right)$ can be divided into k blocks and each block has $\left\lceil \frac{|V(G^k)|}{k} \right\rceil$ vertices. Any vertex v has k - 1 symmetric vertices v' in the other k - 1 blocks.*

The k-automorphic graph is generated by the k-automorphic function defined as follows. These two definitions were proposed by [16].

Definition 4. *K-Automorphic Function [2]. Given a vertex v in a k-automorphic graph G^k, v and its corresponding k - 1 symmetric vertices form an alignment vertex instance (AVI).*

All aligned vertex instances make up the aligned vertex table (AVT). Each row of the AVT table is an AVI, expressed as a circular linked list. For each vertex v in an AVI, we can define k-automorphic functions $F_i(i = 0, \ldots, k - 1)$ in G^k based on the AVI, Fig. 2 shows the AVT and automorphic function corresponding to Fig. 1(a) in the algorithm.

Combining the above-mentioned content, converting an original graph G into a k-automorphic graph mainly consists of three steps: graph partitioning, graph alignment, and edge - copy [21]. First, we adopt the METIS algorithm [31] to divide the vertices in G into k blocks. Second, in the process of graph alignment, the vertices with the highest degrees in each block are selected and aligned. Subsequently, the remaining vertices within the same block are aligned with vertices from other blocks, following a breadth - first search (BFS) traversal order. The result is an alignment vertex table (AVT). Third and finally, based on AVT, symmetry edges are inserted in other (k - 1) blocks for each edge in one block. Crossing edges between two blocks are also copied accordingly.

1	6
2	7
3	8
4	5

$F_1(1) = 6; F_1(6) = 1$

$F_1(2) = 7; F_1(7) = 2$

$F_1(3) = 8; F_1(8) = 3$

$F_1(4) = 5; F_1(5) = 4$

(a) Alignment Vertex Table   (b) Automorphic Function F_1

Fig. 2. AVT and automorphic functions [16]

3.2   Generalization

As we mentioned in the k-automorphism model, the assumption of k-automorphism is a graph without labels, but most graphs in real life contain vertex labels, such as a social network, each vertex contains personal information, so the protection of label information is also essential. In order to protect label information from leakage, k-anonymity [3], ℓ-diversity [4], and t-closeness [5] has been proposed. Generalization is the combination of multiple values into a single value (also known as an equivalence class) and is a popular label anonymization technique. Specifically, each vertex label in the graph is represented by a common set of labels. The output of the label generalization algorithm is a label correspondence table (LCT) that maps this correspondence.

k-anonymity is a privacy protection technique that ensures each equivalence class contains a minimum of k records. This requirement guarantees that each record within the class is indistinguishable from at least k - 1 other records, thereby reducing the risk of re-identification. ℓ-diversity, on the other hand, focuses on the distribution of a sensitive attribute within each equivalence class. It requires that each class exhibits at least ℓ "well-represented" values of the sensitive attribute. This criterion enhances privacy by preventing the over-generalization of sensitive information and providing a more diverse representation within each class. Lastly, t-closeness addresses the distribution of labels within equivalence classes. It ensures that the label distribution in each class is no more than a predefined distance, t, away from the distribution in the entire set of labels. This approach aims to maintain a balanced representation of labels, reducing the risk of statistical inference attacks and preserving privacy.

References [6,16,21] adopt the concepts of k-anonymity, ℓ-diversity, and t-closeness respectively, to enhance privacy preservation in their respective studies. The Fig. 3 below is an example:

4   Privacy-Preserving Subgraph Matching

In this section, we introduce several representative privacy-preserving subgraph matching frameworks, which use different techniques to balance accuracy and efficiency. We introduce [11] first because he first posed the problem we needed to solve and established a strict set of privacy requirements to keep data safe.

Label Group	Labels
A	7000,15000
B	8000,14000
C	Internet, Software
D	Female, Male

Fig. 3. Label Correspondence Table [21]

4.1 Privacy-Preserving Graph Query in Cloud

In [11], the authors define and solve the problem of privacy-preserving query over encrypted graph-structured data in cloud computing (PPGQ) for the first time. To minimize the number of subgraph isomorphic queries, the principle of "filter-validation" is employed to efficiently prune negative data graphs prior to validation. It is important to note that the framework proposed in this paper primarily focuses on the query process using indexes outsourced to cloud servers, rather than addressing the encryption, outsourcing, or access of the data itself. The framework of PPGQ is shown in Fig. 4.

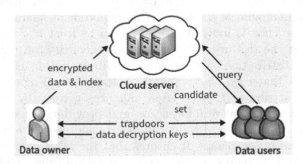

Fig. 4. Encrypted Cloud Data Graph Query Architecture [11]

The main idea is as follows: First, we build a feature-based index to provide feature-related information for each encrypted data graph. Next, select Efficient Inner Product as the pruning tool to perform the filtering process. Both the data graph and the query graph are described as binary vectors, each bit of the former indicates whether the corresponding feature is a subgraph isomorphic to the data graph, and each bit of the latter indicates whether the corresponding feature is included in the query graph. The inner product of query vector and data vector can accurately measure the number of query features contained in the data and is used to filter negative graphs that do not contain query graphs. To preserve privacy, a secure inner product calculation method is proposed, which is adapted from the k-nearest neighbor (KNN) technique [15].

The data owner converts his own data graph into a searchable index according to the BuildIndex($G, \mathcal{K}$) algorithm, and the query algorithm runs on the cloud server as part of the cloud service storage service. The data user runs the trapdoor generation algorithm TDGen($Q, \mathcal{K}$) according to various search control mechanisms. The cloud finally according to the Query algorithm Query($T_Q, \mathcal{I}$) return all candidate hypergraphs.

This paper defines for the first time the problem query over encrypted graph-structured cloud data [11]. It provides a new direction for future research, but it does not involve the encryption, outsourcing, and access technology of the graph itself, and the data graph is a small graph, which may not be suitable for large and complex graphs in practical applications.

4.2   Structural Protection Scheme Based on Cyclic Group

In [7], a method was proposed to protect the edge structure subgraph isomorphism (SPsubIso). This method aims to safeguard the adjacency matrix of both the query graph and the data graph, as well as ensure the privacy of the query process against violations by SP. The approach, based on Ullmann's algorithm, converts the subgraph isomorphism problem into a series of matrix operations denoted as TsubIso. TsubIso consists of three parts:

1. TEnum enumerates all M_i;
2. TMatch verifies the validity of M_i through addition and multiplication between adjacency matrices M_Q and M_G;
3. TRefine reduces the search space of M_i, where SI_Q (SI_G) is the set of $h - hop$ information for each vertex of Q (SI_G) represented by the bit vector.

Then, based on TsubIso, an edge structure protected subIso (SPsubIso) is proposed:

1. First, the paper proposes a CGBE key encryption scheme (cyclic group based encryption scheme) to encrypt the adjacency matrices M_Q and M_G.
2. SPMatch, performs addition and multiplication under CGBE to check the validity of each mapping M_i with negligible false positives.
3. SPEnum, which optimizes mapping enumerations M_i by introducing a protocol involving client participation that filters useless enumerations for SPs.
4. SPRefine, which exploits the privacy inner product of static indexes for optimizations that reduce the number of possible mappings.

The disadvantage of the structure-preserving subgraph matching scheme proposed in this paper is that this algorithm is only suitable for small graphs because Ullmann's algorithm itself has high time complexity when dealing with large-scale graphs. In this case, the time complexity of the algorithm is exponential, so it may not be suitable for the isomorphism judgment of large and complex graphs in practical applications. In addition, the algorithm needs to maintain the state of the node mapping, so it requires a large storage overhead. In terms of security, although the scheme protects the edge structure of the data graph and the query graph, the node labels are still not protected.

4.3 Outsourced Subgraph Matching Based on Star Decomposition

In [16], the author first proposed a simple solution to this problem. First, we use label generalization on the original data graph G to obtain a graph G', whose vertices hide the vertex labels through the label group, and then use the k-automorphism algorithm [2], to obtain the k-automorphic graph G^k and the corresponding vertex symmetry table, and then directly upload G^k to the cloud. Given a query graph Q, each vertex label is represented by the corresponding label group to anonymize Q, and the anonymized graph is denoted as Q^o and submitted to the cloud. The cloud server answers the subgraph query Q^o through the k-automorphic graph G^k. Obviously, $R(Q, G) \subseteq R(Q^o, G^k)$. We introduce more edges and vertices in G to form G^k, and G^k and Q^o use the same LCT to anonymize labels. The formal definition is as follows [16]:

Theorem 1. *For data graph G and query graph Q, $R(Q, G) \subseteq R(Q^o, G^k)$.*

Finally, the server sends $R(Q^o, G^k)$ to the client, and the client filters out false positives in $R(Q^o, G^k)$ according to G, and obtains $R(Q, G)$.

This simple approach is limited by the following [16]: First, we need to upload the k-automorphic graph G^k to the cloud. Due to the addition of edges and vertices, the size of G^k will be significantly larger than G, which will bring communication overhead and higher storage costs. Secondly, it is worth noting that a larger G^k inherently results in a larger search space for subgraph matching. This expanded search space leads to higher query costs, particularly as the value of k increases. Finally, the query graph only generalizes the labels without protecting the edge structure, which will leak the structural information of the graph.

Based on this, the author proposes the following optimization method: The graph G^k uploaded to the cloud is a k-automorphic graph containing k blocks. Only one block of G^k and k-automorphic functions F_i need to be uploaded to the cloud and the cloud can be reconstructed according to G^o and functions F_i. This is also the motivation for defining the outsourced graph G^o.

Definition 5. *Outsourced Graph [16]. For a graph $G(V(G), E(G), T, \Gamma, L)$ and its k-automorphic graph $G^k(V(G^k), E(G^k), T, \Gamma, L)$, an outsourced graph of G is defined as $G^o(V(G^o), E(G^o), T, \Gamma, L)$, where*

1. *$V(G^o)$ is formed by taking the union of the vertices in the first block of G^k, denoted as $V(B_1)$, along with their one-hop neighbors in G^k denoted as $V(N_1)$; and*
2. *$E(G^o)$ consists of a subset of undirected edges from $E(G^k)$ that connect vertices within $V(B_1)$ as well as vertices between $V(B_1)$ and $V(N_1)$.*

According to definition 5, we can get utsourced graph G^o from G^k and upload it to the cloud. G^o contains the vertices in the first block of G^k (the vertices of the first column of AVT) and their 1-hop neighbors and corresponding edges in G^k. Notably, G^o is much smaller than G^k.

Using the label generalization algorithm on G^k to get the outsourced graph G^o, this paper uses a cost model to find a good generalization strategy. Then, subgraph matching is performed on the outsourced graph G^o. Since there is only a concise graph G^o on the cloud, a special star-based matching algorithm is proposed. In the cloud, the algorithm is divided into the following three steps [21]:

step1 Query decomposition. The cloud first decomposes query Q into a set of star shapes $\{S_i\}$. Each star shape consists of a root vertex along with its adjacent edges and neighboring vertices.

step2 Star matching. The cloud then retrieves matchings for each decomposed star $\{S_i\}$ from the succinct graph G^o, denoted as $R(S_i, G^o)$. Leveraging the symmetry of G^k, the cloud futher obtains the matchings for S_i in G^k, denoted as $R(S_i, G^k)$.

step3 Result join. The cloud starts with a $R(S_i, G^k)$ and iteratively performs a natural join operation with $R(S_j, G^k)$ for all $j \neq i$ until all relevant stars are joined. This process continues until the final results, denoted as $R(Q^o, G^k)$, are obtained, representing the matchings for Q^o over G^k.

Finally, on the client side, the algorithm uses the original graph G and the query graph Q to filter false positives in $R(Q^o, G^k)$ and get the final subgraph matching $R(Q, G)$.

The framework proposed in this paper is mainly for privacy-preserving matching on large graphs in the cloud. These designs significantly reduce the size of the graph uploaded to the cloud. The disadvantage is that the edge structure of the query graph is not preserved, and since treat data owners and users as one in the setting, there may be limitations in practicality. Second, star-based subgraph matching is also inefficient because [21] (1): it cannot narrow down the search scope of the query to local regions in the graph due to query decomposition, and (2) natural joins are computationally expensive.

4.4 Partial Graph-Based Outsourced Subgraph Matching

In [21], consistent with most papers [16], the k-automorphism algorithm is used to protect the edge structure, and the generalization method is used to protect the label information of the vertices. The author proposes the (k, t)-privacy of the outsourced graph and claims that this is the strictest generalization-based privacy model for graph structure and label generalization. Additionally, the author proposes a t-closeness label generalization algorithm $TOGGLE$ to optimize the cost of subgraph matching. Furthermore, the author also proposes a partial graph-based subgraph processing algorithm PGP, which does not require query decomposition and utilizes the symmetry of the outsourcing graph to limit the search scope to a local area. It should be noted that this paper regards the data owner and the data queryer (user) as one party, both as clients. The workflow of the whole process is shown in Fig. 5.

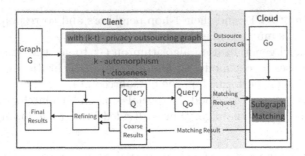

Fig. 5. Workflow of subgraph matching on outsourced graph [21]

To solve the problem that the time complexity of generalization in [16] is exponentially related to the number of labels, this paper proposes the *TOGGLE* algorithm (T-closeness-Optimized Graph Generalization on Label Extension), which aims to summarize labels into groups to minimize the search space. Search space refers to the total number of vertices to explore the query.

Considering that the *TOGGLE* algorithm wants to minimize the search space, the author expresses *TOGGLE* as a combinatorial optimization problem with constraints and then reduces the optimization problem to a General Set Partition Problem (*GSPP*) [14]. Because the asymptotic size of *GSPP* is still an exponent of n, finding the optimal partition is only applicable for small n. Next, the authors propose Algorithm 2 [21], a sub-optimal solution with theoretical guarantees.

Although there has been a lot of research on subgraph matching algorithms [17,19,22,23], only the star-base algorithm proposed by [16] works on outsourced succinct graphs. However, the star decomposition algorithm needs to decompose the original query into multiple subqueries, and the efficiency is low. This paper proposes a partial-graph-based subgraph processing algorithm *PGP* that does not require query decomposition.

The authors of this paper propose a graph label generalization algorithm and an efficient subgraph matching algorithm in the cloud with *t*-closeness and k-automorphism privacy. However, there are also the following limitations: 1. The data owner and the data user are regarded as one in the setting. Although this is simple, it has relatively large limitations in actual application. 2. Although the privacy protection method of the data graph has been optimized, the structure of the query graph has not been protected yet. 3. Mainly work on most of the small graphs, and the actual efficiency on large graphs may be relatively low.

4.5   Oblivious Attributed Subgraph Matching as a Cloud Service

With the continuous development of technology, there have been many works using cryptography and multi-party secure computing technology to realize privacy-preserving subgraph queries. The model architecture is the same as the previous type, except that there are multiple servers on the cloud server side (these servers can be hosted by different cloud providers). In recent years, these multi-server models have grown in popularity in academia and industry to build practical security systems [21,25–27,40].

OblivGM [19] is a system that outsources the oblivious attribute subgraph matching service to the cloud and protects the confidentiality of data content related to attribute graphs and queries. In addition, safe and rich matching functions are supported, where subgraph queries can contain equality predicates and/or range predicates. OblivGM is built with attribute graph modeling and advanced lightweight cryptography such as replicated secret sharing (RSS) [37], functional secret sharing (RSS) [38], and secure shuffling. Figure 6 shows the system architecture of OblivGM.

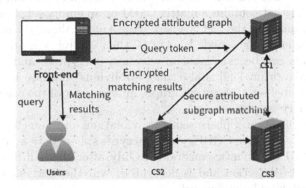

Fig. 6. OblivGM's system architecture [19]

OblivGM has the following four stages: 1. graph and subgraph query modeling, properly modeling the attribute graph and subgraph query to facilitate subsequent oblivious subgraph matching services, 2. attributed graph encryption, fully encrypt the attributed graph, and then send the generated ciphertext to the cloud server, 3. Secure query token generation, SF parses each subgraph query and generates a corresponding secure query token, which is then sent to the cloud server, and 4. Security attribute subgraph matching. The cloud server retrieves an encrypted subgraph isomorphic to the query from the encrypted attribute graph.

Security analysis: In the semi-honest and non-colluding adversary model, OblivGM guarantees that the cloud server will not learn 1) the attribute value and degree of each vertex in the attribute graph, and the connection between these vertices, 2) The subgraph matches the target attribute value associated with each vertex in the query, 3) Search access patterns [43].

This paper implements OblivGM, a new system that outsources oblivious subgraph matching services to the cloud with stronger security and richer features than existing techniques. Although OblivGM supports privacy-preserving subgraph matching processing on a large graph, it does not preserve the structure of the query graph.

4.6   Privacy-Preserving Localized Graph Pattern Query Processing

In [9], subgraph homomorphic query and strong simulation query are implemented, because the performance of subgraph homomorphic query is similar to subgraph homomorphic query. The output of the LGPQ (Localized Graph Pattern Query) algorithm is: the ball defined by the center (node) and the radius (path length) is used as the basic unit of the query result. Among them, the center node of the ball is the node of the same label value in G mapped to the node in Q, and the radius of the ball is equal to the diameter of Q.

The framework Prilo proposed in this paper to provide private query LGPQ services consists of three steps:

step1   Candidate enumeration: Candidate Enumeration Algorithm (hom) finally returns R1 that contains a matrix CMM or an empty set. If it is an empty set, it means that the ball is a false ball. Among them, if the u-node label of Q is equal to the v-node label of G, then C[u][v] = 1, otherwise C[u][v] = 0. Balls can be calculated in advance, encrypted, and stored on the server. Algorithm 1 [9] is also "query-oblivious" because it only relies on the node label set of the query graph, independent of the edge information of the query graph.

step2   Query verification: Sphere semantic violations are detected in a "query-oblivious" way using CGBE, which encrypts sphere data and detects using known LGPQ semantic constraints. Only after the ball data has passed this step of detection and is decrypted, will the next step of plaintext query matching be performed.

step3   Query matching: The final verified ball will be returned to the client in ciphertext, and the user will perform a query on the plaintext after decryption.

The advantage of this paper is that it effectively protects the privacy of user queries by encrypting queries, using Trusted Execution Environment (TEE) [41,44] for pruning, and using secure retrieval schemes. The disadvantage is that the framework may involve complex operations such as encryption, pruning, and secure retrieval. These operations may introduce significant computing and storage overhead when processing large-scale query graphs. This can result in performance degradation of the framework and may necessitate additional computing resources and storage space.

5   Conclusion

This paper surveys research on privacy-preserving subgraph matching in cloud. The results show that privacy-preserving subgraph matching becomes increasingly important when facing the challenges of large-scale graph data and cloud storage. In order to protect the sensitive information of users, researchers have proposed various privacy protection technologies, including anonymization, encryption algorithms, and secure multi-party computation. The application of these technologies makes it possible to perform graph query processing in the cloud.

Existing works cannot simultaneously protect the edge structure and label information privacy of data graphs and query graphs, and only protect some sensitive information. Therefore, how to efficiently protect the privacy information of the data graph and query graph at the same time in future work is a research direction and a challenge.

References

1. http://neo4j.org/
2. Zou, L., Chen, L., Özsu, M.T.: K-automorphism: a general framework for privacy preserving network publication. In: VLDB (2009)
3. Sweeney, L.: k-anonymity: a model for protecting privacy. Int. J. Uncertain. Fuzziness Knowl. Based Syst. **10**(05), 557–570 (2002)
4. Machanavajjhala, A., Gehrke, J., Kifer, D., Venkitasubramaniam, M.: l-diversity: privacy beyond k-anonymity. In: ICDE, p. 24 (2006)
5. Li, N., Li, T., Venkatasubramanian, S.: t-closeness: privacy beyond k-anonymity and l-diversity. In: ICDE, pp. 106–115 (2007)
6. Yuan, M., Chen, L., Philip, S.Y., Yu, T.: Protecting sensitive labels in social network data anonymization. TKDE **25**(3), 633–647 (2013)
7. Fan, Z., Choi, B., Chen, Q., et al.: Structure-preserving subgraph query services. IEEE Trans. Knowl. Data Eng. **27**(8), 2275–2290 (2015)
8. https://tugraph.antgroup.com/
9. Xu, L., Choi, B., Peng, Y., et al.: A framework for privacy preserving localized graph pattern query processing. Proc. ACM Manage. Data **1**(2), 1–27 (2023)
10. Hay, M., Miklau, G., Jensen, D., Towsley, D.F., Weis, P.: Resisting structural re-identification in anonymized social networks. PVLDB **1**(1), 102–114 (2008)
11. Cao, N., Yang, Z., Wang, C., et al.: Privacy-preserving query over encrypted graph-structured data in cloud computing. In: 2011 31st International Conference on Distributed Computing Systems, pp. 393–402. IEEE (2011)
12. Lee, J., Han, W., Kasperovics, R., Lee, J.: An in-depth comparison of subgraph isomorphism algorithms in graph databases. PVLDB **6**(2), 133–144 (2012)
13. Liu, K., Terzi, E.: Towards identity anonymization on graphs. In: SIGMOD, pp. 93–106 (2008)
14. Barnhart, C., Johnson, E.L., Nemhauser, G.L., Savelsbergh, M.W., Vance, P.H.: Branch-and-price: column generation for solving huge integer programs. Oper. Res. **46**(3), 316–329 (1998)
15. Wong, W.K., Cheung, D.W., Kao, B., Mamoulis, N.: Secure KNN computation on encrypted databases. In: Proceedings of SIGMOD (2009)
16. Chang, Z., Zou, L., Li, F.: Privacy preserving subgraph matching on large graphs in cloud. In: Proceedings of ACM SIGMOD (2016)
17. Bi, F., Chang, L., Lin, X., Qin, L., Zhang, W.: Efficient subgraph matching by postponing cartesian products. In: SIGMOD, pp. 1199–1214 (2016)
18. Sun, Z., Wang, H., Wang, H., Shao, B., Li, J.: Efficient subgraph matching on billion node graphs. PVLDB **5**(9), 788–799 (2012)
19. Wang, S., Zheng, Y., Jia, X., et al.: OblivGM: oblivious attributed subgraph matching as a cloud service. IEEE Trans. Inf. Forensics Secur. **17**, 3582–3596 (2022)

20. Tai, C., Tseng, P., Yu, P.S., Chen, M.: Identity protection in sequential releases of dynamic networks. IEEE Trans. Knowl. Data Eng. **26**(3), 635–651 (2014)
21. Huang, K., Hu, H., Zhou, S., Guan, J., Ye, Q., Zhou, X.: Privacy and efficiency guaranteed social subgraph matching. The VLDB Journal, pp. 1–22 (2021)
22. Du, B., Zhang, S., Cao, N., Tong, H.: First: fast interactive attributed subgraph matching. In: SIGKDD, pp. 1447–1456. ACM (2017)
23. Qiao, M., Zhang, H., Cheng, H.: Subgraph matching: on compression and computation. PVLDB **11**(2), 176–188 (2017)
24. Zhou, B., Pei, J.: Preserving privacy in social networks against neighborhood attacks. In 2008 IEEE 24th International Conference on Data Engineering, pp. 506–515 (2008)
25. Tan, S., Knott, B., Tian, Y., Wu, D.J.: CryptGPU: fast privacypreserving machine learning on the GPU. In: Proceedings of IEEE S&P (2021)
26. Dauterman, E., Rathee, M., Popa, R.A., Stoica, I.: Waldo: a private time-series database from function secret sharing. In: Proceedings of IEEE S&P (2022)
27. Wang, S., Zheng, Y., Jia, X., Yi, X.: Privacy-preserving analytics on decentralized social graphs: The case of eigendecomposition. IEEE Trans. Knowl. Data Eng. **35**, 7341–7356 (2022)
28. Cheng, J., Fu, A.W.-c., Liu, J.: K-isomorphism: privacy preserving network publication against structural attacks. In: SIGMOD, pp. 459–470 (2010)
29. Wu, W., Xiao, Y., Wang, W., He, Z., Wang, Z.: K-symmetry model for identity anonymization in social networks. In: EDBT, p. 111122 (2010)
30. Jiang, H., Pei, J., Yu, D., et al.: Applications of differential privacy in social network analysis: a survey. IEEE Trans. Knowl. Data Eng. **35**, 108–127 (2021)
31. Karypis, G., Kumar, V.: Analysis of multilevel graph partitioning. In: ICS, p. 29 (1995)
32. Xu, J., Yi, P., Choi, B., et al.: Privacy-preserving reachability query services for massive networks. In: CIKM, pp. 145–154 (2016)
33. Hu, H., Xu, J., Chen, Q., et al.: Authenticating location-based services without compromising location privacy. In: SIGMOD, pp. 301–312 (2012)
34. Wang, S., Zheng, Y., Jia, X., et al.: PeGraph: a system for privacy-preserving and efficient search over encrypted social graphs. IEEE Trans. Inf. Forensics Secur. **17**, 3179–3194 (2022)
35. Lai, S., Yuan, X., Sun, S.F., et al.: Graphseš: an encrypted graph database for privacy-preserving social search. In: Asia CCS 2019: Proceedings of the 2019 ACM Asia Conference on Computer and Communications Security, pp. 41–54 (2019)
36. Lindell, Y.: Secure multiparty computation (MPC). Cryptology ePrint Archive (2020)
37. Araki, T., Furukawa, J., Lindell, Y., Nof, A., Ohara, K.: High-throughput semi-honest secure three-party computation with an honest majority. In: Proceedings of ACM CCS (2016)
38. Boyle, E., Gilboa, N., Ishai, Y.: Function secret sharing. In: Proceedings of EURO-CRYPT (2015)
39. Zou, L., Chen, L., Özsu, M.T.: Distancejoin: pattern match query in a large graph database. PVLDB **2**(1), 886–897 (2009)
40. Chen, W., Popa, R.A.: Metal: a metadata-hiding file-sharing system. In: Proceedings of NDSS (2020)
41. Sabt, M., Achemlal, M., Bouabdallah, A.: Trusted execution environment: what it is, and what it is not. In: 2015 IEEE Trustcom/BigDataSE/ISPA, pp. 57–64 (2015)

42. Araki, T., Furukawa, J., Ohara, K., Pinkas, B., Rosemarin, H., Tsuchida, H.: Secure graph analysis at scale. In: Proceedings of ACM CCS (2021)
43. Curtmola, R., Garay, J.A., Kamara, S., Ostrovsky, R.: Searchable symmetric encryption: improved definitions and efficient constructions. In: Proceedings of ACM CCS (2006)
44. Costan, V., Devadas, S.: Intel SGX explained, Cryptology ePrint Archive (2016)
45. Ding, X., Wang, C., Choo, K.K.R., et al.: A novel privacy preserving framework for large scale graph data publishing. TKDE **33**(2), 331–343 (2019)

The Analysis of Schnorr Multi-Signatures and the Application to AI

Wenchao Wang[1], Jing Qin[1(✉)], Jinlu Liu[1], Xi Zhang[2,3], Xinyi Hou[1], and Zhongkai Wei[1]

[1] School of Mathematics, Shandong University, Jinan 250100, China
qinjing@sdu.edu.cn
[2] College of Computer and Cyber Security, Hebei Normal University, Shijiazhuang 050024, China
[3] Hebei Provincial Key Laboratory of Network and Information Security, Hebei Normal University, Shijiazhuang 050024, China

Abstract. Artificial intelligence (AI) is having a profound impact on our daily lives. We suggest using digital signatures to protect the user's identity and achieve data accountability. To address high-risk applications, multi-signatures are expected to play an important role in AI. MuSig2 by Nick et al. is an efficient and secure Schnorr multi-signature scheme. MuSig2 implements signature aggregation and key aggregation, and MuSig2 is reduced to the One-More Discrete Logarithms (OMDL) problem in the random oracle model. This comes at the cost that the signer needs four nonces instead of one nonce for each signature. However, MuSig2 ignores the change of nonces in the forking lemma, which leads to the signer signature requiring too many nonces, and makes the proof of the scheme complicated. In this paper, we reduce the number of nonces from 4 to 2 and simplify the security proof of the MuSig2 scheme in the random oracle model. Then by reducing the security requirement slightly, we achieve the MuSig2 scheme's security when the nonces are reused. Finally, we utilize the proof technology of MuSig2 to reduce the MSDL (Discrete-Logarithm based Multi-Signature) scheme by Boneh et al. to the OMDL assumption.

Keywords: Artificial intelligence · Schnorr multi-signature scheme · One-more discrete logarithm · Random oracle model

1 Introduction

1.1 AI and Multi-Signatures

AI is a field that deals with making machines think. Legg and Hutter [10] define AI as a process of imitating human behavior and decision-making capabilities. So, AI is a way to train machines to perform tasks that require intelligence. With the increase in the use of artificial intelligence, it becomes essential to make them reliable and trustworthy. Several requirements, such as fairness, explainability,

J. Vaidya et al. (Eds.): AIS&P 2023, LNCS 14509, pp. 114–130, 2024.
https://doi.org/10.1007/978-981-99-9785-5_9

accountability, privacy, reliability, and acceptance, have been proposed in AI to make these systems trustworthy. A multi-signature scheme [7] is a protocol that enables the n signers to jointly generate a short signature σ on m so that σ convinces a verifier that all n signers signed m. This paper proposes using mult-signatures to ensure accountability, integrity and authenticity of high-value data in AI.

The use of multi-signatures can directly enhance the security of artificial intelligence systems. AI and algorithmic decision-making significantly influence various aspects of our lives, including healthcare, business, government, education, and justice [8]. These applications heavily depend on data for making informed decisions. By utilizing digital signatures, the integrity and authenticity of data can be ensured. The implementation of multi-signatures can further protect high-value data. Moreover, one of the difficulties with AI is to identify who is responsible for the AI, the designer, the user, or the supervisor. Digital signature is to ensure that no good person can be wronged (traceability), nor will any adversaries get into the gap (unforgerability).

Multi-signatures can also indirectly improve the security of AI. Blockchain is being integrated with AI technology, such as through electronic contract multi-documentation and full evidence chain traceability, to protect customer data security and privacy, and make contracts more secure and credible. Digital signatures play a crucial role in transactions on blockchain platforms, particularly in enterprise platforms where multi-signatures are required from a group of peers to endorse a transaction. However, this process is complex and time-consuming. The use of multi-signatures, which allows a group of signers to collaborate and produce a joint signature, has gained significant attention for its ability to improve transaction efficiency [22]. Additionally, The use of multi-signatures can enhance the security of certificate authorities which plays an important role in digital signatures.

1.2   Related Work

MULTI-SIGNATURE BASED ON SCHNORR SIGNATURE. Multi-signature schemes based on Schnorr signatures [2, 4, 6, 11–15, 18, 20] are becoming increasingly popular and practical. Following is a naive way to design a multi-signature scheme that is fully compatible with Schnorr signatures [17]. Say a group of n signers want to sign a message m, and let $L = \{X_1 = g^{x_1}, \ldots, X_n = g^{x_n}\}$ be the multiset of all their public keys. The adversary can choose corrupted public keys arbitrarily and duplicate public keys can appear in L. Each signer generates and communicates at random to others a nonce $\widetilde{R}_i = g^{r_i}$. Each of them then computes $R = \prod_{i=1}^n R_i, c = H(\widetilde{X}, R, m)$ where $\widetilde{X} = \prod_{i=1}^n X_i$ is the product of individual public keys, and a partial signature $s_i = r_i + c x_i$. Partial signatures are then combined into a single signature (R, s) where $s = \prod_{i=1}^n s_i$. The validity of a signature (R, s) on message m for public keys $X_1, \ldots, X_n$ is equivalent to $g^s = R\widetilde{X}^c$ where $\widetilde{X} = \prod_{i=1}^n X_i$ and $c = H(\widetilde{X}, R, m)$. Note that this is exactly the verification equation for an ordinary key-prefixed Schnorr signature with respect to the aggregate public key $\widetilde{X}$. However this simplistic protocol is vulnerable to a rogue-key attack. A corrupted signer is able to set its public key to $X_1 = g^{x_1}(\prod_{i=2}^{n} X_i)^{-1}$, which allows him to produce signatures for public keys $X_1, \ldots, X_n$ by himself.

THE MUSIG SCHEME. An efficient direct defense against rogue-key attacks proposed by Bellare and Neven [4] is to work in the plain public-key model. Public keys can be aggregated without the need to check their validity in this model. The canonical multi-signature scheme provably secure in this model and fully compatible with Schnorr signatures is MuSig (and the variant MuSig-DN [14]) by Maxwell et al. [6], independently proven secure by Boneh, Drijvers, and Neven [5].

To overcome rogue-key attacks in the plain public-key model, we let $\widetilde{X}$ is the aggregate public key corresponding to the multiset of public keys $L = \{X_1, \ldots, X_n\}$. It is defined as $\widetilde{X} = \prod_{i=1}^{n} X_i^{a_i}$ where $a_i = H_{agg}(L, X_i)$ (note that the a_i's only depend on the public keys of the signers). To sign some message m, MuSig computes partial signatures $s_i = r_i + c_i x_i$ with respect to "signer-dependent" challenges $c_i = H_{agg}(L, X_i) \cdot H_{sig}(\widetilde{X}, R, m)$. As a result, the verification equation of a signature (R, s) on message m for public keys $L = \{X_1, \ldots, X_n\}$ becomes $g^s = R\widetilde{X} = R\prod_{i=1}^{n} X_i^{a_i c} = R\widetilde{X}^c$, where $c = H_{sig}(\widetilde{X}, R, m)$. This recovers the key aggregation property enjoyed by the naive scheme, albeit with respect to a more complex aggregate key $\widetilde{X} = \prod_{i=1}^{n} X_i^{a_i}$.

However the above scheme is still not secure. The problem can date to the Bellare and Neven's scheme [4]. When we design provably secure Schnorr-based multi-signature schemes one of the main problems exists in the simulation of the honest signer. To simulate the honest signer, the reduction cannot simply use the zero-knowledge property and program the random oracle. The reason is that the random oracle entry that must be programmed depends on the output of the adversarial signers. Bellare and Neven [4] got around this issue by introducing a preliminary round in the signing protocol where signers exchange commitments in their first rounds.

TWO-ROUND SCHEMES. Following the scheme by Bellare and Neven [4], in which signing requires three rounds of interaction, multiple attempts to reduce this number to two rounds [2,4,12,18] were foiled by Drijvers et al. [6]. More precisely, they prove that if OMDL is hard, then there cannot exist an algebraic black-box reduction that proves the CoSi, MuSig, BCJ, or MWLD schemes secure under the discrete logarithm DL or OMDL assumption. They provide attacks that apply to all schemes based on Wagner's algorithm for the generalized birthday problem [21]. Note that the adversary is allowed to engage in an arbitrary number of concurrent sessions (concurrent security), as required by the standard definition of unforgeability.

MSDL with public-key aggregation has an initial commitment round (like the scheme by Bellare and Neven [4]) to simulate an honest signer in a run of the signing protocol via the standard way of programming the random oracle H_{sig}. In this commitment, each signer commits to its share R_i before receiving the shares of other signers. Altogether, the signing protocol of MSDL requires three communication rounds and only the initial commitment round can be preprocessed without knowing the message to be signed.

It is important that each signer must ensure that their secret nonce r_i changes unpredictably whenever $c = H_{sig}(\widetilde{X}, R, m)$ changes. DWMS [1] and MuSig2 [13] use the idea of a linear combination of multiple nonces to obtain a two-round multi-signature scheme. In terms of provable security, DWMS provides a proof only in the combination of ROM+AGM, whereas MuSig2 additionally provides a proof that only rely on the ROM. MuSig2 is the first multi-signature scheme that simultaneously i) is secure under concurrent signing sessions, ii) supports key aggregation, iii) outputs ordinary Schnorr signatures, iv) needs only two communication rounds, and v) has similar signer complexity as ordinary Schnorr signatures. Furthermore, it is the first scheme in the pure DL setting that supports preprocessing of all but one rounds, effectively enabling non-interactive signing without forgoing security under concurrent sessions. In order to further improve the security of multi-signature, many schemes have been proposed [8, 11, 16, 17, 19, 20] (Table 1).

Table 1. Comparison of multi-signature schemes. We call that signing protocol supports preprocessing, i.e., the first round of the signing protocol is independent of the message being signed. The first two columns show the total number of communication rounds ("tot.") in the signing algorithm and the number of communication rounds that are preprocessed ("pp."). $\mathbb{G}$ is the prime p order group that the schemes work. $k$exp shows k-exponentiation in $\mathbb{G}$. NIZK is non-interactive zero knowledge and NIZK proof corresponds to the number of exp in order to execute the proof and verify algorithm. PK is the aggregated public key. 2-ES is 2-entwined sum.

Scheme	tot.	pp.	Sign	Verify	Sign size	PK size	Security (ROM, . . .)
mBCJ [2]	2	0	5exp	6exp	$|\mathbb{G}| + 2|Z_p|$	$2|\mathbb{G}| + 3|\mathbb{Z}_p|$	DL
MSDL [5]	3	1	1exp	1exp	$|\mathbb{G}|$	$|\mathbb{G}| + |\mathbb{Z}_p|$	co-CDH
MuSig-DN [14]	2	0	NIZK-proof	2exp	$|\mathbb{G}|$	$|\mathbb{G}| + |\mathbb{Z}_p|$	DL, DDH
MuSig2(v=4) [13]	2	1	7exp	2exp	$|\mathbb{G}|$	$|\mathbb{G}| + |\mathbb{Z}_p|$	$4q_s$-OMDL
MuSig2(v=2) [13]	2	1	3exp	2exp	$|\mathbb{G}|$	$|\mathbb{G}| + |\mathbb{Z}_p|$	$2q_s$-OMDL, AGM
DWMS [1]	2	1	(2n+2)exp	2exp	$|\mathbb{G}|$	$|\mathbb{G}| + |\mathbb{Z}_p|$	q_s-OMDL, 2ES, AGM
HBMS [3]	2	0	2exp	3exp	$|\mathbb{G}|$	$|\mathbb{G}| + 2|\mathbb{Z}_p|$	XIDL or DL, AGM
MS(Okamoto) [9]	2	1	4exp	6xp	$2|\mathbb{G}|$	$3|\mathbb{Z}_p|$	DL, AGM, NPROM
MuSig2-H [20]	2	1	3exp	2exp	$|\mathbb{G}|$	$|\mathbb{G}| + |\mathbb{Z}_p|$	DL, RSA, AOMPR
MuSig2(v=2, ours)	2	1	3exp	2exp	$|\mathbb{G}|$	$|\mathbb{G}| + |\mathbb{Z}_p|$	$2q_s$-OMDL

1.3   Our Contribution

To protect the application of AI in high-risk scenarios, this paper suggests using multi-signature in AI. The contributions are summarized as follows:

1. We simplify the proof of the MuSig2 scheme in the ROM model and reduce the nonces from 4 to 2. In this proof, we are keenly aware of the fact that the

calculation of the aggregate private key does not increase the number of DL oracles in the security proof.

2. We affirm that there is a potential attack that can compromise the security of the MuSig2 scheme under the OMDL assumption if the nonce is repeatedly used. To repeatedly use nonces, we require that it is feasible to forge multi-signatures involving more than one honest signer.

3. We observe that the proof of the MSDL scheme is invalid. More specifically, the commitment used in the scheme cannot guarantee that the adversary's nonce is revealed publicly before the simulator. We improve the proof of the scheme using a method analogous to that of MuSig2.

2   Preliminaries

NOTATION. Let S be a finite set. We let $s \overset{\$}{\leftarrow} S$ denote sampling an element uniformly at random from S and assigning it to s. If $\mathcal{A}$ is a randomized algorithm, then $\mathcal{A}(x_1, \ldots; \rho)$ denotes its output on inputs $x_1, \ldots$ and coins ρ, while $y \leftarrow \mathcal{A}(x_1, \ldots)$ means that we choose ρ uniformly at random and let $y := \mathcal{A}(x_1, \ldots; \rho)$. We call the triplet $(\mathbb{G}, p, g)$ the group parameters. We let $\mathbb{G}$ be a cyclic group of order p, where p is a k-bit integer, and g be a generator of $\mathbb{G}$. The group $\mathbb{G}$ will be denoted multiplicatively. We adopt the concrete security approach, i.e., we view $(\mathbb{G}, p, g)$ as fixed. If necessary, the bit length k of p can be considered as a security parameter. The advantage of $\mathcal{A}$ in $Game_{\mathcal{A}}$, which is parameterized by the adversary $\mathcal{A}$, can be defined as $Adv_{\mathcal{A}}^{Game}(\lambda) := Pr[Game_{\mathcal{A}}(\lambda) = true]$.

Definition 1. *Let $(\mathbb{G}, p, g)$ be group parameters. Let $DLOG_g(\cdot)$ be an oracle which takes as input an element $X \in \mathbb{G}$ and returns $x \in \{0, ..., p-1\}$ such that $g^x = X$. An algorithm $\mathcal{A}$ is considered to (q, t, ε)-solve the OMDL problem w.r.t. $(\mathbb{G}, p, g)$ if on input $q + 1$ random group elements $X_1, \ldots, X_{q+1}$, it runs in time at most t, makes at most q queries to $DlOG_g(\cdot)$, and returns $x_1, \ldots, x_{q+1} \in \{0, \ldots, p-1\}$ such that $X_i = g^{x_i}$ for all $1 \leq i \leq q+1$ with probability at least ε, where the probability is taken over the random draw of $X_1, \ldots, X_{q+1}$ and the random coins of $\mathcal{A}$.*

Lemma 1. *(A General Forking Lemma). Fix integers q and m. Let $\mathcal{A}$ be a randomized algorithm taking as input a main input inp generated by some probabilistic algorithm $\mathbf{InpGen}()$, elements $h_1, \ldots, h_q$ from some sampleable set H, elements $v_1, ..., v_m$ from some sampleable set V, and random coins from some sampleable set R, and returning either a distinguished failure symbol $\bot$, or a tuple (i, j, out), where $i \in \{1, \ldots, q\}$, $j \in \{0, \ldots, m\}$, and out is some side output. The probability of accepting A, represented as $acc(A)$, is defined as the probability, over $inp \leftarrow \mathbf{InpGen}(), h_1, \ldots, h_q \leftarrow H, v_1, ..., v_m \leftarrow V$, and the random coins of $\mathcal{A}$, that $\mathcal{A}$ returns a non-$\bot$ output. Consider algorithm Fork $\mathcal{A}$, which takes inp and $v_1, v_1', \ldots, v_m, v_m' \in V$ as input. Let frk be the probability (over $inp \leftarrow \mathbf{InpGen}(), v_1, v_1', \ldots, v_m, v_m' \leftarrow V$, and the random coins of $\mathsf{Fork}^{\mathcal{A}}$) that $\mathsf{Fork}^{\mathcal{A}}$ returns a non-$\bot$ output. Then*

$$frk \geq acc(A)\left(\frac{acc(A)}{q} - \frac{1}{|H|}\right). \tag{1}$$

Algorithm 1: The "forking" algorithm $\mathsf{Fork}^{\mathcal{A}}$ built from $\mathcal{A}$.

$\mathsf{Fork}^{\mathcal{A}}(inp, v_1, v_1', \ldots, v_m, v_m')$
$\rho \leftarrow R$ // pick random coins for $\mathcal{A}$
$h_1, \ldots, h_q \leftarrow H$
$\alpha := \mathcal{A}(inp, (h_1, \ldots, h_q), (v_1, \ldots, v_m); \rho)$
if $\alpha = \bot$ then return $\bot$
$(i, j, out) := \alpha$
$h_1', \ldots, h_q' \leftarrow H$
$\alpha' := \mathcal{A}(inp, (h_1, \ldots, h_{i-1}, h_i', \ldots, h_q'), (v_1, \ldots, v_j, v_{j+1}', \ldots, v_m); \rho)$
if $\alpha' = \bot$ then return $\bot$
$(i', j', out') := \alpha'$
if $i \neq i' \lor h_i = h_i'$ then return $\bot$
return (i, out, out')

---

3 The Multi-Signature Scheme MuSig2

In this paper, the multi-signature scheme MuSig2 is established under the EU-CMA model. A formal security model can be found in the [13]. MuSig2 is parameterized by a group generation algorithm GrGen and an integer v, which specifies the number of nonces sent by each signer. The scheme is defined as follows.

Parameters setup (Setup). The system parameter generation algorithm takes as input a security parameter λ. It runs $(\mathbb{G}, p, g) \leftarrow \mathsf{GrGen}(1^\lambda)$, selects three hash functions $H_{agg}, H_{non},$ and H_{sig} from $\{0,1\}^*$ to $\mathbb{Z}_p$, and returns the system parameters $par = ((\mathbb{G}, p, g), H_{agg}, H_{non}, H_{sig})$.

Key generation (KeyGen). Each signer generates a random secret key $x \leftarrow \mathbb{Z}_p$ and returns the corresponding public key $X = g^x$.

Key aggregation (KeyAgg). The key aggregation algorithm takes as input the system parameters par. Let $L = \{X_1, \ldots, X_n\}$ be a multiset of public keys. The key aggregation coefficient for $L = \{X_1, \ldots, X_n\}$ and a public key $X \in L$ is defined as $\mathsf{KeyAggCoef}(L, X) = H_{agg}(L, X)$. Then the aggregate key corresponding to L is $\tilde{X} := \prod_{i=1}^n X_i^{a_i}$, where $a_i := \mathsf{KeyAggCoef}(L, X_i)$.

First signing round (Sign). Each signer can perform the Sign step before the cosigners and the message to sign have been determined.

Sign: Let n be the number of signers, for each $i \in \{1, \ldots, n\}, j \in \{1, \ldots, v\}$, each signer generates random $r_{i,j} \leftarrow \mathbb{Z}_p$ and computes $R_{i,j} = g^{r_{i,j}}$. It then returns the v nonces $(R_{i,1}, \ldots, R_{i,v})$.

The aggregator receives outputs $(R_{1,1}, \ldots, R_{1,v}), \ldots, (R_{n,1}, \ldots, R_{n,v})$ from all signers. Then, it aggregates them by computing $R_j = \prod_{i=1}^n R_{i,j}$ for each $j \in \{1, \ldots, v\}$ and outputting $(R_1, \ldots, R_v)$.

Second signing round (Sign'). Let X_1 and x_1 be the public and secret key of a specific signer. The Sign' algorithm takes as inputs a message m, the public keys of the cosigners $X_2, \ldots, X_n$, and the multiset of all public keys involved in signing $L = \{X_1, \ldots, X_n\}$.

Sign': The signer uses the key aggregation algorithm to compute $\tilde{X}$ and stores its own key aggregation coefficient $a_1 = \mathsf{KeyAggCoef}(L, X_1)$. When the signer recepts the aggregate first-round output $(R_1, \ldots, R_v)$, it computes $b = H_{non}(\tilde{X}, (R_1, \ldots, R_v), m)$. Then it computes

$$R := \prod_{j=1}^{v} R_j^{b^{j-1}},$$

$$c := H_{sig}(\tilde{X}, R, m),$$

$$s_1 := \sum_{j=1}^{v} r_{1,j} b^{j-1} + c a_1 x_1 \mod p.$$

Upon the aggregator receives outputs $(s_1, \ldots, s_n)$ of all signers, it aggregates them by outputting the sum $s := \sum_{i=1}^{n} s_i \mod p$.

Finally the signer receives s and returns the signature $\sigma := (R, s)$.

Verification (Ver). The verification algorithm takes as input an aggregate public key $\tilde{X}$, a message m, and a signature $\sigma = (R, s)$. It accepts the signature if

$$g^s = R\tilde{X}^c. \tag{2}$$

Correctness is straightforward to verify. Additionaly, the verification is exactly the same as for ordinary key-prefixed Schnorr signatures with respect to the aggregate public key $\tilde{X}$.

3.1 Security of MuSig2 in the ROM

In this section, we establish the security of MuSig2 with $v = 2$ nonces in the random oracle model. In the MuSig scheme, the adversary could see the different hash value c^* and $c^{*'}$ in the two excutions forkong lemma. We keenly observe that this difference makes the $R_2^{(k)}$ turn into $R_2^{(k)'}$ (k is the kth multi-signature) which results the reduction of the MuSig scheme to the OMDL problem is false. While we can't analyze the MuSig2 scheme as above. The reason is that c^* depend on $R_{2j}^{(k)}, j = \{1, 2\}$.

The system parameter generation algorithm generates group parameters $(\mathbb{G}, p, g)$ and the key generation algorithm generates a key pair (x^*, X^*) for the honest signer. The target public key X^* is given as input to the adversary $\mathcal{A}$. Then, the adversary can engage in protocol executions with the honest signer. Precisely he provides a message m to sign and a multiset L of public keys involved in the signing process where X^* occurs at least once, and simulates all signers except one instance of X^*. We set the random oracle model for $H_{agg}, H_{non}, H_{sig} : \{0, 1\}^* \longrightarrow \mathbb{Z}_p$.

THE DOUBLE-FORKING TECHNIQUE. We fork the execution of the adversary twice. First, we retrieve the discrete logarithm of the aggregate public key $\tilde{X}$ with respect to which the adversary returns a forgery by forking the answer to the query $H_{sig}(\tilde{X}, R, m)$. Second, we retrieve the discrete logarithm of X^* by forking the answer to $H_{agg}(L, X^*)$.

Theorem 1. *Let GrGen be a group generation algorithm. If the OMDL problem is hard, the multi-signature scheme MuSig2 [GrGen, v = 2] is EUF-CMA in the random oracle model for $H_{agg}, H_{non}, H_{sig} : \{0,1\} \to \mathbb{Z}_p$. Precisely, for any adversary $\mathcal{A}$ against MuSig2[GrGen, v = 2] running in time at most t, making at most q_s Sign queries and at most q_h queries to each random oracle, and such that the size of L in any signing session and in the forgery is at most N, there exists an algorithm $\mathcal{D}$ taking as input group parameters $(\mathbb{G}, p, g) \leftarrow \text{GrGen}(1^\lambda)$, running in time at most*

$$t' = 2(t + Nq + 2q)t_{exp} + O(qN), \tag{3}$$

where $q = 2q_h + q_s + 1$ and t_{exp} is the time of an exponentiation in $\mathbb{G}$, making at most q_s DLOG$_g$ queries, and solving the OMDL problem with an advantage

$$Adv_{D,GrGen}^{OMDL} \geq \left(Adv_{A,MuSig2[GrGen,v=2]}^{EU-CMA}(\lambda)\right)^4 / q^3 - 22/2^\lambda. \tag{4}$$

PROOF OVERVIEW. To construct a simulated scheme, we construct a "wrapping" algorithm $\mathcal{B}$. It essentially runs the adversary $\mathcal{A}$ and returns a forgery together with some information about the adversary execution, unless some bad events happen. Algorithm $\mathcal{B}$ simulates the random oracles $H_{agg}, H_{non},$ and H_{sig} uniformly at random and the signing oracle by obtaining v DL challenges from the OMDL challenge oracle for each Sign query and by making a single query to the DL oracle for each Sign' query.

Then, we use $\mathcal{B}$ to construct an algorithm $\mathcal{C}$. The answer to the query $H_{sig}(\tilde{X}, R, m)$ allows us to retrieve the discrete logarithm of the aggregate public key $\tilde{X}$ with respect to which the adversary returns a forgery

Finally, we use $\mathcal{C}$ to construct an algorithm $\mathcal{D}$. The answer to $H_{agg}(L, X^*)$ allows us to retrieve the discrete logarithm of X^*.

NORMALIZING ASSUMPTION AND CONVENTIONS. Let a (t, q_s, q_h, N)-adversary be an adversary running in time at most t, making at most q_s Sign queries, at most q_h queries to each random oracle, and such that $|L|$ in any signing session and in the forgery is at most N.

In all the following, we assume that $X^* \in L$ and $X \in L$ for any query $H_{agg}(L, X)$. Other queries that are irrelevant can simply be answered uniformly at random in the simulation. Without loss of generality, we further assume that the adversary makes exactly q_h queries to each random oracle and exactly q_s queries to the Sign oracle, and that the adversary closes every signing session.

Lemma 2. *Given some integer v, let $\mathcal{A}$ be a (t, q_s, q_h, N)-adversary in the random oracle model against the multi-signature scheme MuSig2[GrGen, v],* and let $q = 2q_h + q_s + 1$. Then there exists an algorithm $\mathcal{B}$ that takes as input group parameters $(\mathbb{G}, p, g) \leftarrow \mathsf{GrGen}(1^\lambda)$, uniformly random group elements $X^*, U_1, \ldots, U_{2q_s} \in \mathbb{G}$, and uniformly random scalars $h_{agg,1}, \ldots, h_{agg,q}, h_{non,1}, \ldots, h_{non,q}, h_{sig,1}, \ldots, h_{sig,q} \in \mathbb{Z}_p$, makes at most q_s queries to a discrete logarithm oracle $DLOG_g$, and with accepting probability

$$acc(B) \geq Adv_{A,MuSig2[\mathsf{GrGen},v]}^{EUF-CMA}(\lambda) - \frac{4q^2}{2^\lambda}, \tag{5}$$

outputs a tuple $(i_{agg}, j_{agg}, i_{sig}, L, R, s, \boldsymbol{a})$, where $i_{agg}, i_{sig} \in \{1, \ldots, q\}, j_{agg} \in \{0, \ldots, q\}$, $L = \{X_1, \ldots, X_n\}$ is a multiset of public keys such that $X^* \in L, \boldsymbol{a} = (a_1, \ldots, a_n) \in \mathbb{Z}_p^n$ is a tuple of scalars such that $a_i = h_{agg,i_{agg}}$ for any i such that $X_i = X^*$, and

$$g^s = R \prod_{i=1}^n X_i^{a_i h_{sig,i_{sig}}}. \tag{6}$$

Proof. This proof is deferred to the [13] due to the space constraints.

Lemma 3. *Given some integer v, let $\mathcal{A}$ be a (t, q_s, q_h, N)-adversary in the random oracle model against the multi-signature scheme MuSig2[GrGen, v] and let $q = 2q_h + q_s + 1$. Then there exists an algorithm $\mathcal{C}$ that takes as input group parameters $(\mathbb{G}, p, g) \leftarrow GrGen(1^\lambda)$, uniformly random group elements $X^*, U_1, \ldots, U_{q_s} \in \mathbb{G}$, and uniformly random scalars $h_{agg,1}, \ldots, h_{agg,q}, h_{non,1}, \ldots, h_{non,q} \in \mathbb{Z}_p$, makes at most q_s queries to a discrete logarithm oracle $DLOG_g$, and with accepting probability*

$$acc(\mathcal{C}) \geq \frac{\left(Adv_{A,MuSig2[GrGen,v]}^{EUF-CMA}(\lambda)\right)^2}{q} - \frac{2(4q+1)}{2^\lambda}, \tag{7}$$

outputs a tuple $(i_{agg}, L, \boldsymbol{a}, \tilde{x})$ where $i_{agg} \in \{1, \ldots, q\}$, $L = \{X_1, \ldots, X_n\}$ is a multiset of public keys such that $X^* \in L, \boldsymbol{a} = (a_1, \ldots, a_n) \in \mathbb{Z}_p^n$ is a tuple of scalars such that $a_i = h_{agg,i_{agg}}$ for any i such that $X_i = X^*$, and $\tilde{x}$ is the discrete logarithm of $\tilde{X} = \prod_{i=1}^n X_i^{a_i}$ in base g.

Proof. Algorithm $\mathcal{C}$ runs $Fork_{\mathcal{B}}$ with $\mathcal{B}$ as defined in Lemma 2 and takes additional steps as described below. The mapping with notation of Forking Lemma is as follows:

- $(\mathbb{G}, p, g), X^*, U_1, \ldots, U_{q_s}$ and $h_{agg,1}, \ldots, h_{agg,q}$ play the role of *inp*,
- $h_{sig,1}, \ldots, h_{sig,q}$ play the role of $h_1, \ldots, h_q$,
- i_{sig} plays the role of i,
- $(i_{agg}, L, R, s, \boldsymbol{a}, \tilde{x})$ play the role of *out*.

Concretely, $\mathcal{C}$ picks random coins $\rho_{\mathcal{B}}$ and uniformly random scalars $h_{sig,1}, \ldots, h_{sig,q} \in \mathbb{Z}_p$, and runs algorithm $\mathcal{B}$ on coins $\rho_{\mathcal{B}}$, group description $(\mathbb{G}, p, g)$, group elements $X^*, U_1, \ldots, U_{q_s} \in \mathbb{G}$, and scalars $h_{agg,1}, \ldots, h_{agg,q}, h_{non,1}, \ldots, h_{non,q}, h_{sig,1}, \ldots, h_{sig,q} \in \mathbb{Z}_p$. Recall that scalars $h_{agg,1}, \ldots, h_{agg,q}, h_{non,1}, \ldots, h_{non,q}$

are part of the input of $\mathcal{C}$ and the former will be the same in both runs of $\mathcal{B}$. All $DLOG_g$ oracle queries made by $\mathcal{B}$ are relayed by $\mathcal{C}$ to its own $DLOG_g$ oracle. If $\mathcal{B}$ returns $\perp$, $\mathcal{C}$ returns $\perp$ as well. Otherwise, if $\mathcal{B}$ returns a tuple $(i_{agg}, i_{sig}, L, R, s, \boldsymbol{a}, \tilde{x})$, where $L = \{X_1, \ldots, X_n\}$ and $\boldsymbol{a} = (a_1, \ldots, a_n)$, $\mathcal{C}$ picks uniformly random scalars $h'_{sig,i_{sig}}, \ldots, h'_{sig,q} \in \mathbb{Z}_p$ and runs $\mathcal{B}$ again with the same random coins $\rho_{\mathcal{B}}$ on input $(\mathbb{G}, p, g), X^*, U_1, \ldots, U_{q_s}, h_{agg,1}, \ldots, h_{agg,q}$, $h_{sig,1}, \ldots, h_{sig,i_{sig}-1}, h'_{sig,i_{sig}}, \ldots, h'_{sig,q}$.

Again, all $DLOG_g$ oracle queries made by $\mathcal{B}$ are relayed by $\mathcal{C}$ to its own $DLOG_g$ oracle. If $\mathcal{B}$ returns $\perp$ in this second run, $\mathcal{C}$ returns $\perp$ as well. If $\mathcal{B}$ returns a second tuple $(i'_{agg}, L', R', s', \boldsymbol{a}')$, where $L' = \{X'_1, \ldots, X'_{n'}\}$ and $\boldsymbol{a}' = (a'_1, \ldots, a'_{n'})$, $\mathcal{C}$ proceeds as follows. Let $\tilde{X} = \prod_{i=1}^n X_i^{a_i}$ and $\tilde{X}' = \prod_{i=1}^{n'}(X'_i)^{a'_i}$ denote the aggregate public keys from the two forgeries. If $i_{sig} \neq i'_{sig}$, or $i_{sig} = i'_{sig}$ and $h_{sig,i_{sig}} = h'_{sig,i_{sig}}$, $\mathcal{C}$ returns $\perp$. Otherwise, if $i_{sig} = i'_{sig}$ and $h_{sig,i_{sig}} \neq h'_{sig,i_{sig}}$, we will prove shortly that

$$i_{agg} = i'_{agg}, \; L = L', R = R' \; and \; \boldsymbol{a} = \boldsymbol{a}' \tag{8}$$

which implies in particular that $\tilde{X} = \tilde{X}'$. By lemma 2, the two outputs returned by $\mathcal{B}$ are such that

$$g^s = R\tilde{X}^{h_{sig,i_{sig}}} \; and \; g^{s'} = R'(\tilde{X}')^{h'_{sig,i_{sig}}} = R\tilde{X}^{h'_{sig,i_{sig}}}, \tag{9}$$

which allows $\mathcal{C}$ to compute the discrete logarithm of $\tilde{X}$ as

$$\tilde{x} := (s - s')(h_{sig,i_{sig}} - h'_{sig,i_{sig}})^{-1} \mod p. \tag{10}$$

Then $\mathcal{C}$ returns $(i_{agg}, L, \boldsymbol{a}, \tilde{x})$.

It remains to prove the equalities of Eq. 8. We set $T_{sig}(\tilde{X}, R, m) = h_{sig,i_{sig}}$ in $\mathcal{B}$'s first execution and $T_{sig}(\tilde{X}', R', m') = h'_{sig,i_{sig}}$ in $\mathcal{B}$'s second execution. Since the c^* changes lastly and the adversary's view is consistent up to the c^* changes, we have $R = R'$ and $\tilde{X} = \tilde{X}'$. According to the KeyColl and BadOrder is not ture, $\tilde{X} = \tilde{X}'$ implies $L = L'$, $i_{agg} = i'_{agg}$ and $\boldsymbol{a} = \boldsymbol{a}'$.

$\mathcal{C}$ returns a non-$\perp$ output if $Fork_{\mathcal{B}}$ does, so that by lemmas 2 and 3, and letting $\varepsilon = Adv^{EUF-CMA}_{\mathcal{A}, MuSig2[GrGen,v]}(\lambda)$, $\mathcal{C}'$ accepting probability satisfies

$$
\begin{aligned}
acc(\mathcal{C}) &\geq acc(\mathcal{B})\left(\frac{acc(\mathcal{B})}{q} - \frac{1}{p}\right) \\
&\geq \frac{(\varepsilon - 4q^2/2^\lambda)^2}{q} - \frac{\varepsilon - 4q^2/2^\lambda}{2^{\lambda-1}} \\
&\geq \frac{\varepsilon^2}{q} - \frac{2\varepsilon(4q+1)}{2^\lambda} + \frac{8q^2(2q+1)}{2^{2\lambda}} \\
&\geq \frac{\varepsilon^2}{q} - \frac{2(4q+1)}{2^\lambda}.
\end{aligned}
\tag{11}
$$

We are ready to prove Theorem 1 as follows, which we restate below for convenience, by constructing from $\mathcal{C}$ an algorithm $\mathcal{D}$ solving the OMDL problem.

Proof of Theorem 1. Fix some integer $v \geq 2$. Algorithm $\mathcal{D}$ runs $Fork_{\mathcal{C}}$ with $\mathcal{C}$ as defined in lemma 3 and takes additional steps as described below. The mapping with the notation in our Forking Lemma (lemma 2) is as follows:

- $(\mathbb{G}, p, g), X^*, U_1, \ldots, U_{2q_s}$ play the role of inp,
- $h_{non,1}, h'_{non,1}, \ldots, h_{non,q}, h'_{non,q}$ play the role of $v_1, v'_1, \ldots, v_m, v'_m$,
- $h_{agg,1}, \ldots, h_{agg,q}$ play the role of $h_1, \ldots, h_q$,
- (i_{agg}, j_{agg}) play the role of (i, j),
- $(L, \boldsymbol{a}, \tilde{x})$ play the role of out.

In more details, algorithm $\mathcal{D}$ makes $2q_s + 1$ queries to its challenge oracle $X^*, U_1, \ldots, U_{2q_s} \leftarrow CH()$, picks random coins $\rho_{\mathcal{C}}$ and scalars $h_{agg,1}, \ldots, h_{agg,q}$, $h_{non,1}, \ldots, h_{non,q} \in \mathbb{Z}_p$, and runs $\mathcal{C}$ on coins $\rho_{\mathcal{C}}$, group description $(\mathbb{G}, p, g)$, group elements $X^*, U_1, \ldots, U_{2q_s} \in \mathbb{Z}_p$, and scalars $h_{agg,1}, \ldots, h_{agg,q}, h_{non,1}, \ldots, h_{non,q} \in \mathbb{Z}_p$. It relays all $DLOG_g$ oracle queries made by $\mathcal{C}$ to its own $DLOG_g$ oracle. It use $DLOG_g$ oracle to cache pairs of group elements and responses to avoid making multiple queries for the same group element. If $\mathcal{C}$ returns $\bot$, $\mathcal{D}$ returns $\bot$ as well. Otherwise, if $\mathcal{C}$ returns a tuple $(i_{agg}, j_{agg}, L, \boldsymbol{a}, \tilde{x})$, $\mathcal{D}$ picks uniformly random scalars $h'_{agg,i_{agg}}, \ldots, h'_{agg,q} \in \mathbb{Z}_p$, and runs $\mathcal{C}$ again with the same random coins $\rho_{\mathcal{C}}$ on input $X^*, U_1, \ldots, U_{2q_s}$,

$$h_{agg,1}, \ldots, h_{agg,i_{agg}-1}, h'_{agg,i_{agg}}, \ldots, h'_{agg,q},$$

$$h_{non,1}, \ldots, h_{non,j_{agg}}, h'_{non,j_{agg}+1}, \ldots, h'_{non,q}.$$

It relays all $DLOG_g$ oracle queries made by $\mathcal{C}$ to its own $DLOG_g$ oracle after looking them up in its cache. This avoids making duplicate queries. If $\mathcal{C}$ returns $\bot$ in this second run, $\mathcal{D}$ returns $\bot$ as well. If $\mathcal{C}$ returns a second tuple $(i'_{agg}, j'_{agg}, L', \boldsymbol{a}', \tilde{x}')$, $\mathcal{D}$ proceeds as follows. Let $L = \{X_1, \ldots, X_n\}$, $\boldsymbol{a} = (a_1, \ldots, a_n)$, $L' = \{X'_1, \ldots, X'_{n'}\}$, and $\boldsymbol{a}' = (a'_1, \ldots, a'_n)$. Let n^* be the number of times X^* appears in L. If $i_{agg} \neq i'_{agg}$, or $i_{agg} = i'_{agg}$ and $h_{agg,i_{agg}} = h'_{agg,i_{agg}}$, $\mathcal{D}$ returns $\bot$. Otherwise, if $i_{agg} = i'_{agg}$ and $h_{agg,i_{agg}} \neq h'_{agg,i_{agg}}$. Then according to the forking lemma, we will have that

$$L = L' \text{ and } a_i = a'_i \text{ for each } i \text{ such that } X_i \neq X^*. \tag{12}$$

By lemma 3, we have that

$$g^{\tilde{x}} = \prod_{i=1}^{n} X_i^{a_i} = (X^*)^{n^* h_{agg,i_{agg}}} \prod_{\substack{i \in \{1, \ldots, n\} \\ X_i \neq X^*}} X_i^{a_i},$$

$$g^{\tilde{x}'} = \prod_{i=1}^{n} X_i^{a'_i} = (X^*)^{n^* h'_{agg,i_{agg}}} \prod_{\substack{i \in \{1, \ldots, n\} \\ X_i \neq X^*}} X_i^{a_i}, \tag{13}$$

Thus, $\mathcal{D}$ is able to compute the discrete logarithm of X^* as

$$x^* := (\tilde{x} - \tilde{x}')(n^*)^{-1}(h_{agg,i_{agg}} - h'_{agg,i_{agg}})^{-1} \mod p. \tag{14}$$

According to $r = s - cx$, we can obtain the discrete logarithm $R_{1j}^{(k)}$, $j = \{1, 2\}$, $k = \{1, \ldots, q_s\}$. This is cantradict with $2q_s$OMDL problem.

Neglecting the time needed to compute discrete logarithms , the running time t' of $\mathcal{D}$ is twice the running time of $\mathcal{C}$, which itself is the same as the running time of $\mathcal{B}$. The running time of $\mathcal{B}$ is the running time t of $\mathcal{A}$ plus the time needed to maintain tables T_{agg}, T_{non}, and T_{sig} (we assume each assignment takes unit time) and answer signing and hash queries. The sizes of T_{agg}, T_{non}, and T_{sig} are at most qN, q, and q respectively. Answering signing queries is dominated by the time needed to compute the aggregate key as well as the honest signer's effective nonce, which is at most Nt_{exp} and $(v - 1)t_{exp}$ respectively. Answering hash queries is dominated by the time to compute the aggregate nonce which is at most $(v - 1)t_{exp}$. Therefore, $t' = 2(t + q(N + 2v - 2))t_{exp} + O(qN)$.

Let $\varepsilon = Adv_{\mathcal{A},MuSig2[GrGen,v]}^{EUF-CMA}(\lambda)$. By Lemmas 1 and 3, the success probability of $Fork_{\mathcal{C}}$ is at least

$$acc(Fork_{\mathcal{C}}) \geq acc(\mathcal{C})(\frac{acc(\mathcal{C})}{q} - \frac{1}{p})$$

$$\geq \frac{(\varepsilon^2/q - 2(4q + 1)/2^\lambda)^2}{q} - \frac{\varepsilon^2/q - 2(4q + 1)/2^\lambda}{2^{\lambda-1}}$$

$$\geq \frac{\varepsilon^4}{q^3} - \frac{16 + 4/q - 2}{q \cdot 2^\lambda} \tag{15}$$

$$\geq \frac{\varepsilon^4}{q^3} - \frac{22}{2^\lambda}.$$

The advantage of $\mathcal{D}$ is

$$Adv_{\mathcal{D},GrGen}^{OMDL}(\lambda) \geq acc(Fork_{\mathcal{C}}) \geq \frac{\varepsilon^4}{q^3} - \frac{22}{2^\lambda}. \tag{16}$$

3.2   Practical Considerations

THE CHOICE OF THE NUMBER v OF NONCES. We provide a security proof in the ROM for MuSig2[$v = 2$]. When compared to MuSig2[$v = 4$], the signing algorithm of MuSig2[$v = 2$] saves a multi-exponentiation of size three plus a single exponentiation as well as three group elements of communication in the first round (all per signer).

STATEFULNESS. The issue of ensuring correct state transitions can be challenging in practice if the state is written to persistent storage. In particular, the state may be reused by accident when restoring a backup or through a deliberate attack on the physical storage. After executing Sign' with some state the signer must make sure to never run Sign' again with the same state. Otherwise, the signer will reuse the nonce, allowing trivial extraction of the secret key. Again, similar attacks apply to essentially all Schnorr multi-signature schemes, except for the fully deterministic MuSig-DN. In the MuSig2 scheme, the same state is also not secure. On the one hand, the same state will cause the private key to be leaked, and on the other hand, the same state will make it easy for adversaries to forge new multisignatures from existing multisignatures through the Wagner algorithm.

We reduce the security requirement slightly to achieve the MuSig2 scheme's security even if the nonces are reused. In the MuSig2 scheme's security model, they assume that there is a single honest signer and that the adversary has corrupted all other signers. While we assume there are two or more honest signers in the secure model. This change achieves the MuSig2 scheme's security even if the nonces are reused.

Consider a concrete situation, there are 1000 signers and at least 2 honest signers. Every signer needs two nonce when signing multi-signatures. The probability of nonces being reused is $(1/2)^{60}$. Assume the adversary forges a signature if the number of the same nonces that are repeatedly used are 2^{40} times in the MuSig2 scheme. Altogether the probability that the adversaries will successfully forge the signature is $(1/2)^{60 \times 2 \times 2} \times (1/2)^{40} = (1/2)^{280}$. Furthermore, we assume that $r_i = r_{i1}b_i + r_{i2}b$, $b_i = H_{non}(R_{i1}, (R_1, \ldots, R_v))$. Then the above probabilities become $(1/2)^{60 \times 2 \times 2 \times 2} \times (1/2)^{40} = (1/2)^{520}$. This is negligible. In addition, repeating a random number three times will cause the private key to leak. We note that the probability of repeating a random number three times is negligible, with a probability of $(1/2)^{60 \times 2 \times 2} = (1/2)^{240}$.

4   MSOMDL

Maxell et al. [12] proposes a multi-signature scheme based on Schnorr signatures to realize the aggregate key, and reduces the number of rounds of the multi-signature scheme in Bellare and Neven [4] from 3 to 2. However, this reduction has defects [6]. Boneh et al. [5] proposes MSDL by adding commitments to avoid this defect. Unfortunately, their proof exists an issue related to the commitment. Therefore, MSOMDL is proposed in this paper, and this multi-signature scheme is reduced to the OMDL problem. The MSOMDL scheme is the same as the MSDL scheme [13].

We point out that the MSDL scheme is not reduced to the DL problem. Additionally, we briefly outline the main idea of the proof for Theorem 2, with a detailed demonstration deferred to Appendix A.

We review the commitment in the proof of the MSDL scheme as follows. After receiving values t_j from all other signers, it looks up the corresponding values R_j such that $H_{con}(R_j) = t_j$. If not all such values can be found, then $\mathcal{A}$ sends $R_i \leftarrow \mathbb{G}$ to all signers; unless collision happens, the signing protocol finishes in the next round. If all values R_j are found, then $\mathcal{A}$ chooses $s_i, c \leftarrow \mathbb{Z}_q$, simulates an internal query $a_i \leftarrow H_{agg}(pk_i, PK)$, computes $R_i \leftarrow g^{s_i}pk_i^{-a_i c}$ and $R \leftarrow \prod_{j=1}^n R_j$, assigns $H_{con}(R_i) \leftarrow t_i$ and $H_{sig}(R, \tilde{X}, m) \leftarrow c$, and sends R_i to all signers.

Note that if not all such values can be found, then $\mathcal{A}$ sends $R_i \leftarrow \mathbb{G}$ to all signers; unless collision happens, the signing protocol finishes in the next round. We affirm that the simulator can't use this R_i to simulate the signature successfully. Specifically, the adversary's R_j is public before the simulator's is unreasonable in their proof. When the simulator doesn't know one R_j, the simulator can use commitments to distinguish between true R_j and false R_j (true means the adversary uses R_j in the real scheme). Concretely, the simulator can calculate the adversary's $R_j = R / \prod_{i \neq j} R_i$ when the adversaty query H_{sig} with R. Then we have "R_j is true" iff $H_{con}(R_j) = t_j$. Unfortunately, if the simulator is unaware of more than two R_i, it would be unable to simulate as demonstrated above except the commitment is a homomorphism.

To demonstrate the validity of MSOMDL, we can adopt the same proof methodology used in MuSig2. The key step is to prove that $R = R'$. Since the simulator finally changes c^*, the adversary is unable to modify $\{R_i^{(k)}\}$ in the two executions of the forking lemma. Thus, we have $R = R'$. Additionally, we can easily use Boneh et al's proof to prove the MSDL scheme when the number of signers is 2.

5   Conclusion

In order to maintain the security of AI, especially achieve the accountability, integrity and authenticity of important data, this paper introduces multisignatures in AI. For more efficient use of multi-signatures, we simplify MuSig2's proof under the OMDL assumption in the random oracle model, along with reducing the number of nonces from 4 to 2. Moreover, we achieve the MuSig2 scheme's security when the nonces are reused. and we utilize the proof technology of MuSig2 to reduce the security of the MSDL scheme to the OMDL assumption. We will work on implementing a method with tighter reduction that can withstand concurrent attacks and rogue key attacks.

A   Proof of the MSOMDL

Theorem 2. *MSOMDL is EUF-CMA in the random oracle model for $H_{agg}, H_{con}, H_{sig} : \{0,1\}^* \longrightarrow \mathbb{Z}_p$, if the OMDL problem is hard. More precisely, for any adversary $\mathcal{A}$ against MSOMDL running in time at most t, making at most q_s Sign queries and at most q_h queries to each random oracle, and such that the size of L in any signing session and in the forgery is at most N, there exists an algorithm D taking as input group parameters $(\mathbb{G}, p, g) \leftarrow GrGen(1^\lambda)$, running in time at most*

$$t' = 2(t + Nq)t_{exp} + O(qN), \tag{17}$$

where $q = 2q_h + q_s + 1$ and t_{exp} is the time of an exponentiation in $\mathbb{G}$, making at most $2q_s DLOG_g$ queries, and solving the OMDL problem with an advantage

$$Adv^{OMDL}_{D,GrGen(\lambda)} \geq (Adv^{EUF-CMA}_{A,MSOMDL(\lambda)})4/q^3 - 22/2^\lambda. \tag{18}$$

Proof. We construct algorithm $\mathcal{B}$, algorithm $\mathcal{C}$ and algorithm $\mathcal{D}$. They work the same as described in the proof of Musig2. We first construct a "wrapping" algorithm $\mathcal{B}$ which essentially runs the algorithm $\mathcal{A}$ and returns a forgery together with some information about the adversary execution, unless some bad events happen. For the specific structure of algorithm $\mathcal{B}$, see [13].

Following we construct $\mathcal{C}$ to calculate aggregated secret key $\tilde{x}$. The conditions for the forking lemma are the same as for MuSig2. We have $R = R'$, $\tilde{X} = \tilde{X}'$ since the forger $c^* = H_{sig}(\tilde{X}, R, m)$ change lastly. According to the KeyColl and BadOrder is not ture, $\tilde{X} = \tilde{X}'$ implies $L = L'$, $i_{agg} = i'_{agg}$ and $\boldsymbol{a} = \boldsymbol{a}'$.

Therefore, we have that

$$i_{agg} = i'_{agg}, \ L = L', R = R' \ and \ \boldsymbol{a} = \boldsymbol{a}' \tag{19}$$

which implies in particular that $\tilde{X} = \tilde{X}'$. By lemma 3, the two outputs returned by $\mathcal{B}$ are such that

$$g^s = R\tilde{X}^{h_{sig,i_{sig}}} \ and \ g^{s'} = R'(\tilde{X}')^{h'_{sig,i_{sig}}} = R\tilde{X}^{h'_{sig,i_{sig}}}, \tag{20}$$

which allows $\mathcal{C}$ to compute the discrete logarithm of $\tilde{X}$ as

$$\tilde{x} := (s - s')(h_{sig,i_{sig}} - h'_{sig,i_{sig}})^{-1} \mod p. \tag{21}$$

Then $\mathcal{C}$ returns $(i_{agg}, L, \boldsymbol{a}, \tilde{x})$.

letting $\varepsilon = Adv_{\mathcal{A},MSOMDL}^{EUF-CMA}(\lambda)$, $\mathcal{C}'$ accepting probability satisfies

$$acc(\mathcal{C}) \geq acc(\mathcal{B})(\frac{acc(\mathcal{B})}{q} - \frac{1}{p}) \geq \frac{\varepsilon^2}{q} - \frac{2(4q+1)}{2^\lambda}. \tag{22}$$

Following we constuct $\mathcal{D}$ to calculate secret key x^*. The conditions for the forking lemma are the same as for MuSig2 except requires to replace the query $H_{non}(\tilde{X}, (R_1, \ldots, R_v), m)$ to the query $H_{con}(R_i)$. Since the two executions of $\mathcal{B}$ are identical up to the assignments $T_{agg}(L, X^*) := h_{agg,i_{agg}}$ and $T_{agg}(L', X^*) := h'_{agg,i_{agg}}$, the arguments of the two assignments $T_{agg}(L, X_i)$ and $T_{agg}(L', X'_i)$ must be the same, which implies that $L = L'$, $a_i = a'_i$ for each i such that $X_i \neq X^*$.

The adversaty behaves differently in these two excutions. Concretely, the simulator should use different R_i for different a_i in these two excutions. Thus the simulator requires $2q_s DLOG_g$ oracles.

Therefore, we have that

$$L = L' \ and \ a_i = a'_i \ for \ each \ i \ such \ that \ X_i \neq X^*. \tag{23}$$

By lemma 3, we have that

$$g^{\tilde{x}} = \prod_{i=1}^n X_i^{a_i} = (X^*)^{n^* h_{agg,i_{agg}}} \prod_{\substack{i \in \{1,\ldots,n\} \\ X_i \neq X^*}} X_i^{a_i},$$

$$g^{\tilde{x}'} = \prod_{i=1}^n X_i^{a'_i} = (X^*)^{n^* h'_{agg,i_{agg}}} \prod_{\substack{i \in \{1,\ldots,n\} \\ X_i \neq X^*}} X_i^{a_i}, \tag{24}$$

Thus, $\mathcal{D}$ is able to compute the discrete logarithm of X^* as

$$x^* := (\tilde{x} - \tilde{x}')(n^*)^{-1}(h_{agg,i_{agg}} - h'_{agg,i_{agg}})^{-1} \mod p. \tag{25}$$

We have $r = s - cx$. This is contradict with OMDL problem.

According to $r = s - cx$, we can obtain the discrete logarithm $R_{1j}^{(k)}, j = \{1,2\}, k = \{1, \dots, q_s\}$. Noted that $j = \{1,2\}$ represents that the simulator uses different nonces in the different excutions of the foking lemma. This is cantradict with $2q_s$OMDL problem.

Let $\varepsilon = Adv_{A,MSOMDL}^{EUF-CMA}(\lambda)$. By Lemmas 1 and 3, the success probability of $Fork_{\mathcal{C}}$ is at least

$$acc(Fork_{\mathcal{C}}) \geq acc(\mathcal{C})(\frac{acc(\mathcal{C})}{q} - \frac{1}{p}) \geq \frac{\varepsilon^4}{q^3} - \frac{22}{2^\lambda}. \tag{26}$$

The advantage of $\mathcal{D}$ is

$$Adv_{\mathcal{D},GrGen(\lambda)}^{OMDL} \geq acc(Fork_{\mathcal{C}}) \geq \frac{\varepsilon^4}{q^3} - \frac{22}{2^\lambda}. \tag{27}$$

Time's analyze is the same as described in lemma2 except the time needed to compute the honest signer's effective nonce. Thus the running time t' of $\mathcal{D}$ is $t' = 2(t + qN)t_{exp} + O(qN)$.

References

1. Alper, H.K., Burdges, J.: Two-round trip Schnorr multi-signatures via Delinearized witnesses. IACR Cryptology ePrint Archive (2020)
2. Bagherzandi, A., Cheon, J.H., Jarecki, S.: Multisignatures secure under the discrete logarithm assumption and a generalized forking lemma. In: Proceedings of the 15th ACM Conference on Computer and Communications Security (2008)
3. Bellare, M., Dai, W.: Chain reductions for multi-signatures and the HBMS scheme. In: International Conference on the Theory and Application of Cryptology and Information Security (2021)
4. Bellare, M., and Neven, G. Multi-signatures in the plain public-key model and a general forking lemma. In: Conference on Computer and Communications Security (2006)
5. Boneh, D., Drijvers, M., Neven, G.: Compact multi-signatures for smaller blockchains. IACR Cryptol. ePrint Arch. **2018**, 483 (2018)
6. Drijvers, M., et al.: On the security of two-round multi-signatures. In: 2019 IEEE Symposium on Security and Privacy (SP), pp. 1084–1101 (2019)
7. Itakura, K.: A public-key cryptosystem suitable for digital multisignatures. NEC Res. Dev. **71**, 1–8 (1983)
8. Kaur, D., Uslu, S., Rittichier, K.J., Durresi, A.: Trustworthy artificial intelligence: a review. ACM Comput. Surv. (CSUR) **55**, 1–38 (2022)
9. Lee, K., Kim, H.: Two-round multi-signatures from Okamoto signatures. IACR Cryptol. ePrint Arch. **2022**, 1117 (2023)
10. Legg, S., Hutter, M.: A collection of definitions of intelligence. In: Artificial General Intelligence (2007)

11. Ma, C., Weng, J., Li, Y., Deng, R.H.: Efficient discrete logarithm based multisignature scheme in the plain public key model. Des. Codes Crypt. **54**, 121–133 (2010)
12. Maxwell, G., Poelstra, A., Seurin, Y., Wuille, P.: Simple Schnorr multi-signatures with applications to bitcoin. Des. Codes Cryptogr. **87**, 1–26 (2019)
13. Nick, J.D., Ruffing, T., Seurin, Y.: Musig2: simple two-round Schnorr multisignatures. IACR Cryptology ePrint Archive (2020)
14. Nick, J.D., Ruffing, T., Seurin, Y., Wuille, P.: MuSig-DN: Schnorr multi-signatures with verifiably deterministic nonces. In: Proceedings of the 2020 ACM SIGSAC Conference on Computer and Communications Security (2020)
15. Nicolosi, A., Krohn, M. N., Dodis, Y., Mazières, D.: Proactive two-party signatures for user authentication. In: Network and Distributed System Security Symposium (2003)
16. Pan, J., Wagner, B.: Chopsticks: fork-free two-round multi-signatures from noninteractive assumptions. IACR Cryptol. ePrint Arch. **2023**, 198 (2023)
17. Schnorr, C.-P.: Efficient signature generation by smart cards. J. Cryptol. **4**, 161–174 (2004)
18. Syta, E., et al.: Keeping authorities "honest or bust" with decentralized witness cosigning. In: 2016 IEEE Symposium on Security and Privacy (SP), pp. 526–545 (2015)
19. Szalachowski, P., Matsumoto, S., and Perrig, A. PoliCert: Secure and flexible TLS certificate management. In: Proceedings of the 2014 ACM SIGSAC Conference on Computer and Communications Security (2014)
20. Tessaro, S., Zhu, C.: Threshold and multi-signature schemes from linear hash functions. IACR Cryptol. ePrint Arch. **2023**, 276 (2023)
21. Wagner, D.A.: A generalized birthday problem. In: Annual International Cryptology Conference (2002)
22. Xiao, Y.-L., Zhang, P., Liu, Y.: Secure and efficient multi-signature schemes for fabric: an enterprise blockchain platform. IEEE Trans. Inf. Forensics Secur. **16**, 1782–1794 (2022)

Active Defense Against Image Steganography

Weixuan Tang[(✉)] and Yadong Liu

Institute of Artificial Intelligence, Guangzhou University, Guangzhou, China
tweix@gzhu.edu.cn

Abstract. Image steganalysis is the art and science of detecting whether an image contains secrete messages or not, which can prevent malicious usage of steganography. However, steganalysis belongs to passive defense, in the sense that it can only be applied after the stego image is generated. Therefore, there still exists loophole that secrete messages communication could already been accomplished when the stego image is detected. To eliminate malicious steganography from the source, in this paper, an active defensive framework for deep image steganography called ADPI (Active Defense based on Perturbation Injection) is proposed, wherein a defender competes against a steganographer to learn the active defensive strategy. Specifically, on the side of the defender, a generator is adopted to take the original cover image as input, and learn the imperceptible perturbation map. Such perturbation map is added with the original cover image as the enhanced cover image. On the side of the steganographer, a steganographic network is applied to perform message embedding and extraction on the enhanced cover image. The key of ADPI is that the perturbation map is optimized with the goal of reducing the accuracy of the message recovery while maintaining its invisibility. By this means, the active defense can be launched in an effective and imperceptible manner. Experimental results show that the proposed ADPI can be applied to defend against various steganographic methods.

Keywords: Image Steganography · Image Steganalysis · Active Defense · Generative Adversarial Network

1 Introduction

Image steganography is a technology of information hiding, which conceals secrete messages within cover images. Traditional steganographic methods are designed under distortion minimization framework, which can be formulated as minimizing the distortion function under the payload constraint. Such distortion function is calculated according to the embedding costs of modified pixels, which is the key component of this type of steganographic methods. With the defined embedding costs, practical steganographic codes, such as STC, can be applied for message embedding and extraction. In the past decades, lots of steganographic methods under distortion minimization function have been proposed, including [1–4].

However, limited by the strict constraint of distortion minimization framework, the payload of the above methods are relatively low. To obtain higher embedding capacities, deep steganographic methods are proposed, wherein neural networks are utilized to implement message embedding and extraction, rather than practical steganographic codes.

Hayes et al. [5] first proposed to utilize the encoder-decoder structure within the GAN (Generative Adversarial Network) [6] framework to achieve image steganography. To further improve robustness, Zhu et al. [7] introduced various types of noise layers for model training, which simulate real-world noise attacks and compression artifacts, thereby achieving an end-to-end robust framework. To further increase hiding capacity, Zhang et al. [8] which replaced the discriminator with a critic network to assess image authenticity, and modified the encoder architecture to enable high-capacity hiding. Recently, Tan et al. [9] which incorporates channel-wise attention to enhance security and visual quality. Lu et al. [10] proposed a large capacity image steganography network based on INN (Invertible Neural Network), which uses forward and back propagation operations to realize image hiding and revealing.

Note that steganography is a double-edged technology. It can be utilized by common users to protect their privacy in a good way, but also utilized by criminals to spread malicious and illegal information in a bad way. Specifically, the illegal and malicious use of steganography can cause significant social harm. Therefore, it is essential to develop the technology of image steganalysis from a national security perspective, which aims to detect whether an image contains secrete messages or not. Traditional steganalyzer applies ensemble classifier [11] to classify hand-crafted high dimensional statistical features [12–14]. With the rapid development of deep learning techniques, steganalyzers based on CNN (Convolutional Neural Network) have achieved outstanding performance [15–19]. Note that the technology of steganalysis is the passive defense against image steganography, in the sense that steganalysis can only be applied after the stego image is generated. In the case that the stego image is successfully detected by the steganalyzer, such stego image may already been sent to the receiver and the secrete message communication could be accomplished. Therefore, passive defense could not stop malicious usage of steganography from the source.

To address the above issues, in this paper, we propose active defensive strategy for image steganography, which aims to take action before the behavior of secrete message communication happens. Under the guidance of such strategy, we propose an active defensive framework for deep image steganography called ADPI (Active Defense based on Perturbation Injection). ADPI consists of a defender and a steganographer. From the side of the defender, given an original cover image, a generator is applied to learn a perturbation map, which is added with the original cover image as the enhanced cover image. From the side of the steganographer, such enhanced cover image is fed into the deep image steganographic network for message embedding and extraction. The key idea of ADPI is that the perturbation map is optimized with the goal of reducing the accuracy of the message recovery while maintaining its invisibility. By this means, the active defense can be launched in an effective and imperceptible manner. As the training process terminates, these enhanced cover images are uploaded on the Internet. Once the criminals acquires these images for secrete message communication, it is unable for them to implement successful message embedding and extraction. Therefore, our proposed ADPI can eliminate malicious steganography from the source. The contributions of this paper are as follows.

- The distinctions between active defense and passive defense against image steganography have been analyzed. To the best of our knowledge, this work is the first attempt to apply active defense against image steganography.
- An active defensive framework for deep image steganography called ADPI has been proposed, which can generate the enhanced cover image disabling the steganographic system and undistinguished from its original counterpart.
- Experiments have been conducted to evaluate the performance of the proposed ADPI. Experimental results show that the proposed active defense can be launched in an effective and imperceptible manner.

2   Method

In this section, we propose an active defensive framework for deep image steganography called ADPI (Active Defense based on Perturbation Injection). First, we outline the general framework of ADPI. Then, we introduce the design of the loss function. Finally, we present the network architecture of the generator.

2.1   Notations

In this paper, matrices are denoted by capital bold letters, and elements within matrices are denoted by the corresponding lowercase letters. Specifically, original cover image, enhanced cover image, and stego image are denoted as $\mathbf{X} = (x_{i,j,c})^{H \times W \times 3}$, $\widetilde{\mathbf{X}} = (\widetilde{x}_{i,j,c})^{H \times W \times 3}$, and $\mathbf{X}' = (x_{i,j,c})^{H \times W \times 3}$, respectively. Target messages and revealed messages are denoted as $\mathbf{M} = (m_{i,j,c})^{H \times W \times C}$ and $\mathbf{M}' = (m'_{i,j,c})^{H \times W \times C}$, respectively.

2.2   Overall Framework

In this paper, we propose the active defensive strategy against image steganography for the first time. Its key idea is to proactively process the original cover images such that secrete messages cannot be communicated via regarding these processed images as carriers. By this means, the active defensive strategy can proactively disable the steganographic system. Note that the active defense strategy works in a forward-looking manner, in the sense that it takes action before the behavior of secrete message communication happens. Therefore, it can eliminate malicious steganography from the source.

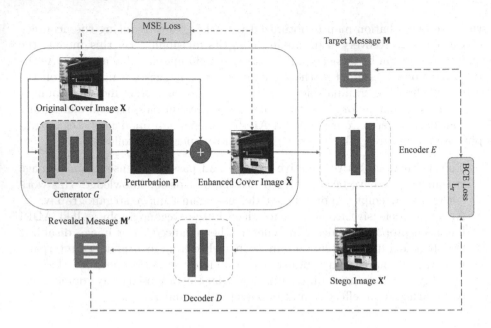

Fig. 1. Illustration of the proposed ADPI.

Following the active defensive strategy, we propose the ADPI framework against image steganography, as shown in Fig. 1. In a nutshell, ADPI consists of a defender and a steganographer. On the side of the defender, the generator G takes original cover image $\mathbf{X} = (x_{i,j,c})^{H \times W \times 3}$ as input, and outputs imperceptible perturbations $\mathbf{P} = (p_{i,j,c})^{H \times W \times 3}$, which can be formulated as

$$\mathbf{P} = G(\mathbf{X}). \tag{1}$$

The perturbations are element-wisely added with the original cover image as the enhanced cover image as

$$\widetilde{\mathbf{X}} = \mathbf{X} + \lambda \mathbf{P}, \tag{2}$$

where λ represents the intensity of the perturbation, and $\widetilde{\mathbf{X}} = (\widetilde{x}_{i,j,c})^{H \times W \times 3}$ is the enhanced cover image. On the other side of the game, the steganographer adopts deep steganographic method to implement message embedding and message extraction entirely based on deep neural networks, which consists of an encoder E and a decoder D. Specifically, the enhanced cover image $\widetilde{\mathbf{X}} = (\widetilde{x}_{i,j,c})^{H \times W \times 3}$ is concatenated with the secrete message $\mathbf{M} = (m_{i,j,c})^{H \times W \times C}$ along the channel axis as $\mathbf{T} = (t_{i,j,c})^{H \times W \times (C+3)}$. And then, to implement message embedding, the encoder E takes in the concatenated tensor $\mathbf{T}$, and outputs the stego image $\mathbf{X}' = (x'_{i,j,c})^{H \times W \times 3}$. Afterwards, to implement message extraction, the decoder D receives the stego image $\mathbf{X}' = (x'_{i,j,c})^{H \times W \times 3}$, and outputs the revealed secrete message $\mathbf{M}' = (m'_{i,j,c})^{H \times W \times C}$. In general, the message embedding and extraction process can be formulated as

$$\mathbf{M}' = D(E(\widetilde{\mathbf{X}}, \mathbf{M})). \tag{3}$$

Note that E and D are jointly optimized to ensure that the revealed messages $\mathbf{M}'$ are close enough to the target messages $\mathbf{M}$ for accurate message communication. Finally, we can calculate the messages' recovered accuracy by comparing each bit in $\mathbf{M}$ and $\mathbf{M}'$. The goal of the defender is to generate perturbation $\mathbf{P}$ such that the recovered accuracy is around 50%. By this means, the ability of transmitting message bits is close to random guessing, indicating that the steganographic system is disabled.

2.3   Loss Functions

In ADPI, we aim to train a generator to produce the enhanced cover image which can simultaneously reduce the recovered accuracy of secrete messages while maintaining satisfied image quality of cover image. Correspondingly, we delicately design the loss function for the generator, which consists of a recovered accuracy loss L_r and a visual quality loss L_v.

As for the recovered accuracy loss L_r, we aim to obtain perturbations which can cause severe decoding error by the deep steganographic method. Considering that this paper makes an early attempt to investigate active defense against image steganography, we make a naive assumption and suppose that the deep neural networks applied for message communication are accessible to the defender. Based on such assumption, L_r is constructed as the negative cross-entropy loss of the message bits between the revealed messages $\mathbf{M}'$ and the target messages $\mathbf{M}$, which can be formulated as

$$L_r = -E_{X \sim \mathbb{P}_C} CrossEntropy(D(E(\mathbf{X} + \lambda G(\mathbf{X}), \mathbf{M})), \mathbf{M}). \qquad (4)$$

As for the visual quality loss L_v, we aim to preserve high visual quality for the enhanced cover image, and try to make the enhanced cover image indistinguished from their original counterparts by human eyes. Therefore, L_v is constructed as the mean square error between the original cover image and the enhanced cover image, which can be formulated as

$$L_v = E_{X \sim \mathbb{P}_C} \frac{1}{3 \times W \times H} \|\mathbf{X}, \mathbf{X} + \lambda G(\mathbf{X})\|_2^2. \qquad (5)$$

Finally, the overall loss L is calculated as the weighted sum of L_r and L_v as

$$L = \alpha L_r + \beta L_v, \qquad (6)$$

where α and β are the hyperparameters balancing the tradeoff between message recovery and visual quality.

2.4   Network Architecture

ADPI consists of a generator and a steganographic network. Specifically, U-Net [20] structure is adopted as the generator. It consists of two main components including an encoder and a decoder, wherein the encoder is responsible for capturing contextual information from the input image, while the decoder part aims to recover the spatial resolution of the segmentation map.

Specifically, the encoder consists of four downsampling blocks, wherein each block is composed of a max pooling layer, two consecutive convolutional layers, a batch normalization layer and a ReLU (Rectified Linear Unit) activation function. As for the decoder, it consists of four upsampling blocks. In each block, a deconvolutional layer is first applied to increase the spatial resolution. The upsampled feature maps are then concatenated with the corresponding feature maps from the encoder for shortcut connection. Afterwards, two consecutive convolutional layers, batch normalization layer, and a ReLU activation function are subsequently applied for further feature extraction. Except for the last convolutional layer which has a kernel size of 1×1, all the other convolutional layers have a kernel size of 3×3 and a padding of 1. All deconvolutional layers have a kernel size of 2×2 and a stride of 2. Tanh activation function is applied in the last layer in UNet.

3 Experiments

In this section, implementation details are described in Sect. 3.1. Evaluation metrics are given in Sect. 3.2. Defensive performance is shown in Sect. 3.3.

3.1 Experimental Setups

Datasets. The experiments are conducted on two datasets, wherein 256×256 ImageNet is applied to train and test the steganographic network, and 256×256 COCO is applied to train and test the generator in the proposed ADPI. Specifically, as for the 256×256 ImageNet, 100,000 images are selected as the training set and 3,000 images are selected as the testing set. As for the 256×256 COCO, 3,000 images are selected as the training set and 1,500 images are selected as the testing set. All images are resized as 256×256.

Steganographic Methods. On the side of the steganographer, three steganographic methods are tested in the experiments, including SteganoGAN, CHAT-GAN and UDH [21]. Each steganographic method embeds messages with 1bpp. Note that UDH is originally applied for hiding secrete images. We properly modify it for hiding secrete messages.

Generators. In the training stage of the generator, the hyperparameter settings for different steganographic methods are different. As for SteganoGAN, α and β are set to $3e^{-4}$ and 1. As for CHAT-GAN, α and β are set to 1 and $2e^{-3}$. As for UDH, α and β are set to $6e^{-3}$ and 1. The perturbation intensity λ is set to 0.5. Adam optimizer with learning rate $1e^{-4}$ is used for optimization. The generator is trained for 15 epochs.

3.2   Evaluation Metrics

We evaluate our active defense framework ADPI from the perspective of image quality of the enhanced cover image by PSNR (Peak Signal to Noise Ratio) and SSIM (Structural Similarity), and the defense performance by RAC (Recovered Accuracy). The definitions of PSNR and SSIM can be referred to [22] and [23], respectively. And RAC is defined as

$$RAC = 1 - \frac{\sum_{i=1}^{H} \sum_{j=1}^{W} \sum_{c=1}^{C} |m_{i,j,c} - m'_{i,j,c}|}{H \times W \times C}, \tag{7}$$

where $\mathbf{M} = (m_{i,j,c})^{H \times W \times C}$ is the target secrete messages and $\mathbf{M}' = (m'_{i,j,c})^{H \times W \times C}$ is the revealed secrete messages.

3.3   Performance of ADPI Against Different Methods

In this part, the proposed ADPI is applied to defend against three different steganographic methods. The results are shown in Table 1 and Fig. 2, and the following observations can be obtained.

– The proposed ADPI can significantly reduce the accuracy of steganographic methods in recovering secret messages. For all three different steganographic methods, the RAC can be decreased below 55%, which is close to random guessing. Such results indicate that the steganographic system is disabled by our proposed defense framework.
– As for objective evaluation in Table 1, the enhanced cover images can still obtain satisfied PSNR and SSIM of 34.29 and 0.8742. As for subjective evaluation in Fig. 2, the enhanced cover images are indistinguishable from the original counterparts. These results indicate that ADPI would not lead to severe visual artifacts on the enhanced cover images, and can be launched in an imperceptible manner.

Table 1. Evaluation of ADPI against different steganographic methods. As for the RAC, the values on the left denote the recovered accuracy when the original cover image is applied, while the values on the right denote the recovered accuracy when the enhanced cover image is applied. As for the PSNR and SSIM, the values on the left denote the metrics for the original cover images, while the values on the right denote the metrics for the enhanced cover images.

Steganographic method	RAC	PSNR	SSIM
SteganoGAN	99.43%/54.29%	38.63/32.13	0.9582/0.8218
CHAT-GAN	99.33%/53.64%	45.62/36.97	0.9931/0.9254
UDH	99.74%/53.89%	41.15/33.79	0.9752/0.8756

Fig. 2. Visualization of the original cover images, enhanced cover images, and their differences multiplied by 10.

3.4 Generalization Performance of ADPI

In this part, the generalization ability of ADPI is investigated. Specifically, ADPI optimized over specific steganographic method is applied to defend against other steganographic methods. The results are shown in Table 2. It can be observed that the images optimized over SteganoGAN can be well generalized to defense against the other two steganographic methods, which can achieve RAC lower than 60%. By contrast, the generalization performance of the images optimized over CHAT-GAN and UDH is inferior to that optimized over SteganoGAN. Their distinction may due to the differences in network architecture.

3.5 Targeted Defense

In Sect. 3.3 and 3.4, the goal of the defender is to disable the steganographic system, such that the steganographer cannot accurately extract the secret messages. In this part, we investigate the scenario of targeted defense, i.e., mislead

Table 2. Investigating the transferability of generators trained with different steganographic models.

Training	Testing		
	SteganoGAN	CHAT-GAN	UDH
SteganoGAN	**54.29%**	55.24%	56.28%
CHAT-GAN	89.49%	**53.64%**	79.45%
UDH	80.24%	73.83%	**53.89%**

the steganographer to extract the messages specified by the defender. In this case, the loss function in Eq. (4) is modified as

$$L_r = E_{X \sim \mathbb{P}_C} CrossEntropy(D(E(\mathbf{X} + \lambda G(\mathbf{X}, \widetilde{\mathbf{M}}), \mathbf{M})), \widetilde{\mathbf{M}}). \qquad (8)$$

The loss function Eq.(5) is modified as

$$L_v = E_{X \sim \mathbb{P}_C} \frac{1}{3 \times W \times H} \left\| \mathbf{X}, \mathbf{X} + \lambda G(\mathbf{X}, \widetilde{\mathbf{M}}) \right\|_2^2, \qquad (9)$$

where $\widetilde{\mathbf{M}}$ is the messages appointed by the defender. In other words, the defender aims to mislead the steganographer extracts the appointed messages $\widetilde{\mathbf{M}}$.

The results are shown in Table 3. It can be observed that the proposed ADPI can still achieve satisfied performance in targeted defense. Specifically, the RAC is higher than 95% in all cases, indicating that ADPI can successfully mislead the steganographic system.

Table 3. Accuracy results of recovering the appointed message during steganography. The RAC is calculated by the revealed messages and appointed messages specified by the defender. As for the PSNR and SSIM, the values on the left denote the metrics for the original cover images, while the values on the right denote the metrics for the enhanced cover images.

Steganographic method	RAC	PSNR	SSIM
SteganoGAN	50.00%/96.80%	38.63/30.65	0.9582/0.8038
CHAT-GAN	50.00%/96.79%	45.62/34.36	0.9931/0.9044
UDH	50.00%/95.25%	41.15/35.74	0.9752/0.9142

4   Conclusion

In this paper, we propose an active defensive framework against image steganography called ADPI, which can take action before the behaviour of secrete message communication happens. Experimental results show that under ADPI, active defense can be launched in an effective and imperceptible manner. Compared with passive steganalysis technique, our proposed active defense is a preventive technique, which can eliminate malicious steganography from the source. In the future, the following aspects may worth further investigations. Firstly, the generalization ability of active defense among different steganographic methods should be improved. Secondly, the scenario of applying active defense against black-box steganographic models should be investigated. Thirdly, more advanced network architecture could be introduced to generate the perturbations.

Acknowledgements. Weixuan Tang is the corresponding author. This work was supported by NSFC (Grant 62002075), Guangdong Basic and Applied Basic Research Foundation (Grant 2023A1515011428), the Science and Technology Foundation of Guangzhou (Grant 2023A04J1723).

References

1. Holub, V., Fridrich, J.J.: Designing steganographic distortion using directional filters. In: Proceedings of the IEEE International Workshop on Information Forensics and Security, pp. 234–239 (2012)
2. Pevný, T., Filler, T., Bas, P.: Using high-dimensional image models to perform highly undetectable steganography. In: Proceedings of the 12th Information Hiding Workshop, pp. 161–177 (2010)
3. Li, B., Wang, M., Huang, J., Li, X.: A new cost function for spatial image steganography. In: Proceedings of the IEEE International Conference on Image Processing, pp. 4026–4210 (2014)
4. Holub, V., Fridrich, J.J.: Digital image steganography using universal distortion. In: Proceedings of the 1st ACM Information Hiding and Multimedia Security Workshop, pp. 59–68 (2013)
5. Hayes, J., Danezis, G.: Generating steganographic images via adversarial training. In: Proceedings of the 30th Conference on Neural Information Processing Systems, pp. 1954–1963 (2017)
6. Goodfellow, I.J., et al.: Generative adversarial nets. In: Advances in Neural Information Processing Systems, pp. 2672–2680 (2014)
7. Zhu, J., Kaplan, R., Johnson, J., Fei-Fei, L.: HiDDeN: hiding data with deep networks. In: Ferrari, V., Hebert, M., Sminchisescu, C., Weiss, Y. (eds.) ECCV 2018. LNCS, vol. 11219, pp. 682–697. Springer, Cham (2018). https://doi.org/10.1007/978-3-030-01267-0_40
8. Zhang, K.A., Cuesta-Infante, A., Xu, L., Veeramachaneni, K.: Steganogan: high capacity image steganography with GANs. arXiv preprint arXiv:1901.03892 (2019)
9. Tan, J., Liao, X., Liu, J., Cao, Y., Jiang, H.: Channel attention image steganography with generative adversarial networks. IEEE Trans. Network Sci. Eng. $9(2)$, 888–903 (2021)
10. Lu, S.P., Wang, R., Zhong, T., Rosin, P.L.: Large-capacity image steganography based on invertible neural networks. In: Proceedings of the IEEE Conference on Computer Vision and Pattern Recognition, pp. 10816–10825 (2021)
11. Kodovsky, J., Fridrich, J., Holub, V.: Ensemble classifiers for steganalysis of digital media. IEEE Trans. Inf. Forensics Secur. $7(2)$, 432–444 (2011)
12. Fridrich, J., Kodovsky, J.: Rich models for steganalysis of digital images. IEEE Trans. Inf. Forensics Secur. $7(3)$, 868–882 (2012)
13. Li, B., Li, Z., Zhou, S., Tan, S., Zhang, X.: New steganalytic features for spatial image steganography based on derivative filters and threshold LBP operator. IEEE Trans. Inf. Forensics Secur. $13(5)$, 1242–1257 (2017)
14. Song, X., Liu, F., Yang, C., Luo, X., Zhang, Y.: Steganalysis of adaptive jpeg steganography using 2D gabor filters. In: Proceedings of the 3rd ACM Workshop on Information Hiding and Multimedia Security, pp. 15–23 (2015)
15. Xu, G., Wu, H.Z., Shi, Y.Q.: Structural design of convolutional neural networks for steganalysis. IEEE Signal Process. Lett. $23(5)$, 708–712 (2016)
16. Zhang, R., Zhu, F., Liu, J., Liu, G.: Depth-wise separable convolutions and multilevel pooling for an efficient spatial CNN-based steganalysis. IEEE Trans. Inf. Forensics Secur. 15, 1138–1150 (2019)
17. You, W., Zhang, H., Zhao, X.: A SIAMESE CNN for image steganalysis. IEEE Trans. Inf. Forensics Secur. 16, 291–306 (2021)
18. Deng, X., Chen, B., Luo, W., Luo, D.: Fast and effective global covariance pooling network for image steganalysis. In: Proceedings of the ACM Workshop on Information Hiding and Multimedia Security, pp. 230–234 (2019)

19. Weng, S., Chen, M., Yu, L., Sun, S.: Lightweight and effective deep image steganalysis network. IEEE Signal Process. Lett. **29**, 1888–1892 (2022)
20. Ronneberger, O., Fischer, P., Brox, T.: U-Net: convolutional networks for biomedical image segmentation. In: Navab, N., Hornegger, J., Wells, W.M., Frangi, A.F. (eds.) MICCAI 2015. LNCS, vol. 9351, pp. 234–241. Springer, Cham (2015). https://doi.org/10.1007/978-3-319-24574-4_28
21. Zhang, C., Benz, P., Karjauv, A., Sun, G., Kweon, I.S.: UDH: universal deep hiding for steganography, watermarking, and light field messaging. In: Proceedings of the 33rd Conference on Neural Information Processing Systems, pp. 10223–10234 (2020)
22. Almohammad, A., Ghinea, G.: Stego image quality and the reliability of PSNR. In: Proceedings of the 2nd International Conference on Image Processing Theory, Tools and Applications, pp. 215–220. IEEE (2010)
23. Wang, Z., Bovik, A.C., Sheikh, H.R., Simoncelli, E.P.: Image quality assessment: from error visibility to structural similarity. IEEE Trans. Image Process. **13**(4), 600–612 (2004)

Strict Differentially Private Support Vector Machines with Dimensionality Reduction

Teng Wang[(✉)][iD], Shuanggen Liu[iD], Jiangguo Liang, Shuai Wang, Lu Wang, and Junying Song

Xi'an University of Posts and Telecommunications, Xi'an, China
{wangteng,liushuanggen201}@xupt.edu.cn,
{liangjiangguo,200019ws,wanglu,songjunying}@stu.xupt.edu.cn

Abstract. With the widespread data collection and processing, privacy-preserving machine learning has become increasingly important in addressing privacy risks related to individuals. Support vector machine (SVM) is one of the most elementary learning models of machine learning. Privacy issues surrounding SVM training classifiers have attracted increasing attention. In this paper, we propose DPDR-DPSVM which is a strict differentially private support vector machine algorithm with high data utility. Aiming at high-dimensional data, we adopt differential privacy in both the dimensionality reduction phase and SVM classifier training phase, which improves model accuracy while achieving strong privacy guarantees. Besides, we train DP-compliant SVM classifiers by adding noise to the objective function itself, thus leading to better data utility. Extensive experiments on three high-dimensional datasets demonstrate that DPDR-DPSVM can achieve high accuracy while ensuring strong privacy protection.

Keywords: Differential privacy · Support vector machine · Data utility

1 Introduction

The rapid development of generative artificial intelligence (Generative AI) and large language models (LLMs) is accelerating changes in our production and living habits [15,19]. As a subfield of (AI), machine learning (ML) has also attracted increasing attention. ML algorithms such as support vector machines and logistic regression can play important roles in text classification, sentiment analysis, information extraction, etc. However, with the proliferation of data collection and processing, privacy concerns have become increasingly important [2,18], especially when dealing with personal or sensitive information. The training process will severely leak the privacy of training data. The adversary may snoop on users' sensitive information through membership inference attacks,

J. Vaidya et al. (Eds.): AIS&P 2023, LNCS 14509, pp. 142–155, 2024.
https://doi.org/10.1007/978-981-99-9785-5_11

attribute inference attacks, or model inversion attacks [9,22], which leads to privacy breaches, identity theft, or other malicious activities.

Privacy-preserving machine learning (PPML) [2] addresses these concerns by allowing the training and inference processes to be performed without exposing the raw data. Support vector machine (SVM) is one of the most elementary learning models of ML. Therefore, there is a huge demand for studying privacy-preserving SVM algorithms. Differential privacy (DP) [5,7] is a rigorous privacy paradigm nowadays and is widely adopted in AI and ML. Concretely, DP has a formal mathematical foundation and therefore prevents the disclosure of any information about the presence or absence of any individual from any statistical operations.

Several approaches have been proposed to train SVM models with differential privacy. These methods typically add noise or perturbation to the training data or model parameters to limit the amount of information that can be learned about any individual data point. Dwork et al. [8] firstly studied the problem of privacy-preserving principal component analysis (PCA) and proved the optimal bounds of DP-compliant PCA, which lays the foundation for applying PCA in PPML. Hereafter, Huang et al. [10] leveraged the Laplace mechanism into PCA-SVM algorithms to achieve differential privacy protection. Sun et al. [17] proposed a differentially private singular value decomposition (SVD) algorithm to provide privacy guarantees for SVM training. To sum end, these methods all consider achieving dimensionality reduction by using PCA, so the algorithms are usually divided into two stages: PCA and SVM. However, the DPPCA-SVM, PCA-DPSVM, and DPSVD mechanisms in [10,17] all only apply differential privacy at one stage in PCA or SVM, resulting in an insufficient degree of privacy protection.

A strict differential privacy protection mechanism should satisfy that DP must be applied whenever the train data is accessed in the algorithm [7]. Therefore, a strict differential privacy SVM mechanism with dimensionality reduction should be further studied. To this end, this paper studies a strict differentially private SVM algorithm with dimensionality reduction, which aims to maintain high data utility while providing strong privacy protection. Our main contributions are summarized as follows.

- We propose a strict privacy-preserving SVM mechanism DPDR-DPSVM which adopts differential privacy in both the dimensionality reduction phase and SVM training phase to provide strong privacy guarantees.
- To overcome the high-dimensional features of data, we introduce a differential privacy-based principal component analysis method to improve model accuracy.
- By leveraging the empirical risk minimization approximations, we train DP-compliant SVM classifiers by adding noise to the objective function itself, leading to better data utility.
- We conduct extensive experiments on three datasets with high-dimensional features. The results demonstrate that our mechanism can achieve high accuracy while ensuring strong privacy protection.

The remainder of the paper is organized as follows. Section 2 provides a literature review. Section 3 introduces the system model, differential privacy, and problem formulation. Section 4 presents our proposed DPDR-DPSVM mechanism. Section 5 shows the experimental results. Section 6 concludes the paper.

2   Related Work

Privacy-preserving machine learning (PPML) [2,16,23] is widely applicable in various domains, including artificial intelligence, large language models, healthcare, finance, and telecommunications. They enable data-driven decision-making and the development of intelligent systems while protecting individuals' sensitive information. Since the introduction of differential privacy (DP) [5,7], DP-based PPML [11] has gained significant attention as a means to ensure privacy while training models on sensitive data.

SVMs are a popular class of machine learning algorithms used for binary classification, regression, and outlier detection tasks. The goal of differential privacy in SVMs is to enable the training process while providing privacy guarantees for the sensitive data used in the training. Moreover, it's important to note that achieving differential privacy in SVMs involves a trade-off between privacy and utility. Chaudhuri et al. [4] proposed to produce privacy-preserving approximations of classifiers learned via (regularized) empirical risk minimization (ERM). They also analyzed the accuracy of proposed mechanisms and the upper bound of the number of training samples, laying the foundation for subsequent research.

Principal component analysis (PCA) is an effective tool to improve the classification accuracy of SVM. Dwork et al. [8] firstly studied the problem of differential privacy-based principal component analysis (PCA) and proved the optimal bounds of DP-compliant PCA, which lays the foundation for applying PCA in PPML. They proposed to perturb the matrix of covariance with Gaussian noise. In contrast, Jiang et al. [12] perturbed the matrix of covariance with Wishart noise, which was able to output a perturbed positive semidefinite matrix. Afterward, Xu et al. [21] applied the Laplace mechanism to introduce perturbation and proposed the Laplace input perturbation and Laplace output perturbation.

What's more, Huang et al. [10] proposed DPPCA-SVM and PCA-DPSVM for privacy-preserving SVM, which perturbed the matrix of covariance with symmetric Laplace noise. However, the DPPCA-SVM and PCA-DPSVM mechanisms only apply differential privacy at one stage in PCA or SVM, resulting in an insufficient degree of privacy protection. It should be claimed that a strict differential privacy protection mechanism should satisfy that DP must be applied whenever the train data is accessed in the algorithm. Besides, Sun et al. [17] proposed DPSVD which uses singular value decomposition (SVD) to project the training instances into the low-dimensional singular subspace. They first added the noise to the raw data D and then obtained the singular values by applying SVD on the perturb data D'. However, the original training dataset is accessed again when computing low-dimensional singular subspace, thus resulting in insufficient privacy protection.

3 Preliminaries

3.1 Safety Model

This paper considers that the server obeys the semi-honest (honest but curious) adversary model. That is, the server adheres to the agreement but also tries to learn more from the received information than the output was unexpected.

3.2 Differential Privacy

Differential privacy (DP) [5,7] is a strict privacy protection model that gives a rigorous and quantified proof of privacy disclosure risk. Thus, DP can provide a probability guarantee for the privacy of any individual's query response and is independent of the prior knowledge of the attacker. Since differential privacy was proposed ten years ago, hundreds of papers based on differential privacy technology have been proposed in security, database, machine learning, and statistical computing applications.

Definition 1 $((\epsilon, \delta)$-Differential Privacy $((\epsilon, \delta)$-DP)). *A randomized mechanism $\mathcal{M}$ satisfies (ϵ, δ)-DP if and only if for any neighboring datasets D and D', and for any possible output $O \subseteq \text{Range}(\mathcal{M})$, it holds*

$$\mathbb{P}[\mathcal{M}(D) \in O] \leq e^{\epsilon} \cdot \mathbb{P}[\mathcal{M}(D') \in O] + \delta, \tag{1}$$

where $\mathbb{P}$ denotes probability.

(ϵ, δ)-DP also called approximated differential privacy. When $\delta = 0$, (ϵ, δ)-DP becomes ϵ-DP, that is, pure differential privacy.

Differential privacy provides a mathematical guarantee of privacy by introducing controlled randomness (i.e., noise) into the data or results of computations. Laplace mechanism [6], exponential mechanism [14], and Gaussian mechanism [7] all can be used to achieve differential privacy. This paper adopts the Gaussian mechanism, which is defined as follows.

Theorem 1 (Gaussian Mechanism). *The Guassian mechanism achieve (ϵ, δ)-DP by adding Gaussian noise with standard deviation $\sigma = \sqrt{2 \ln \frac{1.25}{\delta}} \cdot \frac{\Delta}{\epsilon}$, where Δ is ℓ_2-sensitivity and is computed as the maximal ℓ_2-norm difference of two neighboring datasets D and D'.*

3.3 Problem Formulation

This part introduces the data model, empirical risk minimization, and problem statement.

Data Model. Given a dataset $D = \{(\mathbf{x}_i, y_i) \in \mathcal{X} \times \mathcal{Y} : i = \{1, 2, \cdots, n\}\}$ with n samples, where $\mathbf{x}_i$ and y_i in each sample $(\mathbf{x}_i, y_i)$ denote the data space and label set, respectively. As for binary classification in machine learning, the data space is $\mathcal{X} = \mathbb{R}^d$ and the label set is $\mathcal{Y} = \{-1, 1\}$. That is, each $\mathbf{x}_i = [x_i^1, x_i^2, \cdots, x_i^d]$ is a d-dimensional vector and each $y_i = -1$ or $y_i = 1$. Besides, it assumes that $\|\mathbf{x}_i\|_2 \leq 1$. For convenience, let $X = [\mathbf{x}_1; \mathbf{x}_2; \cdots; \mathbf{x}_n]$ denote the data space of dataset D and let $Y = [y_1; y_2; \cdots; y_n]$ denote the label space of dataset D. That is, $D = (X, Y)$.

Empirical Risk Minimization (ERM). In this paper, we build machine learning models that are expressed as empirical risk minimization. We would like to train a predictor $\beta : \mathbf{x} \to y$. As for machine learning algorithms with empirical risk minimization, the predictor β minimizes the regularized empirical loss. For dataset D, the ERM can be formulated as

$$\mathcal{F}(\boldsymbol{\beta}, D) = \frac{1}{n} \sum_{i=1}^{n} \ell(\boldsymbol{\beta}; \mathbf{x}_i, y_i) + \frac{\lambda}{2} \|\boldsymbol{\beta}\|_2^2 \tag{2}$$

where $\ell(\cdot)$ is the loss function, β is a d-dimensional parameter vector, $\lambda > 0$ is a regularization parameter.

Based on Eq. (2), we aim to compute a d-dimensional parameter vector $\boldsymbol{\beta}^*$ such that

$$\boldsymbol{\beta}^* = \arg\min_{\beta} \mathcal{F}(\boldsymbol{\beta}, D)$$

$$= \arg\min_{\beta} \left[\frac{1}{n} \sum_{i=1}^{n} \ell(\boldsymbol{\beta}_i; \mathbf{x}_i, y_i) + \frac{\lambda}{2} \|\boldsymbol{\beta}\|_2^2 \right] \tag{3}$$

Problem Statement. For input training dataset D, we aim to privately train a machine learning model (i.e., private predictor $\widehat{\boldsymbol{\beta}}^*$) based on ERM with strict differential privacy protection. We will also integrate dimension reduction into all training processes in order to improve model accuracy and reduce computing costs.

4   Our Solution

Considering high dimensional features and insufficient privacy protection, this paper proposes a strict differential privacy support vector machine framework with dimensionality reduction. The framework DPDR-DPSVM mainly includes two phases: the first phase aims to obtain the low-dimensional features with (ϵ_1, δ)-differential privacy, and the second phase aims to train the SVM model with (ϵ_2)-differential privacy.

We mainly use the following two strategies to provide sufficient data privacy and improve the accuracy of SVM models. 1) To provide strict privacy guarantees, we introduce differential privacy in both the dimensionality reduction phase and the SVM training phase. 2) To overcome the high-dimensional features of data, we conduct dimensionality reduction before training by using the principal component analysis (PCA) method, which can improve model accuracy.

In the following, we will describe the main components of DPDR-DPSVM in detail.

Algorithm 1: DPDR: DP-compliant Dimensionality Reduction

Input: dataset $D = (X, Y)$, privacy parameter ϵ_1, δ, expected dimension k.

Output: Private k-dimensional features $\hat{U}^k$.

1 Compute the covariance matrix $M = X^\top X$;

2 Generate Gaussian noise matrix $R \leftarrow \mathcal{N}\left(0, \frac{2\ln(1.25/\delta)\Delta_f^2}{\epsilon_1^2}\right)$;

3 Process R to be a symmetric matrix by each lower triangle entry is copied from its upper triangle counterpart;

4 Compute $\widehat{M} = M + R$;

5 Compute $\hat{U}\hat{S}\hat{V}$ using eigenvalue decomposition of $\widehat{M}$;

6 Grab the first k values of $\hat{U}$ as $\hat{U}^k$;

7 **return** $\hat{U}^k$

4.1 DP-Compliant Dimensionality Reduction

We utilize principal component analysis (PCA) to achieve dimensionality reduction under differential privacy. For d-dimensional dataset $D = (X, Y)$, the $d \times d$ covariance matrix is defined as

$$M = X^\top X = \sum_{i=1}^{n} \mathbf{x}_i^\top \mathbf{x}_i. \tag{4}$$

Thus, we can achieve DP-compliant PCA by applying the Gaussian mechanism first to M_i. Then, the k-principle features of the original dataset are computed by choosing the top-k singular subspace of the noised covariance matrix based on singular value decomposition (SVD).

Algorithm 1 shows the pseudo-code of PCA-based dimensionality reduction while satisfying differential privacy. We simply formalize Algorithm 1 as DPDR(D, ϵ_1, δ, k). Given dataset $D = (X, Y)$, we add Gaussian noise to the covariance matrix to achieve differential privacy. For the function $f(X) = X^\top X$, the ℓ_2-sensitivity of $f(X)$ is $\Delta_f = 1$, as shown in Lemma 1. Thus, the Gaussian noise matrix R is generated from $\mathcal{N}\left(0, \frac{2\Delta_f \ln(1.25/\delta)}{\epsilon_1^2}\right)$ and is processed to be a symmetric matrix by each lower triangle entry is copied from its upper triangle counterpart. Next, we apply SVD to noisy covariance matrix $\widehat{M}$ and thereby grab the top-k singular subspace of $\widehat{M}$, as shown in Lines 5-6.

Lemma 1 *In Algorithm 1 (i.e., DPDR), for all input dataset $D = (X, Y)$, the sensitivity of function $f(X) = X^\top X$ is at most one.*

Proof. Let X' denote the neighboring dataset of X. We consider bounded differential privacy in this paper. That is, X' has the same size as X but only differs in one record. Assuming X' and X differ in the t-th row. Then, the sensitivity can be computed as

$$\|M - M'\|_2$$
$$= \left\| X^\top X - X'^\top X' \right\|_2$$
$$= \left\| \begin{bmatrix} x_t^1 x_t^1 & x_t^1 x_t^2 & \cdots & x_t^1 x_t^d \\ x_t^2 x_t^1 & x_t^2 x_t^2 & \cdots & x_t^2 x_t^d \\ \cdots & \cdots & \ddots & \cdots \\ x_t^d x_t^1 & x_t^d x_t^2 & \cdots & x_t^d x_t^d \end{bmatrix} \right\|_2$$
$$= \sqrt{(x_t^1)^2 \left[(x_t^1)^1 + (x_t^2)^2 + \cdots + (x_t^d)^2 \right] + \cdots + (x_t^d)^2 \left[(x_t^1)^1 + (x_t^2)^2 + \cdots + (x_t^d)^2 \right]}$$
$$= \sqrt{\left[(x_t^1)^1 + (x_t^2)^2 + \cdots + (x_t^d)^2 \right]^2}$$
$$= \sqrt{\|\mathbf{x}_t\|_2}$$
$$\leq 1 \tag{5}$$

where the step of "$\leq$" is achieved since $\|\mathbf{x}_i\|_2 \leq 1$ ($i \in \{1, 2, \cdots, n\}$).

4.2　DPDR-DPSVM: DP-Compliant SVM with Dimensionality Reduction

This part presents the DP-compliant support vector machine (SVM) with dimension reduction. Specifically, we build SVM models based on empirical risk minimization. To improve model accuracy, we apply dimensionality reduction on the original high-dimensional dataset. Besides, to achieve privacy protection, we perturb the objective function to produce the minimizer of the noisy objective function.

Algorithm 2 shows the pseudo-code of our proposed SVM training process under differential privacy. Given the dataset $D = (X, Y)$, we first execute DPDR$(D, \epsilon_1, \delta, k)$ to obtain the private k-dimensional features $\hat{U}^k$ based on Algorithm 1. Then, we can project the data X into k-dimensional space based on $\hat{U}^k$. Therefore, the input dataset for training SVM classifiers is $\widehat{D}^k = \left(\widehat{X}^k, Y \right)$. Next, we compute the privacy parameter which will be used to generate noise in objective function perturbation, as shown in Lines 4-9. Based on the privacy parameter p, the noise vector $\boldsymbol{R}$ can be draw with probability density function $\frac{1}{\alpha} e^{-\frac{p}{2} \|\boldsymbol{R}\|}$. Then, we can perturb the objective function as

$$\mathcal{F}_{\text{priv}}(\boldsymbol{\beta}_i, \widehat{D}_i^k) = \mathcal{F}(\boldsymbol{\beta}, \widehat{D}_i^k) + \frac{1}{n} \boldsymbol{R}^\top \boldsymbol{\beta}_i. \tag{6}$$

At last, we can produce the minimizer of noisy $\mathcal{F}_{\text{priv}}(\boldsymbol{\beta}_i, \widehat{D}^k)$ by

$$\widehat{\boldsymbol{\beta}}^* = \arg\min_{\boldsymbol{\beta}} \left[\mathcal{F}_{\text{priv}}(\boldsymbol{\beta}, \widehat{D}^k) + \frac{\Delta}{2} \|\boldsymbol{\beta}\|_2^2 \right], \tag{7}$$

where $\widehat{\boldsymbol{\beta}}^*$ the optimal parameters of $\mathcal{F}_{\text{priv}}(\boldsymbol{\beta}, \widehat{D}^k)$.

Algorithm 2: DPDR-DPSVM: DP-compliant SVM Training with Dimensionality Reduction

---

Input: dataset $D = (X, Y)$, privacy parameter $\epsilon_1, \epsilon_2, \delta$, regularization
       parameter λ, normalizing constant α.
Output: Differetially private predictor $\widehat{\beta}^*$.

1   Execute DPDR$(D, \epsilon_1, \delta, k)$ to obtain the private k-dimensional features $\hat{U}^k$;

2   Project data space into k dimensions by $\widehat{X}^k = X \cdot \hat{U}^k$;

3   $\widehat{D}^k = \left(\widehat{X}^k, Y \right)$;

4   Compute privacy parameter $p = \epsilon_2 - 2\log(1 + \frac{c}{n\lambda})$;

5   **if** $p > 0$ **then**

6 | $\Delta = 0$;

7   **else**

8 | $\Delta = \frac{c}{n(e^{p/4})} - \lambda$;

9 | $p = \frac{\epsilon_2}{2}$;

10   Draw noise vector $\boldsymbol{R}$ with probability density function $\frac{1}{\alpha} e^{-\frac{p}{2}\|\boldsymbol{R}\|}$;

11   Compute $\mathcal{F}_{\text{priv}}(\boldsymbol{\beta}, \widehat{D}^k) = \mathcal{F}(\boldsymbol{\beta}, \widehat{D}^k) + \frac{1}{n}\boldsymbol{R}^\top \boldsymbol{\beta}$;

12   Minimize $\widehat{\beta}^* = \arg\min_\beta \left[\mathcal{F}_{\text{priv}}(\boldsymbol{\beta}, \widehat{D}^k) + \frac{\Delta}{2} \|\boldsymbol{\beta}\|_2^2 \right]$;

13   **return** $\widehat{\beta}^*$

Based on Eq.(3) in Subsect. 3.3, the minimizer of $\mathcal{F}_{\text{priv}}(\boldsymbol{\beta}, \widehat{D}^k)$ will be computed as

$$
\begin{aligned}
\widehat{\beta}^* &= \arg\min_\beta \left[\mathcal{F}_{\text{priv}}(\boldsymbol{\beta}, \widehat{D}^k) + \frac{\Delta}{2} \|\boldsymbol{\beta}\|_2^2 \right] \\
&= \arg\min_\beta \left[\mathcal{F}(\boldsymbol{\beta}, \widehat{D}^k) + \frac{1}{n}\boldsymbol{R}^\top \boldsymbol{\beta} + \frac{\Delta}{2} \|\boldsymbol{\beta}\|_2^2 \right] \\
&= \arg\min_\beta \left[\frac{1}{n} \sum_{i=1}^{n} \ell\left(\boldsymbol{\beta}_i; \hat{\mathbf{x}}_i^k, y_i\right) + \frac{\lambda}{2} \|\boldsymbol{\beta}\|_2^2 + \frac{1}{n}\boldsymbol{R}^\top \boldsymbol{\beta} + \frac{\Delta}{2} \|\boldsymbol{\beta}\|_2^2 \right]
\end{aligned} \tag{8}
$$

where $\hat{\mathbf{x}}_i^k$ denotes the private k-dimensional data space of $\hat{\mathbf{x}}_i$.

This paper focuses on SVM classification. Thus, the loss function $\ell(\cdot)$ in Eq. (8) is defined as

$$
\ell_{\text{SVM}}(\boldsymbol{\beta}, \mathbf{x}, y) = \max\left\{0, 1 - y\mathbf{x}^\top \boldsymbol{\beta}\right\} \tag{9}
$$

Besides, in Algorithm 2, the parameter c is set as $c = \frac{1}{2h}$ for SVM, where h is huber loss function parameter and picked as $h = 0.5$ for Huber SVM, a typical value [3].

4.3   Privacy Analysis

This part shows the privacy guarantees of our proposed algorithms.

Theorem 2 *Algorithm 1 (i.e., DPDR) satisfies (ϵ_1, δ)-differential privacy.*

Proof. As shown in the 5th line of the Algorithm 1, the Gaussian noise R is introduced into the covariance matrix to provide privacy protection. The noise R is draw from $\mathcal{N}\left(0, \frac{2\ln(1.25/\delta)\Delta_f^2}{\epsilon_1^2}\right)$, that is, the deviation $\sigma = \sqrt{2\ln\frac{1.25}{\delta}} \cdot \frac{\Delta_f}{\epsilon_2}$. Thus, based on Theorem 1, Algorithm 1 (i.e., DPDR) satisfies (ϵ_1, δ)-differential privacy.

Theorem 3 *Algorithm 2 (i.e., DPDR-DPSVM) satisfies (ϵ, δ)-differential privacy, where $\epsilon = \epsilon_1 + \epsilon_2$.*

Proof. Based on Theorem 2, Algorithm 2 executes $DPDR(D, \epsilon_1, \delta, k)$ satisfies (ϵ_1, δ)-differential privacy. Moreover, the privacy guarantee of objective perturbation is shown in lines 4-11, which uses privacy parameter ϵ_2. This can be proved to satisfy ϵ_2-differential privacy by following Theorem 9 in [4]. We omit the details due to space limitations. Therefore, based on the composition theorems [7] of differential privacy, the Algorithm 2 (i.e., DPDR-DPSVM) satisfies $(\epsilon_1 + \epsilon_2, \delta)$-differential privacy.

5 Experiments

This section presents our experimental results of the proposed mechanisms on three popular datasets (Table 1).

5.1 Experiment Setup

Datasets. We select three datasets with different data sizes and dimensions to verify the performance of the mechanism proposed in this paper. MNIST and Fashion-MNIST share the same external characteristics, namely data size and dimension. But Fashion-MNIST is no longer the abstract number symbols, but more concrete clothing images. In contrast, the size of CIFAR-10 is similar to MNIST and Fashion-MNIST in magnitude, but the dimension of CIFAR-10 is much larger than the other two. The details of the three

- MNIST dataset [13] consists of 60,000 training examples and 10,000 testing examples. Each example is a handwritten gray image with 28×28 pixels, associated with a label from 10 classes (i.e., numbers 0 to 9).
- Fashion-MNIST [20] a dataset of Zalando's article images, which consists of a training set of 60,000 examples and a test set of 10,000 examples. Each example is a 28×28 gray-scale image, associated with a label from 10 classes (e.g., coat, dress, bag, etc.).
- CIFAR-10 dataset [1] a computer vision dataset for universal object recognition, which consists of 50,000 training examples and 10,000 testing examples. Each example is a 32×32 color image, associated with a label from 10 classes (e.g., bird, cat, deer, etc.).

Table 1. Datasets used in experiment

Dataset	Data size	Dimension	Target dimension k
MNIST	70,000	784 (28×28 pixels)	{5, 10, 20, 50, 100}
Fashion-MNIST	70,000	784 (28×28 pixels)	{5, 10, 20, 50, 100}
CIFAR-10	60,000	3,072 (32×32×3 pixels)	{5, 10, 20, 50, 100}

Competitors. We evaluate the performance of our proposed DPDR-DPSVM by comparing it with Non-Priv and DPSVM. Non-Priv conducts machine learning with dimensionality reduction but without privacy protection. DPSVM conducts machine learning under differential privacy protection but without dimensionality reduction. Other competitors [10,17] either have insufficient privacy protection or target low-dimensional data learning, thus not comparable.

We will show the accuracy and run time of different methods varying from parameters ϵ, k, n.

5.2   Experimental Results

This section presents our experimental results, including evaluations of accuracy and running time on SVM. By default, we set the parameters $\epsilon = 0.1$, $\delta = 10^{-4}$, $k = 20$, and $n = 10^4$ in all experiments, where $\epsilon_1 = \epsilon_2 = 0.5\epsilon$ are used for DP-compliant dimensionality reduction and DP-compliant machine learning, respectively.

Evaluation of Accuracy. We first validate the performance of dimensionality reduction on SVM classification varying from the target dimension k on three high-dimensional datasets, as shown in Fig. 1. We can see that the SVM classification accuracy of all mechanisms continuously increases with the dimension k increasing from 5 to 100 for all datasets. And, the accuracy does not change much when k is greater than 20. Therefore, we choose the target dimension as $k = 20$ by default in the following experiments. Besides, it can be observed from three datasets that the accuracy of our proposed DPDR-DPSVM is much better than that of DPSVM and is close to Non-Priv when k is large. This demonstrates that DPDR-DPSVM can improve accuracy when dealing with high-dimensional data and can ensure superior data utility while providing strong privacy protection.

As for the CIFAR-10 dataset that has much higher dimensions (i.e., $d = 3,072$), we also utilize the histogram of oriented gradient (HOG) in the experiment to improve accuracy, where the HOG parameters are used as follows: cell size is 4 pixels, number of bins is 9, block size is 2 cell, sliding step is 4 pixels. Nonetheless, the accuracy is not very high compared to MNIST and Fashion-MNIST. Because the SVM used in this paper is a linear model (using hinge loss strategy), and no kernel function is introduced to build a nonlinear model, nor is a convolutional network used. In the follow-up, we will further study the privacy-preserving SVM under the nonlinear model and the convolutional network.

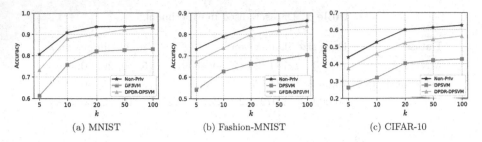

Fig. 1. Accuracy vs. target dimension k on SVM classification ($\epsilon = 0.1$, $n = 10^4$)

Moreover, Fig. 2 shows the high accuracy of our proposed DPDR-DPSVM on three datasets with the privacy parameter ϵ varying from 0.01 to 2.0, where $k = 20, n = 10^4, \delta = 10^{-4}$. Specifically, we consider $\epsilon \in \{0.01, 0.05, 0.1, 0.5, 1.0, 2.0\}$. It can be seen from three figures in Fig. 2 that the accuracy of DPDR-DPSVM is much closer to Non-Priv which has no privacy protection. Thus, this demonstrates again that our proposed DPDR-DPSVM can achieve better accuracy while keeping strong privacy protection. What's more, Fig. 2 shows that the accuracy of DPDR-DPSVM is much superior to DPSVM when applying the same level of privacy protection, which indicates DPDR-DPSVM holds better data utility while keeping the same privacy guarantees.

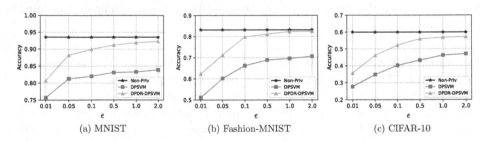

Fig. 2. Accuracy vs. privacy parameter ϵ on SVM classification ($k = 20$, $n = 10^4$)

Furthermore, Fig. 3 shows the comparisons of the impact of data size n on accuracy, where n is set as $n = \{100, 500, 1000, 5000, 10000\}$. It can be seen from Fig. 3 that the accuracy of three mechanisms will increase with the increase of data size n for three datasets. With different data sizes, our proposed DPDR-DPSVM always outperforms DPSVM under the same privacy protection level. This is because DPDR-DPSVM involves the DP-compliant dimensionality reduction to extract the key feature of high-dimensional data, thus leading to a higher accuracy than DPSVM. This also demonstrates that DPDR-DPSVM can improve the data utility in practice even when dealing with high-dimensional data.

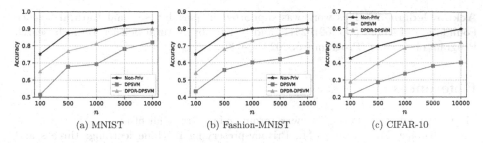

(a) MNIST   (b) Fashion-MNIST   (c) CIFAR-10

Fig. 3. Accuracy vs. data size n on SVM classification ($\epsilon = 0.1$, $k = 20$)

Evaluation of Running Time. We also compared the running time of different mechanisms on SVM training, as shown in Table 2. Here, we set the data size as 10^4 and the target dimension as 20. It can be observed that the running time of Non-Priv and DPDR-DPSVM are much lower than DPSVM, especially when the dataset (i.e., CIFAR-10) is very large. This proves that privacy-preserving dimensionality reduction can surely improve the efficiency of SVM trains while providing privacy protection. Besides, compared with Non-Priv and DPSVM, our proposed DPDR-DPSVM can maintain relatively excellent performance under the premise of providing strong privacy protection.

Table 2. Running time of different mechanisms on different datasets

| Time(s)   Mechanism | Non-Priv | DPSVM | DPDR-DPSVM |
Dataset			
MNIST	3.68	7,873.23	111.70
Fashion-MNIST	5.02	7,988.60	112.70
CIFAR-10	63.45	36,960.70	147.96

6 Conclusion

Support Vector Machine (SVM) is a powerful machine learning algorithm that has been successfully applied to various domains, including text classification, image recognition, bioinformatics, and financial analysis. However, privacy concerns may arise when dealing with sensitive or private data during the training or implementation of SVM. Therefore, this paper proposes a strict differentially private support vector machine algorithm called DPDR-DPSVM with high data utility. We conduct extensive experiments on three high-dimensional data with different characteristics. The experimental results show that our proposed algorithm can maintain good data utility while providing strong privacy guarantees.

Acknowledgment. This work was supported in part by National Natural Science Foundation of China (No. 62102311) and in part by Natural Science Basic Research Program of Shaanxi (Program No. 2022JQ-600).

References

1. Cifar-10 dataset. https://www.cs.toronto.edu/~kriz/cifar.html
2. Al-Rubaie, M., Chang, J.M.: Privacy-preserving machine learning: threats and solutions. IEEE Secur. Priv. **17**(2), 49–58 (2019)
3. Chapelle, O.: Training a support vector machine in the primal. Neural Comput. **19**(5), 1155–1178 (2007)
4. Chaudhuri, K., Monteleoni, C., Sarwate, A.D.: Differentially private empirical risk minimization. J. Mach. Learn. Res. **12**(3), 1069–1109 (2011)
5. Dwork, C.: Differential privacy. In: Bugliesi, M., Preneel, B., Sassone, V., Wegener, I. (eds.) ICALP 2006. LNCS, vol. 4052, pp. 1–12. Springer, Heidelberg (2006). https://doi.org/10.1007/11787006_1
6. Dwork, C., McSherry, F., Nissim, K., Smith, A.: Calibrating noise to sensitivity in private data analysis. In: Halevi, S., Rabin, T. (eds.) TCC 2006. LNCS, vol. 3876, pp. 265–284. Springer, Heidelberg (2006). https://doi.org/10.1007/11681878_14
7. Dwork, C., Roth, A., et al.: The algorithmic foundations of differential privacy. Found. Trends® Theoret. Comput. Sci. **9**(3–4), 211–407 (2014)
8. Dwork, C., Talwar, K., Thakurta, A., Zhang, L.: Analyze gauss: optimal bounds for privacy-preserving principal component analysis. In: Proceedings of the Forty-sixth Annual ACM Symposium on Theory of Computing, pp. 11–20 (2014)
9. Fredrikson, M., Jha, S., Ristenpart, T.: Model inversion attacks that exploit confidence information and basic countermeasures. In: Proceedings of the 22nd ACM SIGSAC Conference on Computer and Communications Security, pp. 1322–1333 (2015)
10. Huang, Y., Yang, G., Xu, Y., Zhou, H.: Differential privacy principal component analysis for support vector machines. Secur. Commun. Netw. **2021**, 1–12 (2021)
11. Ji, Z., Lipton, Z.C., Elkan, C.: Differential privacy and machine learning: a survey and review. arXiv preprint arXiv:1412.7584 (2014)
12. Jiang, W., Xie, C., Zhang, Z.: Wishart mechanism for differentially private principal components analysis. In: Proceedings of the AAAI Conference on Artificial Intelligence, vol. 30 (2016)
13. LeCun, Y., Bottou, L., Bengio, Y., Haffner, P.: Gradient-based learning applied to document recognition. Proc. IEEE **86**(11), 2278–2324 (1998)
14. McSherry, F., Talwar, K.: Mechanism design via differential privacy. In: 48th Annual IEEE Symposium on Foundations of Computer Science (FOCS2007), pp. 94–103. IEEE (2007)
15. Mohamadi, S., Mujtaba, G., Le, N., Doretto, G., Adjeroh, D.A.: ChatGPT in the age of generative AI and large language models: a concise survey. arXiv preprint arXiv:2307.04251 (2023)
16. Ponomareva, N., et al.: How to DP-fy ML: a practical guide to machine learning with differential privacy. J. Artif. Intell. Res. **77**, 1113–1201 (2023)
17. Sun, Z., Yang, J., Li, X., et al.: Differentially private singular value decomposition for training support vector machines. Comput. Intell. Neurosci. **2022**, 2935975 (2022)

18. Tanuwidjaja, H.C., Choi, R., Baek, S., Kim, K.: Privacy-preserving deep learning on machine learning as a service-a comprehensive survey. IEEE Access **8**, 167425–167447 (2020)
19. Wang, Y., Pan, Y., Yan, M., Su, Z., Luan, T.H.: A survey on chatGPT: AI-generated contents, challenges, and solutions. IEEE Open J. Comput. Soc. 1–20 (2023). https://doi.org/10.1109/OJCS.2023.3300321
20. Xiao, H., Rasul, K., Vollgraf, R.: Fashion-MNIST: a novel image dataset for benchmarking machine learning algorithms (2017)
21. Xu, Y., Yang, G., Bai, S.: Laplace input and output perturbation for differentially private principal components analysis. Secur. Commun. Netw. **2019**, 1–10 (2019)
22. Zhang, X., Chen, C., Xie, Y., Chen, X., Zhang, J., Xiang, Y.: A survey on privacy inference attacks and defenses in cloud-based deep neural network. Comput. Stand. Interfaces **83**, 103672 (2023)
23. Zhu, T., Ye, D., Wang, W., Zhou, W., Philip, S.Y.: More than privacy: applying differential privacy in key areas of artificial intelligence. IEEE Trans. Knowl. Data Eng. **34**(6), 2824–2843 (2020)

Converging Blockchain and Deep Learning in UAV Network Defense Strategy: Ensuring Data Security During Flight

Zhihao Li[1,2], Qi Chen[1,2(✉)], Weichuan Mo[1], Xiaolin Wang[1], Li Hu[1], and Yongzhi Cao[3]

[1] Institute of Artificial Intelligence, Guangzhou University, Guangzhou 510006, China
{zhihaoli,motongxue,hl_27}@e.gzhu.edu.cn, chenqi.math@gmail.com
[2] State Key Laboratory of Integrated Service Networks (Xidian University),
Xi'an 710071, China
[3] Key Laboratory of High Confidence Software Technologies (MOE),
School of Computer Science, Peking University, Beijing 100871, China
caoyz@pku.edu.cn

Abstract. Unmanned Aerial Vehicles (UAVs) serve as highly versatile and efficient tools utilized across diverse industries for data collection purposes. However, they face vulnerabilities associated with wireless communication and data exchange, such as unauthorized access, data theft, and cyberattacks. These risks pose significant challenges to the establishment of reliable UAV network services. This study introduces a comprehensive blockchain-based architecture for UAV network services, designed to address these challenges. The proposed architecture tackles concerns related to identity authentication and privacy protection through the seamless integration of blockchain technology. Moreover, it incorporates advanced deep learning techniques to enhance UAV safety during operations and provide robust protection against cyber threats. A series of experimental tests were conducted, simulating various UAV network attack scenarios. The results of these experiments unequivocally demonstrate the feasibility and effectiveness of the blockchain-driven UAV network service architecture.

Keywords: Unmanned Aerial Vehicles · Blockchain · Deep Learning

1 Introduction

Unmanned Aerial Vehicles (UAVs) have made a transition from military to civilian applications, showcasing rapid deployment capabilities and adaptability in challenging terrains. Equipped with advanced sensors, these UAVs serve various purposes, including surveillance, environmental monitoring, and disaster response [1–5]. They play a pivotal role in extending network coverage and

J. Vaidya et al. (Eds.): AIS&P 2023, LNCS 14509, pp. 156–171, 2024.
https://doi.org/10.1007/978-981-99-9785-5_12

enhancing functionality in remote areas, thus contributing to the Internet of Things (IoT). While previous studies have often considered UAVs as supplementary tools to address networking challenges, such as extending vehicular network connectivity [6], providing WiFi services in open spaces [7], and enabling mission-critical IoT communication [8], this perspective has limited the full potential of UAV flight automation. To expedite UAV flight automation, with a focus on secure, real-time services, it is imperative to develop a comprehensive UAV network architecture.

Traditional networks that rely on centralized cloud servers pose challenges such as computational loads and latency, which are particularly unsuitable for UAV networks, especially for tasks sensitive to latency. The importance of ensuring UAV network service security cannot be overstated, given the extensive data transactions involved in flight automation. Inadequate security measures risk breaches and compromises in data integrity [9]. Furthermore, the integration of edge services amplifies these concerns, necessitating the implementation of robust privacy and integrity mechanisms. Promisingly, blockchain technology, with its secure consensus mechanisms, offers potential solutions for addressing these challenges [10–12].

1.1 Our Contributions

In summary, we presents a comprehensive framework for trusted UAV network services, leveraging blockchain and deep learning technology to enhance identity authentication, privacy, and security within the UAV operating environment. Our key contributions can be outlined as follows:

Blockchain-Based Trusted UAV Network Service Architecture: We introduce an innovative trusted UAV network (IoD) service architecture that integrates blockchain, deep learning and edge cloud technologies. This integration ensures the provision of secure, reliable, and efficient network services for UAVs, supporting the automation of UAV fractionalization tasks.

Addressing Trusted Identity Authentication for UAV Clusters: Our proposal includes a blockchain-based identity management module for UAVs, addressing identity authentication, authorization, and end-device management. Utilizing blockchain distributed ledger technology, OpenSSL certificates, hash encryption algorithms, and digital signature technology, this module ensures robust identity management.

Mitigating Cybersecurity Risks for UAVs: We present a real-time UAV cluster terminal device security situational awareness system based on WAF traffic detection technology and malicious process detection technology. Additionally, we introduce a UAV terminal reputation value. Combining these elements with UAV terminal security situational awareness, automated security policies are implemented for both the ground control station and the UAV cluster, ensuring a secure operating environment for UAVs.

Simulation and Validation Through Docker-Based Experiments: We utilize Docker for simulating a UAV cluster and conducting a series of experiments. These experiments not only completed the development of a trusted UAV network service architecture platform but also include simulations of UAV network attacks, validating the feasibility and effectiveness of our proposed scheme and architecture.

2   Related Work

2.1   Emerging UAV Identity Authentication Mechanisms

Recent advancements in UAV technology have spurred the exploration of UAV swarms in military research, garnering significant attention. However, traditional UAV management techniques have raised security concerns, including the vulnerability of centralized systems to single-point failures and the lack of robust authentication, making them susceptible to potential threats from malicious actors [13,14]. Addressing these challenges, scholars have integrated blockchain technology to facilitate UAV identity authentication and decentralized decision-making processes [15]. Blockchain empowers UAVs to engage in encrypted identity validation, secure data sharing, and distributed voting, thereby enhancing security and collaborative aspects in UAV swarm systems.

Additionally, another the reference [16] focuses on enhancing UAV security by preserving the integrity of the UAV service environment. This research introduces two crucial mechanisms: node authentication and credit evaluation. The former validates incoming UAVs, while the latter assesses the credibility of edge nodes and their services, recording evaluations on the blockchain. Negative evaluations can restrict user UAV access, introducing a form of public oversight. Both mechanisms operate within the consensus-driven framework of blockchain, eliminating the need for intermediaries. This democratic approach significantly improves reliability, transparency, and mitigates vulnerabilities arising from third-party involvement. The research presents innovative mechanisms that advance security and trust in the dynamic UAV service environment.

2.2   Blockchain-Driven Services for the Internet of UAVs

In the realm of network applications, a pivotal study conducted by [17] introduces a tailored blockchain-based architecture for the Internet of Unmanned Aerial Vehicles. This innovative framework integrates UAVs, edge servers, and a super-ledger blockchain network, transcending the limitations of conventional UAV systems. Consequently, blockchain technology emerges as a potent solution for addressing the intricate security needs associated with UAV data collection [18].

However, the expansion of UAV networks brings forth communication challenges and complex security issues. To mitigate these challenges, researchers advocate for the integration of blockchain into peer-to-peer UAV networks, thereby enhancing communication security and scalability. This novel approach fosters collective communication integrity among diverse entities. While blockchain was initially designed to tackle double-spending in cryptocurrency, a more nuanced exploration is imperative for ensuring robust UAV network security [19]. Concurrently, in wireless sensor networks (WSNs) within monitoring paradigms, the work by Li et al. [20] stands out. This research proposes a blockchain-enhanced data aggregation strategy harnessing the capabilities of UAVs in WSNs to minimize data redundancy. This cutting-edge solution significantly enhances data security and the trustworthiness of UAVs in assisted WSNs.

2.3   UAV Attack and Defense Strategies

Various configurations of UAV-based networks are gaining prominence in critical domains, including emergency response, environmental monitoring, defense, security, and commercial endeavors [22]. However, these UAVs are susceptible to network-based intrusions due to their involvement in transporting sensitive real-time data. Moreover, the wireless communication infrastructure supporting UAV networks remains vulnerable to a range of network attacks [21].

In the realm of real-time UAV systems, Chen et al. [23] present an innovative concept. Their solution introduces a software framework named "ContainerDrone," meticulously designed to enhance the system's resilience against Denial of Service (DoS) attacks. This framework encompasses three pivotal system resources: CPU, memory, and communication channels, each fortified with tailored defense mechanisms. This strategic approach not only bolsters the stability of real-time UAV systems but also enhances their security, rendering them more resistant to disruptive attacks.

3   System Model

3.1   System Architecture

This section outlines the architecture of an Unmanned Aerial Vehicle Network (IoD) service system, emphasizing security, reliability, and efficiency, achieved through the integration of blockchain, deep learning and edge cloud technologies. The proposed system architecture, depicted in Fig. 1, consists of five essential components, each fulfilling a crucial role in ensuring the seamless operation of the system.

Unmanned Aerial Vehicle Cluster: This vital component comprises unmanned aerial vehicles equipped with various Internet of Things (IoT) devices, including cameras, infrared sensors, and more. Guided by instructions from the ground control station, these UAVs meticulously collect a range of data, encompassing UAV parameters such as speed and battery levels, flight altitude, RGB and thermal images, and other relevant metrics. The collaborative sharing of these data within the UAV cluster facilitates efficient task execution in accordance with predefined directives.

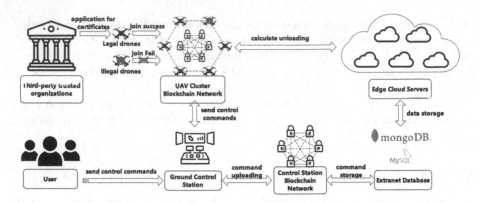

Fig. 1. Overview of System Architecture.

Blockchain Network: At the core of the architecture's security and communication framework lies the blockchain network, enabling seamless interactions among the UAV cluster, ground control station, and users. This blockchain network is meticulously tailored to accommodate the unique characteristics of UAV network data. Notably, two distinct blockchain networks are instantiated to cater to the needs of the UAV cluster and the control station.

- **UAV Cluster Blockchain Network:** Grounded in a distributed, immutable database, this network orchestrates fluid data exchange within the UAV cluster, ensuring both security and reliability. Each UAV serves as a node, inherently linking its UAV UUID to the blockchain's identity authentication framework. When a new UAV node desires to join this blockchain network, a trusted third-party agent administers an admission certificate. Subsequently, the legitimacy of this certificate undergoes scrutiny within the identity authentication system. Ultimately, a collective consensus within the UAV cluster determines the UAV's inclusion in the blockchain network.
- **Control Station Blockchain Network:** This facet manages a secure and dependable distributed immutable database, designated for user operations and interactions between ground control stations and UAVs.

Trusted Third-Party Agent: This essential component plays a pivotal role in facilitating the identity authentication mechanism of UAVs. By providing services related to reliable certificate authorization and proficient certificate management, the trusted third-party agent significantly enhances the security framework of the system.

Ground Control Station: Serving as the hub for data reception and processing, the ground control station fulfills critical functions during task execution. Dynamic adjustments to the UAV cluster's task instructions are made based on executed tasks, involving the dissemination of fresh task directives and the updating of control parameters. To support these functions, the ground control station utilizes both MySQL and MongoDB distributed databases as off-chain repositories. These repositories store structured and unstructured data collected from the UAV cluster. Importantly, essential attributes of off-chain data are recorded within the blockchain network to mitigate data storage overhead on the blockchain.

Edge Cloud Servers: This component is positioned to address the computational limitations inherent in UAVs by offloading computational tasks. The edge cloud servers optimize task execution by conducting data preprocessing either on the UAVs themselves or upon receiving transmitted data. The availability of substantial computational resources within the edge cloud servers enables judicious task offloading, guided by factors such as data volume and UAV computational capacity.

3.2   Threat Model: Security Challenges in UAV Network Services

The current architecture of UAV network services emphasizes the critical need to address security and vulnerability concerns, especially within UAV clusters. Collaborative task execution among UAV clusters, ground control stations, and edge cloud servers requires nearly real-time data interactions for efficient task fulfillment. However, the presence of unreliable wireless network communications and data exchanges introduces a range of risks. These risks include malevolent actors pilfering private data, unauthorized UAVs infiltrating the network, and the potential for malicious attacks targeting the UAV clusters. In the context of these challenges, a comprehensive threat model is outlined below, delineating the security threats and challenges inherent in the system.

UAV Attacks: Within the UAV system, a variety of attacks pose significant threats, including:

- **UAV Identity Spoofing:** This refers to attackers evading identity authentication mechanisms or masquerading as legitimate UAVs, deceiving the system into granting unauthorized access. The consequences of such attacks can be severe, ranging from interference with other UAVs' flights to tampering with mission instructions or data theft. Strategies for these attacks may involve identity forgery, password cracking, or the exploitation of stolen credentials.
- **UAV Hijacking Attacks:** Attackers target UAVs lacking adequate safeguards or communication security, gaining control of legitimate UAVs to manipulate their flight paths and operations. This control manipulation can lead to dire outcomes, such as data leaks or mission failures, significantly jeopardizing the overall system security.
- **Denial of Service (DoS/DDoS) Attacks:** Adversaries inundate the system with malicious requests or exhaust UAV resources to incapacitate a UAV, disrupting tasks or communication. This can lead to loss of control, severed communication links, or system crashes.

- **Port Scanning Attacks:** Attackers utilize tools to scan the system's ports, searching for open network access points for potential exploitation. Knowledge of these open ports can facilitate subsequent malicious activities.
- **Incorrect Position Updates:** Attackers manipulate UAV position calculations through erroneous GPS signals or communication disruption, affecting navigation, potentially causing deviation from intended trajectories, and undermining mission success.

Ground Control Station Attacks: Adversaries exploit vulnerabilities in ground control stations to inject malicious inputs, launching injection attacks:

- **SQL Injection Attack:** Malicious SQL statements are inserted via user inputs, tricking the system into executing unauthorized database queries, potentially leading to data leaks or manipulation.
- **NoSQL Injection Attack:** Attackers inject malicious code into NoSQL databases, jeopardizing data integrity, causing information leaks, and disrupting control station operations.

In light of these multifaceted threats, a comprehensive security strategy is necessary to strengthen the architecture and mitigate risks in UAV network services.

3.3   System Objectives: A Comprehensive Overview

The system's primary objectives encompass both functional and security requirements, resulting in a meticulously designed framework that addresses the multifaceted demands of UAV network services.

Functional Requirements

- **Blockchain-based UAV Identity Management Module:** This module demonstrates the system's commitment to establishing secure and trustworthy identity mechanisms. Through the integration of blockchain distributed ledger technology, PKI public-private key technology, digital certificates, and digital signatures, the module achieves the task of authenticating, authorizing, and meticulously managing terminal device identities, spanning UAV clusters and ground control stations.
- **Security Situational Awareness Defense Module:** The system proactively fosters real-time security situational awareness within terminal devices of UAV clusters and ground control stations. Employing advanced WAF traffic detection technology and malicious process detection technology, this module safeguards the system against potential threats. Additionally, the module utilizes smart contracts to continually track the reputation value of terminal devices, facilitating the deployment of automated security strategies, strengthening both ground control stations and UAV clusters.

- **Blockchain-based Trusted Data Interconnection Module:** Ensuring the integrity of data transmission and interaction is fundamental to this module's function. By harnessing blockchain technology and cryptographic algorithms, the module enables data encryption, transmission, and storage across various interactions—ranging from UAV cluster interactions, interactions between UAV clusters and edge cloud servers, to interactions between ground control stations and UAV clusters. This reinforces the principles of data verifiability, traceability, and accountability within the system.

Security Requirements

- **Confidentiality and Privacy:** The system rigorously upholds the confidentiality and privacy of transmitted data. Through the meticulous application of encryption algorithms, data exchanges—such as those occurring between UAV clusters, interactions between UAV clusters and edge cloud servers, and command data between ground control stations and UAV clusters—are securely encrypted, accessible only to duly authorized users validated by the system.
- **System Availability:** The system's unwavering commitment to high availability enables it to persistently function, even in the face of adversarial attacks, technical glitches, or network disruptions. The system remains operational, responsive, and adept at executing tasks, embodying uninterrupted service.
- **Data Integrity:** The system stands as an indomitable bulwark against data tampering. It diligently safeguards data integrity during transmission and storage, thwarting malevolent efforts to manipulate or compromise the authenticity of the data.
- **Identity Authentication Mechanism:** The system enforces a rigorous identity authentication mechanism that scrutinizes UAVs seeking to join the UAV cluster meticulously. Moreover, identity verification for user access to UAVs and the issuance of control commands through ground control stations are fortified by blockchain and digital signature encryption technologies.

4   System Design

4.1   Blockchain-Based UAV Identity Management: Design and Implementation

The system architecture design of blockchain-based UAV identity authentication is depicted in Fig. 2. This framework is built upon three fundamental components: UAV identity certificate construction, UAV identity privacy protection, and UAV identity authentication. These components synergistically work together to establish a comprehensive and robust identity management ecosystem.

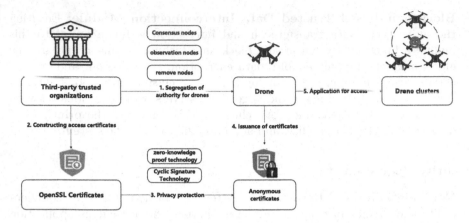

Fig. 2. UAV Identity Authentication.

UAV Identity Certificate Construction: The construction of UAV identity certificates involves a two-fold process: permission division and identity validity proof. The system leverages FiscoBcos-based permission division to allocate UAV permissions judiciously, overcoming challenges associated with permission delineation. OpenSSL, a versatile tool, is utilized for creating UAV identity certificates. Identity validity proof challenges are effectively addressed through this OpenSSL-based technology.

The operational sequence begins with the generation of a private key denoted as node.key using OpenSSL technology. This private key forms the foundational basis for creating a certificate request file, denoted as node.csr. The private key, along with relevant information, is used to construct the certificate request file. This file undergoes thorough scrutiny by the institution administrator, who conducts a meticulous verification process encompassing both legitimacy and the corresponding key. Upon successful validation, the institution administrator utilizes the institution's private key, denoted as agency.key, to issue a certificate to the respective node.

UAV Identity Privacy Protection: The decentralized nature of the blockchain ledger, while transformative, raises significant concerns about identity privacy among the participating nodes. The distributed consensus paradigm introduces apprehensions about privacy exposure, potentially leading to vulnerabilities like tracking and malevolent actions orchestrated by unscrupulous nodes. Within the context of blockchain-infused UAV clusters, safeguarding identity privacy becomes of paramount importance. This project aims to employ sophisticated technologies such as ring signatures and zero-knowledge proofs to proactively achieve UAV identity privacy protection. The application process involves concealing the sender's address and public key through the ingenious orchestration of ring signatures. The mechanics are outlined as follows:

The aspiring node generates an array of public keys, being aware of the private key corresponding to one public key in the array (possession of just one is sufficient). Using this set of public keys alongside the corresponding private key, the node creates a ring signature. Validators can verify that the ring signature originates from the possessor of the private key aligned with one of the public keys in the array. However, the identity of the specific public key linked to the private key remains concealed.

UAV Identity Authentication: The core of identity authentication lies in validating the authenticity of communication counterparts, fortified against adversarial exploits such as forgery and impersonation. The architecture's foundation for identity authentication is supported by cryptographic methodologies, including symmetric encryption algorithms, public-key cryptography algorithms, and digital signature algorithms.

The aspiring applicant employs a digital signature algorithm to encrypt identity information, resulting in the generation of an identity information hash digest. Subsequently, the applicant utilizes the AES encryption algorithm to create an AES key, named Key, to encrypt identity information. This process produces ciphertext, referred to as CipherData. The AES key, meticulously encrypted through the RSA private key (designated as PrivateKey), forms a key block known as CipherKey. Both CipherData and CipherKey are transmitted to the verifier. The verifier, utilizing the RSA public key (denoted as PublicKey), decrypts the RSA-encrypted CipherKey, acquiring the AES key, Key. Using Key, the verifier decrypts CipherData, completing the process of identity authentication.

4.2 Security Situational Awareness Defense Module

The system's security situational awareness defense module architecture is depicted in Fig. 3. At the core of the system's reinforcement lies the Security Situational Awareness Defense Module, a vital structure comprising three essential components: Security Situational Awareness, Malicious UAV Identification based on Reputation, and Automated Security Policies. Together, these elements establish a dynamic and robust defense mechanism, meticulously crafted to counter emerging security challenges.

Security Situational Awareness: At the core of this module lies the seamless integration of behavioral sensing and anomaly reporting, forming a multifaceted mechanism for real-time security awareness and responsiveness. This integration creates a robust and versatile framework, enabling real-time security awareness and effective responses to emerging threats.

Behavioral Perception, utilizing advanced traffic capture technology and machine learning-based traffic detection methods, meticulously analyzes malevolent traffic attributes and extracts significant features from UAVs' behavioral data. This component facilitates meticulous monitoring, analysis, and prediction

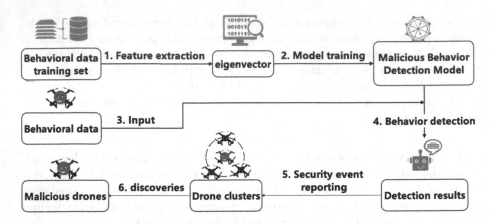

Fig. 3. Security Situational Awareness Defense.

processes covering UAV cluster and ground control station terminal traffic. By discerning patterns of malicious behavior, this layer enhances early detection of potential threats.

Anomaly Reporting, leveraging insights from Behavioral Perception, identifies UAVs exhibiting malicious behavior. This information is relayed to the UAV cluster using blockchain technology, ensuring immutable recording of such malevolent activities. The collective decision-making process facilitated by the blockchain's consensus algorithm allows the expulsion of malicious UAVs from the cluster, fostering a secure and dependable environment.

Additionally, the module incorporates a reputation-based scoring mechanism for precise identification of malicious UAVs. Initial scores assigned to terminal devices dynamically adjust based on instances of anomalous conduct, forming a layered defense against threats. Through these efforts, Security Situational Awareness adopts a holistic approach, safeguarding UAV clusters, promoting real-time security vigilance, and enabling swift responses to potential risks. Key guidelines for score deduction include:

- High-confidence classification of attack traffic (confidence >60%) originating from a cluster UAV results in immediate labeling of the UAV as malicious, while confidence levels below 50% trigger reputation deductions.
- Vigilant monitoring of outbound traffic from UAVs, combined with the identification of malevolent software patterns, classifies UAVs as malicious if detection confidence surpasses 60%. Confidence levels below 50% trigger reputation deductions.
- During UAV cluster identity verification and consensus, detection of counterfeit identities promptly categorizes the node as malicious.

– In the context of blockchain data validation, discovery of errors, including verification failures and missing data, leads to corresponding deductions in the trustworthiness score.

Automated Security Policies: Upon the identification of malicious Unmanned Aerial Vehicles (UAVs), the Automated Security Policies component triggers a swift and coordinated response. This response mechanism operates through blockchain-embedded smart contracts and consensus procedures. Detection outcomes are seamlessly broadcasted via smart contracts, initiating an automated cascade that ultimately severs connections with malevolent UAVs. The synergy of blockchain-backed mechanisms and consensus protocols empowers the UAV cluster with the ability to independently verify messages and reach a consensus, thereby orchestrating automated measures to mitigate anomalies.

4.3 Data Trustworthy Interconnection Module Based on Blockchain Technology

The module for data trust interconnection, utilizing blockchain technology, plays a pivotal role in establishing secure and reliable interconnections within the system, as depicted in Fig. 4. This module encompasses two fundamental aspects: the sharing of data within the UAV cluster and the verification of the shared UAV data. Each unmanned aerial vehicle (UAV) terminal, conceptualized as an autonomous blockchain node, undergoes rigorous identity verification through blockchain protocols upon joining the cluster. During task execution, UAVs equipped with IoT devices engage in data exchange to enhance collaborative efficiency. The UAV cluster receives shared data, and a consensus algorithm, employing blockchain, smart contracts, and cryptography, ensures the integrity and consistency of this data.

In the pursuit of establishing a reliable interconnection module for UAV data, it is crucial to account for the diverse performance attributes inherent in different UAV models and the varied operational requirements they fulfill. Consequently, the data requiring blockchain-based verification within the UAV cluster system is classified into two distinct categories: essential and non-essential data. Essential data comprises information of paramount importance for task execution, necessitating immediate blockchain verification. On the other hand, non-essential data includes information with minimal impact on tasks, allowing for deferred blockchain integration.

To address the varying levels of criticality associated with shared verification data, a robust middleware system is developed to oversee attestation for UAV cluster data sharing and validation for shared UAV data. This system relies on two fundamental functionalities: data sharing attestation facilitated by the middleware system and the validation of shared UAV data, ensuring an audit trail.

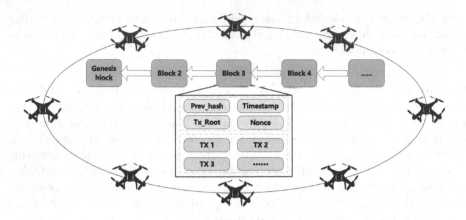

Fig. 4. UAV Uplinking.

As illustrated in Fig. 5, for pivotal key data, the application system proactively submits data to the middleware system through a RESTful API. Upon data submission, the middleware system processes and validates the received data through standardized procedures and signature authentication. This process concludes with the invocation of the blockchain system's API, seamlessly integrating the data into the blockchain. Non-essential data is managed using a passive message queue mechanism. When new data arrives in the message queue, the middleware system autonomously retrieves and processes the data, enabling its seamless incorporation into the blockchain. This approach significantly enhances operational efficiency, promoting an elevated level of system performance.

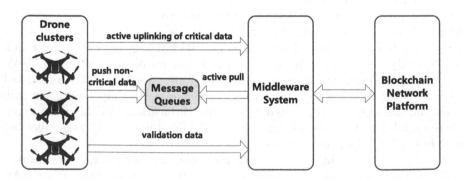

Fig. 5. Middleware System.

5   Experiments and Analyses

5.1   Experimental Design

Unmanned Aerial Vehicle (UAV) Network Service System: In our study, we implemented the blockchain platform using the FISCO BCOS framework and constructed the UAV network service framework system with the Spring Boot framework. The UAV network service system was deployed using Docker Compose and Kubernetes. Each UAV was configured as a blockchain node associated with a unique UUID. Docker containers were utilized to emulate third-party trusted agencies, providing reliable certificate authorization and management services for UAV identity authentication.

Unmanned Aerial Vehicle (UAV) Node Simulation: To simulate ground control stations sending task commands to UAV clusters, we developed code using Spring Boot. Docker containers were employed to simulate UAV clusters and capture flight status data, including speed, battery level, flight altitude, RGB images, and thermal images. Data exchange among UAV clusters was governed by smart contracts on the blockchain platform. Additionally, edge cloud servers managed UAV computation offloading and storage, storing both raw UAV data and computed results in a distributed database.

Unmanned Aerial Vehicle (UAV) Attack Simulation: We constructed a series of adversarial scenarios using Docker Compose to simulate attacks on our system. These scenarios included UAV identity spoofing, UAV hijacking, denial-of-service attacks, and port scanning attacks. The security performance of the system was evaluated by analyzing the frequency of successfully detected attacks and the detection success rate.

5.2   Experimental Analyses

Figure 6 illustrates the network latency of the unmanned aerial vehicle (UAV) network system. We employed a stress-testing tool to invoke interfaces, simulating data transmission between the UAV cluster and the ground control station. Timestamps for data transmission and reception were recorded to calculate network latency. Experimental results indicate that the average network latency of the UAV network system is less than 106.75 milliseconds, demonstrating favorable latency performance in real-time communication.

Figure 7 demonstrates the effectiveness of the unmanned aerial vehicle network service system against malicious attacks. Various types of attacks were simulated, including UAV identity spoofing, hijacking attacks, and denial of service attacks. During the evaluation process, Docker was used to simulate 100 UAVs and 20 malicious UAVs. The results reveal that the security posture-based defense module detected approximately 92% of malicious UAV identity spoofing attacks, around 67% of malicious UAV hijacking attacks, about 82% of denial of service attacks, approximately 87% of SQL injection attacks, and roughly 86% of NoSQL injection attacks. The system demonstrates a high level of security performance.

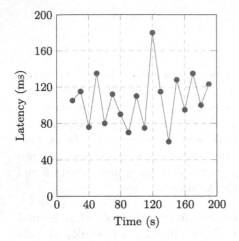

Fig. 6. Network Latency

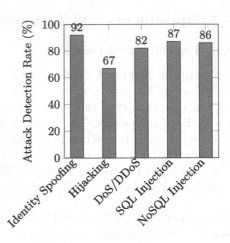

Fig. 7. Attack Detection Rate

6   Conclusion

In this paper, we presented a robust UAV network service architecture built on blockchain technology, seamlessly integrating blockchain, edge cloud computing, and UAV networks. This integration establishes a strong foundation for customized, secure, reliable, and efficient network services for UAVs. By incorporating innovative mechanisms such as blockchain-based identity management, real-time security situation awareness, and the introduction of reputation values for UAV terminals, we effectively mitigate cyber-attack vulnerabilities faced by UAVs. This results in a trusted interconnection of data within UAV clusters. We extensively validated our system through simulations and experiments utilizing Docker container technology, confirming its high level of security, effectiveness, and usability.

Acknowledgment. The authors would like to thank the reviewers for their helpful comments and suggestions. This work was supported by the National Key Project of China (No. 2020YFB1005700).

References

1. Gupta, L., Jain, R., Vaszkun, G.: Survey of important issues in UAV communication networks. IEEE Commun. Surv. Tutorials **18**(2), 1123–1152 (2015)
2. Hayat, S., Yanmaz, E., Muzaffar, R.: Survey on unmanned aerial vehicle networks for civil applications: a communications viewpoint. IEEE Commun. Surv. Tutorials **18**(4), 2624–2661 (2016)
3. Zhang, Q., Jiang, M., Feng, Z., et al.: IoT enabled UAV: network architecture and routing algorithm. IEEE Internet Things J. **6**(2), 3727–3742 (2019)
4. Srivastava, A., Prakash, J.: Internet of low-altitude UAVs (IoLoUA): a methodical modeling on integration of internet of "Things" with "UAV" possibilities and tests. Artif. Intell. Rev. **56**(3), 2279–2324 (2023)

5. Datta, S.K., Dugelay, J.L., Bonnet, C.: IoT based UAV platform for emergency services. In: Proceedings of the International Conference on Information and Communication Technology Convergence, ICTC 2018, pp. 144–147 (2018)
6. Lihua, Z., Jianfeng, D., Yu, W., et al.: An online priority configuration algorithm for the UAV swarm in complex context. Procedia Comput. Sci. **150**, 567–578 (2019)
7. Dai, M., Huang, N., Wu, Y., et al.: Unmanned-aerial-vehicle-assisted wireless networks: advancements, challenges, and solutions. IEEE Internet Things J. **10**(5), 4117–4147 (2022)
8. Orsino, A., Ometov, A., Fodor, G., et al.: Effects of heterogeneous mobility on D2D-and drone-assisted mission-critical MTC in 5G. IEEE Commun. Mag. **55**(2), 79–87 (2017)
9. Peng, S., Zhou, F., Li, J., et al.: Efficient, dynamic and identity-based remote data integrity checking for multiple replicas. J. Netw. Comput. Appl. **134**, 72–88 (2019)
10. Strobel, V., Castelló, F.E., Dorigo, M.: Blockchain technology secures robot swarms: a comparison of consensus protocols and their resilience to byzantine robots. Front. Robot. AI **7**, 54 (2020)
11. Wen, Q., Gao, Y., Chen, Z., et al.: A blockchain-based data sharing scheme in the supply chain by IIoT. In: Proceedings of the IEEE International Conference on Industrial Cyber Physical Systems, ICPS 2019, pp. 695–700 (2019)
12. Gao, Y., Chen, Y., Hu, X., et al.: Blockchain based IIoT data sharing framework for SDN-enabled pervasive edge computing. IEEE Trans. Industr. Inf. **17**(7), 5041–5049 (2020)
13. Han, P., Sui, A., Wu, J.: Identity management and authentication of a UAV swarm based on a blockchain. Appl. Sci. **12**(20), 10524 (2022)
14. Millard, A.G., Timmis, J., Winfield, A.F.T.: Towards exogenous fault detection in swarm robotic systems. In: Natraj, A., Cameron, S., Melhuish, C., Witkowski, M. (eds.) TAROS 2013. LNCS, vol. 8069, pp. 429–430. Springer, Heidelberg (2013)
15. Castelló Ferrer, E.: The blockchain: a new framework for robotic swarm systems. In: Arai, K., Bhatia, R., Kapoor, S. (eds.) FTC 2018. AISC, vol. 881, pp. 1037–1058. Springer, Cham (2019). https://doi.org/10.1007/978-3-030-02683-7_77
16. Wang, J., Liu, Y., Niu, S., et al.: Lightweight blockchain assisted secure routing of swarm UAS networking. Comput. Commun. **165**, 131–140 (2021)
17. Nguyen, T., Katila, R., Gia, T.N.: A novel internet-of-drones and blockchain-based system architecture for search and rescue. In: Proceedings of the International Conference on Mobile Ad Hoc and Smart Systems, MASS 2021, pp. 278–288 (2021)
18. Lee, S., Kim, M., Kim, M., et al.: Timely update probability analysis of blockchain ledger in UAV-assisted data collection networks. In: Proceedings of the International Conference on Communications, ICC 2022, pp. 4559–4564 (2022)
19. Ghribi, E., Khoei, T.T., Gorji, H.T., et al.: A secure blockchain-based communication approach for UAV networks. In: Proceedings of the International Conference on Electro Information Technology, EIT 2020, pp. 411–415 (2020)
20. Li, G., He, B., Wang, Z., et al.: Blockchain-enhanced spatiotemporal data aggregation for UAV-assisted wireless sensor networks. IEEE Trans. Industr. Inf. **18**(7), 4520–4530 (2021)
21. Safavat, S., Rawat, D.B.: OptiML: an enhanced ML approach towards design of SDN based UAV networks. In: Proceedings of the International Conference on Communications, ICC 2022, pp. 1–6 (2022)
22. Mershad, K.: PROACT: parallel multi-miner proof of accumulated trust protocol for internet of drones. Veh. Commun. **36**, 100495 (2022)
23. Chen, J., Feng, Z., Wen, J.Y., et al.: A container-based DoS attack-resilient control framework for real-time UAV systems. In: Proceedings of the Design, Automation & Test in Europe Conference & Exhibition, DATE 2019, pp. 1222–1227 (2019)

Towards Heterogeneous Federated Learning: Analysis, Solutions, and Future Directions

Yongwei Lin(ID), Yucheng Long(ID), Zhili Zhou$^{(\boxtimes)}$(ID), Yan Pang(ID), and Chunsheng Yang(ID)

Artificial Intelligence Research Institute, Guangzhou University, Guangzhou 510006, China
{YongweiLin,YuchengLong}@e.gzhu.edu.cn, zhou_zhili@163.com, chunsheng.yang@gzhu.edu.cn

Abstract. With the rapid growth of edge devices such as smartphones, wearables, and mobile networks, how to effectively utilize a large amount of private data stored on these devices has become a challenging issue. To address this issue, federated learning has emerged as a promising solution. Federated learning allows multiple devices to train machine learning models collaboratively while keeping the data decentralized and following local privacy policies. However, the heterogeneous differences in data distributions, model structures, network environments, and devices pose challenges in realizing collaboration. In this paper, we reviewed the heterogeneous federated learning (HFL) approaches and classified them into data heterogeneity, device heterogeneity, communication heterogeneity, and model heterogeneity. Also, we concluded their advantages and disadvantages and gave the solutions to the limitations in detail. Meanwhile, this paper introduces the commonly used methods for evaluating the performance of federated learning and suggests the future directions of the HFL framework.

Keywords: Heterogeneous Federated Learning · Trustworthy AI · Federated Learning

1 Introduction

In today's technologically advanced society, edge devices such as smartphones, wearable devices, and mobile networks are widely used in real-world applications. However, it is challenging to use the large amount of personal data stored on these devices without compromising privacy. To address this challenge, federated learning has emerged as a promising solution. Federated learning allows multiple devices to train machine learning models collaboratively while keeping

This work is supported in part by the National Natural Science Foundation of China under Grant 62372125, in part by the Guangdong Natural Science Funds for Distinguished Young Scholar under Grant 2023B1515020041.

J. Vaidya et al. (Eds.): AIS&P 2023, LNCS 14509, pp. 172–189, 2024.
https://doi.org/10.1007/978-981-99-9785-5_13

the data decentralized, adhering to the data locality principle. In this paradigm, the devices participating in the federated learning system are called clients. Federated learning is a secure and distributed machine learning framework based on encryption techniques, enabling organizations to engage in collaborative model training while safeguarding data privacy. Federated learning(FL) [1,2], a collaborative learning model paradigm, has attracted increasing attention from industry and academia. Extensive research on this approach has been conducted in various real scenarios, including healthcare [3], recommendation systems [4], anti-money laundering [5], and data security [6].

Federated learning has achieved significant success. However, since most existing research in federated learning is based on the assumption of homogeneous data that can be easily aggregated. there are numerous challenges [7], including variations in data distributions, model structures, network environments, and edge devices, making federated collaboration hard to implement. The heterogeneity issues have existed in various aspects of the learning process, including data heterogeneity, device heterogeneity, communication heterogeneity, and model heterogeneity. As shown in Fig. 1, the specific challenges are summarized as follows.

(1) Data heterogeneity: Due to the Non-Independent Identical Distribution (Non-IID) problem of the client's local data, the results obtained from model training on one client may not be able to be generalized to other client's data, resulting in a decline in the overall performance of the model.

(2) Device heterogeneity: Due to the differences in client's storage, computation, and communication capabilities, the computational power of some clients is weak, and they cannot perform complex model training or gradient calculation, resulting in an imbalance between the devices involved in federated learning, affecting the overall training effect.

(3) Communication heterogeneity: Due to the differences in the network environment in which the client is located, there may be communication delays and bandwidth constraints, resulting in a blockage of the aggregation process of the model parameters, which affects the model's updating and convergence speed.

(4) Model heterogeneity: In various application scenarios, different tasks require different models, so customers need to effectively integrate different types of models. However, this is a challenging task that requires solving the problem of model fusion and integration.

To address the above heterogeneity problems, we provide a comprehensive survey of research work on HFL in this paper. We conduct a comprehensive investigation into the fundamental causes of heterogeneity in federated learning and subsequently classify HFL approaches into four categories: data heterogeneity, device heterogeneity, communication heterogeneity, and model heterogeneity. This study sufficiently analyzes the solutions to address these challenges. Additionally, it employs commonly employed performance evaluation methodologies to evaluate the performances of existing HFL approaches and also gives four potential research directions of the HFL framework.

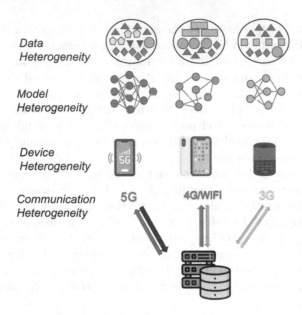

Data
Heterogeneity

Model
Heterogeneity

Device
Heterogeneity

Communication   5G   4G/WIFI   3G
Heterogeneity

Fig. 1. Schematic of heterogeneous federated learning

1.1 Related Surveys

A survey conducted by Yang et al. [2] was quite influential in establishing the fundamental principles and concepts of Federated Learning. It also proposed an extensive and robust FL framework. Kairouz et al. [8] expanded the applications of FL to various scenarios. Wahab et al. [9] conducted a thorough investigation and synthesis, presenting a multi-level classification methodology and evaluation criteria, and exploring the prospects of federated learning within communication and network systems. In a more recent survey [10], the domain of Personalized Federated Learning (PFL) was introduced, accompanied by an exploration of the fundamental challenges of privacy-preserved machine learning on heterogeneous data. The survey described PFL techniques, pivotal concepts, and future research directions.

However, it is worth noting that several surveys focus on HFL. The study [11] offered an all-encompassing assessment of the profound impact of heterogeneity on quality and fairness in federated learning, highlighting significant effects on model performance and fairness in mixed heterogeneity scenarios. The concept of HFL was initially introduced by Gao et al. [12], who endeavored to tackle the intricate challenges posed by heterogeneity in federated learning through the comprehensive investigation of various aspects, including data space, statistics, systems, and model heterogeneity. In a recent survey, Ye et al. [13] provided a systematic examination and comprehensive review of the practical challenges and innovative solutions of HFL. The survey research challenges in HFL, a thorough review of recent advancements, analysis of existing approaches, and an insightful on future research directions.

2   A Taxonomy of Heterogeneous Federated Learning

2.1   Definition

The concept of HFL aims to address the inherent heterogeneity among participants in terms of data, devices, communication, and models. The primary objective of HFL is to facilitate the integration of knowledge across diverse participants, thereby enhancing model performance and generalization capabilities.

2.2   Analysis

Data Heterogeneity. Data Heterogeneity is often regarded as statistical heterogeneity, where the data deviates from complete independence and identical distribution, commonly known as non-independent and identically distributed (non-i.i.d.).

Model Heterogeneity. Model heterogeneity presents numerous technological and algorithmic challenges in the field of federated learning. Primarily, models with different architectures may have different quantities of parameters and follow distinct update rules. Consequently, this makes the aggregation of model parameters notably complex during the federated learning process. Moreover, model heterogeneity gives rise to disparities in model performance, as divergent model types may exhibit variances in data processing and learning tasks.

Device Heterogeneity. In a pristine federation environment, clients demonstrate a diverse range of device configurations, including variations in GPU, CPU, software, and network conditions. This heterogeneity leads to significant discrepancies in device overhead, such as compute time and resource utilization, when striving to accomplish the same task. Consequently, this exacerbates the performance degradation of a global model.

Communication Heterogeneity. In real-world implementations of the Internet of Things (IoT), devices are commonly deployed in diverse network environments, each characterized by distinct network connectivity settings. Consequently, that leads to variations in communication attributes such as bandwidth, latency, and reliability, resulting in what is widely known as communication heterogeneity.

3   Heterogeneous Federated Learning Taxonomy

3.1   Federal Learning Strategies with Heterogeneous Data

The presence of non-i.i.d. data among clients poses a challenge known as data heterogeneity. Addressing the detrimental effects of non-i.i.d. data remains an ongoing research endeavor. Non-i.i.d. data exhibits distributional skews, often observed as label distribution skew, feature distribution skew, and quantity skew.

Label Distribution Skew: In the context of label distribution skew [14], the heterogeneous distribution of target labels (or classes) can lead to significant disparities among diverse clients.

Feature Distribution Skew: Feature distribution skew refers to discrepancies in the distributions of input features among clients [15]. The variation in feature distributions could increase due to divergent data collection processes employed by different clients.

Quantity Skew: Quantity skew refers to disparities in the available data volume among clients. Certain clients may possess a substantial amount of data [16], while others may have a limited number of data samples. Consequently, clients with abundant data can exert undue influence on the training process. Effectively managing quantity skew requires techniques to mitigate the impact of data imbalance and prevent clients with limited data from being overshadowed.

Based on the classification of data heterogeneity problems, there are several potential solutions to consider. One such solution is data augmentation, which involves enriching or amplifying data by incorporating supplementary information or features. Fedmix [17] aims to integrate Mixup techniques into federated learning to enhance the mean-based federated learning paradigm. This innovative approach introduces mean-enhancement techniques within the federated learning framework, approximating the benefits of Mixup. As a result, Fedmix effectively addresses challenges such as overfitting, enhances model generalization, and mitigates issues associated with imbalanced data distribution. However, it is important to note that Fedmix has its limitations. The collection of local data distributions may introduce potential information leakages, raising concerns about privacy and security. Astraea [18] tackles the challenges of data imbalance and model bias through adaptive sample selection and uncertainty-driven model updating strategies. The efficacy of these approaches is rigorously demonstrated across various datasets encompassing mobile deep-learning applications. However, it is worth noting that some researchers have expressed concerns about Astraea's method, suggesting that the disclosure of local data distributions during upload could inadvertently expose vulnerabilities and make it susceptible to malicious intrusion.

This section delves into the data heterogeneity issues encountered in federated learning, encompassing label distribution skewness, feature distribution skewness, and quantity skewness. To address these challenges, the section examines potential solutions, including data augmentation and the Fedmix method. In future research, it would be valuable to explore approaches that effectively mitigate the limitations of these methods and further enhance their efficacy and security.

3.2   Federal Learning Strategies with Heterogeneous Model

Model heterogeneity presents a challenge when attempting to transfer knowledge between the clients that employ different models with a model-independent approach. To tackle this issue, we classify model heterogeneity into two categories: partial heterogeneity and full heterogeneity.

Partial Heterogeneity. Partial heterogeneity refers to the differences in the model architectures adopted by various clients, while certain components or layers of the models remain consistent across clients. In other words, there exists a partial overlap in the model architectures. This variability may arise due to hardware limitations, individual requirements, or variations in the task properties that clients aim to address.

Complete Heterogeneity. Complete heterogeneity in federated learning occurs when different clients use model architectures that have significant differences, leading to a wide variety of models. To effectively tackle this challenge, sophisticated strategies like meta-learning [19] or model-agnostic mechanisms are necessary. These strategies facilitate the generalization and transfer of knowledge while accommodating the diverse model structures.

Several solutions are proposed based on the above categorization of model heterogeneity into partial and complete heterogeneity. One of these solutions is knowledge distillation, which relaxes the stringent requirements for homogeneous local models by using logarithms as a representation of knowledge transfer. This approach allows for the creation of federated learning systems that can accommodate different model architectures [19]. Wang introduced the VFedTrans framework for facilitating privacy-preserving data sharing and knowledge transfer among healthcare organizations [20]. This framework utilizes a joint modeling approach to extract a federated representation of shared samples by combining their features. However, researchers have expressed concerns regarding the effectiveness, scalability, and applicability of this approach in different scenarios of vertical federated learning. In the field of federated learning, Le et al. proposed FedLKD, an approach that utilizes layer-wise knowledge distillation [21]. The goal of FedLKD is to enhance the local training process by applying knowledge distillation between global and local models, using a small proxy dataset. However, it is important to carefully consider the potential impact of this method on privacy preservation, as emphasized by several researchers. Yu et al. proposed an innovative approach to address the inherent heterogeneity in joint learning through local adaptation [22]. This technique aims to enhance model efficiency and convergence by incorporating local model adaptation and parameter tuning. It enables each client to personalize model training based on its local data characteristics and device capabilities. However, it is important to consider the potential limitations of this method when utilizing logits, as it may result in insufficient integration of local information.

In summary, dealing with model heterogeneity presents a significant challenge in the field of federated learning, which can be categorized as partial heterogeneity and complete heterogeneity. To tackle this challenge, researchers have proposed several effective solutions, such as knowledge distillation, federated inter-layer distillation, local model adaptation, and parameter tuning. However, it is crucial to conduct further research and practical validation to improve the efficiency and feasibility of these approaches. Moreover, it is important to carefully consider essential aspects like privacy protection and potential limitations when implementing these methods.

3.3   Federal Learning Strategies with Heterogeneous Communication

Within the intricate landscape of the Internet of Things, the prevalence of communication heterogeneity poses significant challenges, characterized by high communication costs and suboptimal efficiency [23], thereby diminishing the efficacy of federated learning. Several methodologies have emerged as joint strategies to address the pervasive challenge of communication heterogeneity. These encompass the optimization of compression parameters and gradients, the reduction of communication rounds, and the implementation of asynchronous training techniques.

Compression Parameters and Gradients. Model parameter compression is an effective strategy for dealing with variations in communication during federated learning. It reduces the amount of data transmitted by compressing model parameters and can be personalized based on the characteristics and limitations of individual devices. By selectively transmitting gradient updates according to device characteristics and communication conditions, we can minimize communication overhead and improve efficiency.

From the perspective of Compression Parameters and Gradients, there are several methods to address the communication heterogeneity in federated learning. For example, Communication-Mitigated Federated Learning [24] addresses the transmission of inconsequential updates to the central server by evaluating the compliance of local updates with global updates. This method is effective in reducing the workload of communication transmission. However, it is crucial to take into account the limitations of this approach when dealing with networks that are highly diverse or unreliable. The Federated Deep Neural Networks Framework [25]introduces a transformative approach by substituting every fully connected (FC) layer with a pair of low-rank projection matrices, thereby achieving model compression within the DNNs architecture. The framework establishes a comprehensive global error function to reconstruct the output of the compressed DNNs model, ensuring fidelity in the compression process. In addition, FedSkel [26] enhances federated learning by improving computational efficiency and optimizing communication on edge devices. This is achieved through selective model updates that solely target the essential components, thereby reducing resource requirements. However, it is important to note that the scalability of the system, particularly concerning privacy and security concerns, has not been extensively analyzed.

Reducing Communication Rounds. Reducing the number of communication rounds is an effective strategy to handle communication heterogeneity. It helps to minimize the overall communication overhead between participants. In the context of federated learning, FedMMD [27] improves the optimization process by introducing the Maximum Mean Discrepancy constraint into the loss function. This integration leads to a reduction in the required communication rounds. Another approach, FedSeq [28] enhances the algorithm's performance and convergence rate by setting a predefined communication round budget. This approach effectively manages resource allocation and streamlines the learning process.

Reducing the number of communication rounds is an effective approach to address communication differences in federated learning. FedMMD integrates the Maximum Mean Discrepancy constraint into the loss function, aiming to minimize communication rounds. Meanwhile, FedSeq enhances performance and convergence by setting a predetermined limit on the communication rounds allowed.

Asynchronous Training. In asynchronous training, participants have the freedom to update model parameters independently without waiting for others to finish their updates. This concurrent process has the potential to improve communication efficiency, especially in situations with high communication latency [7]. In the context of asynchronous training, several methods have been proposed to handle communication differences in federated learning. For example, FedSeC [29] introduces a framework for differential privacy that incorporates an optimization technique based on updates. On the other hand, FedSA [30] uses a semi-asynchronous mechanism that relies on the sequential order of model updates. Additionally, FedHe [31] applies a knowledge distillation-like approach to reduce communication overhead.

Communication heterogeneity in federated learning poses significant challenges, including high communication costs and suboptimal efficiency. To address this issue, researchers have developed several methodologies, including compression parameters and gradients, reducing communication rounds, and asynchronous training techniques. Methods such as Communication-Mitigated Federated Learning, Federated Deep Neural Networks Framework, and FedSkel optimize compression and computational efficiency, while FedMMD and FedSeq reduce communication rounds. Asynchronous training methods such as FedSeC, FedSA, and FedHe also address communication heterogeneity.

3.4   Federal Learning Strategies with Heterogeneous Devices

Device heterogeneity in federated learning arises due to disparities in device configurations, including hardware, software, and network conditions [7]. To address this challenge, methodologies such as fine-tuning training tasks and client selection are utilized to allocate suitable tasks to edge devices, aiming to optimize overall efficiency.

Training Tasks Adjustment. To optimize global efficiency in federated learning, it is essential to allocate appropriate tasks based on the computational capabilities of each device, while also considering factors like fairness, privacy, and data diversity. Intelligent algorithms play a crucial role in navigating device heterogeneity and enabling effective collaboration. For instance, Abdellatif et al. [32] propose an efficient user and resource allocation scheme for horizontal federated learning. This system leverages the vast volumes of data generated by Internet of Things (IoT) devices to train deep learning models, addressing the challenges and requirements posed by data privacy and resource-constrained environments.FedSAE [33] tackles the issue of performance degradation in federated learning through two key mechanisms: automatic adjustment of device training task capabilities and participant selection. This approach utilizes comprehensive information about a device's history of training tasks to predict its training load capacity, enabling adaptive participant selection. However, it is important to note that refining workload allocation based on client training history may introduce temporal delays.

Client Selection. Client selection is a critical aspect of federated learning, aiming to identify suitable clients for each iteration based on their constraints, such as network bandwidth, computation capability, and local resources. Selection strategy plays a crucial role in accelerating convergence and improving model accuracy. Wang et al. [34] have made significant contributions in this field. Their research tackles the challenges posed by non-IID data in federated learning. They propose a reinforcement learning framework specifically designed for this scenario, including an effective data representation method, an optimized task allocation strategy, and a model aggregation mechanism. It is important to note that reinforcement learning models require a substantial amount of data for effective training. Furthermore, in addressing the challenges arising from device heterogeneity, client selection is often combined with task adjustment. Researchers have developed methodologies like CFL-HC [35]and HeteroSAg [36]to handle the varying computational capabilities of edge devices. These approaches effectively mitigate the impact of device heterogeneity, ensuring optimal performance and efficiency in federated learning settings.

In federated learning, addressing device heterogeneity requires the allocation of suitable tasks to edge devices and the selection of appropriate clients based on their constraints. Fine-tuning training tasks and client selection strategies are crucial for maximizing overall efficiency while considering factors like fairness, privacy, and data diversity. To tackle the challenges posed by device heterogeneity and ensure optimal performance and efficiency in federated learning settings, researchers have developed methods such as FedSAE, CFL-HC, and HeteroSAg. These methodologies effectively handle the varying computational capabilities of edge devices. Additionally, reinforcement learning models show promise in addressing the issue of non-IID data.

More details of the contributions and limitations of the existing Heterogeneous FL method can be found in the supplemental material. Given the evolving nature of this field, it is essential to establish widely ac-knowledged benchmarking and evaluation frameworks for heterogeneous scenarios with different complexities.

4 Heterogeneous Federal Learning Evaluation Methods

The concept of collaborative learning was first introduced by McMahan et al. [1]. In this rapidly evolving field, it is crucial to establish widely recognized benchmarking and evaluation frameworks for different scenarios with varying complexities. Empirical evaluation plays a vital role in examining a verifiable federated learning approach in simulated or real-world error-prone environments. It allows for a comprehensive exploration of its effectiveness in complex computational landscapes while ensuring the reliability of the findings.

Model Performance and Communication Overhead. The evaluation of federated learning methods takes into account precision, convergence velocity, and factors such as client heterogeneity and disparate data distributions [37]. It is crucial to strike a balance among precision, communication overhead, and model performance when assessing these methods. Researchers commonly use metrics such as accuracy, precision, recall, F1 score, and convergence velocity to measure the effectiveness of federated learning approaches. By analyzing both model performance and communication overhead, potential areas for optimization can be identified.

Robustness. Robustness is an essential metric for assessing the resilience of federated learning methods against adversarial scenarios [38,39]. It ensures that the model maintains its performance and accuracy in the federated learning environment. Evaluation techniques commonly include adversarial attacks such as model inversion, membership inference, and data contamination. Metrics such as accuracy degradation, model divergence, and anomaly detection are used to quantify robustness.

Privacy Protection. Privacy protection is an important consideration when evaluating the effectiveness of HFL methods. In federated learning, participants often have sensitive data, so ensuring the security of this data is paramount. Researchers evaluate the effectiveness of methods in preserving individual privacy using metrics such as differential privacy [40], information entropy, and data aggregation. These metrics allow for quantifying the level of privacy protection provided. By enhancing privacy protection measures, researchers aim to ensure data security and privacy during the federated learning process.

Customer Contribution. Analyzing client contributions is a crucial aspect of federated learning. It involves quantifying individual contributions by considering factors such as the quality and quantity of data, computational capabilities, and reliability. Metrics like data quality, data quantity, and computational resources are used to assess client contributions. Understanding the varying degrees of contribution is important for optimizing the federated learning process, improving model performance, and addressing data heterogeneity. For instance, FedCav [41]introduces an algorithm for model aggregation that takes into account client contributions in the presence of heterogeneous data.

5   Future Directions

Our empirical investigation unequivocally demonstrates the burgeoning prominence of HFL research. Nonetheless, many difficulties persist, necessitating their resolution to empower this technology to confront the difficulties encountered in real-world applications. We will look over the next steps for future inquiry to enhance the efficacy of addressing heterogeneous predicaments within forthcoming Federated Learning systems.

5.1   Privacy Protection

In HFL, participants often have sensitive personal data, so privacy protection measures are necessary. Future research should prioritize the development of efficient and secure privacy-preserving mechanisms to ensure participants have control over their privacy while sharing data. Differential privacy offers a mathematical guarantee that statistical results can be publicly released while safeguarding individual privacy [46]. By combining differential privacy with federated learning, it becomes possible to prevent the disclosure of sensitive information during model training and aggregation. Future research should focus on improving differential privacy algorithms and mechanisms that can accommodate diverse data types and privacy requirements in HFL. Another important research direction is investigating the use of Secure Multi-Party Computation in the context of federated learning [47]. However, it is important to note that these solutions may need to be adapted to account for system heterogeneity.

5.2   Improving Communication Efficiency

Communication plays a pivotal role in coordinating the collaborative learning process among heterogeneous participants, but it often incurs substantial costs in terms of bandwidth, latency, and energy consumption [42,43,45]. To enhance communication efficiency in HFL, researchers can explore the following aspects. (1)Integration of differential privacy techniques: By incorporating differential privacy techniques, the amount of information exchanged during the federated learning process can be effectively reduced. (2)Gradient compression techniques: These techniques aim to minimize the size of gradients communicated during federated learning, thereby reducing the communication overhead. (3)Leveraging edge computing capabilities and enabling local model updates: By utilizing the computational capabilities of edge devices and facilitating local model updates, the dependency on frequent communication with the central server can be decreased, leading to improved communication efficiency. (4)Knowledge transfer techniques: Exploring techniques that enable knowledge transfer among clients can significantly reduce the extensive communication requirements. Methods such as transfer learning, model personalization, and parameter sharing facilitate the transfer of learned knowledge from high-resource clients to low-resource clients, thereby mitigating overall communication needs.

In conclusion, enhancing communication efficiency is a critical area for future research in HFL. By employing techniques such as differential privacy, gradient compression, edge computing, and federated learning with knowledge transfer, we can effectively reduce communication overhead and enhance the scalability and efficiency of federated learning in heterogeneous settings.

5.3   Federated Fairness

Federal equity is a crucial consideration in the design and implementation of federated learning systems.The presence of diverse and distributed data sources among heterogeneous participants introduces biases and inequalities, and thus effective mitigation is required urgently [6]. Future research should prioritize the development of strong frameworks and algorithms that actively promote fairness, equality, and nondiscrimination in federated learning. One way to enhance fairness in federated learning is to focus on privacy-preserving methods that safeguard sensitive data and prevent unauthorized access or misuse [44]. By exploring privacy-enhancing technologies, we can facilitate collaborative learning while respecting individual privacy rights. This not only mitigates the risk of biased model updates but also fosters fairness in the process of aggregating data. Additionally, it is crucial to design federated learning algorithms explicitly to tackle the challenges posed by data heterogeneity and fairness requirements [45]. Traditional federated learning methods may unintentionally favor participants with more extensive or representative data, resulting in biased models and persistent inequality. To address this, future research should concentrate on innovative techniques such as sample weighting, domain adaptation, and model regularization. These approaches effectively account for data heterogeneity and ensure fairness throughout the model training and aggregation processes.

In conclusion, it is crucial to prioritize addressing equity at the federal level in federated learning. Researchers can promote fairness, equality, and nondiscrimination in federated learning systems by developing methods that protect privacy, exploring technologies that enhance privacy, and designing algorithms that explicitly address the diversity of data and fairness.

5.4   Uniform Benchmarks

The growing fascination with HFL is evident based on the results of our recent survey. However, as we delve further into this domain, numerous challenges arise that require immediate attention to make this technology suitable for practical applications. A crucial aspect for future research directions in addressing heterogeneity in FL systems revolves around the establishment of standardized benchmarks.

Improved Datasets. To accurately represent the diverse nature of real-world federated learning scenarios, it is crucial to develop comprehensive and realistic datasets. Improving datasets is a key area for future advancements in the field of heterogeneous federated learning. Researchers should focus on different aspects of dataset construction, including being aware of heterogeneity, using representative data sampling techniques, assessing and enhancing data quality, generating privacy-preserving datasets, creating benchmark datasets, and incorporating real-world data. By addressing the challenges associated with data heterogeneity using these strategies, researchers can enhance the performance and effectiveness of federated learning models in diverse settings. The availability of these improved datasets will enable more realistic and impactful research in the field of HFL.

Enhanced Evaluation Metrics. Establishing clear and consistent evaluation metrics is crucial for effectively measuring the performance of Horizontal Federated Learning. It is essential to advance the field by developing enhanced evaluation metrics that can provide a comprehensive understanding of the strengths and limitations of federated learning systems. A key focus of future research should be on expanding existing models such as FedEval [48]. The objective should be to create metrics that consider heterogeneity awareness, privacy preservation, fairness orientation, robustness emphasis, resource efficiency, and real-world performance. These enhanced evaluation metrics will drive progress in the field and contribute to the development of more effective and equitable federated learning systems.

6   Conclusion

This paper aims to provide a comprehensive definition and analysis of HFL. It categorizes HFL into four types of heterogeneity: data, model, device, and communication, based on the underlying causes of heterogeneity in federated learning. The study offers a meticulous examination of potential solutions to address these challenges, ultimately enhancing the reader's comprehension of the impact of heterogeneity on federated learning. Furthermore, it succinctly summarizes commonly employed performance evaluation methods and proposes future directions for the development of the HFL framework. These insightful discussions hold significant value in contributing to the advancement of the HFL community. HFL presents itself as an engaging research avenue, necessitating collaborative efforts from the machine learning, systems, and data privacy communities (Tables 1, 2, 3 and 4).

Appendix

Table 1. Heterogeneous data methods

Methods	Key Contributions	Limitations
Zhang et al. [14]	FedIC solves the problem of skewed label distribution in federated learning by calibrating logits and introducing label boundaries	The effectiveness in dealing with extreme labeling distribution skewness still needs further research and improvement.
Luo et al. [15]	DFL solves the problem of uneven attribute distribution on the performance and convergence stability of federated learning	Challenges remain in dealing with complex relationships between domain-specific and cross-invariant attributes.
Yoon et al. [17]	FedMix for improving the performance of federated learning with non-independent Identically distributed Data and Addressing Privacy Preservation	Collecting local data distributions may bring potential information leakage.
Duan et al. [18]	Astraea for Improving Classification Accuracy in Mobile Deep Learning Applications	Disclosure of local data distribution during upload may inadvertently expose vulnerabilities and make it susceptible to malicious intrusion

Table 2. Heterogeneous model methods

Methods	Key Contributions	Limitations
Fallah et al. [19]	MAML uses a personalized version of joint averaging algorithm and evaluates its performance against gradient specification of the non-convexloss function	verlooking other potential approaches or techniques that could enhance personalization in federated learning.
Wang et al. [20]	VFKF proposes a vertical federated knowledge transfer mechanism for feature enrichment in cross-party machine learning systems	The scalability and applicability of vertical federated learning in different scenarios are not apparent.
Le et al. [21]	FedLKD effectively addresses the statistical heterogeneity challenge by leveraging knowledge istillation between global and local models	Its effectiveness and privacy preservation may vary depending on the specific characteristics of the dataset and the selection of proxy data.
Yu et al. [22]	They alleviate the issue of overfitting in personalized updates by augmenting the coherence of logits between the global and local models	The exploitation of logits may engender inadequate assimilation of local information

Table 3. Heterogeneous communication methods

Methods	Key Contributions	Limitations
Hou et al. [23]	FedChain combines the advantages of local and global update methods infederated learning,achieving fast convergenco while leveraging data similarity	Devices may connect slowly, rendering them expensive and unreliable communicate.
Lu et al. [24]	CMFL avoids transmitting irrelevant updates to the server by measuring the consistency of local updates with global updates	Difficult to handle highly heterogeneous or unreliable network environments.
Li et al. [25]	They presents a concise and efficient federated learning framework fortraining deep neural networks on resource-constrained mobile device	Lack of in-depth analysis of potential privacy or security implications of proposed frameworks.
Luo et al. [26]	Fedskel enables federated learning for efficient computation and efficient communication on edge devices by updating the essential parts of the mode	Limited scalability analysis of the system with privacy or security concerns

Table 4. Heterogeneous device methods

Methods	Key Contributions	Limitations
Abdellatif et al. [32]	Allow massive amounts of data generated by IoT devices to train deep learning models	Failure to minimize communication overhead in hierarchical joint learning.
Li et al. [33]	FedSAE effectively addresses systems heterogeneity by adjusting the training tasks of devices and actively selecting participants	Refining the allocation of workloads in accordance with the client straining history may encounter temporal delays.
Wang et al. [34]	Favor dynamically curates the optimal cohort of clients to engage in iterations of federated learning	Raining reinforcement learning models necessitates a substantial volume of data

The four tables above summarize the solutions to federated learning data heterogeneity, model heterogeneity, communication heterogeneity, and device heterogeneity, and analyze the main contributions and limitations of each approach. These valuable discussions can contribute to the high-quality development of the heterogeneous federated learning community.

References

1. McMahan, B., et al.: Communication-efficient learning of deep networks from decentralized data. In: Artificial Intelligence and Statistics. PMLR (2017)
2. Yang, Q., et al.: Federated machine learning: Concept and applications. ACM Trans. Intell. Syst. Technol. (TIST) **10**(2), 1–19 (2019)
3. Dayan, I., et al.: Federated learning for predicting clinical outcomes in patients with COVID-19. Nat. Med. **27**(10), 1735–1743 (2021)
4. Wu, C., et al.: FedGNN: federated graph neural network for a privacy-preserving recommendation. arXiv preprint arXiv:2102.04925 (2021)
5. Suzumura, T., et al.: Towards federated graph learning for collaborative financial crimes detection. arXiv preprint arXiv:1909.12946 (2019)

6. Usynin, D., et al.: Adversarial interference and its mitigations in privacy-preserving collaborative machine learning. Nat. Mach. Intell. **3**(9), 749–758 (2021)
7. Li, T., et al.: Federated learning: challenges, methods, and future directions. IEEE Signal Process. Mag. **37**(3), 50–60 (2020)
8. Kairouz, P., et al.: Advances and open problems in federated learning. Found. Trends® Mach. Learn. **14**(1-2), 1–210 (2021)
9. Wahab, O.A., et al.: Federated machine learning: survey, multi-level classification, desirable criteria and future directions in communication and networking systems. IEEE Commun. Surv. Tutor. **23**(2), 1342–1397 (2021)
10. Tan, A.Z., et al.: Towards personalized federated learning. IEEE Trans. Neural Netw. Learn. Syst. **34**, 9587–9603 (2022)
11. Abdelmoniem, A.M., et al.: A comprehensive empirical study of heterogeneity in federated learning. IEEE Internet Things J. **10**, 14071–14083 (2023)
12. Gao, D., Yao, X., Yang, Q.: A survey on heterogeneous federated learning. arXiv preprint arXiv:2210.04505 (2022)
13. Ye, M., et al.: Heterogeneous Federated Learning: State-of-the-art and Research Challenges. arXiv preprint arXiv:2307.10616 (2023)
14. Zhang, J., et al.: Federated learning with label distribution skew via logits calibration. In: International Conference on Machine Learning. PMLR (2022)
15. Luo, Z., et al.: Disentangled federated learning for tackling attributes skew via invariant aggregation and diversity transferring. arXiv preprint arXiv:2206.06818 (2022)
16. Zhu, H., et al.: Federated learning on non-IID data: a survey. Neurocomputing **465**, 371–390 (2021)
17. Yoon, T., et al.: Fedmix: Approximation of mixup under mean augmented federated learning. arXiv preprint arXiv:2107.00233 (2021)
18. Duan, M., et al.: Astraea: self-balancing federated learning for improving classification accuracy of mobile deep learning applications. In: 2019 IEEE 37th International Conference on Computer Design (ICCD). IEEE (2019)
19. Fallah, A., Mokhtari, A., Ozdaglar, A.: Personalized federated learning with theoretical guarantees: a model-agnostic meta-learning approach. Adv. Neural. Inf. Process. Syst. **33**, 3557–3568 (2020)
20. Wang, L., Huang, C., Han, X.: Vertical federated knowledge transfer via representation distillation. In: FL-IJCAI Workshop (2022)
21. Le, H.Q., et al.: Layer-wise Knowledge Distillation for Cross-Device Federated Learning. In: 2023 International Conference on Information Networking (ICOIN). IEEE (2023)
22. Yu, T., Bagdasaryan, E., Shmatikov, V.: Salvaging federated learning by local adaptation. arXiv preprint arXiv:2002.04758 (2020)
23. Hou, C., et al.: FeDChain: Chained algorithms for near-optimal communication cost in federated learning. arXiv preprint arXiv:2108.06869 (2021)
24. Luping, W., Wei, W., Bo, L.: CMFL: mitigating communication overhead for federated learning. In: 2019 IEEE 39th International Conference on Distributed Computing Systems (ICDCS). IEEE (2019)
25. Li, X., et al.: A unified federated DNNs framework for heterogeneous mobile devices. IEEE Internet Things J. **9**(3), 1737–1748 (2021)
26. Luo, J., et al.: Fedskel: efficient federated learning on heterogeneous systems with skeleton gradients update. In: Proceedings of the 30th ACM International Conference on Information & Knowledge Management (2021)
27. Yao, X., et al.: Federated learning with additional mechanisms on clients to reduce communication costs. arXiv preprint arXiv:1908.05891 (2019)

28. Zaccone, R., et al.: Speeding up heterogeneous federated learning with sequentially trained superclients. In: 2022 26th International Conference on Pattern Recognition (ICPR). IEEE (2022)
29. Gao, Z., et al.: FedSeC: a robust differential private federated learning framework in heterogeneous networks. In: 2022 IEEE Wireless Communications and Networking Conference (WCNC). IEEE (2022)
30. Ma, Q., et al.: FedSA: a semi-asynchronous federated learning mechanism in heterogeneous edge computing. IEEE J. Sel. Areas Commun. **39**(12), 3654–3672 (2021)
31. Chan, Y.H., Edith, C.H.N.: Fedhe: heterogeneous models and communication-efficient federated learning. In: 2021 17th International Conference on Mobility, Sensing and Networking (MSN). IEEE (2021)
32. Abdellatif, A.A., et al.: Communication-efficient hierarchical federated learning for IoT heterogeneous systems with imbalanced data. Future Gener. Comput. Syst. **128**, 406–419 (2022)
33. Li, L., et al.: FedSAE: a novel self-adaptive federated learning framework in heterogeneous systems. In: 2021 International Joint Conference on Neural Networks (IJCNN). IEEE (2021)
34. Wang, H., et al.: Optimizing federated learning on non-IID data with reinforcement learning. In: IEEE INFOCOM 2020-IEEE Conference on Computer Communications. IEEE (2020)
35. Wang, D., et al.: CFL-HC: a coded federated learning framework for heterogeneous computing scenarios. In: 2021 IEEE Global Communications Conference (GLOBECOM). IEEE (2021)
36. Elkordy, A.R., Salman Avestimehr, A.: Heterosag: secure aggregation with heterogeneous quantization in federated learning. IEEE Trans. Commun. **70**(4), 2372–2386 (2022)
37. Li, Y., et al.: FedH2L: Federated learning with model and statistical heterogeneity. arXiv preprint arXiv:2101.11296 (2021)
38. Takahashi, H., Liu, J., Liu, Y.: Breaching FedMD: image recovery via paired-logits inversion attack. In: Proceedings of the IEEE/CVF Conference on Computer Vision and Pattern Recognition (2023)
39. Liu, Y., et al.: A secure federated learning framework for 5G networks. IEEE Wirel. Commun. **27**(4), 24–31 (2020)
40. Ding, J., et al.: Differentially private and communication efficient collaborative learning. In: Proceedings of the AAAI Conference on Artificial Intelligence, vol. 35. No. 8. (2021)
41. Zeng, H., et al.: FedCAV: contribution-aware model aggregation on distributed heterogeneous data in federated learning. In: Proceedings of the 50th International Conference on Parallel Processing (2021)
42. Bibikar, S., et al.: Federated dynamic sparse training: computing less, communicating less, yet learning better. In: Proceedings of the AAAI Conference on Artificial Intelligence, vol. 36. No. 6 (2022)
43. Hardt, M., Price, E., Srebro, N.: Equality of opportunity in supervised learning. In: Advances in Neural Information Processing Systems, vol. 29 (2016)
44. Lyu, L., et al.: Towards fair and privacy-preserving federated deep models. IEEE Trans. Parallel Distrib. Syst. **31**(11), 2524–2541 (2020)
45. Gálvez, B.R., et al.: Enforcing fairness in private federated learning via the modified method of differential multipliers. In: NeurIPS 2021 Workshop Privacy in Machine Learning (2021)
46. Sun, L., Lyu, L.: Federated model distillation with noise-free differential privacy. arXiv preprint arXiv:2009.05537 (2020)

47. Bonawitz, K., et al.: Practical secure aggregation for privacy-preserving machine learning. In: Proceedings of the 2017 ACM SIGSAC Conference on Computer and Communications Security (2017)
48. Chai, D., et al.: FedEval: A Holistic Evaluation Framework for Federated Learning. arXiv preprint arXiv:2011.09655 (2020)

From Passive Defense to Proactive Defence: Strategies and Technologies

Chong Shi[1], Jiahao Peng[1], Shuying Zhu[1], and Xiaojun Ren[1,2](✉)

[1] Instituite of Artificial Intelligence, Guangzhou University,
510006 Guangzhou, China
shichong@e.gzhu.edu.cn
[2] Guangdong Provincial Key Laboratory of Blockchain Security,
510006 Guangzhou, China
renxiaojun@gzhu.edu.cn

Abstract. The goal of network defense mechanisms is to enable systems to actively detect and withstand attacks, reduce reliance on external security measures, and quickly recover and repair. This paper elaborates on relevant works from both passive defense and proactive defense perspectives. Our first contribution is to introduce strategies and technologies related to passive defense, discussing in detail access control strategies, identity authentication technologies, and firewall technologies. These technologies play a significant role in protecting computer systems and networks from unauthorized access and malicious activities. Addressing the limitations of passive defense, such as: difficult to resolve uncertainty attacks and passive self-defense, our second contribution is to introduce strategies and technologies related to proactive defense. Firstly, we provide a comparative introduction to moving target strategies, intrusion tolerance strategies, and mimic defense strategies. Secondly, based on the mimic defense strategy, we provide a detailed introduction to mimic routers and mimic server technologies, which simulate normal network traffic and service behavior to enhance system security. Moreover, we provide future prospects and suggest potential directions. These approaches can help protect computer systems and networks from various security threats and provide valuable insights for researchers and security professionals on how to address evolving threats.

Keywords: Cybersecurity · Passive Defense · Proactive defense

1 Introduction

Cyberspace security is the comprehensive concept of protecting network systems, data, and online information, encompassing information security and network security. With the increasing popularity of the Internet, cyberspace has become the fifth domain [1] after land, sea, air, and space, holding significant strategic importance for maintaining national security and driving economic development. The rapid advancement of technology poses numerous challenges to cyberspace

J. Vaidya et al. (Eds.): AIS&P 2023, LNCS 14509, pp. 190–205, 2024.
https://doi.org/10.1007/978-981-99-9785-5_14

security, such as hacker attacks, virus intrusions, and phishing attempts, which threaten the security of individuals, businesses, and nations. The main issues in network security are concentrated in the following aspects:

- **Software and hardware backdoors.** Malicious code left in network information systems allows hackers to bypass security controls and gain access to the information system. Unfortunately, backdoor issues persist in the era of globalized supply chains.
- **Unknown security vulnerabilities.** Current network security defense technologies cannot address vulnerabilities or backdoors caused by information system software and hardware design flaws. This problem is one of the main targets for future resolution.
- **Low cost of attack.** Any group or organization capable of discovering and exploiting vulnerabilities or backdoors can easily disrupt and undermine the principles of space network security, significantly increasing the cost and consequences of network security defense.

In response to the issues above, our first contribution is to introduce the traditional strategies and techniques (Passive Defense) for network security defense. Traditional network security defense is primarily achieved through the comprehensive application of various security measures. Among them, traditional network security defense strategies, mainly employing identity authentication, access control, data encryption, and firewalls, play a memorable role in network security, laying the foundation for defense.

The essence of a passive defense system lies in achieving effective and precise defense by pre-gathering threat characteristic information such as attack sources and attack patterns. However, standard passive defense security techniques may have the following drawbacks:

- **Uncertainty attacks are difficult to address.** Existing security defense technologies worldwide have difficulty ensuring the trustworthiness of encryption and authentication devices.
- **Passive self-defense.** Traditional passive defense theories, relying on prior knowledge, cannot effectively mitigate the threats of dynamic network attacks. They can only dynamically search for system vulnerabilities or flaws and then improve through passive defense techniques such as antivirus measures and patching.
- **Singular defense system architecture.** The transparent architecture of network information systems makes it challenging to close the security chain, resulting in a singular defense system architecture.

Tradition network security defense measures are technologically passive. Thus, the future trend of network security defense will shift from passive defense technologies towards active defense. our second contribution is introducing the existing proactive defense techniques. Proactive defense empowers cybersecurity personnel with proactive control over system defense, reversing the traditional reactive approach. Shifting from "passive" to "active", it establishes an asymmetric strategic defense advantage and propels network security protection into a more advanced stage. It represents the future direction of development for network security defense technologies.

2   Passive Defense

Network security static defense employs a variety of strategies and technologies to protect networks from security threats. These strategies include access control, network segmentation, patch management, configuration hardening, and secure coding practices, aiming to restrict access to network resources, reduce the attack surface, address known vulnerabilities, and develop secure code. In terms of technologies, firewalls are used to filter unauthorized network traffic, IDS/IPS systems monitor and prevent malicious activities and attacks, secure protocols and data encryption ensure secure data transmission, and SIEM integrates logs and event data for real-time monitoring and threat detection. By applying these strategies and technologies in a comprehensive manner, organizations can establish a robust network security static defense framework, enhance network security, and effectively mitigate various threats.

2.1   Access Control

The goal of access control is to minimize the security risks of unauthorized access to physical or logical systems. It utilizes access control policies, and authentication technology to protect data confidentiality and security. Traditional access control models include Discretionary Access Control [17–21], Mandatory Access Control [22,23], Role-Based Access Control [24–28], and Attribute-Based Access Control models [29–33]. Discretionary Access Control (DAC) is an access control model that is based on the ownership of resources by subjects to control access permissions. In this model, users can freely determine the access permissions for other users to their resources. This model offers high flexibility but also carries the risk of permission misuse, requiring system administrators and users to negotiate and manage permissions themselves. cc In this model, security policies are defined by system administrators or policymakers and are applied to all subjects and resources. This model provides higher security but lower flexibility, making it less adaptable to dynamic environments. Role-Based Access Control (RBAC) is an access control model that is based on user roles and responsibilities to control access permissions for resources. In this model, access permissions are assigned to roles, and users obtain corresponding permissions through the roles assigned to them. This model simplifies permission management and maintenance, improving system scalability and manageability. Attribute-Based Access Control (ABAC) is an access control model that is based on attributes (such as user attributes, resource attributes, environmental attributes, etc.) to control access permissions. In this model, access control rules are defined based on attribute values and conditions using a policy language. This model has flexible access control rules, allowing access control decisions based on dynamic attributes. As Table 1, we discuss the differences of the aforementioned policies.

Table 1. Comparison of different types of access control policies

Type	DAC	RBAC	ABAC	MAC
Features	Autonomous	Role-based	Attribute-based	Label-based
Focus	Object's c permission list	Subject's permission list	Attribution of suject, sobject, and request	Labels of subject, object and request
Use Cases	User-controlled permissions	Administtrator controlled permissions	Scenarios where role definition is unclear	Scenarios where labeling of all data is possible and request
Example	Consumer-facing where users have control over the their own content	Internal company systems where administrators design roles and assign users to their roles	Network reguests when there are multiple subjects and objects involved and it is challenging to clearly define roles	Government systems where each piece if data and every individual has a specpfic level of confidentiality classfication

2.2 Identity Authentication Technology

Identity authentication technologies are used to verify the identity of users
or entities, ensuring that only authorized individuals can access systems or
resources. These technologies play a crucial role in safeguarding system and
data security by confirming the legitimacy of users and preventing unauthorized access. The most common identity authentication technology is password
authentication, where users verify their legitimacy and validity through a user-
name/password combination. Two-factor authentication technology [54,55] adds
an authentication token beyond a static password. Users need to authenticate
both static and dynamic passwords during the login process to confirm their
identity. This approach offers higher security and dynamism. With technological advancements, biometric-based identity authentication technology [56–60]
has emerged. It utilizes unique biological features of individuals to authenticate their identities, extracting physiological characteristics or specific behaviors
as verification methods. It combines digital identity with a person's real identity. Currently, widely used biometric authentication methods include fingerprint
recognition [61,62], iris recognition [63,64], and behavioral recognition [65]. However, compared to password authentication, where passwords are static and susceptible to interception by Trojan programs in computer memory or network
monitoring devices, biometric authentication also has lower security. Additionally, the accuracy and stability of biometric recognition are greatly influenced by
environmental conditions, especially for injured or ill users. Furthermore, biometric authentication systems tend to have higher costs and are suitable for
scenarios with high-security requirements. Moreover, digital certificates [66–68]
are also an essential means of user identity authentication. Digital certificates
serve as proof of user identity and contain identity-related data, similar to physical identification cards in real life. Digital certificate authentication requires

the support of a trusted third-party certification authority (CA) responsible for issuing digital certificates and ensuring the authority and authenticity of user identity authentication.

Moreover, data encryption techniques have become indispensable with the advancement of identity authentication technologies. As shown in Fig. 1, Data Encryption Technology transforms information into meaningless ciphertext using encryption keys and encryption algorithms, and the recipient can decrypt the ciphertext back into plaintext using decryption keys and decryption algorithms. Therefore, data encryption technology effectively prevents the leakage of computer system information. Traditional data encryption technologies include symmetric encryption [34–38], asymmetric encryption [39–42], and transparent encryption [43,44]. Symmetric encryption uses the same key for encryption and decryption, providing fast speed but complex key management. Asymmetric encryption uses a pair of keys, with the public key used for encryption and the private key for decryption, offering high security but slower speed. Transparent encryption automatically encrypts and decrypts data during transmission or storage, providing a seamless and convenient experience for users, but it may impact system performance to some extent. Today, symmetric encryption algorithms [45,46] and triple Data Encryption Algorithm (3DES) [47,48] are widely used, primarily due to their superior security levels and computational speeds. Additionally, there are asymmetric encryption algorithms such as RSA [49], ElGamal [50], and elliptic curve algorithms [51]. Among them, the RSA algorithm has a higher security level. It provides solid guarantees for the security and integrity of data information transmission, making it the most commonly used asymmetric encryption algorithm. Finally, there are two common types of transparent encryption techniques: hook-based transparent encryption (application-layer transparent encryption) [52] and driver-layer transparent encryption [53].

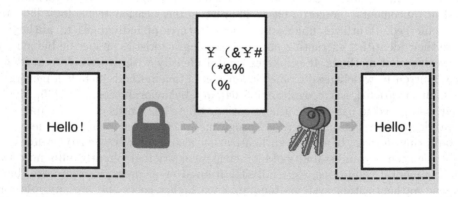

Fig. 1. Data Encryption Technology

2.3   Firewall Technology

A firewall is a traffic control device that controls secure access at different network boundaries. It is a system that exists between networks and executes security policies. The firewall is a protective gateway between the enterprise intranet and the public network. It can be categorized into hardware and software, with the main principle being the filtration and blocking traffic according to predefined security policies. Packet filtering, application proxy, and stateful inspection are three main types of firewall technologies. Packet filtering firewall [2–7], the earliest firewall technology, blocks or allows packets based on packet header information and filtering rules. Application proxy firewall [8–10] examines the application data of packets and maintains a complete connection state. It can analyze complete command sets of different protocols and allow or block specific commands based on security rules. It also has features like URL filtering, data modification, identity authentication, and logging. Intrusion state detection firewall [11–16], also known as an adaptive firewall or dynamic packet filtering firewall, which generates filtering rules dynamically based on past communication information and other application-derived state information. It filters new communication based on the newly generated filtering rules. Traditional firewall technologies have the advantages of high security, complete user and application authentication, and effective isolation of direct communication between internal and external networks, but they cannot detect unknown attacks. Based on protocol analysis, a stateful inspection firewall reads packets at the network layer, filters abnormal connections based on existing filtering rules (or attacks signature libraries based on packet headers), generates corresponding alerts and logs, and enhances security, achieving a balance between performance and security. As Table 2, we compare and analyze the differences among these three technologies.

Table 2. Comparison of different firewall technologies.

Type	Packet Filter	Application Proxy	Dynamic Packet Filter
OSI Layer	Transport Layer	Application Layer	Transport Layer
Strength	1. High Performance 2. User-friendly 3. Wide applicability	1. In-depth inspection 2. Provide access control and authentication 3. Hide internel network structure	1. Combining the advantages of the previous 2. Flexible configuration
Weakness	1. Limited application layer control 2. Prone to deception 3. Difficult to handle complex protoclos and applications	1. Performance degradation 2. Complex configuration 3. Limited flexibility	1. Complex configuration 2. Requires real-time updates and adjustments

3   Proactive Defence

Traditional security defense strategies tend to be passive and static, relying on prior knowledge of known attacks and needing more ability to address new types of network attacks and threats, resulting in poor generalization. However, with the advancement of attack techniques, network attacks have become automated, intelligent, intense, diverse, and highly covert. Additionally, the extreme asymmetry in the current network's attack and defense situation makes it difficult for traditional network security defense techniques to effectively respond to complex and diverse network attacks and threats, let alone attacks based on unknown exploitable vulnerabilities and backdoors.

Proactive defense primarily involves real-time monitoring of the entire computer system, enabling quick detection of changes in network traffic, analysis of program behavior, and prohibition of any suspicious activities to protect computer system security. At the same time, proactive defense systems also collect information on how suspicious behaviors connect to the computer (potential intrusion behavior) and other helpful information (knowledge about the intruders). Users can utilize this information to employ "counterattack" measures, making it difficult for intruders to carry out their attacks. The main strategies include: Moving Target Defense (MTD) [70], Intrusion Tolerant System (ITS) [71] and Mimic Defense (MD) [69]. We compares and analyzes the differences among these strategies as Table 3.

Table 3. Comparison of three proactive defense techniques.

Strategy	IST			MTD			MD	
Architecture	SITAR	MAFTIA	ITUA	SCIT	MAS	TALENT	Mimic Router	MimicServer
Diversity	✓		✓	✓	✓	✓	✓	✓
Redundancy	✓	✓		✓			✓	✓
Voting	✓	✓			✓		✓	✓
Migration			✓	✓	✓	✓		✓
Focus	Mitigate attack impact			Increace attack difficulty			1.Disrupt attack chains 2.Increace the difficulty of sustained attack	
Strength	1.Implement node-protection 2. High security			1.Achieve stealeffect 2.High confidentiality			1.High cost-effectiveness 2. High availability, integrity and confidentiality	
Weakness	High cost			1.Limited pattern diversity 2.Poor stability			Requires futher maturation	

MTD is a network security strategy and technology designed to increase the difficulty for attackers and reduce the success rate of attacks. The core idea of MTD is to constantly change the system's attack surface, making it challenging for attackers to discover and exploit vulnerabilities in the system. In addition, the three significant architectures based on intrusion tolerance technology, namely SalableIntrusion-tolerant Architecture for Distributed Services (SITAR) [72] and, Malicious-and AccidentalFault Tolerance for Internet Applications (MAFTIA) [73], and Intrusion Tolerance by Unpredictable Adaptation (ITUA) [74] have attracted significant attention from researchers.

ITS is a computer system designed to maintain regular operation in the face of persistent and sophisticated attacks. It achieves high system availability and security by incorporating fault-tolerant techniques and multiple layers of defense. Specifically, the Self-Cleansing Intrusion Tolerance (SCIT) [75] architecture applied to DNS and Web servers, the Moving Attack Surface (MAS) [76] architecture based on different software-hardware combinations of virtual server stacks, and the Trusted Dynamic Logical Heterogeneity System (TALENT) [77] architecture applied to kernel-level, operating layer, and hardware layer have all received significant attention from researchers.

MD is a computer security strategy that protects systems from malicious activities by imitating attacker behavior, hiding system features, and expanding the attack surface. Its core idea is to make the system unpredictable, making it difficult for attackers to identify and exploit vulnerabilities, thus enhancing system security and resilience. We primarily focus on discussing the concepts of Mimic Routers and Mimic Server.

3.1 Mimic Defense Strategy

MD proposed by academician Wu Jiangxing, aims to dynamically and pseudo-randomly select and execute various software and hardware variants under active and passive triggering conditions. This approach aims to create a hardware execution environment and software behavior that internal and external attackers find unobservable or difficult to construct an attack chain based on vulnerabilities or backdoors. The goal is to reduce system security risks. Both MTD and MD share the concept of dynamic, diverse, and random proactive defense. MD inherits MTD's dynamic, proactive defense concept while enriching its content. It achieves an inherent proactive defense capability by applying heterogeneous redundant architecture and mimic computing. The theoretical framework of mimic defense consists of three main aspects: dynamic, heterogeneous, and redundant (DHR). The typical architecture [79] is illustrated in Fig. 2. The heterogeneous execution units are composed of heterogeneous elements with the same functionality but different internal structures from heterogeneous pool, serving as the first step in achieving dynamic scheduling and policy distribution. The execution unit refers to the computing systems responsible for task execution. The policy distribution module dispatches tasks to the online execution unit set, while the policy arbitration module evaluates the processing results of the execution unit set and provides feedback to the policy scheduling module. The output module handles the arbitration results and performs the output.

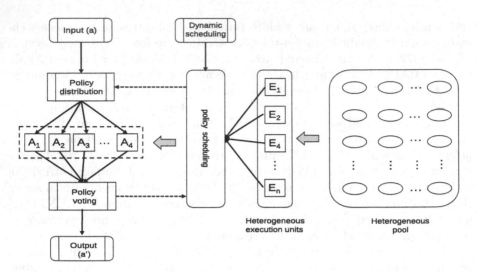

Fig. 2. DHR Architecture

The mimic defense model achieves its effectiveness by dynamically selecting a set of execution bodies and adaptively changing the system composition, rendering the attack information obtained by attackers ineffective and making it difficult to sustain or reproduce the same attack. The mimic defense model has been continuously developed in recent years with a more enriched and robust model system.

3.2 Mimic Router Technology

Mimic router technology [78] combines network routing with mimic techniques to dynamically rewrite and forward network traffic, achieving obfuscation and concealment of the network traffic. It can simulate multiple virtual network nodes and perform deceptive routing operations between nodes, making it difficult for attackers to obtain accurate network topology and target node information. Mimic router technology prevents attackers from accurately tracking and identifying real targets, enhancing network security and privacy protection capabilities. The critical elements of mimic router technology are as follows:

- **Virtual node simulation:** Mimic router technology utilizes virtual node simulation to emulate multiple fake network nodes. The presence of these virtual nodes makes it challenging for attackers to determine the real target nodes, increasing the difficulty of attacks.
- **Deceptive routing operations:** Mimic router technology performs deceptive routing operations between virtual nodes. This means that data packets transmitted in the network may follow a series of fake routing paths, making it difficult for attackers to determine the real data flow and target nodes.

- **Traffic obfuscation and concealment:** Mimic router technology obfuscates and hides network traffic, making it difficult for attackers to distinguish between real and fake traffic. This can be achieved by modifying packet header information, encrypting data, or using randomization techniques.

Based on the above, mimic router technology effectively disrupts and confuses attackers' behavior, reduces the probability of successful attacks, and improves network security. It plays a crucial role in defending against network attacks, protecting network privacy, and countering network reconnaissance. Therefore, it has received extensive attention and application in the field of computer security.

3.3   Mimic Server Technology

The main idea of mimic server technology [79] is to create one or more virtual mimic servers that appear nearly identical to real servers from the outside, but are actually fake or protective. Mimic servers can employ various techniques and mechanisms to simulate the behavior of real servers, including fake service ports, forged operating system and application fingerprints, false log records, and more. By introducing mimic servers, attackers will face the following difficulties and challenges:

- **Confusing attack targets:** Mimic server technology makes it difficult for attackers to determine the location and characteristics of real servers. Attackers may be redirected to mimic servers, reducing the threat to real servers.
- **Increasing attack complexity:** Mimic servers can simulate various responses and behaviors, making it challenging for attackers to differentiate between real and fake servers. Attackers need to invest more time and resources in analyzing, testing, and attacking mimic servers, thereby increasing the complexity and cost of attacks.
- **Providing early threat detection:** Mimic servers can monitor and record the behavior of attackers. By analyzing the attack activities on mimic servers, potential threats can be detected and identified early, allowing for appropriate defensive measures.

Mimic server technology has wide-ranging applications in various scenarios, including network security defense, intrusion detection and response, penetration testing, and more. It can enhance system security and privacy protection capabilities, helping organizations effectively address various network attacks and threats.

4   Outlook

Proactive defense in cyberspace empowers cybersecurity personnel with the initiative in system defense, effectively reversing the passive defense dilemma of traditional networks, transforming from "passive" to "active", and establishing an asymmetrical strategic defensive advantage. This advancement propels network security protection into a more advanced stage and represents the future direction of security technologies. However, in comparison to the increasingly complex methods of cyber attacks, further promotion of emerging security technologies, innovative defense concepts, and the enhancement of proactive, collaborative, and holistic network defense techniques are necessary. This will contribute to reducing the imbalance between network attacks and defenses, mitigating risks and losses caused by various types of cyber attacks, and ensuring the security of computer networks at all levels. However, as follows, there are still aspects worth exploring.

1) **Mlti-source Open Heterogeneous:** The mimic defense technology based on the DHR architecture faces challenges in achieving fine-grained heterogeneity across the entire software stack and underlying hardware. Moreover, due to the limited space of intrinsic security component mimic entropy, there is an urgent need to develop multi-source open heterogeneous implementation mode designs and novel secure software and hardware creations with native intrinsic security.

2) **Heterogeneity Gain and Execution Entity Synchronization:** On one hand, to enhance the security gain of system transformation and prevent multi-modal collaborative attacks based on redundant inter-component port links, shared spaces, and other channels or synchronization mechanisms, it is necessary for execution entities and components to achieve a high degree of heterogeneity. On the other hand, when synchronizing the working states and input-output vectors of heterogeneous execution entities, although relatively secure plugin modes or message queue modes can be used, the high heterogeneity of execution entities increases deployment difficulties and synchronization delays. The trade-off between the gain in heterogeneity and the associated costs constrains the improvement of network performance. Furthermore, there is a lack of mature quantifiable validation and measurement mechanisms, making it challenging to achieve a trusted balance condition.

3) **The Balance of Security and Functionality:** Firstly, the introduction of various security components for mimic defense technology will inevitably increase costs and pose challenges in the deployment of heterogeneous design for execution entities. Secondly, in order to enhance the credibility of scheduling and decision-making strategies, the complexity of scheduling and decision-making strategies is increasing. The current research on diverse dynamic indicators and complex algorithms needs to address issues related to costs, deployment difficulties, and network support in a timely manner.

4.1   Conclusion

The rapid development of the Internet poses new challenges to network security. In the face of the fundamentally insecure nature of the cyberspace, which is characterized by being "easy to attack and difficult to defend". This paper introduces the origin of network security imbalance and analyzes the deficiencies and limitations of existing passive defense security systems. Furthermore, it introduces the development framework of proactive defense systems, focusing on the essential core and technological features of mimic defense, and extracting a comparison between intrusion tolerance and moving target techniques. Lastly, two technical implementations of mimic defense strategy are proposed. Moreover, we propose the presentation of the future development vision for defense technologies. The aim is to promote further advancements in defense technologies and provide a globally trusted theoretical and technological foundation for achieving the "cybersecurity rebalancing strategy".

References

1. Mijwil, M., et al.: Cybersecurity challenges in smart cities: an overview and future prospects. Mesop. J. Cybersecur. **2022**, 1–4 (2022)
2. Sahana, Y.P., Gotkhindikar, A., Tiwari, S.K.: Survey on can-bus packet filtering firewall. In: 2022 International Conference on Edge Computing and Applications (ICECAA). IEEE (2022)
3. Sreelaja, N.K.: A fireworks-based approach for efficient packet filtering in firewall. In: Handbook of Research on Fireworks Algorithms and Swarm Intelligence. IGI Global, pp. 315–333 (2020)
4. Durante, L., Seno, L., Valenzano, A.: A formal model and technique to redistribute the packet filtering load in multiple firewall networks. IEEE Trans. Inf. Forensics Secur. **16**, 2637–2651 (2021)
5. Malikovich, K.M., Rajaboevich, G.S., Karamatovich, Y.B.: Method of constructing packet filtering rules. In: 2019 International Conference on Information Science and Communications Technologies (ICISCT). IEEE (2019)
6. Ari Muzakir, A.: Analisis Kinerja Packet Filtering Berbasis Mikrotik Routerboard Pada Sistem Keamanan Jaringan. Analisis Kinerja Packet Filtering Berbasis Mikrotik Routerboard pada Sistem Keamanan Jaringan (2022)
7. Liang, J., Kim, Y.: Evolution of firewalls: toward securer network using next generation firewall. In: 2022 IEEE 12th Annual Computing and Communication Workshop and Conference (CCWC). IEEE (2022)
8. Jingyao, S., Chandel, S., Yunnan, Yu., Jingji, Z., Zhipeng, Z.: Securing a network: how effective using firewalls and VPNs are? In: Arai, K., Bhatia, R. (eds.) FICC 2019. LNNS, vol. 70, pp. 1050–1068. Springer, Cham (2020). https://doi.org/10.1007/978-3-030-12385-7_71
9. Muzaki, R.A., et al.: Improving security of web-based application using ModSecurity and reverse proxy in web application firewall. In: 2020 International Workshop on Big Data and Information Security (IWBIS). IEEE (2020)
10. Yina, Q.: Discussion on computer network security technology and firewall technology. Int. J. New Dev. Eng. Soc. **6**(4), 1–5 (2022)
11. Amouei, M., Rezvani, M., Fateh, M.: RAT: reinforcement-learning-driven and adaptive testing for vulnerability discovery in web application firewalls. IEEE Trans. Dependable Secure Comput. **19**(5), 3371–3386 (2021)
12. Praise, J., Jeya, R., Raj, J.S., Bibal Benifa, J.V.: Development of reinforcement learning and pattern matching (RLPM) based firewall for secured cloud infrastructure. Wirel. Personal Commun. **115**, 993–1018 (2020)
13. Bagheri, S., Shameli-Sendi, A.: Dynamic firewall decomposition and composition in the cloud. IEEE Trans. Inf. Forensics Secur. **15**, 3526–3539 (2020)

14. Chebrolu, C.S., Chung-Horng, L., Ajila, S.A.: Dynamic packet filtering using machine learning. In: 2022 IEEE 23rd International Conference on Information Reuse and Integration for Data Science (IRI). IEEE (2022)
15. Kailanya, E., Mwadulo, M., Omamo, A.: Dynamic deep stateful firewall packet analysis model. Afr. J. Sci. Technol. Soc. Sci. 1(2), 116–123 (2022)
16. Malikovich, K.M., Rajaboevich, G.S., Karamatovich, Y.B.: Method of constructing packet filtering rules. In: 2019 International Conference on Information Science and Communications Technologies (ICISCT). IEEE (2019)
17. Sandhu, R., Munawer, Q.: How to do discretionary access control using roles. In: Proceedings of the Third ACM Workshop on Role-Based Access Control (1998)
18. Dranger, S., Sloan, R.H., Solworth, J.A.: The complexity of discretionary access control. In: Yoshiura, H., Sakurai, K., Rannenberg, K., Murayama, Y., Kawamura, S. (eds.) IWSEC 2006. LNCS, vol. 4266, pp. 405–420. Springer, Heidelberg (2006). https://doi.org/10.1007/11908739_29
19. Solworth, J.A., Sloan, R.H.: A layered design of discretionary access controls with decidable safety properties. In: Proceedings of IEEE Symposium on Security and Privacy, 2004. IEEE (2004)
20. Vijayalakshmi, K., Jayalakshmi, V.: A study on current research and challenges in attribute-based access control model. Intell. Data Commun. Technol. Internet Things Proc. ICICI **2022**, 17–31 (2021)
21. Aftab, M.U., et al.: Traditional and hybrid access control models: a detailed survey. Secur. Commun. Netw. **2022**, 1–5 (2022)
22. Gihleb, R., Giuntella, O., Zhang, N.: The effect of mandatory-access prescription drug monitoring programs on foster care admissions. J. Human Resourc. **57**(1), 217–240 (2022)
23. Namane, S., Dhaou, I.B.: Blockchain-based access control techniques for IoT applications. Electronics **11**(14), 2225 (2022)
24. Fragkos, G., Johnson, J., Tsiropoulou, E.E.: Dynamic role-based access control policy for smart grid applications: an offline deep reinforcement learning approach. IEEE Trans. Human-Mach. Syst. **52**(4), 761–773 (2022)
25. Ameer, S., Benson, J., Sandhu, R.: An attribute-based approach toward a secured smart-home IoT access control and a comparison with a role-based approach. Information **13**(2), 60 (2022)
26. Kormpakis, G., et al.: An advanced visualisation engine with role-based access control for building energy visual analytics. In: 2022 13th International Conference on Information, Intelligence, Systems Applications (IISA). IEEE (2022)
27. Ghazal, R., et al.: Intelligent role-based access control model and framework using semantic business roles in multi-domain environments. IEEE Access **8**, 12253–12267 (2020)
28. Alshammari, S.T., Albeshri, A., Alsubhi, K.: Integrating a high-reliability multicriteria trust evaluation model with task role-based access control for cloud services. Symmetry **13**(3), 492 (2021)
29. Ding, S., et al.: A novel attribute-based access control scheme using blockchain for IoT. IEEE Access **7**, 38431–38441 (2019)
30. Bhatt, S., et al.: Attribute-based access control for AWS internet of things and secure industries of the future. IEEE Access **9**, 107200–107223 (2021)
31. Aghili, S.F., et al.: MLS-ABAC: efficient multi-level security attribute-based access control scheme. Future Gener. Comput. Syst. **131**, 75–90 (2022)
32. Guo, H., Meamari, E., Shen, C.-C.: Multi-authority attribute-based access control with smart contract. In: Proceedings of the 2019 International Conference on Blockchain Technology (2019)

33. Zhong, H., et al.: An efficient and outsourcing-supported attribute-based access control scheme for edge-enabled smart healthcare. Future Gener. Comput. Syst. **115**, 486–496 (2021)
34. Alenezi, M.N., Alabdulrazzaq, H., Mohammad, N.Q.: Symmetric encryption algorithms: review and evaluation study. Int. J. Commun. Netw. Inf. Secur. **12**(2), 256–272 (2020)
35. He, K., et al.: Secure dynamic searchable symmetric encryption with constant client storage cost. IEEE Trans. Inf. Forensics Secur. **16**, 1538–1549 (2020)
36. Li, J., et al.: Searchable symmetric encryption with forward search privacy. IEEE Trans. Dependable Secure Comput. **18**(1), 460–474 (2019)
37. Patranabis, S., Mukhopadhyay, D.: Forward and backward private conjunctive searchable symmetric encryption. Cryptology ePrint Archive (2020)
38. Gui, Z., Paterson, K.G., Patranabis, S.: Rethinking searchable symmetric encryption. In: 2023 IEEE Symposium on Security and Privacy (SP). IEEE (2023)
39. Zhang, Q.: An overview and analysis of hybrid encryption: the combination of symmetric encryption and asymmetric encryption. In: 2021 2nd International Conference on Computing and Data Science (CDS). IEEE (2021)
40. Sharifovich, A.S., Maxmudovich, H.X., Mansurovich, B.M.: Protocol for electronic digital signature of asymmetric encryption algorithm, based on asymmetric encryption algorithm based on the complexity of prime decomposition of a sufficiently large natural number. Texas J. Multidiscip. Stud. **7**, 238–241 (2022)
41. Verma, G., et al.: An optical asymmetric encryption scheme with biometric keys. Optics Lasers Eng. **116**, 32–40 (2019)
42. Bao, Z., Xue, R., Jin, Y.: Image scrambling adversarial autoencoder based on the asymmetric encryption. Multimed. Tools App. **80**(18), 28265–28301 (2021)
43. Hu, Z., et al.: Reversible 3D optical data storage and information encryption in photo-modulated transparent glass medium. Light Sci. App. **10**(1), 140 (2021)
44. Jiang, F., et al.: Research on the application of transparent encryption in distributed file system HDFS. In: 2020 19th International Symposium on Distributed Computing and Applications for Business Engineering and Science (DCABES). IEEE (2020)
45. Su, N., Zhang, Y., Li, M.: Research on data encryption standard based on AES algorithm in internet of things environment. In: 2019 IEEE 3rd Information Technology, Networking, Electronic and Automation Control Conference (ITNEC). IEEE (2019)
46. Yazdeen, A.A., et al.: FPGA implementations for data encryption and decryption via concurrent and parallel computation: a review. Qubahan Acad. J. **1**(2), 8–16 (2021)
47. Ramachandra, M.N., et al.: An efficient and secure big data storage in cloud environment by using triple data encryption standard. Big Data Cogn. Comput. **6**(4), 101 (2022)
48. Akande, O.N., Abikoye, O.C., Kayode, A.A., Aro, O.T., Ogundokun, O.R.: A dynamic round triple data encryption standard cryptographic technique for data security. In: Gervasi, O., et al. (eds.) ICCSA 2020. LNCS, vol. 12254, pp. 487–499. Springer, Cham (2020). https://doi.org/10.1007/978-3-030-58817-5_36
49. Rivest, R., et al.: A method for obtaining digital signatures and public-key cryptosystems. Commun. ACM **21**(2), 120–126 (1978)
50. Elgamal, T.: A public key cryptosystem and a signature scheme based on discrete logarithms. IEEE Trans. Inf. Theory **31**(4), 469–472 (1985)
51. Ye, G., Liu, M., Mingfa, W.: Double image encryption algorithm based on compressive sensing and elliptic curve. Alex. Eng. J. **61**(9), 6785–6795 (2022)

52. Cui, H., et al.: TraceDroid: A Robust Network Traffic Analysis Framework for Privacy Leakage in Android Apps. In: Su, C., Sakurai, K., Liu, F. (eds.) Science of Cyber Security. SciSec 2022. LNCS, vol. 13580, pp. 541–556. Springer, Cham (2022). https://doi.org/10.1007/978-3-031-17551-0_35

53. Singh, S.K., Yi, P., Park, J.H.: Blockchain-enabled secure framework for energy-efficient smart parking in sustainable city environment. Sustainable Cities Soc. **76**, 103364 (2022)

54. Kaur, S., Kaur, G., Shabaz, M.: A secure two-factor authentication framework in cloud computing. Secur. Commun. Netw. **2022**, 1–9 (2022)

55. Watters, P., et al.: This would work perfectly if it weren't for all the humans: two factor authentication in late modern societies. First Monday (2019)

56. Palma, D., Montessoro, P.L.: Biometric-based human recognition systems: an overview. In: Recent Advances Biometrics, pp. 1–21 (2022)

57. Singh, V., Kant, C.: Biometric-based authentication in Internet of Things (IoT): a review. Adv. Inf. Commun. Technol. Comput. Proc. AICTC **2022**, 309–317 (2021)

58. Bera, B., et al.: On the design of biometric-based identity authentication protocol in smart city environment. Pattern Recogn. Lett. **138**, 439–446 (2020)

59. Gupta, S., Buriro, A., Crispo, B.: DriverAuth: a risk-based multi-modal biometric-based driver authentication scheme for ride-sharing platforms. Comput. Secur. **83**, 122–139 (2019)

60. Sengupta, S.: A secured biometric-based authentication scheme in IoT-based patient monitoring system. In: Mandal, J.K., Bhattacharya, D. (eds.) Emerging Technology in Modelling and Graphics. AISC, vol. 937, pp. 501–518. Springer, Singapore (2020). https://doi.org/10.1007/978-981-13-7403-6_44

61. Priesnitz, J., et al.: An overview of touchless 2D fingerprint recognition. EURASIP J. Image Video Process. **2021**(1), 1–28 (2021)

62. Rajasekar, V., et al.: Enhanced multimodal biometric recognition approach for smart cities based on an optimized fuzzy genetic algorithm. Sci. Rep. **12**(1), 622 (2022)

63. Boyd, A., et al.: Post-mortem iris recognition-a survey and assessment of the state of the art. IEEE Access **8**, 136570–136593 (2020)

64. Wang, C., et al.: Towards complete and accurate iris segmentation using deep multi-task attention network for non-cooperative iris recognition. IEEE Trans. Inf. Forensics Secur. **15**, 2944–2959 (2020)

65. Dargan, S., Kumar, M.: A comprehensive survey on the biometric recognition systems based on physiological and behavioral modalities. Expert Syst. Appl. **143**, 113114 (2020)

66. Capece, G., Ghiron, N.L., Pasquale, F.: Blockchain technology: redefining trust for digital certificates. Sustainability **12**(21), 8952 (2020)

67. Rahardja, U., et al.: Immutable ubiquitous digital certificate authentication using blockchain protocol. J. Appl. Res. Technol. **19**(4), 308–321 (2021)

68. Maulani, G., et al.: Digital certificate authority with blockchain cybersecurity in education. Int. J. Cyber IT Serv. Manage. **1**(1), 136–150 (2021)

69. Hu, H., et al.: Mimic defense: a designed-in cybersecurity defense framework. IET Inf. Secur. **12**(3), 226–237 (2018)

70. Zhuang, R., et al.: A theory of cyber attacks: a step towards analyzing MTD systems. In: Proceedings of the Second ACM Workshop on Moving Target Defense (2015)

71. Reynolds, J., et al.: The design and implementation of an intrusion tolerant system. In: Proceedings International Conference on Dependable Systems and Networks. IEEE (2002)

72. Wang, F., et al.: SITAR: a scalable intrusion-tolerant architecture for distributed services. In: Workshop on Information Assurance and Security, vol. 1 (2003)

73. Cachin, C., et al.: Malicious-and Accidental-Fault Tolerance in Internet Applications: reference model and use cases (2000)

74. Pal, P., et al.: Intrusion tolerance by unpredictable adaptation (ITUA). Technical report. AFRL-IF-RS-TR-2005-119 (2005)

75. Bangalore, A.K., Sood, A.K.: Securing web servers using self cleansing intrusion tolerance (SCIT). In: 2009 Second International Conference on Dependability. IEEE (2009)

76. Huang, Y., Anup K. Ghosh. "Introducing diversity and uncertainty to create moving attack surfaces for web services. In: Moving Target Defense: Creating Asymmetric Uncertainty for Cyber Threats, pp. 131–151.Springer, New York, NY (2011)

77. Okhravi, H., et al.: Creating a cyber moving target for critical infrastructure applications using platform diversity. Int. J. Critical Infrastruct. Protect. 5(1), 30–39 (2012)

78. Li, X., et al.: A router abnormal traffic detection strategy based on active defense. In: Journal of Physics: Conference Series. Vol. 1738. No. 1. IOP Publishing (2021)

79. Tong, Q., et al.: Design and implementation of mimic defense Web server. J. Softw. 28(4), 883–897 (2017)

Research on Surface Defect Detection System of Chip Inductors Based on Machine Vision

Xiao Li[1,2] ⓘ, Xunxun Pi[3] ⓘ, Hong Tang[1,2] ⓘ, and Junhang Qiu[1,2(✉)] ⓘ

[1] School of Information Engineering, East China Jiao Tong University, Nanchang 33000, China
2020031010000221@ecjtu.edu.cn
[2] Jiangxi Artificial Intelligence Production-Education Integration Innovation Center, Nanchang, China
[3] School of Information Engineering, Jiangxi University of Science and Technology, No.115 Yaohu University Park, Nanchang, Jiangxi, China

Abstract. Artificial inspection of surface defects in chip inductors faces issues such as low efficiency and poor accuracy. To enhance production efficiency, increase intelligence, and reduce production costs, this paper proposes the use of machine vision technology for chip inductor surface defect detection. Specifically, the paper builds upon the DETR model, improving its feature extraction network and attention mechanism. The approach involves transferring the pre-trained detection model to generalize it for chip inductor surface defect datasets. Consequently, the improved DETR model is applied to chip inductor surface defect detection. Experimental results demonstrate that the enhanced DETR model successfully detects chip inductor surface defects, improving the feature extraction and object localization capabilities of the network while reducing training time. The application of machine vision in chip inductor surface defect detection enhances efficiency, addresses the issue of lengthy DETR model training and poor small object detection performance, achieves classification and localization of chip inductor surface defects, and validates the feasibility of the detection method.

Keywords: Machine Vision · Surface Defect Detection · Chip Inductor Surface Defect Dataset · DETR Model

1 Image Preprocessing and Dataset Creation

To achieve surface defect detection in chip inductors, it is essential to train the detection model to learn defect characteristics. Therefore, the creation of a defect dataset is a crucial step in achieving successful detection.

Supported by organization x.

1.1 Analysis of Surface Defects in Chip Inductors

Introduction to the Inspection Subject. Chip inductor components are electronic components primarily used for circuit functions such as filtering, isolation, and matching. Due to their characteristics of minia-turization, high quality, high energy storage, and low resistance, they are widely employed in various electronic products. The chip inductors used for detection in this paper are wire-wound chip inductors made from insulated conductors, collectively referred to as chip inductors, as shown in Fig. 1 below.

Fig. 1. Chip Inductive Element

Defect Analysis. During the production process of chip inductors, every factor may cause defects on the surface of chip inductors. These defects not only affect the appearance of the product, but also affect the performance of chip inductors. These four defects are mainly detected, as shown in Fig. 2 below.

1.2 Image Preprocessing

During the process of image capture and transmission, there are various sources of interference that affect the quality of chip inductor component images, rendering them unsuitable for direct use in image detection. Further preprocessing

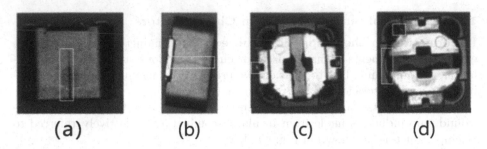

(a)   (b)   (c)   (d)

Fig. 2. (a) Damaged Magnetic Rings, (b) Concealed Cracks in Magnetic Rings, (c) Exposed Copper on Electrodes, (d) Exposed Wire on Electrodes

of the images is required. Image preprocessing serves to enhance image quality, eliminate interference, and, in cases of limited data, employ image enhancement techniques to augment the dataset.

Distortion Correction. Initially, this paper selected a 9×9 calibration board with an accuracy of 0.01 mm and captured images of the calibration board from six different positions, as shown in Fig. 3.

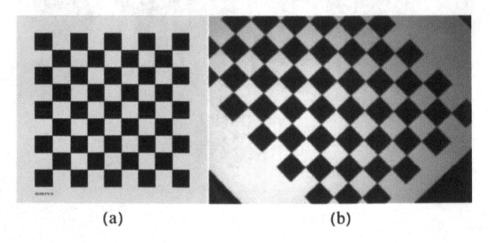

(a)   (b)

Fig. 3. Chip Inductive Element

After performing distortion correction on the images, the resulting capture quality is shown in the figure below (Fig. 4):

(a)   (b)

Fig. 4. Effect Chart of Distortion Correction (a) Before Correction (b) After Correction

Image Denoising. Gaussian Filter

After Gaussian filtering, the effect on the chip inductor component image is shown in Fig. 5, where (a) is the original image, and (b) is the filtered image.

Median Filter

In this paper, we used [specific filter name] as the filter to process the patch inductor component images, and the filtering results are shown in Fig. 6. Figure 6(a) represents the original image of the chip inductor component captured, while Fig. 7(b) shows the chip inductor component image after undergoing median filtering.

Image Enhancement. Data augmentation methods [1] refer to a set of transformation techniques applied to original data to generate new training data, thereby increasing the dataset's size, diversity, and generalization capability. In this project, a total of 1280 × 1024 pixel color images were captured for chip inductor components. The dataset includes images of good-quality components as well as those with dark cracks and broken magnetic cores. To address the issue of limited sample data collected in this study, the decision was made to employ data augmentation techniques to supplement the dataset and enhance the performance of the neural network model. Common data augmentation methods include random cropping, flipping, rotation, scaling, translation, deformation, arbitrary adjustments to image brightness, contrast, saturation, and adding

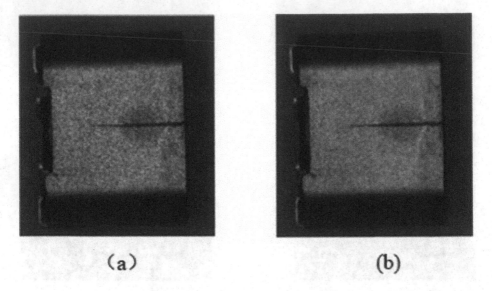

Fig. 5. Effect of Gaussian Filtering

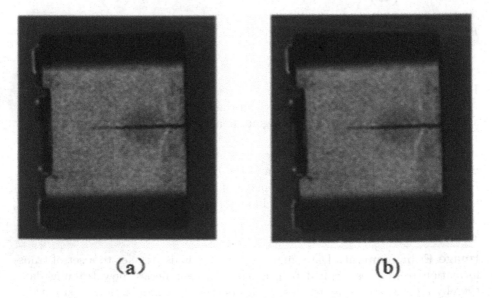

Fig. 6. Effect Chart of Median Filtering

noise, among others. These data augmentation methods can be used individually or in combination to generate a more diverse and robust training dataset.

1.3   Creation of Chip Inductor Surface Defect Dataset

Data Cleaning. Data cleaning refers to the process of handling and transforming data to eliminate redundant information, correct erroneous data, deal with missing values, and make the data more suitable for analysis and utilization. The following are the steps involved in cleaning the collected patch inductor image data: removing duplicate data, handling missing values, correcting erroneous data, and formatting image data.

Data Labeling. The format of the dataset in this project follows the format of the commonly used MS COCO [2] dataset in object detection, making it convenient for pre-training of neural network models. For the surface defect detection of chip inductors studied in this paper, the dataset is created following the format of the MS COCO dataset.

In this defect detection task, data annotation was carried out using the open-source software called labelme [3]. Upon completion of the annotation, annotation files quired for network training are generated. The labelme software interface and the structure of annotation files are illustrated in Fig. 8.

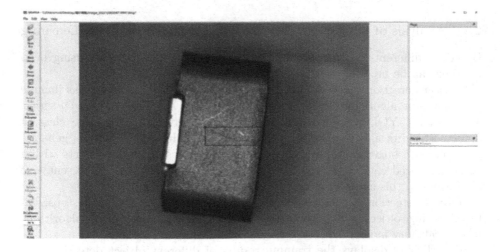

Fig. 7. Annotation Assistant lebelme Annotation Interface

The Division of the Detection Dataset. TA complete dataset consists of three parts: the training set, validation set, and test set. Splitting the dataset is a crucial step as it helps us evaluate the model's performance and prevents overfitting. Generally, the majority of the data in the dataset is used for training the model, a smaller portion is used to validate whether the model has learned successfully, and another portion is used to test the model's generalization ability. In this detection project, 618 images of chip inductors were captured, and after data augmentation, the dataset was expanded to 970 images. The dataset was divided into training, testing, and validation sets in an 8:1:1 ratio, with the data in the validation and test sets being mutually exclusive from the training set. The specific division is shown in Table 1 below.

Table 1. Dataset Split.

Surface Mount Inductor Dataset	Training Set	Validation Set	Test Set
970	776	97	97

2   The Relevant Design and Detection of Surface Defects in SMT Inductor Inspection Algorithm

To achieve the detection of surface defects on chip inductors, the core component of the image processing system, the defect detection algorithm, is designed. In cases with limited data, an improved object detection model is generalized to the chip inductor surface defect detection project using transfer learning, thereby enabling the detection of surface defects on chip inductors.

2.1   Analysis of Object Detection Methods Based on Deep Learning

Based on different learning frameworks, object detection methods using deep learning can be broadly categorized into two types:

Object detection methods based on Convolutional Neural Networks include representative algorithms such as the YOLO (You Only Look Once) [4] series, YOLOv4 [5], YOLOv7 [6], etc. These methods directly regress the class of detected objects and predict bounding boxes. Another type involves generating candidate boxes containing objects and then classifying the objects within these generated candidate boxes and regressing their positions. Representative algorithms in this category include Faster R-CNN [7], among others.

Object detection methods based on Transformers follow three main steps: feature learning, object estimation, and label matching. Representative algorithms in this category include DETR [8], VIT FRCNN [9], and others.

The Table 2 displays the training results of different object detection network models on the COCO 2017 dataset [10] and our chip inductor surface defect dataset. In the table, APs represents the average precision for small object detection, APM represents the average precision for medium object detection, and APL represents the average precision for large object detection. Analysis of the table reveals that, on average precision, object detection methods based on Convolutional Neural Networks tend to have higher AP values overall. However, both types of methods show less-than-ideal results in detecting small objects.

Table 2. Performance Comparison of Different Object Detection Network Models

Object detection model	Dataset	AP/%	APs/%	APM/%	APL/%
Faster RCNN	COCO 2017	40.2	24.2	43.5	52
		37.8	21.7	39.3	46.8
YOLOv4	COCO 2017	43.5	26.7	46.7	57.3
		37.7	20.5	41.6	52.7
YOLOv7	COCO 2017	55.9	31.8	55.5	65
		49.6	25.7	50.6	56.7
VIT FRCNN	COCO 2017	37.8	17.8	41.4	57.3
		34.1	12.3	37.7	54.8
DETR	COCO 2017	44.9	23.7	49.5	64.1
		41.7	21.8	46.7	60.3

For medium and large object detection, the difference in APM and APL values between the two methods is not significant. Despite being slightly less performant than object detection methods based on Convolutional Neural Networks, Transformer-based object detection algorithms have a relatively short research history in the field. They hold significant research value and room for improvement. Hence, in this paper, we choose to build upon the DETR network model and improve the DETR algorithm for chip inductor surface defect detection.

2.2 Improved DETR Algorithm

Improvement of the Feature Extraction Network. The improvement of the feature extraction network in the DETR model relies on convolutional neural networks (CNNs). When CNNs extract image features, their convolutional kernels have a fixed size. When using the same convolutional kernel for convolution operations, feature extraction can only occur within a fixed region of the image. This means that the receptive field of the convolutional kernel remains fixed and cannot adapt to changes in the shape of the detected objects, resulting in a limited ability to extract image features. To address this issue, this paper introduces deformable convolutional neural networks (DCNNs) to enhance the model's feature extraction capability.

Improvement of the Attention Mechanism. Traditional self-attention mechanisms spend a significant amount of time locating object position information within redundant background information in images. To address this issue, a Spatial Prior module is introduced on top of the self-attention mechanism, allowing the attention mechanism to select crucial information from a vast amount of data quickly and efficiently locate object positions. This aims to reduce the model's training time. The computational principle of the improved self-attention mechanism is illustrated in Fig. 10.

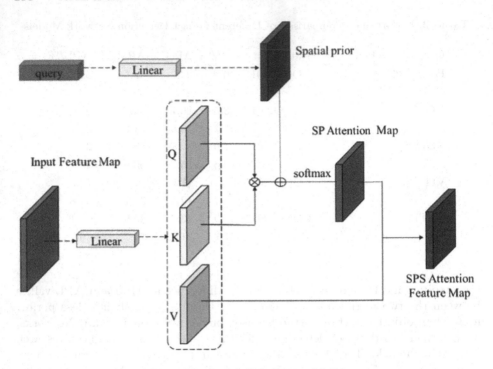

Fig. 8. Spatial Self Attention Model

The computational formula for the spatial self-attention model is as shown in Equation (6). Compared to the calculation formula of the self-attention mechanism, it incorporates spatial weights for the spatial prior. The spatial weights are computed using a two-dimensional Gaussian function, as illustrated in Equation (7).

$$SPSattention = \text{softmax}(\frac{K_i^T Q}{\sqrt{d}} + \log G)V_i \tag{1}$$

$$G(i,j) = e^{(-\frac{(i-x)^2}{\varphi w^2} - \frac{(j-y)^2}{\varphi h^2})} \tag{2}$$

In Equation (7), where x and y respectively represent the center coordinates of the predicted bounding boxes in the query, w and h represent the width and height of the predicted boxes, and are used to adjust the bandwidth of the Gaussian distribution. As depicted in Fig. 10, the input to the spatial prior module is the query values after linear transformation. The spatial weights, denoted as G, are computed using a Gaussian function. These spatial weights are then added to the result of the dot product between Q and K. Afterward, a softmax operation is applied to normalize the result, yielding the SP Attention Map. Finally, this map is multiplied by to obtain the spatial self-attention feature map. The principle of the spatial prior module is to adjust the search range for each object query vector in the self-attention mechanism to be around the object's center and the region near the predicted bounding box. It disregards low-importance areas that are far from the object center, thereby accelerating the convergence of DETR.

The Enhanced DETR Network Architecture. The improved DETR network structure retains the overall framework of DETR and is primarily composed of the following components, as shown in Fig. 11 and (Fig. 9).

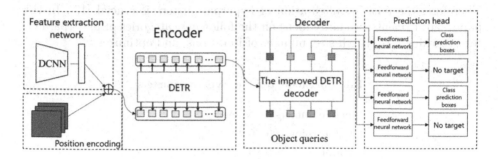

Fig. 0. Improved DETR network structure

2.3   Surface Defect Detection of Chip Inductors Based on Improved DETR Network

Setting up the Testing Environment. The selection of hardware and software for the image processing system is crucial for establishing the operational environment of the detection algorithm. As the original DTER algorithm was developed in the Ubuntu operating system using the Python language, we opted to install the Ubuntu operating system on the computer to facilitate later training and testing of the detection algorithm. Additionally, we selected PyCharm as the software environment.

PyCharm serves as a Python IDE and offers excellent support for Python language development. Its robust interactive environment, along with features such as project management, autocompletion, code modification, syntax highlighting, code navigation, error display, version control, and more, significantly enhance development efficiency. This allows us to create project environments that match the detection algorithm's requirements effectively.

Detection Model Training. In the field of object detection, when dealing with cross-category detection, many object detection methods often employ transfer learning [11]. This approach reduces the algorithm's reliance on large datasets and significantly enhances the accuracy of object detection and recognition when the model is generalized to the classes it needs to detect. Due to the limited dataset for detecting surface defects on patch inductors, in order to improve the accuracy of the improved DETR model for object detection and recognition, it was decided to use the transfer learning approach. This involved pre-training the improved DETR algorithm on the COCO 2017 dataset and then generalizing it to the dataset for surface defects on patch inductors to train the improved DTER model.

During the training process, it is typically assessed by monitoring changes in loss and accuracy to evaluate the current training status of the model. Adjustments to hyperparameters are made promptly to achieve better training results. Therefore, the initial configuration of hyperparameters is crucial [12]. In this paper, adjustments primarily focus on the following categories of hyperparameters: learning rate, batch size, number of iterations, and optimizer.

Table 3. Model training hyperparameters

Hyperparameters	Numerical
Initial learning rate	0.01
Batch Size	3
Iteration count	300
Optimizer	SGD with momentum

To observe the training performance of the improved DETR model on the surface defect dataset of patch inductors before and after the enhancement, a line chart was used to visualize the changes in the loss function during the training process, as shown in Fig. 12 The horizontal axis in the chart represents epochs, while the vertical axis represents loss. It can be observed that the improved DETR model significantly reduced the training time compared to DETR. In the case of the improved DETR model, the loss on the training set gradually decreases around the 50th epoch and then quickly converges. In contrast, the DETR model's loss only begins to converge around the 250th epoch.

To assess the recall of the DETR model before and after the improvement, a line chart was utilized to visualize the changes in recall during the training process, as depicted in Fig. 13. From the graph, it is evident that the improved DETR model exhibited an enhanced recall compared to the original DETR model. The recall increased from approximately 0.69 to around 0.8, and the improved DETR model also converged faster than DETR.

To assess the accuracy of the DETR model before and after the improvement, a line chart was used to visualize the changes in accuracy during the training process, as shown in Fig. 14. From the graph, it is evident that the improved DETR model achieved an increased accuracy, rising from around 0.67 to approximately 0.8. Based on the training results mentioned above, the improved DETR model significantly reduced training time and improved both recall and accuracy in the detection of surface defects on patch inductors. This improvement is likely attributed to the deformable feature extraction network and spatial self-attention mechanism of the improved DETR model, which are advantageous for obtaining better feature perception and object localization capabilities during the training phase, thus surpassing the training performance of the original DETR model.

Detection Results and Performance Analysis. Figure 15 displays the PR performance curves of the improved DETR model for various defect categories, showing the Precision and Recall values at different confidence thresholds for each defect category. When the confidence threshold is close to 0, all annotated boxes are detected, resulting in high Recall and low Precision, corresponding to the lower-right corner of the PR curve. Conversely, when the confidence threshold is close to 1, most predicted boxes include the target, leading to high Precision and low Recall, corresponding to the upper-left corner of the PR curve. The area enclosed by each curve represents the Average Precision (AP) for that category. Since the PR performance curves traverse various confidence thresholds, AP is a performance evaluation metric that is independent of confidence.

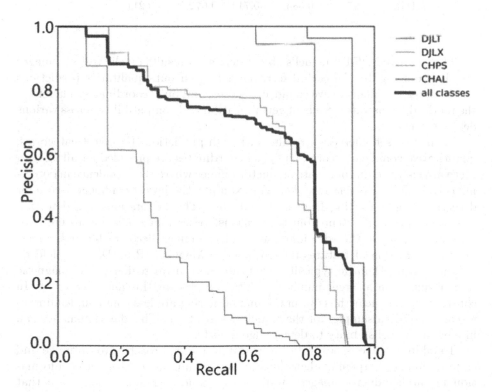

Fig. 10. PR performance curve of improved DETR model for each defect category

From Fig. 15, it can be observed that when the threshold is set to 0.5, the Average Precision (AP) for the detection of all defect categories reaches 0.75.

Table 4 presents the performance evaluation metrics for the model on the surface defect dataset of patch inductors, including Electrode Exposed Copper (DJLT), Magnetic Ring Cracks (CHAL), Exposed Wire of Electrode (DJLX), and Magnetic Ring Damage (CHPS). These correspond to the areas under the PR performance curves for each class in the figure. Specifically, Magnetic Ring Cracks achieve an AP of 0.931, Electrode Exposed Copper has an AP of 0.804, Exposed Wire of Electrode has an AP of 0.672, and Magnetic Ring Damage has an AP of 0.297.

Table 4. Performance Evaluation Metrics for the Model on the Surface Defect Dataset of Patch Inductors

Category	Images	precision	Recall	mAP@0.5	mAP@0.5:0.95
CHPS	97	0.589	0.228	0.297	0.116
CHAL	97	0.918	0.929	0.931	0.513
DJLT	97	0.831	0.81	0.804	0.392
DJLX	97	0.686	0.714	0.672	0.211

The improved DETR model's object detection results on the test set images are shown in Fig. 16. For each defect, the model outputs qualitative predictions for the category, quantitative confidence scores, and pixel coordinates. Therefore, the model demonstrates excellent generalization and compatibility across various defect categories.

In the test set's detection results, although predictions for defect categories, quantitative confidence scores, and pixel coordinates are provided for all images, there are instances in some patch inductor images where the confidence in detecting certain defects is relatively low. For example, the lowest confidence score for detecting Magnetic Ring Damage is only 0.3. The factors affecting detection confidence primarily stem from two reasons. After image filtering in inductor component images, the edge information of the image becomes blurred, affecting the extraction of features related to edge Magnetic Ring Damage defects. Magnetic Ring Damage typically occurs in the central region of the magnetic ring, resulting in a larger number of data samples for this defect category. In contrast, defects near the edge of the magnetic ring are less common, leading to a scarcity of data samples for the model to learn from. This data imbalance can hinder the model's ability to detect edge defects.

To address these issues in future research, one approach is to optimize and enhance image sharpening algorithms to better emphasize edge defect information in patch inductor images. Additionally, collecting more image data that includes instances of edge Magnetic Ring Damage will make it easier for the model to learn the features of such defects.

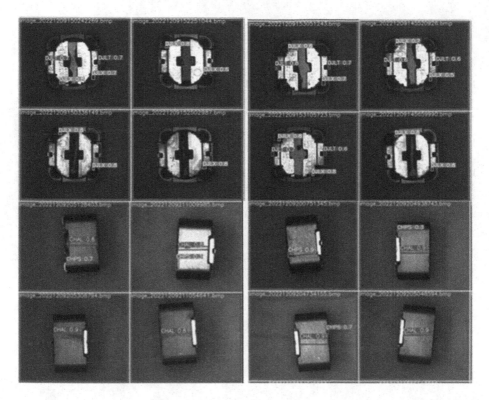

Fig. 11. Improved DETR Model for Test Set Image Target Detection Results

References

1. Aaij, R., et al.: Test of lepton universality with b 0→ k* 0 + - decays. J. High Energy Phys. **2017**(8), 1–31 (2017)
2. Beal, J., Kim, E., Tzeng, E., Park, D.H., Zhai, A., Kislyuk, D.: Toward transformer-based object detection (2020). arXiv preprint arXiv:2012.09958
3. Beal, J., Kim, E., Tzeng, E., Park, D.H., Zhai, A., Kislyuk, D.: Toward transformer-based object detection (2020). arXiv preprint arXiv:2012.09958
4. Chai, M., Xia, F., Hao, S., Peng, D., Cui, C., Liu, W.: PV power prediction based on LSTM with adaptive hyperparameter adjustment. IEEE Access **7**, 115473–115486 (2019)
5. Mahto, P., Garg, P., Seth, P., Panda, J.: Refining YOLOv4 for vehicle detection. Int. J. Adv. Res. Eng. Technology (IJARET) **11**(5), 409–419 (2020)
6. Parekh, Z., Baldridge, J., Cer, D., Waters, A., Yang, Y.: Crisscrossed captions: Extended intramodal and intermodal semantic similarity judgments for MS-COCO (2020). arXiv preprint arXiv:2004.15020
7. Redmon, J., Divvala, S., Girshick, R., Farhadi, A.: You only look once: unified, real-time object detection. In: Proceedings of the IEEE Conference on Computer Vision and Pattern Recognition, pp. 779–788 (2016)

8. Ren, S., He, K., Girshick, R., Sun, J.: Faster R-CNN: towards real-time object detection with region proposal networks. In: Advances in Neural Information Processing Systems. vol. 28 (2015)
9. Wada, K., et al.: Labelme: Image polygonal annotation with python (2016)
10. Wang, C.Y., Bochkovskiy, A., Liao, H.Y.M.: YOLOv7: trainable bag-of-freebies sets new state-of-the-art for real time object detectors. In: Proceedings of the IEEE/CVF Conference on Computer Vision and Pattern Recognition, pp. 7464–7475 (2023)
11. Weiss, K., Khoshgoftaar, T.M., Wang, D.: A survey of transfer learning. J. Big Data **3**(1), 1–40 (2016)
12. Yan, C., Gong, B., Wei, Y., Gao, Y.: Deep multi-view enhancement hashing for image retrieval. IEEE Trans. Pattern Anal. Mach. Intell. **43**(4), 1445–1451 (2020)

Multimodal Fatigue Detection in Drivers via Physiological and Visual Signals

Weijia Li[1,2], Xunxun Pi[3], Hong Tang[1,2], and Junhang Qiu[1,2]($\boxtimes$)

[1] School of Information Engineering, East China Jiao Tong University, Nanchang 33000, China
2020031010000221@ecjtu.edu.cn
[2] Jiangxi Artificial Intelligence Production-Education Integration Innovation Center, Nanchang, China
[3] School of Information Engineering, Jiangxi University of Science and Technology, No. 115, Yaohu University Park, Nanchang, Jiangxi, China

Abstract. Driving conditions such as stress, sleepiness and fatigue can easily lead to traffic accidents, and the detection of these unsafe conditions is an important means of ensuring driving safety. The fatigue detection system currently on the market suffers from a single detection feature and low detection accuracy. In view of the powerful feature extraction and fusion capabilities of neural networks, a multimodal driver fatigue detection method based on physiological and visual signals is proposed for processing physiological signals from wearable devices and visual signals from cameras during the driving process. Firstly, the model processes the physiological signal sequences and locates significant changes in physiological state, and extracts visual features based on visual signals, mainly eye features and complementary mouth and head features. Secondly, the fatigue features are obtained by fusing the physiological and visual features in the temporal dimension, based on which an effective fatigue detection model can be trained. We use publicly available datasets to segment physiological signal sequences such as heart rate and quantitatively capture change points. After testing in different visual environments such as day and night and with or without face occlusion, this model can meet the requirements of basic real-time fatigue detection with a high detection accuracy.

Keywords: Fatigue Detection · Multimodal Fusion · Time-Series Segmentation

1 Related Work on Fatigue Testing

The automotive industry and researchers have developed various protocols to perceive and detect the state of humans while driving. It can be mainly divided into methods based on facial features and methods based on human physiological

---

Supported by Jiangxi Province 03 Special Project (No. 20203ABC03W07).

J. Vaidya et al. (Eds.): AIS&P 2023, LNCS 14509, pp. 221–236, 2024.
https://doi.org/10.1007/978-981-99-9785-5_16

features. The method based on facial features has low accuracy detection, low cost, and is easy to achieve, with a low probability of affecting driver operations. The method based on human physiological characteristics has high detection accuracy, but it is costly and difficult to popularize. This section mainly introduces the relevant methods and research status of fatigue testing.

1.1   Based on Facial Features

The existing work on fatigue, drowsiness, or stress detection mainly focuses on the occurrence of extreme fatigue and obvious signs, such as yawning, drowsiness, and prolonged eye closure [3]. For example, Zhang et al. [4] used convolutional neural networks (CNN) to detect yawning by using features from the nose region rather than the mouth region caused by the vehicle driver turning their head. Ouyang et al. [23] extracted ROI based on facial feature points of drivers, migrated deep networks that performed well in other computer vision tasks to fatigue detection tasks, and combined LSTM's ability to process temporal data for fatigue detection. Wang Hongjun et al. [24] used a multi-threaded optimized Dlib (Image Processing Open Source Library) to locate and track the driver's face. The facial key point detector in the Dlib open source library was used to extract key feature points of the driver's face, and the aspect ratio and mouth aspect ratio of the driver's eyes were calculated in real-time. At the same time, four indicators were calculated: blink frequency, closure frequency, percentage of eye closure time, and yawn frequency, Using mathematical methods for real-time fusion, driver fatigue status is graded based on the values of fusion indicators.

However, for the driving operator, this obvious sign may not appear until a moment before the accident occurs. Therefore, it is necessary to detect fatigue as early as possible and provide drivers with more time to make appropriate responses. Therefore, in our literature, we utilize temporal information to detect unsafe states as soon as possible.

1.2   Based on the Characteristics of Human Physiological Parameters

The first issue that needs to be addressed in fatigue detection methods based on human physiological parameter characteristics is the issue of experimental data. In most cases, experimental data is collected by recruiting subjects for driving. Firstly, in the real world, some pre-defined routes [6] can be used, or driving simulator can be used to configure driving settings on the driving simulator [7] and collect information based on the subject's driving situation. In most cases, we not only need to pay attention to the driver's physiological signals, but also need to pay attention to the vehicle's status. Because vehicle usage data, such as pedal and steering wheel usage, or meta parameters such as vehicle speed and acceleration, can provide information about the driver's state and behavior [8]. However, for wake-up estimation tasks, the measurement of physiological signals remains the most important feature. And it is usually necessary to record multiple signals simultaneously, as a single measurement may result in significant differences between subjects.

Some widely used sensing data methods include electrocardiogram (ECG), electromyography (EMG), blood pressure, respiratory rate, electrodermal activity (EDA), and skin temperature. The heart rate (HR) and heart rate variability (HRV) extracted from electrocardiogram data are commonly used analytical data. There is already literature proving that in stress events, heart rate increases [9] and heart rate variability decreases [10]. HRV can be extracted in the time or frequency domain using various linear or nonlinear methods [11]. Respiratory rate is another indicator of unsafe driving conditions, such as significant changes in respiratory rate under stress or fatigue. In reference [12], it can be seen that changes in respiratory rate are related to increased pressure. Electromyography and temperature measurements are more susceptible to noise and artifacts [13], therefore their applicability is limited. Meanwhile, EDA is considered one of the most representative measurement data as it is related to the autonomic nervous system activity that causes physiological arousal [14]. EDA can capture in vitro conduction changes caused by sweating. Due to many factors that affect sweat gland activity, such as environmental temperature, different treatment methods are usually used. The commonly used method is to extract two signal components corresponding to low-frequency trends and high-frequency oscillations: skin conductivity level (SCL) and skin conductivity response (SCR). In our study, we followed reference [15] and regarded EDA as the reference ("gold") standard for driver wake-up status.

2   Introduction of Our Method

Given the powerful feature extraction and fusion capabilities of neural networks, this paper proposes a multimodal driver fatigue detection method based on physiological and visual signals, which is used to process physiological signals from wearable devices and visual signals from cameras during driving. This method consists of three parts, as shown in Fig. 1. Firstly, the model processes physiological signal sequences and locates significant changes in physiological states, extracting visual features based on visual signals, primarily eye features and supplemented by mouth and head features. Secondly, physiological and visual features are fused in the temporal dimension to obtain fatigue features. Based on this feature representation, an effective fatigue detection model can be trained. Finally, driver fatigue detection and discrimination are carried out based on the obtained fatigue characteristics.

For visual signals, we need to extract facial features. Obtain key eye, mouth, and head regions through facial recognition and key point localization of images, and extract features from each image to generate a time series. For physiological signals, high-frequency noise is removed from the physiological signal time series through a low-pass filter, and then the filtered time series is downsampled to ensure detection and analysis at a common sampling rate. Use Gaussian segmentation algorithm to segment time series data and apply time series clustering to the obtained segmentation. Then, physiological and visual features are fused in the time dimension to obtain fatigue features, and based on this feature representation, an effective fatigue detection model can be trained. Finally,

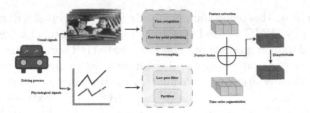

Fig. 1. Multimodal driver fatigue detection method based on physiological and visual signals.

driver fatigue detection and discrimination are carried out based on the obtained fatigue characteristics.

2.1 Extraction of Facial Features

For the processing of visual signals, we use a hybrid deep neural network architecture to extract facial features from images. As shown in Fig. 2, the architecture of the proposed hybrid deep neural network consists of three main modules: (1) face detector, (2) spatial feature extractor, and (3) temporal feature modeling. They are connected together through multiple learning networks. Firstly, the face detector uses a multi task cascaded convolutional neural network (MTCNN) [16] to allocate facial region bounding boxes and corresponding facial feature points in each frame of the video. Further extract the eye, mouth, and head regions of the facial region. Secondly, a customized mobile vision application efficient convolutional neural network (MobileNet) [17] is used as a spatial feature extractor to extract facial features from images of each frame. Finally, since fatigue features follow a pattern over time, an LSTM network is used to utilize the time patterns of a series of features within a specific time interval. The following sections provide a detailed introduction to each module.

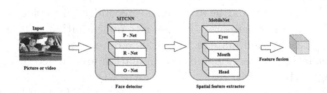

Fig. 2. Facial Feature Extraction.

2.2 Face Detector

Driver fatigue detection based on video images has many problems, as facial area detection and alignment are influenced by many factors such as lighting conditions, driver gestures, video resolution, facial angle, and occlusion. Therefore,

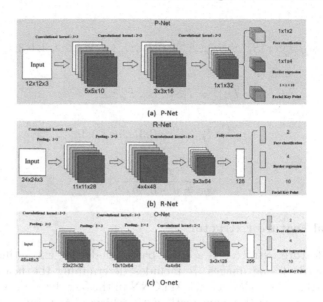

Fig. 3. Three architectures of MTCNN.

the design of facial detectors is crucial for achieving accurate facial region detection before facial feature extraction and fatigue detection. During driving, the driver may change their posture due to vehicle movement, extreme lighting or darkness inside the vehicle, and abnormal occlusion. Due to significant changes in the driver's posture and the fact that the camera often has the same viewing angle, extracting the specific positions of the mouth and eye areas becomes more difficult. The MTCNN proposed by Zhang et al. [16] is known as one of the fastest and most accurate facial detectors. To address the above issues, MTCNN uses several different stages for face detection and alignment tasks. As shown in Fig. 3, MTCNN consists of three network architectures (P-Net, R-Net, and O-Net) to obtain facial bounding boxes and feature points of three different scales. In MTCNN, h is the result parameter set and the input image, as shown in formula 1. Through three networks (P-Net, R-Net, and O-Net), the predicted facial bounding box positions are. In addition, the five coordinates including left eye, right eye, nose, left mouth corner, and right mouth corner are represented as lx0, ly0, lx1, ly1, lx2, ly2, lx3, ly3, lx4, and ly4, respectively. To obtain a more accurate image for subsequent cropping, the head area of the image is cropped according to formula 2 based on the position of the facial bounding box. Based on five coordinates, the eye and mouth regions are also cropped and extracted, using a 30% face bounding box size (w represents the width of the bounding box, and h represents the height of the bounding box) and the center coordinates, as shown in Eqs. 3 and 4.

$$h_{\theta MTCNN}(I) = [S_x; S_y; E_x; E_y; l_x0; l_x1; l_y1; l_x2; l_y2; l_x3; l_y3; l_x4; l_y5] \qquad (1)$$

$$I_{face} = I[S_x : S_y, E_x : E_y] \qquad (2)$$

$$I_{crop} = I_{face}[xc : 0.3w, yc : 0.3h] \qquad (3)$$

$$\begin{aligned} xc = x - \frac{0.3w}{2} \\ yc = y - 0.3h/2 \end{aligned} \qquad (4)$$

2.3 Spatial Feature Extractor

The spatial feature extractor is a CNN based model for extracting facial features from single frame images. It includes determining the head, eyes, and mouth as facial markers through the MTCNN in the face detector. In this study, MobileNet was used as the main method to achieve a fast and stable training process to generate feature extraction models. This model has achieved good performance in image recognition on various datasets. MobileNet and its variants were introduced as a solution primarily for speed optimization. Figure 4 shows the improved MobileNet architecture, which includes thirteen convolutional layers (grouped as Conv 1–13), five maximum pooling layers (Max Pool 1–5), one average pooling layer (Ave Pool), and one fully connected feedforward network layer (FC).

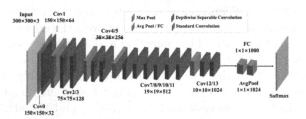

Fig. 4. Architecture of a CNN based feature extraction model

2.4 Time Series Feature Extraction

Although feature extractors can predict the fatigue level of each image frame based on spatial features, sometimes it is still difficult to distinguish slight dynamic changes with strong temporal dependencies, such as yawning and speaking. Therefore, it is beneficial to consider the time information in consecutive frames. To this end, deep LSTM [18] was applied to model temporal features.

LSTM is a special type of recurrent neural network (RNN) used to analyze hidden sequence patterns in temporal and spatial sequence data. It can learn long-term dependencies because it has a unique structure of input, output, and forgetting gates, which can control long-term sequence pattern recognition. The LSTM used in the proposed hybrid network aims to avoid long-term dependence by controlling the amount of information provided within each time frame through gate composition. The working principle of a door is to try to forget some unimportant information from the previous frame. At the same time, it also analyzes information within the current time frame and makes assumptions based on current information and previous important information.

2.5   Time Series Feature Extraction

Although feature extractors can predict the fatigue level of each image frame based on spatial features, sometimes it is still difficult to distinguish slight dynamic changes with strong temporal dependencies, such as yawning and speaking. Therefore, it is beneficial to consider the time information in consecutive frames. To this end, deep LSTM [18] was applied to model temporal features. LSTM is a special type of recurrent neural network (RNN) used to analyze hidden sequence patterns in temporal and spatial sequence data. It can learn long-term dependencies because it has a unique structure of input, output, and forgetting gates, which can control long-term sequence pattern recognition. The LSTM used in the proposed hybrid network aims to avoid long-term dependence by controlling the amount of information provided within each time frame through gate composition. The working principle of a door is to try to forget some unimportant information from the previous frame. At the same time, it also analyzes information within the current time frame and makes assumptions based on current information and previous important information (Fig. 5).

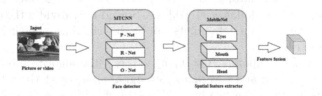

Fig. 5. Human Physiological Parameter Feature Extraction.

2.6   Data Preprocessing

For our research, we will limit the scope to HR and BR, as these are easily obtainable or extractable in all the aforementioned datasets and have previously been proven to be effective indicators of human stress. In the absence of respiratory measurements in the dataset, we continue to extract such information using the ECG derived respiratory algorithm proposed in [20]. Then, we use

[21] to extract the respiratory rate from the respiratory time series. Physiological indicators were initially sampled at various sampling rates, so we performed downsampling. Specifically, we first apply low-pass, third-order Butterworth filters to reduce high-frequency components, and then extract the filtered signal to a universal rate of 0.5 Hz. In this way, we can focus on significant changes in the signal while eliminating possible artifacts. In Table 1, we summarize the sampling parameters used in each dataset.

As for the basic facts, we have attempted three different methods to explain the variability of the provided data. Electrodermal activity (EDA) has been used as the gold standard for emotional state estimation, so we use it as a basic factual measure in all cases. We will also evaluate our subjective stress level methods when available. In order to improve the integrity of our work, we also consider considering reference standards by averaging the time series and subjective ratings of EDA as much as possible. This strategy was adopted in reference [34] as a method for determining subjective stress levels. All ground truth measures are further filtered to eliminate any rapid oscillations, as shown in Table 1.

Table 1. Physiological Signal Experimental Dataset Processing.

Dataset	Physiological signal	Driver	Distance length	Sampling rate
DriveDB ECG	EDA	24	>30 km	15.5 Hz
HCL Lab Driving	ECG, HR, EDA	10	24 km	1024 Hz
AffectiveROAD	HR, EDA	14	31 km	1–4 Hz

2.7   Gaussian Segmentation

To segment time series data, we used the greedy Gaussian segmentation (GGS) algorithm proposed by Hallac et al. [19]. GGS gradually divides the data stream into multiple parts, and its data points can be described as independent samples of Gaussian distribution.

Given a set of breakpoints $B = (b_1, b_2, \ldots, b_k)$, the algorithm only considers the distribution (mean and covariance) of signal changes at these breakpoints. More specifically, given the two breakpoints bi and bi+1 in B, GGS estimates the empirical covariance $S^{(i)}$ of that segment using the following formula:

$$S^{(i)} = \frac{1}{b_{i+1} - bi} \sum_{t=b_i}^{b_{i+1}} (x_t - \mu^{(i)})(x_t - \mu^{(i)})^T \tag{5}$$

where x_t is the data sample at the t_{th} time point, $\mu^{(i)}$ and $\sum(i)$ are the mean and covariance of the segments. Using the covariance calculated from the above equation, GGS tends to estimate B to maximize the likelihood as follows:

$$-\frac{1}{2}\sum_{i=1}^{K+1}[(b_{i+1}-bi)\log(S^{(i)}+\frac{\lambda}{b_{i+1}-bi}\mathbf{I})-\lambda\mathbf{Tr}(S^{(i)}+\frac{\lambda}{b_{i+1}-bi}\mathbf{I})^{-1}]\quad(6)$$

Here λ is a regularization term that sets the importance of covariance. The use of dynamic programming has solved the search and decision-making problem of multiple breakpoints. We intentionally chose GGS as the segmentation algorithm because it can effectively work in multimodal scenarios by considering the multivariate distribution shown in [22]. Our goal is to detect such breakpoints in physiological signals and evaluate their robustness in estimating corresponding changes in driver pressure levels.

2.8 Time Series Clustering

The main objective of this study is to detect the most prominent changes in stress levels, but not to identify any type of change in the data. To quantify this concept, we apply time clustering based on the breakpoints proposed by the GGS algorithm to the ground real time series. After running GGS, we treat each fragment as an independent time series sample and perform a simple time series k-means to cluster the fragments into k sets. Subsequently, we use the clustering results to discard all breakpoints located between segments in the same cluster. In Fig. 6, you can see the visualization of the result representation of the example EDA signal.

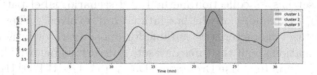

Fig. 6. Clustering example of EDA ground truth signals from the HCI dataset.

3 Experimental Analysis

3.1 Experimental Data

Due to privacy issues with physiological data, there are currently no publicly available datasets containing multiple physiological signals, nor are there datasets related to driving status. Therefore, in our study, we used three public datasets that best met our goals.

The visual signal data includes videos from two sources: the available public datasets NTHU-DDD, UTA-RLDD, and YawnDD, as well as videos captured by the authors interviewing expert operators while performing a crane operation simulation in the Unity3D game environment. Details of the videos are shown in Table 3. They were filmed in different scenarios, including working in front of a computer, simulated or real driving environments, and simulated crane operations. They have different facial features, behaviors, races, lighting conditions, acquisition scenes, and facial poses (different camera positions). Videos are also captured at different resolutions, e.g. 640 × 480, 1280 × 720, etc. (Table 2)

Table 2. Experimental dataset.

Dataset	Driver behavior	Lighting	Scenario introduction
NTHU-DDD	Quiet		Wear glasses during the day
	Yawn		No accessories during the day
	Nod	Day and night	Wear sunglasses
	Look to the side		No jewelry at night
	Look to the side		Wear glasses at night
YawnDD	Usual		Wear glasses
	Speak	Daytime	Wear sunglasses
	Yawn		No accessories
	Sing		Have a beard
UTA-RLDD	Alert	Morning	Wear glasses
	Low vigilant	Noon	Wear sunglasses
	Drowsy	Midnight	Have a beard

Each dataset has its collection method and scenario, label pattern, dataset size, and facial expressions of whether fatigue has "evolved". They are used to understand dataset characteristics suitable for crane operator fatigue detection.

More information about the three datasets is described below: The University of Texas at Arlington Real Life Drowsiness Dataset (UTA-RLDD) [1] was created for a multi-stage drowsiness detection task. The goals of the dataset focused on discriminants such as subtle micro-expressions in fatigue situations, not just extreme and easily observable expressions. 60 healthy participants recorded 30 h of RGB video in the dataset. By using the participants' cell phones or webcams, they recorded facial videos themselves in real life.

Therefore, it is expected to detect fatigue or drowsiness at an early stage, activating mechanisms to prevent drowsiness through these subtle conditions. Due to the physiology and instincts of the participants, it was difficult for them to pretend to be lethargic or tired by mimicking subtle micro-expressions.

The NTHU Driver Drowsiness Detection dataset (NTHU-DDD) [3] is a public dataset collected by the computer vision laboratory of National Tsinghua University, which contains 36 IR videos under various simulated driving scenarios. These scenes include normal driving, yawning, slow blinking, falling asleep, laughing, etc. The videos were shot under daytime and night lighting conditions. However, they are all themed around fake fatigue.

The Yawning Detection Dataset (YawDD) [4] was collected by the Distributed Collaborative Virtual Environment Research Lab (DISCOVER Lab) at the University of Ottawa. It contains two available sub-datasets: the first contains 322 RGB videos of normal facial expressions, and the second contains 29 RGB videos of drivers yawning. Both sub-datasets consist of bespectacled and non-bespectacled/sunglasses male and female drivers from different races. In addition, there are three different mouth situations in the dataset: (1) shut up and drive normally (no talking), (2) talk or sing while driving, and (3) yawn while driving. In other applications, it can also be used for yawning and fatigue detection, such as simulating communication between an operator and a rigging.

During the experiment, due to the long-term dependence on specific datasets, a problem arose that, within a few seconds, alert facial expressions on a series of frames, if just restored, would still be considered signs of drowsiness. Warning expressions after drowsiness. Furthermore, the level of detail of existing labels in this dataset does not allow for the identification of drowsiness states with high precision in the time dimension. Compared with other datasets, there is also no uniform evaluation criteria and labeling principles among them. To address these issues, the authors relabeled three available datasets, NTHU-DDD, UTA-RLDD, and YawnDD, in segmented units per frame and per minute. Those typical facial states or behaviors, such as eyes closed, yawning, head bowed, are still considered evidence for judging whether a frame contributes to fatigue awareness. To describe the transition states between alarm and fatigue, and to establish uniform evaluation criteria, we propose a relabeling workflow and uniform relabeling principles.

3.2 Data Preprocessing

For all videos, MTCNN is used to detect faces in all frames. The detected face bounding box with five landmark points is cropped together with the boundary pixels, and the cropped face area is adjusted to a fixed size of 64 × 64. Due to the high frame rate of the dataset (e.g., 30 fps or 15 fps), this study re-samples the video frames by a factor of 6 or 3 and inputs the face sequence to the proposed hybrid neural networks at a frame rate of 5 fps. The classification results (prediction level) can be upsampled back to the original video length. In addition, some videos in the dataset are grey release videos. Therefore, each frame should be replicated three times as a 3-channel image in order to generalize the proposed method when processing color or grey release inputs.

3.3 Training and Testing

In experiments, two classifiers of the proposed learning architecture, called Spatial Feature Extractor (MobileNet) and Temporal Feature Modeling (LSTM), are trained and evaluated respectively. The entire architecture is then tested by combining two trained classifiers. General training and testing results are as follows:

1) All videos in the public dataset are cropped into fixed-length video clips that can start from any frame of the original video. Sequential features calculated from the human eye, mouth, and head regions in one video clip are considered. From all available data obtained from the video clip, 70% are randomly selected to train the classifier, which is then evaluated using the other 30% of the remaining data.
2) For the spatial feature extractor (MobileNet), the eye, mouth and head regions detected from the customized MTCNN have three fatigue levels: alertness, low vigilance and fatigue for training and validation. Additionally, this custom MobileNet is used to extract sequence features for further training in the next step.
3) For temporal feature modeling (LSTM), since fatigue features follow a certain pattern over time, it is used to exploit a temporal pattern of a series of features over a specific time interval. It is also trained and evaluated by randomly selecting the same set of 70% and 30% data from all available datasets.
4) After the training of the two classifiers is completed, for the three datasets, select the data to test the trained model to integrate the two classifiers. Then evaluate all final performance based on real labels.

3.4 Evaluation Indicators

The performance of the proposed fatigue detection architecture on multiple datasets is quantitatively evaluated in terms of accuracy and loss. To achieve more detailed and valuable fatigue level predictions, the mean absolute error (MAE) is used as a loss measure because fatigue level detection is a multi-class ordinal classification problem. It takes into account the intermediate problem between regression and classification. Furthermore, for multi-class ordinal classification, Gaudette and Japkowicz compared various indicators of the accuracy of ordinal classification. They found that, as a single statistic, MAE (mean absolute error) or MSE (mean squared error) performed better than other indicators they found in the literature. Although MAE/MSE is designed for continuous data, its more severe penalty for deviation from the mean applies to ordered data converted to small integers.

Evaluation metrics are defined in the Eqs. 5 and 6. Accuracy is the main metric in this study. It refers to the percentage of entire videos that have been correctly classified, not individual video clips.

$$Accuracy = \frac{TP + TN}{TP + TN + FP + FN} \tag{7}$$

TP, TN, FP, and FN represent true positives, true negatives, false positives, and false negatives, respectively, based on comparisons between fatigue test results and underlying facts.

The loss (i.e. mean absolute error, MAE) is the mean absolute difference between the estimated value and the actual value:

$$Loss = \frac{\sum_{i=1}^{N} \left| Y_i - \hat{Y}_i \right|}{N} \tag{8}$$

Y_i represents the predicted degree of fatigue and $\hat{Y}_i$ represents the actual label value. N is the video frame number for fatigue detection.

3.5   Experimental Results and Analysis

In contrast, when explicit stress annotations are taken as basic facts, both signals are robust and have comparable accuracy in locating state changes. In the third version of our experiment, we considered the fused gold standard for reporting ratings and EDA, and AffectiveROAD achieved a significant improvement of 27% compared to EDA-only performance and a 13% improvement compared to rating-only performance, indicating that this fusion effectively lays the foundation for arousal estimation in AffectiveROAD. We emphasize that the performance of the Multi-modality model doubles its score compared to the model using EDA-only (Fig. 7).

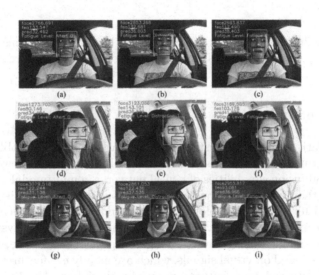

Fig. 7. Data Test

Select a YawDD dataset containing videos captured at different camera positions for testing to determine the appropriate angle for facial video capture. As shown in Fig. 8, YawDD contains two sets of driver video sets with various facial features for yawn detection. In the first set, the cameras are mounted under the vehicle's front windshield at an angle to the driver (side view). The second set of cameras is mounted on the vehicle's dashboard and faces directly toward the driver (front view). In the dataset, each driver has three or four videos. Each video contains facial expressions in different mouth states, such as still, talking/singing, and yawning. The dataset provides 322 videos, including male and female drivers, with and without glasses/sunglasses, different races, and three different scenarios: (1) driving normally (without talking); (2) talking and singing while driving; (3) yawning while driving.

As shown in Table 3, the fatigue detection accuracy of the driver's forward-looking video is 88.97%, which is higher than the 83.01% of the side-looking driver's video. It meets the natural expectation that the more parts of the face to be captured, the more features to be detected to improve the accuracy of fatigue detection. Nonetheless, the trained model achieved high accuracy in this experiment through the detection performance of side-looking video.

Table 3. Experimental dataset.

Dataset	Camera Localization	Loss	Accuracy
	Vehicle dashboard	0.2356	0.8897
YawnDD	Under the front windshield	0.4105	0.8301

Therefore, in general, our proposed method has great advantages over other single fatigue detection methods, and has more guiding significance and practical value.

4   Conclusion

At present, fatigue detection systems on the market have the problem of single detection features and low detection accuracy. In view of the powerful feature extraction and fusion capabilities of neural networks, a multimodal driver fatigue detection method based on physiological signals and visual signals is proposed, which is used to process physiological signals from wearable devices and visual signals from cameras during driving. First, the model processes physiological signal sequences and locates significant changes in physiological state, and extracts visual features based on visual signals, which are mainly eye features and supplemented by mouth and head features. Secondly, physiological and visual features are fused in the time dimension to obtain fatigue features. An effective fatigue detection model can be trained based on the feature representation. We use public datasets to segment physiological signal sequences such as heart rate and quantify and capture change points. After testing in different visual environments such as day and night and whether the face is occluded, this model can meet the requirements of basic real-time detection of fatigue, and the detection accuracy is high.

References

1. Sarker, H., et al.: Finding significant stress episodes in a discontinuous time series of rapidly varying mobile sensor data. In: Proceedings of the 2016 CHI Conference on Human Factors in Computing Systems, pp. 4489–4501 (2016)

2. Fogarty, J., Hudson, S.E., Lai, J.: Examining the robustness of sensor-based statistical models of human interruptibility. In: Proceedings of SIGCHI, pp. 207–214 (2004)
3. Ghoddoosian, R., Galib, M., Athitsos, V.: A realistic dataset and baseline temporal model for early drowsiness detection. In: Proceedings of the IEEE Conference on Computer Vision and Pattern Recognition Workshops, pp. 178–187 (2019). https://doi.org/10.1109/CVPRW.2019.00027'
4. Zhang, W., Murphey, Y.L., Wang, T., Xu, Q.: Driver yawning detection based on deep convolutional neural learning and robust nose tracking. In: 2015 International Joint Conference on Neural Networks (IJCNN), IEEE, pp. 1–8 (2015)
5. Arbelaez, P., Maire, M., Fowlkes, C., Malik, J.: Contour detection and hierarchical image segmentation. IEEE Trans. Pattern Anal. Mach. Intell. **33**(5), 898–916 (2010)
6. Schneegass, S., Pfleging, B., Broy, N., et al.: A data set of real world driving to assess driver workload. In: Proceedings of the 5th International Conference on Automotive User Interfaces and Interactive Vehicular Applications, pp. 150–157 (2013)
7. Saeed, A., Trajanovski, S., Van Keulen, M., et al.: Deep physiological arousal detection in a driving simulator using wearable sensors. In: 2017 IEEE International Conference on Data Mining Workshops (ICDMW). IEEE, pp. 486–493 (2017)
8. Sahayadhas, A., Sundaraj, K., Murugappan, M.: Detecting driver drowsiness based on sensors: a review. Sensors **12**(12), 16937–16953 (2012)
9. Kim, H.G., Cheon, E.J., Bai, D.S., et al.: Stress and heart rate variability: a meta-analysis and review of the literature. Psychiatry Investig. **15**(3), 235 (2018)
10. Munla, N., Khalil, M., Shahin, A., et al.: Driver stress level detection using HRV analysis. In: 2015 International Conference on Advances in Biomedical Engineering (ICABME). IEEE, pp. 61–64 (2015)
11. Malik, M.: Heart rate variability: standards of measurement, physiological interpretation, and clinical use: task force of the European society of cardiology and the north American society for pacing and electrophysiology. Ann. Noninvasive Electrocardiol. **1**(2), 151–181 (1996)
12. Widjaja, D., Orini, M., Vlemincx, E., et al.: Cardiorespiratory dynamic response to mental stress: a multivariate time-frequency analysis. Comput. Math. Methods Med. **2013**, 451857 (2013)
13. Chowdhury, A., Shankaran, R., Kavakli, M., et al.: Sensor applications and physiological features in drivers' drowsiness detection: a review. IEEE Sens. J. **18**(8), 3055–3067 (2018)
14. Affanni, A., Bernardini, R., Piras, A., et al.: Driver's stress detection using skin potential response signals. Measurement **122**, 264–274 (2018)
15. Stappen, L., Schumann, L., Sertolli, B., et al.: MuSe-Toolbox: the multimodal sentiment analysis continuous annotation fusion and discrete class transformation toolbox. In: Proceedings of the 2nd on Multimodal Sentiment Analysis Challenge, pp. 75–82 (2021)
16. Lapuschkin, S., Binder, A., Muller, K.R., et al.: Understanding and comparing deep neural networks for age and gender classification. In: Proceedings of the IEEE International Conference on Computer Vision Workshops, pp. 1629–1638 (2017)
17. Howard, A.G., Zhu, M., Chen, B., et al.: MobileNets: efficient convolutional neural networks for mobile vision applications. arXiv preprint arXiv:1704.04861 (2017)
18. Greff, K., Srivastava, R.K., Koutnik, J., et al.: LSTM: a search space odyssey (2015). arXiv preprint arXiv:1503.04069 (2016)

19. Hallac, D., Nystrup, P., Boyd, S.: Greedy Gaussian segmentation of multivariate time series. Adv. Data Anal. Classif. **13**(3), 727–751 (2019)
20. Van Gent, P., Farah, H., Van Nes, N., et al.: HeartPy: a novel heart rate algorithm for the analysis of noisy signals. Transport. Res. F: Traffic Psychol. Behav. **66**, 368–378 (2019)
21. Makowski, D., et al.: NeuroKit2: a Python toolbox for neurophysiological signal processing. Behav. Res. Methods **53**(4), 1689–1696 (2021). https://doi.org/10.3758/s13428-020-01516-y
22. Feng, T., Booth, B.M., Narayanan, S.S.: Modeling behavior as mutual dependency between physiological signals and indoor location in large-scale wearable sensor study. In: ICASSP 2020–2020 IEEE International Conference on Acoustics, Speech and Signal Processing (ICASSP). IEEE, pp. 1016–1020 (2020)
23. Chen, X., Gu, Q., Liu, W., Liu, S., Ni, C.: Research on static software defect prediction methods. J. Softw. 27(1), 1–25 (2016). https://www.jos.org.cn/1000-9825/4923.htm, https://doi.org/10.13328/j.cnki.jos.004923 https://doi.org/10.13328/j.cnki.jos.004923
24. Hongjun, W., Hao, B., Hui, Z., et al.: Driver fatigue state detection and warning technology based on computer vision. Sci. Technol. Eng. **22**(12), 4887–4894 (2022). https://doi.org/10.3969/j.issn.1671-1815.2022.12.027

Protecting Bilateral Privacy in Machine Learning-as-a-Service: A Differential Privacy Based Defense

Le Wang[1], Haonan Yan[1,2], Xiaodong Lin[1(✉)], and Pulei Xiong[3]

[1] University of Guelph, Guelph, Canada
{lwang20,xlin08}@uoguelph.ca, yanhaonan.sec@gmail.com
[2] Xidian University, Xi'an, China
[3] National Research Council Canada, Ottawa, Canada
Pulei.Xiong@nrc-cnrc.gc.ca

Abstract. With the continuous promotion and deepened application of Machine Learning-as-a-Service (MLaaS) across various societal domains, its privacy problems occur frequently and receive more and more attention from researchers. However, existing research focuses only on the client-side query privacy problem or only focuses on the server-side model privacy problem, and lacks a simultaneous focus on bilateral privacy defense schemes. In this paper, we design privacy-preserving mechanisms based on differential privacy for the client and server side respectively for the first time. By injecting noise into query requests and model responses, both the client and server sides in MLaaS are privacy-protected. Experimental results also demonstrate the effectiveness of the proposed solution in ensuring accuracy and providing privacy protection for both the clients and servers in MLaaS.

Keywords: Machine Learning as a Service · Bilateral Privacy · Privacy Leakage · Model Extraction · Differential Privacy

1 Introduction

Since the advent of the internet, the pace at which human society produces data has continuously accelerated, and the complexity of this data has increasingly grown. By 2025, according to IDC, the total amount of data is expected to reach a staggering 175 ZB [1]. Traditional business intelligence tools have become inadequate for handling such vast quantities and varieties of data, requiring more efficient analytical tools. The rapid advancement of machine learning (ML) technology aptly fills this gap, while machine learning-as-a-service (MLaaS) came into being. Due to its remarkable service capabilities and consistently decreasing costs, it quickly garnered favor among customers worldwide, with the global market size projected to reach $16.7 billion by 2027 [2]. Concurrently, major global tech corporations have established their commercial foothold in the MLaaS sector, including platforms like Amazon Marketplace [3], Google Cloud AI [4], and Azure Machine Learning [5].

J. Vaidya et al. (Eds.): AIS&P 2023, LNCS 14509, pp. 237–252, 2024.
https://doi.org/10.1007/978-981-99-9785-5_17

However, with the incremental adoption of MLaaS, researchers gradually find that there are numerous security problems in MLaaS. Specifically, during clients enjoy the convenient MLaaS services, their proprietary data must be uploaded to service providers, leading to significant potential risks associated with client privacy breaches [6–9]. Additionally, the service models of providers are also susceptible to model extraction attacks [10], leading to model privacy (i.e., parameters, hyperparameters, training dataset) exposure. Meanwhile, malicious clients can even exploit these parameters to construct fake models with similar performance or launch inversion attacks [13] and membership inference attacks [14].

Addressing these concerns, numerous studies have been conducted, resulting in a myriad of solutions. In defense against model extraction attacks, using technologies such as rounding confidence and differential privacy, researchers have introduced a variety of methods or mechanisms, achieving satisfactory performance in terms of privacy and/or utility [15–18]. To mitigate client data privacy leaks, numerous strategies and methods have been proposed leveraging k-anonymity, ℓ-diversity, differential privacy, homomorphic encryption, and secure multi-party computation [19–21]. To summarize, even though these solutions are effective for their respective privacy challenges, the MLaaS domain currently lacks a universally practical scheme that safeguards both client and server model privacy.

In this study, we propose a strategy that employs differential privacy techniques for both client and server ends, offering dual-sided defense capabilities. On the client side, we amalgamate the exponential mechanism with the Laplace mechanism to provide data privacy release capabilities. On the server side, we utilize a defined model decision space to introduce effective differential perturbations for sensitive queries. Furthermore, experiment results prove the effectiveness of the schemes we proposed. The contributions of this paper are summarized as follows:

1) We first focus on and propose a generalized scheme with the ability to protect the privacy of both the client and the server side at the same time.
2) We propose a method to provide client query data privacy release by integrating organically the exponential and Laplacian mechanisms.
3) We propose a method for server-side privacy protection by modeling the decision space to identify and perturb responses.
4) We perform experiments to demonstrate the effectiveness of the proposed scheme in providing both client- and server-side privacy protection capabilities.

The remainder of this paper is organized as follows: Section 2 introduces the privacy leakage risks faced by both the server and client side and the reserve knowledge of the proposed schemes; Sect. 3 reviews the previous research results; Sect. 4 describes the details of the schemes that satisfy the privacy preservation needs of both sides simultaneously; Sect. 5 demonstrates our evaluation methodology and the experimental results, and Sect. 6 concludes the proposed schemes and gives an outlook on the application directions.

2   Preliminaries

2.1   MLaaS

MLaaS is a cloud-based solution designed to simplify the adoption of machine learning for developers and businesses [22]. By offering pre-trained models, easy-to-use APIs, and scalable infrastructure, MLaaS removes the complexities of managing machine learning environments. This enables organizations to seamlessly integrate machine learning capabilities into their applications, regardless of their expertise in AI [23]. With major cloud providers offering these services, businesses can now harness the potential of machine learning without the need for extensive in-house resources or technical knowledge.

2.2   Privacy Leakage in MLaaS

The utilization of MLaaS presents not only privacy challenges for ML model owners (service providers) but also data privacy concerns for data owners (clients). Specifically, clients using MLaaS are concerned about the privacy and security of data submitted to the MLaaS platform for use in making predictions [6]. In contrast, MLaaS platform proprietors are worry about the potential theft of their models by adversaries masquerading as clients or other malicious attackers [10].

Client / Data Owner. Typically, MLaaS represents a suite of services. Prominent companies like IBM, Google, Microsoft, and FICO, in pursuit of augmented profits and user friendliness, commonly retail machine learning tools as a component of their cloud computing offerings. This array of services includes but is not limited to, data visualization, APIs, facial recognition, natural language processing, predictive analytics, and deep learning. Once the clients (data owners) purchase any of the above services, depending on the type of service selected, the client is required to upload the relevant data to the service provider where the actual computation of the data takes place in the service provider's data center, and it can be seen that the clients' data is geared towards a very high risk of leakage and malicious use in the above process.

   More specifically, the business nature of MLaaS requires clients to share data with service provider. Within various societal sectors, this type of data might be highly sensitive. Examples include personal information aggregated for social science investigations, patient records garnered for medical studies, and individual viewing preferences amassed for advertising research. This poses a risk of accidental data leakage or misuse, especially when service providers fall short in safeguarding data or divert it for extraneous objectives. Even more disconcerting is the scenario where a service provider, commanding a significant market share, accumulates data across diverse societal domains from distinct clientele. The inherent associations and concealed patterns within this data may be deciphered by the provider, culminating in more profound intrusions into client privacy.

Furthermore, under the existing business model of MLaaS, once a client uploads data to an MLaaS provider, the client typically has little control over where the data goes and how it is used, i.e., it loses ownership of the data. This engenders two potential hazards: first, the client has no way to know and verify concerns such as data access rights, data storage locations, and data retention periods, and the protection of the client's data relies entirely on the service provider's ethical self-discipline, which history has shown us to be completely unreliable. Second, because the client loses ownership of the data and has no way of monitoring and controlling how the service provider uses the data, there is no way for the client to enjoy the rewards of trading the data as an asset.

Server / Model Owner. In the context of MLaaS, each model can be perceived as a specific function that maps input data to output data. Generally, MLaaS providers offer users two categories of models: generative models and discriminative models. The primary distinction between these two lies in their treatment of input data and their respective objectives. Generative models aim to learn the joint probability distribution of the inputs and infer the conditional probability given the input data, and they can be employed to describe or generate samples from a dataset. Discriminative models, on the other hand, directly learn the mapping or decision boundary from input data to output labels, focusing on distinguishing between different classes. In essence, while generative models are concerned with how data is generated, discriminative models emphasize how to differentiate or classify outputs based on the input.

Ever since Tramèr et al. [10] introduced model extraction attacks in 2016, the field of model extraction has been undergoing rapid advancements. At present, based on the objectives of the adversary, these attacks are generally categorized into two types: fidelity extraction attacks and accuracy extraction attacks [24]. To elucidate further, if one regards a victim model (target model) as a function f_{vic} that maps input data x to output data y, i.e., $y = f_{vic}(x)$. For discriminative models, the goal of an accuracy extraction attack is to construct a new model f_{acc}, even if f_{acc} differs in structure and parameters from f_{vic}, it still ensures that the accuracy of f_{acc} on a test dataset remains highly consistent with that of the target model f_{vic}. Fidelity extraction attacks, however, emphasize a comprehensive replication of the target model f_{vic}, encompassing its structure, parameters, and functionality, ensuring that the new model f_{fid} not only matches the accuracy of the target model f_{vic} but also replicates its errors. In a nutshell, while accuracy extraction attacks focus on functional performance, fidelity extraction attacks are dedicated to the complete replication of the model.

As illustrated in Fig. 1, the attack process of common model extraction attacks is depicted, and they are categorized based on the attacker's objectives: to steal either accuracy model f_{acc} or fidelity model f_{fid}. Specifically, ❶ the attacker systematically sends input samples $X = \{x_1, x_2, \cdots, x_n\}$ to the target model f_{vic} and observes the corresponding outputs $O_{dataset} = \{f_{vic}(x_1), \cdots, f_{vic}(x_n)\}$. ❷ With accumulated input-output pairs (X and $O_{dataset}$), the adversary commences the training of the surrogate model. This sur-

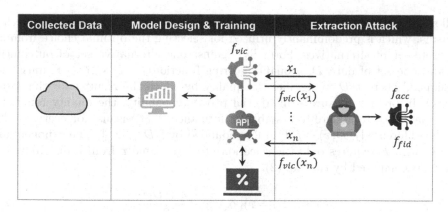

Fig. 1. The diagram of model extraction attack.

rogate endeavors to mirror the behavior exhibited by the original. ❸ To optimize the accuracy or fidelity of the surrogate model f_{acc} or f_{fid}, iterative feedback loops, involving additional queries, may be necessary.

2.3 Differential Privacy

It is well understood that differential privacy (DP) is a privacy-preserving technique [11,12]. The principle behind it is to impose a constraint on randomized computations, ensuring they do not reveal particular details of individual records in the input. This restriction is realized by mandating that the algorithm behaves nearly identically on any two datasets that are closely related.

Let's assume a dataset D that each record originated from an abstract domain (AD). This dataset can be depicted as a function mapping from AD to the natural numbers $\mathbb{N}$, where $D(x)$ denotes the frequency of record x within dataset D. Given this, $\| D - D' \|$ can be utilized to indicate the total absolute difference in frequencies between datasets D and D'. This represents the number of records required to be added or subtracted to transition D to D'.

Definition 1 (Differential Privacy). *Having F be a mechanism that mapping datasets D and D' to distributions on the output space R, which satisfies (ϵ, δ)-differential privacy if for all possible outputs $S \subseteq R$ and for datasets D, D' where $\| D - D' \| \leq 1$,*

$$Pr[F(D) \in S] \leq e^{\epsilon} Pr[F(D') \in S] + \delta, \tag{1}$$

if $\delta = 0$ we say that F provides ϵ-differential privacy.

The DP frequently employs mechanisms such as the exponential mechanism, Laplace mechanism, random response mechanism, and Gaussian mechanism. The solution proposed in this paper is designed on both the exponential and Laplace mechanisms, and a detailed description of these two mechanisms is as follows. The exponential mechanism E is one of the ϵ-differentially private mechanisms, which is predominantly utilized for selecting the optimal choice from a discrete set of alternatives. Formally, against one alternative set of outcomes R and one set of data D, a quality scoring function $f : D \times R \to \mathbb{R}$ must be designed, where $f(D, r)$ signifies the quality by which the result r for the data set D. In order to ensure ϵ-differential private capability, the quality function f is mandated to adhere to a stable performance, that means, for every result r, the difference $|f(D, r) - f(D', r)|$ is bounded by $\| D - D' \|$. The exponential mechanism E requires only the selection of an outcome r from a distribution, which is computed by the Eq. (2).

$$Pr[(E(D) = r)] \propto \exp \frac{\epsilon \times f(D, r)}{2}. \tag{2}$$

Within the context of DP, a linear query is typically defined as a function q that maps data records in the dataset to the interval $[-1, +1]$. Formally, for a data set D, the result of a linear query can be represented as $q(D) = \sum_{x \in D} q(x) D(x)$.

As previously mentioned, the Laplace mechanism is another type of ϵ-DP mechanism, primarily employed to compute the approximate sum of bounded functions within a data set. For a data set D, if q denotes a linear query, the Laplace mechanism L is defined by the Eq. (3),

$$Pr[(L(D) = r)] \propto \exp(-\epsilon \times |r - q(D)|). \tag{3}$$

whereas the Laplace mechanism can be regarded as a specific instance of the exponential mechanism, it can be more efficiently realized by introducing Laplace noise with parameter $\frac{1}{\epsilon}$ in the $q(D)$ value. Furthermore, given that the Laplace distribution is an exponentially concentrated distribution, it is possible for the Laplace mechanism to approximate the true sum fairly accurately.

To counteract model extraction attacks, our primary approach is based on the concept of model decision space, employed to safeguard the query responses near the model's decision boundary [32]. The specific definition of the model decision space is as the Eq. (4).

Definition 2 (Model Decision Space). *In the given feature space* **S**, *there exists a model f and parameters Λ chosen by the model provider. All feature vectors x that are adjacent to the model decision space form a zone $\mathbf{S}_\Lambda$ within* **S**,

$$\mathbf{S}_\Lambda = \{x \in \mathbb{R}^d \mid dist(x, f) < \Lambda\}, \tag{4}$$

where the distance between a feature vector x and the decision boundary of model f is quantified by $dist(\cdot)$. All queries falling within the $\mathbf{S}_\Lambda$ space are deemed sensitive, carrying the risk of exposing the decision space of model f.

To this end, in order for the model to successfully resist extraction attack, it is necessary to perturb all responses within the model decision space. The aim is to perturb the responses of any two sensitive queries, preventing the attacker from ascertaining the true decision boundary within that space. To achieve the aforementioned goal, we introduce the concept of model decision space differential privacy. Its formal definition is as follows:

Definition 3 (ϵ-Decision Space Differential Privacy, ϵ-DSDP). *If and only if for any two queries ℓ_1 and ℓ_2 within the model decision space $\mathbf{S}_\Lambda$, an perturbation algorithm $O(\cdot)$ achieves ϵ-decision space differential privacy, then for the true responses ξ_1 and ξ_2 and the perturbed responses $O(\xi_1)$ and $O(\xi_2)$, the following inequality consistently holds true,*

$$e^{-\epsilon} \leq \frac{P_r[\xi_1 = \xi_2 | O(\xi_1), O(\xi_2)]}{P_r[\xi_1 \neq \xi_2 | O(\xi_1), O(\xi_2)]} \leq e^{\epsilon}. \tag{5}$$

Wherein, Eq. (5) ensures that attackers cannot ascertain whether the perturbed responses $O(\xi_1)$ and $O(\xi_2)$ are derived from the same query response $\xi_1 = \xi_2$ or different query responses $\xi_1 \neq \xi_2$, with a high level of confidence (governed by ϵ). As a result, irrespective of the number of meticulously crafted queries the attacker initiates, they cannot discern the authentic decision boundary within the model decision space $\mathbf{S}_\Lambda$.

3 Related Works

3.1 Defenses for Client in MLaaS

To the best of our knowledge, there's limited research on how to protect client data privacy within the MLaaS. Nonetheless, in the field of data publishing, researchers have published several works. Hardt et al. [25] introduced a novel differential privacy data release method named MWEM, which achieves theoretical guarantees by integrating the multiplicative weights update rule with the exponential mechanism. A major limitation of this method is that maintaining a complete distribution becomes infeasible when the data domain is very large. Addressing numerous queries on high-dimensional data sets, Gaboardi et al. [26] proposed an immensely practical privacy-preserving algorithm, named Dual Query. This algorithm encapsulates computationally challenging steps into an indirect integer program, significantly enhancing computational efficiency. However, both of the aforementioned methods suffer from task bias, unable to process tasks of any type with satisfactory accuracy. Subsequently, by replacing core components of MWEM and Dual Query, Vietri et al. [27] introduced three oracle-efficient algorithms for constructing differential privacy synthetic data, yet these did not overcome their primary limitations. As a result, Zhang et al. [28] introduced the PrivSyn method, a general differential privacy data release mechanism that can autonomously and confidentially identify correlations in original data and generate sample data from dense graph models.

3.2 Defenses for Server in MLaaS

The main object of this work is to design defense methods against model extraction attacks, i.e., model output perturbation [29]. Initially, Tramèr et al. [10]

delineated a fundamental form of output perturbation, which entails rounding off prediction results. However, they concurrently noted its impracticality. Lee et al. [30] introduced a deceptive perturbation method, intended to disrupt the activation layer located at the model's last position, thereby amplifying the difficulty for attackers to retrieve comparable models. Specifically, this technique perturbs the model's raw response by appending an inverse sigmoid function, subsequently employing a normalizer to ensure the sum remains unity. Nonetheless, this approach falls short of model extraction attacks that merely operate on prediction labels. Orekondy et al. [31] proposed a utility-constrained defensive framework, termed "prediction poisoning", which harmonizes model privacy and utility objectives by perturbing predictions. Explicitly, the perturbation maximizes the gradient deviation of prediction posteriors from those of the original model. However, this method requires a large number of gradient computations to implement. In the same year, Zheng et al. [32] introduced the BDPL method, leveraging differential privacy to perturb model output responses. It appends a boundary differential privacy layer after the model output layer, obfuscating responses to queries within the decision boundary zone. Explicitly, BDPL discerns query sensitivity via pre-defined model decision boundary-sensitive zones. Upon identifying a query as sensitive, the boundary differential privacy layer invokes a boundary random response algorithm controlled by the privacy budget to return the obfuscated response. Yan et al. [33] proposed a monitoring-based differential privacy technique to resist adaptive query-flooding parameter duplication attacks. Especially, this mechanism realizes real-time evaluation of the model state and adaptively adjusts the privacy budget based on the evaluation results, which in turn dynamically adjusts the amount of noise added to the model response. Li et al. [34] introduced a personalized local differential privacy mechanism to defend against the equation-solving model extraction attacks on regression models. This mechanism makes the model adaptively noisy by adding high latitude Gaussian noise to the model coefficients.

To summarize, although researchers have carried out a lot of research work in the respective areas of DP data release and model protection, there is still a lack of work that integrates the privacy needs of both the server and the client and conducts research on holistic privacy protection mechanisms in MLaaS.

4 Proposed Method

4.1 Overview

As mentioned above, both service providers and clients grapple with a myriad of privacy leakage risks and threats in the domain of MLaaS. As depicted in Fig. 2, our research pioneers an integrated solution aimed at simultaneously preserving the privacy of both the service providers and clients. When handling private query data uploaded by clients to service providers, we have innovated a data release method that cleverly combines Laplace noise with the exponential mechanism (LNEM). Further, in relation to the response results that the service provider disseminates to the client, we have architected a model decision space DP (ϵ-DSDP) that is appended to the model outputs, with the primary objective of effectively perturbing responses to sensitive queries originating from the client.

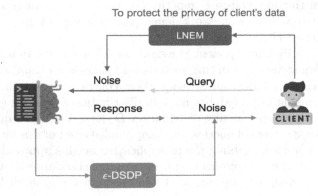

Fig. 2. The framework of the proposed method.

4.2 Privacy for Client

In this part, we utilize the multiplicative weights framework introduced by Hardt and Rothblum [35] to iteratively refine an approximative distribution to more accurately resemble the intrinsic true distribution. The fundamental observation is that, upon detecting a query where the approximative distribution returns a markedly higher value compared to the true distribution, it is pertinent to augment the weights of records that positively contribute to this query, while simultaneously diminishing the weights of those that negatively contribute. In the inverse scenario, where the approximative distribution procures a significantly reduced query result, adjustments to the weights are made in the reverse order.

To elucidate with greater formality, consider q as a linear query structured over a domain $\mathbf{D}$ encompassing records. Let's assume that distribution D endeavors to emulate the true distribution D' in relation to query q. The multiplicative weights update protocol dictates the modification of the weight D allots to a specific record x as Eq. (6):

$$\frac{D_{new}(x)}{D(x)} = \exp(\frac{q(x) \times (q(D') - q(D))}{2}).$$ (6)

Upon standardization of these weights, Hardt and Rothblum substantiated that such an update decremental impacts the relative entropy between D and D' by an additive factor of $(q(D) - q(D'))^2$ in each iteration. Hence, pinpointing queries exhibiting disparities between D and D' facilitates the iterative refinement of our approximation.

In essence, this update rule adaptively modulates the weights contingent on each record's influence on the discrepant query, culminating in a heightened probability density on records that resonate more harmoniously with distribution D'. When this mechanism is repetitively applied to in-congruent queries, the multiplicative weights framework demonstrably navigates distribution D in closer proximity to D'.

To further clarify, the exponential weight modification function accentuates weights of affirmatively influencing records and diminishes those of adversely influencing records, commensurate with their contributory significance. This strategic realignment of probability mass in distribution D ensures greater congruence with records epitomizing distribution D' for a stipulated query. When enacted across a spectrum of queries, the compounded effect of this multiplicative re-calibration amplifies, enabling the multiplicative weights approach to assimilate D into D', even when commencing from a rudimentary approximation. This methodology epitomizes a sophisticated and theoretically grounded stratagem for distribution alignment.

4.3 Privacy for Server

In this section, we primarily introduce solutions designed to combat model extraction attacks. The key idea is to incorporate ϵ-decision space differential privacy (ϵ-DSDP) into the model output, that is, appending an ϵ-DSDP layer following the model's output layer. Especially, this ϵ-DSDP layer consists of two pivotal steps: ❶ identifying sensitive queries from either attackers or users based on the corner-point technique; ❷ employing a perturbation algorithm to perturb the responses of these sensitive queries to adhere to ϵ-DSDP. Moving forward, we will focus on elaborating the proposed perturbation algorithm that satisfies ϵ-DSDP.

Warner et al. introduced the randomized response technique [36] as a survey methodology aimed at mitigating potential biases stemming from non-responses and societal expectations when posing questions related to sensitive behaviors and beliefs. At the core of this technique is the intent to safeguard privacy in the original data by leveraging uncertainty in responses to sensitive inquiries. It's primarily tailored for dichotomous data, which assumes two distinct values. Notably, the perturbation algorithm defined within our boundary differential privacy framework in this paper offers just two output options. Consequently, we've devised the BWRR (Boundary Warner Randomized Response) algorithm, drawing inspiration from the randomized response technique, to fulfill the requirements of ϵ-DSDP.

For a given model within MLaaS, given a query sample ℓ_q and its true response $\xi_q \in \{0,1\}$, the BWRR algorithm $B(\xi_q)$ perturbs ξ_q according to the Eq. (7), ensuring that $B(\xi_q)$ adheres to ϵ-DSDP.

$$B(\xi_q) = \begin{cases} \xi_q, & \text{with probability } \frac{1}{2} + \frac{\sqrt{e^{2\epsilon}-1}}{2+2e^\epsilon} \\ 1-\xi_q, & \text{with probability } \frac{1}{2} - \frac{\sqrt{e^{2\epsilon}-1}}{2+2e^\epsilon} \end{cases} \tag{7}$$

To this point, we proceed to detail the procedure of the ϵ-DSDP layer that is affixed subsequent to the output layer of model f. Specifically, upon receiving a fresh query ℓ_q, if a preliminary lookup discerns that it has been previously queried by other users, the ϵ-DSDP layer directly furnishes the querier (client) with the cached response ξ'_q. This strategy is employed to thwart adversaries from gleaning multiple perturbed responses for an identical query, which might consequently undermine the privacy guarantees of ϵ-DSDP. If this is not the case, the process is as follows: ❶ the ϵ-DSDP layer procures the authentic query outcome ξ_q as returned by model f. ❷ this layer subsequently scrutinizes all corner points to ascertain whether ℓ_q resides within the model decision space. Should any corner point be deemed a flip point, the query is promptly flagged as sensitive. In such an instance, the BWRR($\cdot$) algorithm described in Eq. (6) equipped with a predefined privacy budget ϵ, is utilized to ensure privacy preservation, i.e., $\xi'_q = \text{BWRR}(\xi_q, \epsilon)$. Conversely, should ℓ_q be adjudged non-sensitive post comprehensive corner point evaluation, the querier is provided with the authentic query outcome ξ_q as derived from model f. ❸ the resultant output of the BWRR($\cdot$) algorithm is then harnessed. This perturbed query result ξ'_q is relayed to the querier. Concurrently, it's cached locally, laying the groundwork for potential subsequent inquiries.

5   Experiment

5.1   Setup

• **Datasets.** We evaluate our approach on four publicly available datasets commonly used for machine learning research: the Email Spam dataset contains emails labeled as spam or not spam based on the presence of certain words. The Mushrooms dataset from scikit-learn [37] classifies mushrooms as poisonous or edible based on their characteristics. The other two datasets are obtained from Kaggle and represent a diverse range of ML tasks. Together, these four datasets cover several orders of magnitude in size and complexity, enabling comprehensive analysis.

For all datasets, categorical features are encoded using one-hot encoding to avoid making assumptions about ordinal relationships between categories. Missing values in the data are imputed by replacing them with the mean value for that feature, a simple and widely used approach for handling missing data. Table 1 summarizes key statistics of the evaluation datasets.

Table 1. Datasets

Dataset	Instances	Dimensions
SocialAds	401	5
Titanic	1310	28
Email Spam	4601	46
Mushrooms	8124	112

To evaluate client-side privacy, we add DP noise directly to the user requests before sending them to the server.

For server-side evaluation, we randomly split each dataset into a training set (70%) and a test set (30%). The training set serves as private data to train the target model. The test set is used to evaluate the accuracy of the target model as well as the accuracy of the model extracted by the adversary (i.e., to calculate *Accuracy*). The test set also measures how closely the extracted model matches the target model (i.e., to calculate $1 - R_{test}$).

In our adversary model, the adversary queries the target model exhaustively overall feature spaces. It then uses the prediction results to train the extracted model. Note that the test set is kept private and not exposed during extracted model training. Evaluating the extracted model on the private test set illustrates its generalization ability beyond the query dataset.

• **Evaluation Metrics.** For client-side evaluation, we use mean squared error (MSE) to evaluate range queries and relative entropy (Kullback-Leibler divergence, KLD) for binary contingency table queries.

For server-side evaluation, we measure the utility of the LR and NN models on the test set using accuracy. To evaluate how closely the extracted model matches the original, we use the R_{test} metric from [33] calculated on the test set. Therefore, $1 - R_{test}$ represents the extraction status in terms of test error. Since the test set follows the same distribution as the training set, the extracted model can also be evaluated on random datasets with different distributions to estimate its fidelity across the full feature space uniformly, denoted as R_{unif}. *Test error R_{test}* measures the similarity between the extracted model and the original model on the test set. A lower R_{test} indicates higher similarity and a more effective extraction attack. For clarity, we define *Extraction_status* = $1 - R_{test}$ and use this metric in our experiments. Formally, given the extracted model $\widetilde{f}(x)$ and the test dataset D_{test},

$$R_{test} = \frac{1}{|D_{test}|} \sum_{i \in D_{test}} d\left(f(x^{(i)}) \neq \widetilde{f}(x^{(i)}) \right) \tag{8}$$

where d is an indicator function that equals 1 if $f(x^{(i)}) = \widetilde{f}(x^{(i)})$, otherwise 0.

5.2   Evaluate Client's Mechanism

Here we estimate the error of our proposed defense. The baseline is to add noise using only the Laplace mechanism, a common simple, and efficient scheme. The results are shown in Fig. 3.

When accuracy is emphasized, selecting the necessary dimensions to add noise can significantly improve accuracy. When the privacy budget is very sufficient, or the query set is very simple, directly perturbing all dimensions produces better results than spending a fraction of the privacy budget to determine what to perturb. In more challenging cases such as complex data and limited privacy budget, our algorithm outperforms previous schemes.

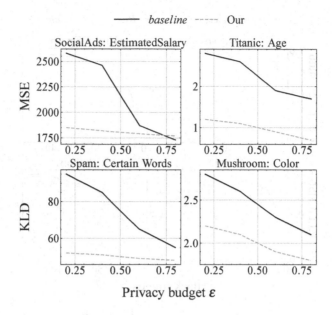

Fig. 3. Comparison of the proposed mechanism with the Laplace mechanism on four datasets. The y-axis measures MSE and KLD for each query, averaged over five independent experimental replications as ϵ changes. The smaller the ϵ value, the more noise is added and the better the effect of our mechanism performs.

5.3   Evaluate Server's Mechanism

We use the novel extraction attack to examine our proposed defense, along with the Rounding Confidences (RC) mechanism as the baseline. The results are shown in Fig. 4.

   In order to evaluate the effectiveness of the protection of the model parameters, we compare the proposed mechanism with the RC mechanism along with the effectiveness of the attack in the unprotected state. We chose the strongest attack scheme available, the QPD model extraction attack. It can be seen that in various experimental setups, the scheme proposed in this paper shows a significant protection effect compared to no defense. And the proposed scheme is consistently the best performer while the intensity of the attack increases.

6   Conclusions

In this work, we focus on privacy leakage issues pertaining to the client and server within the MLaaS. Utilizing differential privacy techniques, we safeguard both the query requests from the client and the model responses from the server. Specifically, the proposed method offers effective privacy protection for the client-side query and the server-side model parameters. This is also the first initiative that concurrently addresses the privacy of both the client and the server in

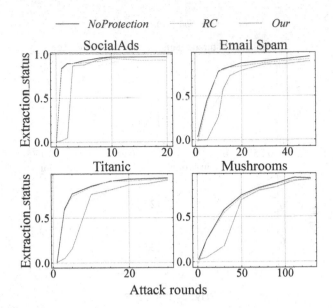

Fig. 4. The proposed mechanism is compared with the RC mechanism in terms of model extraction defense effects on the four datasets. At the beginning of the attack, our defense was significantly better than RC's.

the MLaaS setting. We aspire that our solution will tangibly address privacy protection concerns in MLaaS, enhance its privacy security in real-world applications, and promote the application and evolution of trustworthy AI technology in MLaaS.

Acknowledgement. This project was supported in part by collaborative research funding from the National Research Council of Canada's Artificial Intelligence for Logistics Program. Part of Haonan Yan's work is done when he visits the School of Computer Science at the University of Guelph.

References

1. Reinsel, D., Gantz, J., Rydning, J.: Data Age 2025: the evolution of data to life-critical. https://www.seagate.com/files/www-content/our-story/trends/files/Seagate-WP-DataAge2025-March-2017.pdf. Accessed Aug 2023
2. mordor intelligence, machine learning as a service (MLaaS) market size & share analysis-growth trends & forecasts (2023–2028). https://www.mordorintelligence.com/industry-reports/global-machine-learning-as-a-service-mlaas-market
3. Amazon marketplace. https://aws.amazon.com/marketplace. Accessed Aug 2023
4. Google cloud AI. https://cloud.google.com/solutions/ai. Accessed Aug 2023
5. Azure machine learning. https://azure.microsoft.com/en-ca/free/machine-learning. Accessed Aug 2023

6. Tanuwidjaja, H.C., Choi, R., Baek, S., et al.: Privacy-preserving deep learning on machine learning as a service-a comprehensive survey. IEEE Access **8**, 167425–167447 (2020)
7. De Cristofaro, E.: A critical overview of privacy in machine learning. IEEE Secur. Priv. **19**(4), 19–27 (2021)
8. Qayyum, A., Ijaz, A., Usama, M., et al.: Securing machine learning in the cloud: a systematic review of cloud machine learning security. Front. Big Data **3**, 587139 (2020)
9. Acar, G., Eubank, C., Englehardt, S., et al.: The web never forgets: persistent tracking mechanisms in the wild. In: Proceedings of the. ACM SIGSAC Conference on Computer and Communications Security, vol. 2014, pp. 674–689 (2014)
10. Tramèr, F., Zhang, F., Juels, A., et al.: Stealing machine learning models via prediction APIs. In: 25th USENIX Security Symposium (USENIX Security 16), pp. 601–618 (2016)
11. Chen, Q., Chai, Z., Wang, Z., et al.: QP-LDP for better global model performance in federated learning. In: 2022 18th International Conference on Mobility, Sensing and Networking (MSN). IEEE, pp. 422–426 (2022)
12. Chen, Q., Wang, H., Wang, Z., et al.: LLDP: a layer-wise local differential privacy in federated learning. In: 2022 IEEE International Conference on Trust, Security and Privacy in Computing and Communications (TrustCom). IEEE, pp. 631–637 (2022)
13. Fredrikson, M., Jha, S., Ristenpart, T.: Model inversion attacks that exploit confidence information and basic countermeasures. In: Proceedings of the 22nd ACM SIGSAC Conference on Computer and Communications Security, pp. 1322–1333 (2015)
14. Shokri, R., Stronati, M., Song, C., et al.: Membership inference attacks against machine learning models. In: 2017 IEEE Symposium on Security and Privacy (SP). IEEE, pp. 3–18 (2017)
15. Shokri, R., Shmatikov, V.: Privacy-preserving deep learning. In: Proceedings of the 22nd ACM SIGSAC Conference on Computer and Communications Security, pp. 1310–1321 (2015)
16. Mohassel, P., Zhang, Y.: SecureML: a system for scalable privacy-preserving machine learning. In: 2017 IEEE Symposium on Security and Privacy (SP). IEEE, pp. 19–38 (2017)
17. Hesamifard, E., Takabi, H., Ghasemi, M., et al.: Privacy-preserving machine learning in cloud. In: Proceedings of the 2017 on Cloud Computing Security Workshop, pp. 39–43 (2017)
18. Zheng, W., Popa, R.A., Gonzalez, J.E., et al.: Helen: maliciously secure cooperitive learning for linear models. In: 2019 IEEE Symposium on Security and Privacy (SP). IEEE, pp. 724–738 (2019)
19. Ristenpart, T., Tromer, E., Shacham, H., et al.: Hey, you, get off of my cloud: exploring information leakage in third-party compute clouds. In: Proceedings of the 16th ACM Conference on Computer and Communications Security, pp. 199–212 (2009)
20. Sweeney, L.: k-anonymity: a model for protecting privacy. Internat. J. Uncertain. Fuzziness Knowl. Based Syst. **10**(05), 557–570 (2002)
21. Machanavajjhala, A., Kifer, D., Gehrke, J., et al.: L-diversity: privacy beyond k-anonymity. ACM Trans. Knowl. Discov. (TKDD), **1**(1), 3-es (2007)
22. Ribeiro, M., Grolinger, K., Capretz, M.A.M.: MLaaS: machine learning as a service. In: 2015 IEEE 14th International Conference on Machine Learning and Applications (ICMLA). IEEE, pp. 896–902 (2015)

23. Weng, J., Weng, J., Cai, C., et al.: Golden grain: building a secure and decentralized model marketplace for MLaaS. IEEE Trans. Dependable Secure Comput. **19**(5), 3149–3167 (2021)

24. Jagielski, M., Carlini, N., Berthelot, D., et al.: High accuracy and high fidelity extraction of neural networks. In: 29th USENIX Security Symposium (USENIX Security 20), pp. 1345–1362 (2020)

25. Hardt, M., Ligett, K., McSherry, F.: A simple and practical algorithm for differentially private data release. In: Advances in Neural Information Processing Systems, vol. 25 (2012)

26. Gaboardi, M., Arias, E.J.G., Hsu, J., et al.: Dual query: practical private query release for high dimensional data. In: International Conference on Machine Learning. PMLR, pp. 1170–1178 (2014)

27. Vietri, G., Tian, G., Bun, M., et al.: New oracle-efficient algorithms for private synthetic data release. In: International Conference on Machine Learning. PMLR, pp. 9765–9774 (2020)

28. Zhang, Z., Wang, T., Li, N., et al.: PrivSyn: differentially private data synthesis. In: 30th USENIX Security Symposium (USENIX Security 21), pp. 929–946 (2021)

29. Gong, X., Wang, Q., Chen, Y., et al.: Model extraction attacks and defenses on cloud-based machine learning models. IEEE Commun. Mag. **58**(12), 83–89 (2020)

30. Lee, T., Edwards, B., Molloy, I., et al.: Defending against neural network model stealing attacks using deceptive perturbations. In: 2019 IEEE Security and Privacy Workshops (SPW). IEEE, pp. 43–49 (2019)

31. Orekondy, T., Schiele, B., Fritz, M.: Prediction poisoning: utility-constrained defenses against model stealing attacks. In: International Conference on Representation Learning (ICLR), vol. 2020 (2020)

32. Zheng, H., Ye, Q., Hu, H., Fang, C., Shi, J.: BDPL: a boundary differentially private layer against machine learning model extraction attacks. In: Sako, K., Schneider, S., Ryan, P.Y.A. (eds.) ESORICS 2019. LNCS, vol. 11735, pp. 66–83. Springer, Cham (2019). https://doi.org/10.1007/978-3-030-29959-0_4

33. Yan, H., Li, X., Li, H., et al.: Monitoring-based differential privacy mechanism against query flooding-based model extraction attack. IEEE Trans. Dependable Secure Comput. **19**(4), 2680–2694 (2021)

34. Li, X., Yan, H., Cheng, Z., et al.: Protecting regression models with personalized local differential privacy. IEEE Trans. Dependable Secure Comput. **20**(2), 960–974 (2022)

35. Hardt, M., Rothblum, G.N.: A multiplicative weights mechanism for privacy-preserving data analysis. In: 2010 IEEE 51st Annual Symposium on Foundations of Computer Science. IEEE, pp. 61–70 (2010)

36. Warner, S.L.: Randomized response: a survey technique for eliminating evasive answer bias. J. Am. Stat. Assoc. **60**(309), 63–69 (1965)

37. Pedregosa, F., Varoquaux, G., Gramfort, A., et al.: Scikit-learn: machine learning in Python. J. Mach. Learn. Res. **12**, 2825–2830 (2011)

FedCMK: An Efficient Privacy-Preserving Federated Learning Framework

Pengyu Lu[1] , Xianjia Meng[1,2(✉)] , and Ximeng Liu[2]

[1] Northwest University, Xi'an, Shaanxi 710069, China
xianjiam@nwu.edu.cn
[2] Fuzhou University, University Town Fuzhou, Fujian 350108, China

Abstract. Federated learning emerged to solve the privacy leakage problem of traditional centralized machine learning methods. Although traditional federated learning updates the global model by updating the gradient, an attacker may still infer the model update through backward inference, which may lead to privacy leakage problems. In order to enhance the security of federated learning, we propose a solution to this challenge by presenting a multi-key Cheon-Kim-Kim-Song (CKKS) scheme for privacy protection in federated learning. Our approach can enable each participant to use local datasets for federated learning while maintaining data security and model accuracy, and we also introduce FedCMK, a more efficient and secure federated learning framework. Fed-CMK uses an improved client selection strategy to improve the training speed of the framework, redesigns the key aggregation process according to the improved client selection strategy, and proposes a scheme vMK-CKKS, to ensure the security of the framework within a certain threshold. In particular, the vMK-CKKS scheme adds a secret verification mechanism to prevent participants from malicious attacks through false information. The experiments show that our proposed vMK-CKKS schemes significantly improve security and efficiency compared with the previous encryption schemes. FedCMK reduces training time by 21% on average while guaranteeing model accuracy, and it provides robustness by allowing participants to join or leave during the process.

Keywords: Homomorphic encryption · Federated learning · Multi-key · CKKS · Machine learning · Secret sharing

1 Introduction

With improved computing power and increased data volume, deep learning has achieved remarkable success in computer vision, natural language processing, and other fields. However, large-scale data collection and storage often lead to privacy leakage and data security issues. To solve the problem of privacy leakage

Supported by National Natural Science Foundation of China under grant number 62276211.

in machine learning with a large amount of data and the problem of data island that a large amount of data cannot be applied, the concept of federated learning comes into being, aiming at distributed machine learning under the premise of protecting data privacy [14,19]. In contrast, federated learning empowers individual devices to train models locally and share only the model updates with a central server, preserving the users' privacy. In recent years, federated learning has received much attention from academia and industry due to its potential in various applications such as healthcare, finance, and the smart internet of things [4,15,23].

Nevertheless, federated learning still faces some challenges in practical applications [16]. In order to enhance the security of federated learning, many scholars have carried out research on homomorphic encryption, differential privacy, and secure multi-party computation [9,11,24]. Additionally, federated learning can also be attacked by malicious actors who may transmit faulty model updates or manipulate training processes, damaging overall model performance [1]. Therefore, it is crucial to design an effective security mechanism to detect and resist such attacks. Furthermore, federated learning involves multiple participants conducting model training locally, which may result in a slower overall convergence speed. Balancing the process of local training and global aggregation to improve training speed and model performance is also a problem worth studying.

In this study, we use improved client selection strategies and multi-key homomorphic encryption schemes to solve the problems of data leakage, malicious party attacks, model training, and convergence rate optimization. Specifically, we improve the client selection strategy to improve model training and convergence speed [20]. Based on the improved client selection strategy, we designed a multi-key homomorphic encryption scheme, namely vMK-CKKS, to protect data privacy and solve the problem of collusion attacks of malicious actors. We propose a complete federated learning framework, FedCMK, and verify its security and robustness through experiments. Our contribution is as follows:

(1) We propose a client selection strategy to solve the device heterogeneity problem and ensure the model's accuracy while maximizing the training efficiency. Experiments show that our client-selected federated learning framework can improve the training speed by about 21% compared with the traditional federated learning framework without more than 1% accuracy loss. For a specific model, the training speed can be improved by up to 33%.

(2) We design a MK-CKKS scheme: vMK-CKKS. Based on the client selection strategy, we redesigned the way of public key aggregation for different clients. The vMK-CKKS scheme is based on verifiable secret sharing, ensuring security within a limited threshold. It prevents the participants from maliciously sending false information to destroy the decryption. Experiments show that our multi-key CKKS scheme is more efficient and secure than the traditional encryption schemes.

(3) We propose a cross-device federated learning framework FedCMK based on the client selection strategy with vMK-CKKS encryption scheme. Under the premise of enhancing privacy security and training efficiency, the

framework also supports arbitrary training strategy and has high scalability and robustness. Experiments show that FedCMK can effectively complete different cross-device federated learning tasks and resist malicious attacks from participants within a certain threshold.

(4) We theoretically prove the security of our federated learning framework and compare its communication cost and computation cost with some existing homomorphic encryption schemes, evaluate the efficiency and performance of our federated learning framework, and discuss some potential risks.

The remainder of this paper is organized as follows. In Sects. 2 and 3, we introduce related works and the preliminaries. In Sect. 4, we introduce the Fed-CMK framework, improved client selection strategies, and vMK-CKKS scheme. Then we present the experimental environment and parameters in Sect. 5, and we evaluate the experimental results. Finally, we provide proof of the framework's security in Sect. 6 and conclude the paper in Sect. 7.

2   Related Work

Our research is mainly aimed at federated learning and multi-key homomorphic encryption. In this section, we will summarize the current federated learning framework based on multi-key homomorphic encryption. At the same time, we will introduce traditional single-key homomorphic encryption schemes.

2.1   Homomorphic Encryption Based FL

Federated learning based on homomorphic encryption offers more robust security without affecting model accuracy. Homomorphic encryption has become the most common privacy protection method in federated learning. Recently, many federated learning frameworks using homomorphic encryption have been proposed. For example, Dimitris et al. proposed MetisFL, a homomorphic encryption-based federated learning model for training neural models and predicting certain diseases [21]. However, their research did not optimize homomorphic encryption schemes or consider potential model leakage. Moreover, their focus was on personalized FL [22]. In recent years, many federated learning frameworks have used the Paillier semi-homomorphic encryption scheme [24–26]. However, the Paillier scheme's nature is unsuitable for large-scale machine-learning gradient encryption. In recent years, the CKKS scheme has become the mainstream homomorphic encryption framework for federated learning. Microsoft has implemented CKKS in the SEAL library, and the CKKS scheme has been widely researched and applied in recent years.

2.2   Multi-key Homomorphic Encryption Based FL

Multi-key homomorphic encryption is more suitable for large-scale multi-party federated learning scenarios than traditional single-key homomorphic encryption schemes [5]. Ma et al. have proposed the xMK-CKKS scheme, which simplifies the aggregation of public keys by adding them to form an aggregated public key for encrypting model updates [17]. This approach has the advantage of being easy to implement. It can be extended to multiple participants while maintaining a certain level of security against collusion between $K - 1$ participants and the server. However, this approach also has some limitations. Firstly, when there are many participants, aggregating all public keys may lead to excessive noise, which can negatively impact the accuracy of the ciphertext and increase computation and communication costs. Secondly, the entire aggregate public key must be reset if a participant drops out. Finally, although it effectively prevents collusion between the server and other actors, it is assumed that the server is a trusted third party in federated learning. In extreme collusion cases, the server can send false information to obtain private data. In contrast, Du et al. have proposed the tMK-CKKS scheme, which uses Shamir's secret sharing to reduce overhead while ensuring security [7]. However, Shamir's secret-sharing scheme is not effective in preventing malicious secret sharing between participants, which could result in decryption failure. Some studies focus on vertical federation learning scenarios, such as CryptoBoost, an XGBoost framework proposed by Jin and Wang et al. based on multi-party homomorphic encryption technology [12]. CryptoBoost is end-to-end secure, and it proposes a new set of communication protocols to reduce costs. Applying multi-key homomorphic encryption under vertical federation learning is also a primary direction for future research [18]. Based on this, we improve the above algorithm and design a multi-key CKKS variant that addresses efficiency and security issues in existing schemes.

3   Preliminaries

In this section, we outline part of the notations used in the paper and introduce the FedCS client selection protocol, which is the basis of our improved client selection protocol, while we present the multi-key homomorphic encryption-related techniques.

We set the secret distribution chi to be the uniform distribution over the set of polynomials in R with coefficients $\{0, \pm1\}$. Each coefficient of the error $e \leftarrow \psi$ is plotted according to a discrete Gaussian distribution centered at zero and standard deviation $\sigma = 3.2$. The model's weight used in the experiment is represented by 32 bits of floating point numbers.

3.1   Federated Learning

Federated learning is a cutting-edge artificial intelligence technology that prioritizes user privacy and data security by ensuring participants cannot access each other's data. As a mainstream algorithm in federated learning, FedAvg allocates a fixed number of training steps to each participant and aggregates locally trained models to compute a new global model. This approach allows for model updates without requiring the exchange of raw data between participants, ensuring privacy and security. The FedAvg algorithm has been widely adopted in practical applications of federated learning.

3.2 Client Selection

In cross-device federated learning, the participation of numerous edge devices and mobile terminals with limited performance capabilities can lead to a significant waste of computing resources or the exclusion of numerous devices that cannot perform multiple epochs within a given time frame. Nishiod et al. proposed the FedCS protocol to solve the heterogeneous problems in federated learning [20]. We improved on this to fit our framework.

3.3 MK-CKKS Scheme

Song and Dai proposed a multi-key homomorphic encryption (MK-HE) scheme based on CKKS in their work [3,5,6,8]. They designed multi-key variants of Brakerski-Fan-Vercauteren (BFV) and CKKS and provided a new relinearization scheme. Moreover, they applied the MKHE scheme to evaluate convolutional neural network (CNN) models.

However, directly applying the MK-CKKS scheme to federated learning may result in privacy risks since the server can decrypt model updates and access personal data during decryption. While the server is typically trustworthy, this contradicts the goals of federated learning. Therefore, our MK-CKKS scheme limits the decryption ability of the server to the ciphertext of a single client, preventing the potential privacy leakage risk on the server side. Specifically, for encrypted model updates, the server can only decrypt the sum of all model updates in ciphertext and cannot decrypt the model updates of individual participants separately.

4   FedCMK

In this section, we introduce our cross-device federated learning framework Fed-CMK, including its system model, our simulated attack model, improved client selection algorithm, and its encryption algorithm and overall system flow.

4.1   Problem Statement

Suppose there is an encrypted cross-device federated learning framework, a total of K clients participate in the training process, and the whole training process is based on the Federated Average (FedAvg) algorithm. The server randomly selects several clients for this round of learning and sends them the global model. The selected clients use the local data set for local training and upload the updated gradient information to the server after encryption. In such a system, we might face the following challenges:

Device Heterogeneity: In a cross-device federated learning system, the performance between devices and the amount of data is different. If the client is randomly selected, it may cause a lot of computing power or data waste. Therefore, how to choose the client will affect the accuracy and time of training.

Encryption and Decryption of Gradients: In this cross-device federated learning system, each participating client may have its key. If each client encrypts the gradient with its key and uploads it, the decryption process will be difficult for the server. If a uniform key is used, the security of the key is not satisfactory to every client, and the parties may be malicious and conspire to steal data. At the same time, if the server can decrypt the ciphertext of each client separately, then a not fully trusted server will easily steal all the data, which is also unacceptable. Therefore, selecting an appropriate encryption scheme is essential to ensure local data security.

Robustness of the System: In this cross-device federated learning system, any client may join or leave during the process, and their actions should not affect the entire training.

4.2   Threat Model

In our threat model, we default the federation launcher to be a trusted entity. In the vMK-CKKS scheme, it will generate the secret to building the aggregate public key used for encryption. While the federation controller and federation learners are honest and curious, they will strictly follow the protocol but will be curious to infer the data of other learners. In order to better reflect the security, we introduce an active adversary $\mathcal{A}$ into the model. In the vMK-CKKS scheme, we set $A = t$. $\mathcal{A}$ should be several learners smaller than A, or a federation controller and several learners smaller than A. The goal of $\mathcal{A}$ is to obtain as many ciphertext decryption results as possible to steal the local data of learners. The following are some possible inferences:

1. $\mathcal{A}$ may consist of at most $A - 1$ learners who obtain each other's ciphertexts or decrypt shares by collusive attacks and wish to decrypt the ciphertexts to steal data from the remaining learners.
2. $\mathcal{A}$ is some maliciously participating learner who broadcasts the wrong secret shares to other learners.
3. $\mathcal{A}$ maliciously sends the wrong decrypt share and hopes to cause an error in the decryption process.
   We notice that such an opponent is very typical in the threat model [2].

4.3   Our Client Selection Design

The original FedCS scheme uses a greedy algorithm to strive for as many high-performance participants as possible to participate in the training task, but this also has some shortcomings. Secondly, it may lead to many devices being unable to participate in the training task. In the case of uneven data distribution, it may lead to certain data waste problems. Finally, some high-performance devices may be malicious, leading to persistent malicious attacks. Therefore, we make some optimizations based on Nishio et al. 's FedCS framework to be more suitable for our federated learning framework and homomorphic encryption scheme. Firstly, we notice that the choice of T_r in the original scheme has a significant impact on the final model update, so we compare the training time and accuracy of FedCMK under different T_r, and choose a more appropriate T_r value. We also add a safety margin $T_{r_s} = \frac{1}{60}T_r$ to account for the network fluctuations that communication may face in real-world applications. Second, we clustered all the clients and selected clients for each clustered client set to make use of as many devices as possible. Finally, we observe that clients with larger datasets tend to be underutilized, leading to severe data wastage and potentially lower model accuracy. To solve this problem, we introduce a weighted selection scheme in which we add several clients with large datasets according to a weight W_s in the S set. In addition, we cache the clients with excellent performance in subset S' and can directly schedule them for subsequent training if they are idle. The Settings of weight W_s and subset S' vary from device to device. Our experiments compare the accuracy and time of training under different W_s. Considering the balance of performance and efficiency, we chose $W_s = 0.2$ and $|S'| = 0.1|S|$.

4.4   Our MK-CKKS Scheme Design

We have optimized and proposed a MK-CKKS variant that will be used to build our federated learning framework FedCMK. The vMK-CKKS scheme is based on verifiable secret sharing (VSS), which is similar to Shamir's secret sharing but with an additional verification mechanism [10]. This mechanism enables participants to verify the correctness of the received secret fragments, helping to prevent malicious actors from tampering with or forging shares during the sharing process, thus enhancing the system's security. Moreover, the verification function of all parties of VSS improves the system's fault tolerance and robustness since even if some participants provide incorrect shares when restoring the aggregate public key, the final result will not be affected. We will discuss this method in more detail below.

- **SecretShare** : A trusted third party performs secret generation, and we refer to this third party as the generator hereinafter(GH). GH randomly selects $a_i \in \mathbb{Z}_p$, and construct a polynomial of degree t-1, satisfying $f(x) = a_0 + a_1x + \cdots + a_{t-1}x^{t-1}$(mod$p$), sets $a_0 = z$. For a client k_i, its secret share is $z_i = f(i)$. Any t participants can jointly reconstruct the secret.
- **Setup** : For a given security parameter λ, set the RLWE dimension n, ciphertext modulus q, key distribution χ and error distribution ψ over R. Then, takes all the security parameters as input and returns the public parameterization(n,q,χ,ψ,R).
- **KeyGen** : For the generator hereinafter, randomly selects a secret $z \in \mathbb{Z}_p$, this secret will be split into n shares, and each of which will be held by one participant. Simultaneously GH computes $A_j = g^{a_j}$, where $j = 0, 1 \ldots, t-1$,

and exposes those parameters. Therefore, the aggregated public key can be expressed as $\tilde{b} = -s_i \cdot a + e \,(\mathrm{mod}\ q)$, where $s_i = z \cdot s \,(\mathrm{mod}\ q)$, $s \leftarrow \chi$.

- **Verify** : For a client k_i verifies the correctness of the secret after receiving it and refuses to perform subsequent operations if the equality $g^{z_i} = \Pi_{j=0}^{t-1} A_j^{x_i^j}$ is not met.

- **Encryption** : Let $a = \mathbf{a}[0]$, $b = \mathbf{b}[0]$. Sample $v \leftarrow \chi$ and $e_0, e_1 \leftarrow \psi$, For a client k_i, encoding a plaintext $m_i \in \mathcal{M}$ and outputs a ciphertext $ct_i \in \{0, 1\}$, where

$$ct_i = (c_0^{k_i}, c_1^{k_i}) = (v'^{k_i} \cdot \tilde{b} + m_i + e_0^{k_i}, v'^{k_i} \cdot a + e_1^{k_i}) \,(\mathrm{mod}\ q) \qquad (1)$$

- **Add** : The sum of ciphertext is as follows:

$$C_{sum} = \sum_{i=1}^{K} ct_i \triangleq (C_{sum_0}, C_{sum_1},) = \left(\sum_{i=1}^{K} (v'^{k_i} \cdot \tilde{b} + m_i + e_0^{k_i}), \sum_{i=1}^{K} (v'^{k_i} \cdot a + e_1^{k_i}) \right)$$
$$(\mathrm{mod}\ q)$$
$$(2)$$

- **Decryption** : Any K participants can jointly reconstruct the ciphertext, and the decryption share is calculated as follows:

$$D_i = s_i \cdot C_{sum_1} + e_i^* = s_i \cdot \sum_{i=1}^{K} (v'^{k_i} \cdot a + e_1^{k_i}) + e_i^* \,(\mathrm{mod}\ q), e^* \leftarrow \psi \qquad (3)$$

Then the sum of all plaintexts C_{sum} can be decrypted as the same.

4.5 FedCMK Design

Based on the above discussion, we have designed a federated privacy-preserving learning framework using the vMK-CKKS scheme. In this framework, a trusted server acts as the federation launcher, serving as the entry point for the entire federated learning process. Before the federated learning process begins, the federation launcher initializes the model and defines the required machine learning architecture. It also generates hyperparameters and a secret for aggregating public keys distributed to all learners. The federation controller is responsible for scheduling learners to perform federated learning tasks and aggregating the local model updates of each learner to compute a new global model. Prior to each round of training, the federation controller selects K' learners, who transmit their performance status and resource information to the federation controller. The federation controller then selects S learners to participate in the current round of training. Each participant in the federated learning process is referred to as a federated learner, and communication between learners is limited to the broadcast phase during secret verification. The learner receives the global model from the federation controller and trains it locally using their private dataset and the tasks assigned by the federation controller. After completing one round of training, the learner sends the ciphertext of their model update to the federation controller.

Therefore, a complete round of federated learning process will be expressed as follows:

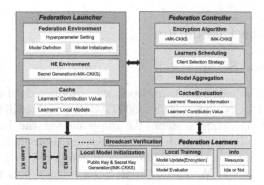

Fig. 1. The Federated Learning Framework Based on our MK-CKKS Homomorphic Encryption (FedCMK)

Initialization: The federation launcher completes the setting of hyperparameters, such as the dimension of RLWE, the ciphertext modulus, and the sampling distribution, and sets up the federation environment. At the same time, the federation launcher generates secrets for aggregated public keys.

Client Selection: The federation controller randomly selects K' learners, and the selected K' learners send their current resource information to the federation controller, such as whether the CPU/GPU is occupied, the approximate size of the local dataset. The federation controller then selects S learners according to the client selection strategy for this round of training.

Local Training: After determining the learners for this round, the federation controller selects the training model, and the selected learners download the global model and conduct local training on their private dataset, generating the local model.

Model Update Encryption: The learners encrypt their local model updates using the secret and public key distributed by the federation launcher and upload the encrypted model updates to the federation controller.

Ciphertext Aggregation: After the federation controller receives the model update ciphertexts of all participating learners, it adds all the ciphertexts into C_{sum}.

Decryption: The federation controller sends C_{sum_1} to learners in this round (if learners in this round S are less than the decryption threshold t, learners in K' are selected in turn), then the selected learners calculate their decryption shares and upload to federation controller. After the federation controller gets all the decrypted shares, it uses C_{sum} and the decrypted shares to restore the plaintext and then updates the global model for $w + 1$ rounds. Then federation controller distributes the new global model to learners participating in the next round (Fig. 1).

5   Performance Evaluation

5.1   Experimental Setup

Our evaluation of the federated framework was conducted on a server with an Intel i5-11400F CPU, NVIDIA RTX 3060Ti GPU, and 16GB RAM, running the Ubuntu 22.04 operating system. We implemented the FedAvg algorithm using Pytorch to evaluate our federated framework. Our multi-key encryption scheme was built using the HEAAN library and compared with several previous multi-key CKKS schemes. We also compared the privacy-preserving learning of Paillier's scheme.

5.2   Results

To evaluate our federated framework, we first measured the accuracy and time cost of one round of FedAvg without multi-key homomorphic encryption. Next, we measured the accuracy and time cost of the round of communication after adding the vMK-CKKS scheme. We used three datasets, MNIST, Shakespeare, and CIFAR100, to conduct four experiments. Additionally, we compared the performance of our federated learning framework with several recent federated frameworks.

Client Selection: We first compared the classical FedAvg federated learning scheme without introducing the client selection strategy and the federated learning scheme with the introduction of the client selection strategy. After comparison, the average time to reach convergence of the federated learning scheme with the introduction of the client selection strategy is significantly reduced. Due to different data sets and different parameter Settings, the convergence will be greatly affected. $T_r = 1\,\text{min}$, and the number of clients selected in each round $S = 0.1K'$. Under this parameter setting, the average accuracy difference between the experiment and the federal learning scheme without client selection is less than 1%, but the training time is reduced by 23% on average.

Accuracy: To compare the accuracy and security of our federated learning framework, we evaluated the performance of different models and different datasets and compared it with several other federated learning frameworks.

Table 1. Comparison of convergence time of different FL schemes

FL Scheme	Total training time	Accuracy
Paillier based FL	105 min	78.9%
xMK-CKKS based FL	73 min	79.5%
tMK-CKKS based FL	61 min	79.6%
vMK-CKKS based FL	52 min	78.8%

We contrast our federated framework with a federated learning framework without privacy protection. Four experiments were conducted for each scheme, using the MNIST, Shakespeares, and CIFAR100 datasets. After comparing the experimental results, before adding homomorphic encryption, the federated learning framework has an accuracy rate of 79.2%, 79.8%, 65.5% and 53.0% in the four experiments, and then added our two multi-key After the CKKS scheme, the accuracy rates are 78.9%, 79.1%, 65.2%, 52.8%, and 78.8%, 78.9%, 64.2%, 52.9%. It means that our multi-key CKKS scheme keeps the accuracy of federated learning model training the same. We list some parameters of the experiment and the final result curve, and we can see that our federated learning framework curve is very similar to the original framework without privacy protection. We detailed our experimental parameter Settings in Table 1, and the results obtained are shown in Fig. 2.

Efficiency: In order to evaluate the performance of the two MK-CKKS schemes, we compared it with the mainstream federal environment homomorphic encryption scheme Paillier, and we also compared it with the xMK-CKKS scheme and the tMK-CKKS scheme. Our experiments mainly compare the following aspects: first, the model update ciphertext size under the Paillier scheme and the model update ciphertext size under different multi-key CKKS schemes, which are compared in detail in Table 2, and second, the time cost of encryption and decryption in the process of other encryption schemes and our encryption scheme, that is, the computational cost.

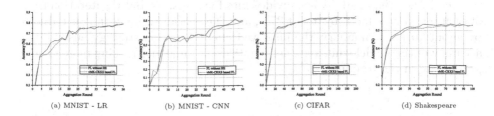

(a) MNIST - LR   (b) MNIST - CNN   (c) CIFAR   (d) Shakespeare

Fig. 2. Performance of different datasets and models under several FL frameworks

According to Table 2, although the encryption and decryption speed of the Paillier scheme is faster than that of the MK-CKKS scheme within a specific range, the average encryption time of the CKKS scheme(0.04ms) is much smaller than that of the Paillier scheme(0.34ms) because of more plaintext can be packaged in the ciphertext. For several MK-CKKS schemes, xMK-CKKS, tMK-CKKS, and vMK-CKKS are homomorphic schemes that meet a threshold. The threshold of MK-CKKS is K, and the threshold of tMK-CKKS and vMK-CKKS can be unified into t so that the decryption time will change according to the values of t and K. Theoretically, when t is less than K, The xMK-CKKS scheme takes longer to decrypt. In general, S will be much smaller than t and K, so the decryption time per round will be shorter than several other MK-CKKS schemes.

Table 2. Different Homomorphic Scheme Parameters And Time Costs

Scheme	Library	Security level	Packing Size	Key size	Ciphertext size	Enc(ms)	Dec(ms)	Add(ms)
Paillier	Python-Paillier	128	60	3072	6144	31.3	15.7	0.1
xMK-CKKS	HEAAN	128	2048	4096	8192	77.1	19.2	2.5
tMK-CKKS	HEAAN	128	2048	4096	8192	77.1	19.2	2.5
vMK-CKKS	HEAAN	128	1024	2048	4096	33.7	12.4	1.8
vMK-CKKS	HEAAN	128	2048	4096	8192	77.1	19.2	2.5

Figure 3 shows the effect of different client weights and clustering on federated learning training time and accuracy. We notice that when the considerable data weight reaches 0.25, which means that there is at least one-quarter of big data clients, our federated learning model can guarantee almost the same accuracy as the original model, but the training time decreases by about 17%. When the weight reaches 0.3, the accuracy of the model is improved by 0.01%, but the training time is only decreased by 10% compared to the original model. Considering the efficiency requirement in practical applications, we set the weight to 0.25 to ensure the balance between accuracy and time overhead. The model can obtain high accuracy quickly for the clustering strategy when the number of clusters is 4. That is, 1000 clients are grouped into 4 clusters of 250 clients per group. Compared with the unclustered algorithm, the accuracy of the model is reduced by less than 0.01% when divided into four clusters, but the iteration time of each round is reduced by about 27%, and the overall training time is reduced by about 10%. In summary, the improved client selection strategy ensures the accuracy of the model while reducing the training time and making more clients participate in the training.

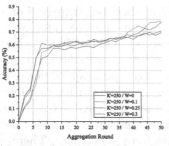

(a) Accuracy and time cost under different client weights

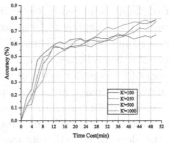

(b) Accuracy and time cost under different client clusters

Fig. 3. Accuracy and time cost under different client clusters and weights

Figure 4 shows the computational cost of encryption, decryption, ciphertext addition, and ciphertext decryption under several schemes. We compare the computational cost of our two schemes with Paillier's scheme, the xMK-CKKS scheme, and the tMK-CKKS scheme. For Paillier's scheme, as the CKKS scheme packs more ciphertexts simultaneously (based on polynomial dimension), as the amount of data increases, It is faster than Paillier's scheme in encryption and decryption, and the gap increases linearly with the number of models to be encrypted. For xMK-CKKS and tMK-CKKS schemes, there is no obvious difference in the speed of encryption, decryption, and ciphertext addition under the same parameters. However, in the decryption phase, the xMK-CKKS scheme requires all K clients to calculate the decryption share, so the computational cost is high. For the tMK-CKKS scheme and vMK-CKKS scheme, more than t clients must jointly decrypt the calculation because t is usually less than K, and the computational cost is low. At the same time, due to client selection, clients involved in vMK-CKKS decryption share calculation often have better performance. Therefore, the decryption speed is faster than tMK-CKKS. We show a more detailed comparison of several MK-CKKS schemes in Fig. 5.

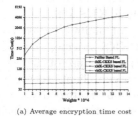

(a) Average encryption time cost

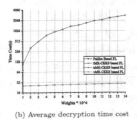

(b) Average decryption time cost

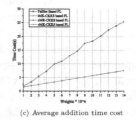

(c) Average addition time cost

Fig. 4. Comparison of calculation time cost between Paillier and different MK-CKKS schemes

Figure 6 shows the computational overhead on the server side. In Fig. 6(a), we compare the influence of different client numbers K, threshold size t, and the number of clients selected in each round of the client selection strategy S on the decryption of the aggregated ciphertext. It can be seen that the value of K is much larger than t and S in general. The decryption cost of xMK-CKKS is higher than that of other schemes because it requires all clients to aggregate. For different choices of t, a smaller value of t will bring faster decryption speed but reduce the security of collusion attacks. The threshold-based secret sharing method for decryption is still faster than the aggregation method in xMK-CKKS. For our multi-key homomorphic scheme, since S in the client selection strategy is smaller than t, the decryption still requires at least t clients to participate. Therefore, the whole is still faster than the tMK-CKKS scheme under the same threshold t. In Fig. 6(b), we compare the time cost of different models. It can be seen that the vMK-CKKS scheme reduces the time cost by about 6% compared with tMK-CKKS. In practical application, considering the balance between security and efficiency, the weight of vMK-CKKS can further reduce the time cost.

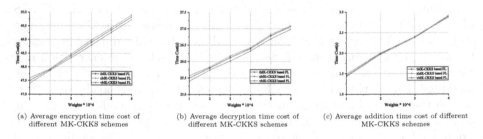

(a) Average encryption time cost of different MK-CKKS schemes   (b) Average decryption time cost of different MK-CKKS schemes   (c) Average addition time cost of different MK-CKKS schemes

Fig. 5. Calculation cost of different MK-CKKS schemes

Figure 7 shows the communication cost under different encryption schemes, and we take the ciphertext size simplicity of different schemes as the cost of the communication overhead. In the Paillier scheme, the cost of the ciphertext grows linearly much more than the other CKKS schemes. The xMK-CKKS scheme ($K = 1000$) always has a higher ciphertext cost than tMK-CKKS and vMK-CKKS based on a threshold ($t = 300$). In the tMK-CKKS scheme, due to the client selection strategy of vMK-CKKS), the size of plaintext used for encryption in each round is less than that of the tMK-CKKS scheme, and the ciphertext size is also slightly reduced. However, in practice, because of the verification mechanism, vMK-CKKS needs to broadcast between clients to verify the correctness of secret fragments, and the client selection strategy requires the client to inform the server of its corresponding resource information, which increases the communication requirement of each round by about 11KB compared with other schemes. Suppose the total number of aggregated rounds is 50. It will incur about 0.53 MB of communication overhead, which is still an order of magnitude smaller than ciphertext. Therefore, the communication overhead of the vMK-CKKS scheme is still smaller than that of the other two MK-CKKS schemes when the number of aggregation rounds is small.

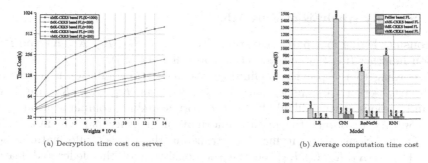

(a) Decryption time cost on server   (b) Average computation time cost

Fig. 6. Decryption time cost on server

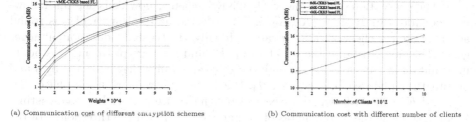

(a) Communication cost of different encryption schemes   (b) Communication cost with different number of clients

Fig. 7. Communication cost of different encryption schemes

6 Security and Functionality Analysis

6.1 Analysis of vMK-CKKS

In Sect. 6.1, we analyze the multi-key CKKS scheme and prove its security, based on which we will analyze the security of the vMK-CKKS scheme.

Theorem 1. *The vMK-CKKS scheme is semantically secure, based on the hardness of the RLWE problem [3].*

Proof. The security of the vMK-CKKS scheme follows directly from the security of the CKKS homomorphic encryption scheme. We can see that $ct_i = (c_0^{k_i}, c_1^{k_i}) = (v'^{k_i} \cdot \tilde{b} + m_i + e_0^{k_i}, v'^{k_i} \cdot a + e_1^{k_i}) \pmod{q}$, and the decryption share $D_i = s_i \cdot C_{sum_1} + e^* = s_i \cdot \sum_{i=1}^{K} (v'^{k_i} \cdot a + e_1^{k_i}) + e^* \pmod{q}$, these messages are all added with errors. The security of the CKKS scheme relies on the hardness of the Ring Learning with Errors (RLWE) problem. RLWE problem is believed to be hard in the worst-case sense, even in the presence of quantum computers.

The aggregate public key security of the vMK-CKKS scheme is guaranteed by the Feldman threshold secret sharing scheme, which is information-theoretic secure. If the secret is divided into s shares, any $s-1$ shares can not recover the original secret so that the vMK-CKKS scheme can resist a certain threshold of collusion attack.

6.2   Security Analysis of FedCMK

We consider here a four-party scenario, with learners K_1, K_2, K_3, K_4 and federation launcher (i.e. S_1) and federation controller (i.e. S_2).

According to the previous definition, the federation initiator is a trusted third-party server, assuming that the participants K_1, K_2, K_3 are honest and curious, they will abide by the corresponding communication protocol but try to obtain the private information of other participants, while K_4 is a malicious party, it may not transmit the correct information. Let $F = \{S_1, S_2, K_1, K_2, K_3, K_4\}$ be the set consisting of this federated learning framework. We consider potentially several kinds of adversaries $\mathcal{A} = \{\mathcal{A}_{s_2}, \mathcal{A}_{K_1}, \mathcal{A}_{K_2}, \mathcal{A}_{K_3}, \mathcal{A}_{K_4}\}$, where $\mathcal{A}_{K_1}$ represents a possible inference attack by learner K_1, and so on.

If the encryption adopts the vMK-CKKS scheme, now consider the following scenario. Firstly, $S = \{K_1, K_2, K_4\}$ is selected as the learner of this round according to the client selection strategy. Ideally, the trusted federation launcher generates and distributes the secret s. K_1, K_2 and K_4 encrypt the model update information m_1, m_2 and m_4 through the aggregate public key $\tilde{b}$ formed by s and output the ciphertexts ct_1, ct_2 and ct_4. But in the secret distribution phase, K_1, K_2, and K_4 will verify whether their secret shares are correct by broadcasting. Finally, suppose the secret shares of all participants reach the threshold t of the secret sharing scheme. In that case, at least t participants have the correct secret shares, and the federation controller decrypts and outputs the sum of plaintexts m on the premise that at least t participants jointly decrypt. We consider the algorithm to be secure.

6.3   The Security of FedCMK

Here, we perform the security proof of the federated learning framework FedCMK based on the analysis in Sect. 6.2.

Theorem 2. *Any private privacy information of the parties involved in Fed-CMK will not be inferred, in the presence of honest and curious adversaries* $\mathcal{A} = \{\mathcal{A}_{s_2}, \mathcal{A}_{K_1}, \mathcal{A}_{K_2}, \mathcal{A}_{K_3}\}$.

Proof. We here analyze the effect of inference attacks by semi-honest adversaries on the overall system. In the vMK-CKKS scheme, since the aggregate public key used for encryption is based on the threshold secret sharing technique, neither individual semi-honest federation controllers nor learners can decrypt the ciphertext independently because they cannot reconstruct the secret independently. Therefore, a semi-honest adversary cannot steal the private data of other learners alone.

Theorem 3. *Even if at most n - 1 learners perform a collusion attack, any private privacy information of the parties involved in FedCMK will not be inferred, in the presence of honest and curious adversaries* $\mathcal{A} = \{\mathcal{A}_{s_2}, \mathcal{A}_{K_1}, \mathcal{A}_{K_2}, \mathcal{A}_{K_3}\}$. *(In the vMK-CKKS scheme, n represents the threshold t of secret sharing)*

Proof. We here analyze the impact of collusion attacks among multiple members on the overall system. In the vMK-CKKS scheme, considering the worst case, $t-1$ learners conduct a collusion attack with the federation controller, hoping to infer the private information of the remaining learner. We introduce the model in Sect. 6.4 for illustration, that is, the federation controller S_2 conspires with learners K_2 and K_4 to infer the private information of K_1. Since the vMK-CKKS scheme builds on the VSS scheme, any holder of $t+1$ secret shares can recover it by polynomial modulo q, while the holder of $t-1$ shares cannot. The VSS scheme is based on the discrete logarithm problem, and there is no probability of cracking through the polynomial complexity algorithm, so it is computationally secure. Therefore, $t-1$ malicious attackers cannot obtain the data of other learners through joint collusion.

Theorem 4. *Even if some malicious adversary shares the wrong secret share, it will not derive any private information of the parties involved in FedCMK or break the decryption, in the presence of honest and curious and malicious adversaries* $\mathcal{A} = \{\mathcal{A}_{s_2}, \mathcal{A}_{K_1}, \mathcal{A}_{K_2}, \mathcal{A}_{K_3}\}$.

Proof. The vMK-CKKS scheme is based on Feldman's verifiable secret sharing scheme. After each client receives the secret share, it needs to verify whether z_i satisfies the equation $g^{z_i} = \Pi_{j=0}^{t-1} A_j^{x_i^j}$. The equality is derived as follows:

$$\Pi_{j=0}^{t-1} A_j^{x_i^j} \,(\mathrm{mod}\ q) = g^{\left(a_j x_i^{t-1} + b_j x_i^{t-1} + c_j x_i^{t-1} + v_j\right)} \,(\mathrm{mod}\ q) \\ = g^{z_i} \,(\mathrm{mod}\ q) \tag{4}$$

According to the difficulty of discrete logarithm calculation, all parameters are hard to be calculated, so if one party provides the wrong secret share, it can not participate in the final decryption calculation, nor can he steal the data. Since only a threshold of t parties with secret shares is needed for secret reconstruction, the wrong secret shares sent by malicious parties do not affect the final decryption.

7   Conclusion

We have improved xMK-CKKS and tMK-CKKS, the two previous multi-key encryption schemes, and optimized the algorithm for efficiency and security. We have improved the aggregation mode of public keys to better adapt to the characteristics of distributed training of federated learning. The overall federated learning framework is more secure, robust, and efficient.

We evaluate our scheme in terms of accuracy, computation cost, and communication cost and compare our scheme with the mainstream Paillier encryption scheme and several different multi-key CKKS schemes. Experiments show that our federated learning framework is more efficient than the traditional federated learning framework while ensuring accuracy and can conduct secure federated training under the condition of having a trusted server.

However, when the number of clients is large, the verifiable secret sharing mechanism requires broadcasting between clients to verify the correctness of the obtained secret snippets, which can incur significant additional overhead, and the communication between clients may cause potential security problems.

In future work, we may optimize scenarios for large-scale federated learning for large-scale clients to ensure that there is not a large amount of additional communication overhead. Optimizing the client selection mechanism may be an option [13]. At the same time, we hope that our encryption scheme can be more suitable for distributed scenarios, such as federated learning without the participation of a trusted third party and vertical federated learning, where each participant holds different keys.

References

1. Bagdasaryan, E., et al.: How to backdoor federated learning. In: International Conference on Artificial Intelligence and Statistics. PMLR, pp. 2938–2948 (2020)
2. Bonawitz, K., et al.: Practical secure aggregation for privacy-preserving machine learning. In: Proceedings of the 2017 ACM SIGSAC Conference on Computer and Communications Security, pp. 1175–1191 (2017)
3. Brakerski, Z., Vaikuntanathan, V.: Fully homomorphic encryption from ring-LWE and security for key dependent messages. In: Rogaway, P. (ed.) CRYPTO 2011. LNCS, vol. 6841, pp. 505–524. Springer, Heidelberg (2011). https://doi.org/10.1007/978-3-642-22792-9_29
4. Brisimi, T.S., et al.: Federated learning of predictive models from federated electronic health records. Int. J. Med. Inf. **112**, 59–67 (2018)
5. Chen, H., et al.: Efficient multi-key homomorphic encryption with packed ciphertexts with application to oblivious neural network inference. In: Proceedings of the 2019 ACM SIGSAC Conference on Computer and Communications Security, pp. 395–412 (2019)
6. Cheon, J.H., Kim, A., Kim, M., Song, Y.: Homomorphic encryption for arithmetic of approximate numbers. In: Takagi, T., Peyrin, T. (eds.) ASIACRYPT 2017. LNCS, vol. 10624, pp. 409–437. Springer, Cham (2017). https://doi.org/10.1007/978-3-319-70694-8_15
7. Du, W., et al.: A efficient and robust privacy-preserving framework for cross-device federated learning. In: Complex & Intelligent Systems, pp. 1–15 (2023)
8. Fan, J., Vercauteren, F.: Somewhat practical fully homomorphic encryption. Cryptology ePrint Archive (2012)
9. Fang, H., Qian, Q.: Privacy preserving machine learning with homomorphic encryption and federated learning. Future Internet **13**(4), 94 (2021)
10. Feldman, P.: A practical scheme for non-interactive verifiable secret sharing. In: 28th Annual Symposium on Foundations of Computer Science (SFCS 1987), pp. 427–438. IEEE (1987)
11. Geyer, R.C., Klein, T., Nabi, M.: Differentially private federated learning: a client level perspective. arXiv preprint arXiv:1712.07557 (2017)
12. Jin, C., et al.: Towards End-to-end secure and efficient federated learning for XGBoost (2022)
13. Konečný, J., et al.: Federated learning: strategies for improving communication efficiency. arXiv preprint arXiv:1610.05492 (2016)

14. Federated Learning: Collaborative machine learning without centralized training data. Publication date: Thursday, April 6 (2017)
15. Li, T., et al.: Federated learning: challenges, methods, and future directions. IEEE Signal Process. Mag. **37**(3), 50–60 (2020)
16. Lyu, L., Yu, H., Yang, Q.: Threats to federated learning: a survey. arXiv preprint arXiv:2003.02133 (2020)
17. Ma, J., et al.: Privacy-preserving federated learning based on multi-key homomorphic encryption. Int. J. Intell. Syst. **37**(9), 5880–5901 (2022)
18. Matsumoto, M., Oguchi, M.: IoT device friendly leveled homomorphic encryption protocols. In: IEEE International Conferences on Internet of Things (iThings) and IEEE Green Computing & Communications (GreenCom) and IEEE Cyber, Physical & Social Computing (CPSCom) and IEEE Smart Data (SmartData) and IEEE Congress on Cybermatics (Cybermatics), pp. 525–532. IEEE (2022)
19. McMahan, B., et al.: Communication-efficient learning of deep networks from decentralized data. In: Artificial Intelligence and Statistics, pp. 1273–1282. PMLR (2017)
20. Nishio, T., Yonetani, R.: Client selection for federated learning with heterogeneous resources in mobile edge. In: ICC 2019–2019 IEEE International Conference on Communications (ICC), pp. 1–7. IEEE (2019)
21. Stripelis, D., et al.: Secure federated learning for neuroimaging. arXiv preprint arXiv:2205.05249 (2022)
22. Tan, A.Z., et al.: Towards personalized federated learning. IEEE Trans. Neural Networks Learn. Syst. **32**, 9587–9603 (2022)
23. Yuan, B., Ge, S., Xing, W.: A federated learning framework for healthcare IoT devices. arXiv preprint arXiv:2005.05083 (2020)
24. Zhang, C., et al.: Batchcrypt: efficient homomorphic encryption for cross-silo federated learning. In: Proceedings of the 2020 USENIX Annual Technical Conference (USENIX ATC 2020) (2020)
25. Zhang, J., et al.: PEFL: a privacy-enhanced federated learning scheme for big data analytics. In: IEEE Global Communications Conference (GLOBECOM), pp. 1–6. IEEE (2019)
26. Zhang, X., et al.: A privacy-preserving and verifiable federated learning scheme. In: ICC 2020–2020 IEEE International Conference on Communications (ICC), pp. 1–6. IEEE (2020)

An Embedded Cost Learning Framework Based on Cumulative Gradient Rewards

Weixuan Tang$^{(\boxtimes)}$ and Yingjie Xie

Institute of Artificial Intelligence, Guangzhou University, Guangzhou, China
`tweix@gzhu.edu.cn`

Abstract. The structure of the Generative Adversarial Network (GAN) has demonstrated good performance in various tasks, mainly comprising two competing sub-networks. The GAN has the potential to effectively generate artificial samples that closely resemble the actual sample distribution. The field of steganography utilizing the Generative Adversarial Network (GAN) structure has witnessed a wealth of research with highly successful outcomes. This paper proposes a steganography framework that integrates reinforcement learning and introduces a new reward function to analyze the embedding cost of images in the steganography problem. In this framework, the reward function assigns distortion values to each pixel of the image and relates the security performance of steganography. Based on the conducted experiments, an enhanced steganographic embedding scheme can ultimately be achieved.

Keywords: Steganography · steganalysis · reinforcement learning · embedding policy · automatic cost learning

1 Introduction

1.1 Steganography

Image steganography is a technique for hiding secret information in an image. The current state-of-the-art steganography methods are primarily implemented using a distortion minimization framework [1], which aims to modify as few detectable elements in the image as possible while achieving a specified embedding capacity. One approach involves the use of tuned filters. Methods such as HUGO [2], WOW [3], HILL [4], S-UNIWARD [5], and MiPod [6] are all based on the distortion minimization framework, where various filters are employed to estimate distortion costs, resulting in excellent anti-detection performance while accommodating a large information payload.

One steganography method involves a combination of a distortion minimization framework and deep learning. The utilization of deep learning to enhance both steganography and steganalysis represents a cutting-edge approach in this field. ASDL-GAN [7] employs a generative adversarial network to autonomously learn the probability of embedding changes for individual pixels in a provided cover. The discriminator in ASDL-GAN utilizes XuNet [8], and it estimates the corresponding gradient by constructing an auxiliary neural network. SPAR-RL [9] learns an attack steganography strategy using reinforcement learning and adversarial attacks against a given analyzer based on convolutional neural networks, aiming to achieve optimal security. MCTSteg [10] enhances the order and extent of modifications to the state and pixel embedding of the cover, with more detailed operations resulting in improved steganography. ReLOAD [11] optimizes the asymmetric distortion applied in additive steganography, leading to a substantial enhancement in the performance of this steganographic method.

There are also methods that exclusively rely on deep learning techniques to embed information, significantly enhancing the image's embedding capacity while sacrificing a portion of its security performance. SSGAN [12] uses information embedding as the generative part and visual image recognition and steganography detection as the discriminative part. Some studies [13–15] have successfully concealed information within images using generative adversarial networks. StegGAN [16] generates steganographic images through unsupervised adversarial training. SteganoGAN [17] employs deep convolutional steganographic structures to attain improved results in both visual quality and steganographic security. HiDDeN and others [18,19] embed a substantial amount of information while enhancing security and visual quality through the joint training of encoder and decoder networks.

Most existing research on deep learning-based steganographic structures focuses on the learning objective of embedding probability. However, the work presented in this paper will concentrate on learning the embedding cost. Our work in this paper employs a reinforcement learning approach for image steganography. Specifically, the agent network takes a state as input, and its output corresponds to the embedding cost. Using the embedding cost, we apply the additive distortion framework to derive the embedding probability, which, in turn, is used to generate the final stego. Additionally, we consider the performance of steganalysis on the generated steganographic image when providing the reward function for the agent network and configuring the environment. The primary objective of employing reinforcement learning as a framework is to estimate the approximate cost associated with embedding the cover image. During this process, we face several primary challenges. Firstly, multiple distinct embedding schemes exist for a cover image, leading to a one-to-many problem. Determining the associated embedding cost through specific embedding results is challenging due to the imperfect nature of embedding analysis. Addressing this issue requires a substantial number of embedding samples, which inevitably increases the costs and complexity of the learning process. Another issue arises from incorporating the discriminator as part of the environment. The robustness of the discriminator network becomes a major concern since the information from the discriminator network significantly influences the learning capacity of the agent network.

2 Preliminaries

Since the framework proposed in this paper combines two fundamental techniques, namely the minimum distortion framework and reinforcement learning, we will begin by introducing them in this section. For the remainder of this article, matrices will be denoted by bold capital letters, while the individual elements of the matrix will be represented by the corresponding lowercase letters.

2.1 Reinforcement Learning

Reinforcement learning is a machine learning paradigm that models a problem as a Markov decision process. In this process, the agent interacts with the environment to learn a strategy that maximizes the expected return. A Markov decision process comprises three essential elements: state, action, and reward. The state signifies the condition of the environment, the action is the output of the agent based on the state, and the reward is the signal from the environment in response to the agent's action. At any given time t, a Markov decision process defines the Markov nodes and is represented as follows.

$$\{s_t, a_t, r_t\}, s_t \in S, a_t \in A, r_t \in R,$$

Here, S, A, and R all represent finite sets. The transition from a state at any given time t to the state at the next time $t + 1$ can be expressed as follows:

$$s_t \xrightarrow[R(a_t, s_t)]{a_t} s_{t+1}.$$

The state value function, described by the following equation, is calculated as the expectation of discounted rewards in the current state s:

$$V_\pi(s) = \mathbb{E}_\pi \left[\sum_{t=0}^{t=T-1} \gamma^t R(a_t, s_t) \right], s_0 = s, \tag{1}$$

the loss function for reinforcement learning can be expressed as

$$\boldsymbol{\theta} \leftarrow \boldsymbol{\theta} + \eta \nabla_\theta \log \pi^\theta (a|s) A(a|s), \tag{2}$$

where A denotes the advantage function

$$A(a|s) = R(a|s) - V(a|s). \tag{3}$$

2.2 Distortion Minimization Framework

The framework commonly used in steganography for minimising distortion involves modelling the process of information embedding as a constrained optimization problem. This problem can be expressed as follows, with a given information embedding capacity of C.

$$\min_{\mathbf{Y}} D(\mathbf{X}, \mathbf{Y}), \quad \text{s.t. } \psi(\mathbf{Y}) = C, \tag{4}$$

where $\mathbf{X}$ denotes cover and $\mathbf{Y}$ denotes stego. Function ψ measures the embedded information capacity C and is usually measured in bits.

It can be assumed that the distortion cost between each pixel is independent of each other, and the overall distortion can be defined by the additive distortion function, as in the following equation

$$D(\mathbf{X},\mathbf{Y}) = \sum_{i=1}^{H}\sum_{j=1}^{W} \rho_{i,j}|x_{i,j} - y_{i,j}|, \tag{5}$$

where H and W represent the height and width of the image, respectively, and $\rho_{i,j}$ represents the embedding distortion value for the i-th row and j-th column. The function D is used to measure the overall distortion value.

In the ternary embedding scheme, the modification point is $\mathcal{M} \in (-1, 0, 1)$, and the embedding probability is determined by the given embedding capacity as follows:

$$p_{i,j}^{(m)} = \frac{e^{-\lambda \rho_{i,j}^{(m)}}}{\sum_{\tilde{m}\in\mathcal{M}} e^{-\lambda \rho_{i,j}^{(\tilde{m})}}}, \quad m \in \mathcal{M}, \tag{6}$$

where λ is a parameter determined by the following constraint:

$$-\sum_{i=1}^{H}\sum_{j=1}^{W}\sum_{m\in\mathcal{M}} p_{i,j}^{(m)}\log_2 p_{i,j}^{(m)} = C. \tag{7}$$

3   Framework

3.1   Overview of the Overall Framework

A reinforcement learning framework is introduced to autonomously learn the embedding cost of cover images through interactions between the agent and the environment. The framework for this work comprises two main components: one is the U-Net, which serves as the agent, and the other is the environment. The environment consists of two subcomponents, a distortion minimization framework and a steganalysis discriminator, which can be a commonly used steganography analyzer. The overall framework is illustrated in Fig. 1. The Agent network's parameters are denoted as θ, and the network takes an image $\mathbf{C} = (c_{i,j})^{H \times W}$ as input. The Agent network will directly output the corresponding embedding cost, denoted as $\rho = \mathbf{\Pi}^{\theta} = \left(\pi_{i,j}^{\theta}\right)^{H \times W}$, for the image, where $\pi_{i,j}^{\theta}$ represents the embedding cost of each pixel. The embedding cost will be transformed through an additive distortion framework into an embedding probability map, denoted as $\mathbf{p}^{\theta} = \left(p_{i,j}^{\theta}\right)^{H \times W}$, which will be sampled to obtain a simulated embedding point, denoted as $\mathbf{M}$. The modification point obtained in this manner can be represented as $\mathbf{M} = \left(m_{i,j}^{\theta}\right)^{H \times W}$, where $m_{i,j}^{\theta}$ represents the modification value of the pixel at row i, column j. In this manner, we can obtain the stego as represented in the following equation:

$$\mathbf{S} = \mathbf{C} + \mathbf{M}, \tag{8}$$

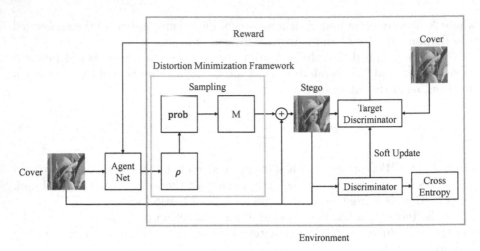

Fig. 1. Schematic diagram of the overall framework.

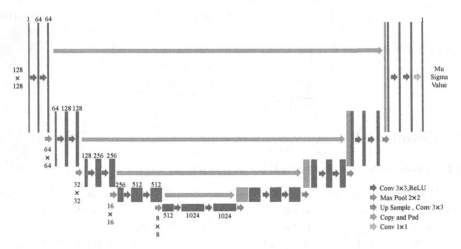

Fig. 2. The above figure represents the U-Net structure used as agent, where the features extracted by one encoder will be used as shared features for three decoders. The three decoders have the same structure.

and the second deep neural network is called the environment network, denoted by the parameter ω, which is used by the steganalysis network to provide reward signals. Typical CNN steganalysers, such as XuNet and SRM [20], can be employed for this purpose. The environment network is trained using the corresponding steganographic images generated by the covers and agents. To calculate the reward, we design a reward function that considers the distortion of the steganographic information at each pixel.

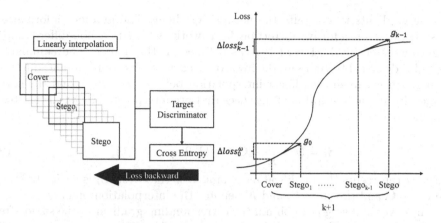

Fig. 3. Information on cumulative rewards. The stego and cover are linearly interpolated into many slices. Each slice can be used to obtain the inverse gradient of each point in the image using a fixed discriminator. Ideally, the higher the number of slices obtained by linear interpolation, the cumulative gradient will allow the accurate calculation of the distortion value of each modified point relative to the cover image.

It's important to note that the proposed framework operates in a manner where the sampled actions depend on the reward signals provided by the environment for policy updates. Actions with higher rewards will have a greater likelihood of being sampled in the next iteration, while actions with lower rewards will still have a certain probability of being sampled. This framework represents a steganographic single-step multi-agent Markov Decision Process (MDP) problem. The details of the U-Net structure used as an agent are illustrated in Fig. 2. It is worth noting that in the architecture of this paper, the discriminator located in the environment part adopts a soft update strategy. On one hand, in the early stages of training, the performance of the discriminator may be unstable. The inclusion of the soft update strategy can stabilize the gradient information provided by the discriminator within the environment. On the other hand, the concept of soft update is inspired by DDPG [21], where the idea of deterministic gradient effectively addresses the issue of excessively large state spaces.

3.2 Details of Cumulative Rewards

In reinforcement learning, agents learn by responding to reward signals, with the ultimate goal of maximizing their cumulative rewards. The paper models the process of steganographic embedding using multiple agents in an environment where the steganalyzer plays a dominant role. The reward signal is generated by the reverse gradient passing through the steganalyzer, with the ultimate learning goal of minimizing the overall distortion cost. Building on the concept above, our objective is to compute the distortion value for every corresponding pixel in both the cover and the stego images. Typically, adversarial networks employ reverse gradients to compute the network gradient. The gradient information must traverse an additive distortion framework, which is a non-differentiable structure, after the backward loss, as detailed in this paper. The cumulative reward is designed to compute the reward for each pixel-specific action. Initially, we perform evenly spaced linear interpolation between the stego and the cover images. Here is a description of the slices that are interpolated between the stego and the cover:

$$\mathbf{S}_i = \mathbf{M} * \frac{i}{K} + \mathbf{C}, i \in \{1, 2, \cdots, k-1\}, \tag{9}$$

Here, $\mathbf{M}$ represents the modification points, $\mathbf{C}$ is the cover, $\mathbf{S}_i$ is a slice between the cover $\mathbf{C}$ and the stego $\mathbf{S}$, and K denotes the interpolation number.

The cover and stego can obtain the corresponding gradient of the stego after the discriminator undergoes the loss backpropagation. The corresponding gradient of the stego can be represented as

$$g_i^\omega = \frac{\partial loss_\omega}{\partial \mathbf{S}_i}. \tag{10}$$

Multiplying the corresponding gradient of each slice by the amount of modification gives the loss of the modification point to the discriminator loss function. This process can be expressed as follows:

$$\triangle loss_i^\omega \approx g_i^\omega \frac{\mathbf{M}}{K}, \tag{11}$$

the approximate loss can be obtained by accumulating the g_i^ω obtained from each slice as follows

$$\triangle L^\omega = \sum_{i=1}^K \triangle loss_i^\omega = \sum_{i=1}^K g_i^\omega \frac{\mathbf{M}}{K} = \sum_{i=1}^K \frac{\partial loss_\omega}{\partial \mathbf{S}_i} \frac{\mathbf{M}}{K}, \tag{12}$$

the final reward function is expressed as

$$\mathbf{R} = -\mathbf{P} \odot \triangle L^\omega, \tag{13}$$

where ω represents a steganalysis network. Ideally, as i tends to infinity, we can obtain a precise measurement of the pixel-level distortion between the stego and the cover. The gradient corresponding to the modification point is obtained through backward gradient propagation for each equally spaced interpolated slice. The distortion value resulting from each modification point pair is obtained by accumulating the gradients of the slices. The specific implementation is depicted in Fig. 3 and (Fig. 4). The ultimate learning objective of the policy network can be expressed as the following optimization problem, which is defined by the loss function

$$\theta \leftarrow \theta + \eta \nabla_\theta \log \pi^\theta(a|s) \mathbf{R}. \tag{14}$$

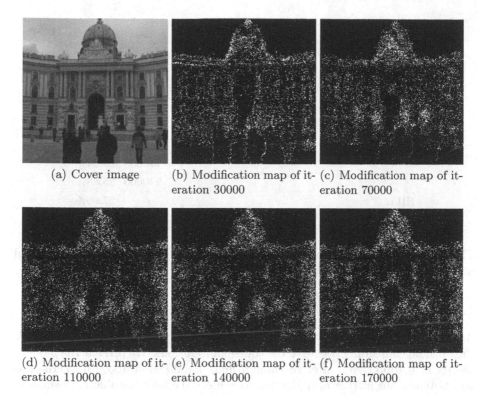

(a) Cover image   (b) Modification map of iteration 30000   (c) Modification map of iteration 70000

(d) Modification map of iteration 110000   (e) Modification map of iteration 140000   (f) Modification map of iteration 170000

Fig. 4. Illustration of the cover image "01013.pgm" from 256 × 256 *BOSSBase* and its modification map of different steganographic methods on 0.4 bpp, where white points indicate modifications of ±1.

4   Experiment

4.1   Setting

The framework presented in this paper is trained on the Alaska256 dataset, which comprises 37,511 grayscale images with dimensions of 256 × 256. The testing phase employs the BossBase dataset, which includes 10,000 grayscale images with dimensions of 256 × 256. The generator is used to create the distortion map for the cover images in the testing phase. This is accomplished by minimizing the distortion framework to obtain simulated embedding points, thereby generating stego images. The security performance of the framework is assessed using a deep learning-based SRM feature set as described in this paper. The test error of the steganalyzer is employed to evaluate security performance and is calculated as the average of the false alarm rate and the miss detection rate. This is represented by the equation:

$$P_E = \frac{1}{2}(P_{FA} + P_{MD}).   \tag{15}$$

Table 1. Security performance of Our work.

Steganographic scheme		0.4 bpp
		SRM
Our Scheme with different Training Iterations	30000	21.20%
	70000	23.87%
	110000	24.87%
	140000	25.42%
	170000	25.24%
	210000	25.88%
HILL		27.00%

The framework proposed in this paper is implemented on PyTorch and trained using a sheet of NVIDIA Tesla V100s (Table 1).

4.2 Analysis

As demonstrated, the generator in this paper is optimized and converges towards the optimal embedding cost over the course of iterations. We present the model's results at 0.4bpp. Even after reaching a final stabilization above 0.25, there is still potential for further optimization to achieve the best possible performance. Simultaneously, we must acknowledge that the final performance falls short of the expected results. We believe this may be attributed to a combination of factors, as outlined below.

- As this framework is based on generative adversarial networks, the generator's performance is significantly influenced by the discriminator's strength, and hence the discriminator's robustness impacts the inverse gradient information.
- Regarding the reward function design, while this paper proposes a pixel-level reward function, the overall embedding security performance has not been fully accounted for and leaves room for further exploration.

5   Conclusion

The steganography performance of the generator improves as the model iterates. However, we also observe that the final results fail to surpass the performance of the traditional filter-based approach. We believe that there may be unaccounted factors in the design of the reward function that contribute to the final performance falling short of our expectations.

Acknowledgements. This work was supported by NSFC (Grant 62002075), Guangdong Basic and Applied Basic Research Foundation (Grant 2023A1515011428), the Science and Technology Foundation of Guangzhou (Grant 2023A04J1723).

References

1. Filler, T., Judas, J., Fridrich, J.J.: Minimizing additive distortion in steganography using syndrome-trellis codes. IEEE Trans. Inf. Forensics Secur. **6**(3–2), 920–935 (2011)
2. Pevný, T., Filler, T., Bas, P.: Using high-dimensional image models to perform highly undetectable steganography. In: Böhme, R., Fong, P.W.L., Safavi-Naini, R. (eds.) IH 2010. LNCS, vol. 6387, pp. 161–177. Springer, Heidelberg (2010). https://doi.org/10.1007/978-3-642-16435-4_13
3. Holub, V., Fridrich, J.J.: Designing steganographic distortion using directional filters. In: 2012 IEEE International Workshop on Information Forensics and Security, WIFS 2012, Costa Adeje, Tenerife, Spain, 2–5 December 2012, pp. 234–239. IEEE (2012)
4. Li, B., Wang, M., Huang, J., Li, X.: A new cost function for spatial image steganography. In: 2014 IEEE International Conference on Image Processing, ICIP 2014, Paris, France, 27–30 October 2014, pp. 4206–4210. IEEE (2014)
5. Li, B., Wang, M., Huang, J., Li, X.: A new cost function for spatial image steganography, pp. 4206–4210 (2014)
6. Sedighi, V., Cogranne, R., Fridrich, J.J.: Content-adaptive steganography by minimizing statistical detectability. IEEE Trans. Inf. Forensics Secur. **11**(2), 221–234 (2016)
7. Tang, W., Tan, S., Li, B., Huang, J.: Automatic steganographic distortion learning using a generative adversarial network. IEEE Sig. Process. Lett. **24**(10), 1547–1551 (2017)
8. Xu, G., Wu, H., Shi, Y.: Structural design of convolutional neural networks for steganalysis. IEEE Sig. Process. Lett. **23**(5), 708–712 (2016)
9. Tang, W., Li, B., Barni, M., Li, J., Huang, J.: An automatic cost learning framework for image steganography using deep reinforcement learning. IEEE Trans. Inf. Forensics Secur. **16**, 952–967 (2021)
10. Mo, X., Tan, S., Li, B., Huang, J.: MCTSteg: a Monte Carlo tree search-based reinforcement learning framework for universal non-additive steganography. IEEE Trans. Inf. Forensics Secur. **16**, 4306–4320 (2021)
11. Mo, X., Tan, S., Tang, W., Li, B., Huang, J.: ReLOAD: using reinforcement learning to optimize asymmetric distortion for additive steganography. IEEE Trans. Inf. Forensics Secur. **18**, 1524–1538 (2023)
12. Shi, H., Dong, J., Wang, W., Qian, Y., Zhang, X.: SSGAN: secure steganography based on generative adversarial networks. In: Zeng, B., Huang, Q., El Saddik, A., Li, H., Jiang, S., Fan, X. (eds.) PCM 2017, Revised Selected Papers, Part I. LNCS, vol. 10735, pp. 534–544. Springer, Cham (2018). https://doi.org/10.1007/978-3-319-77380-3_51
13. Volkhonskiy, D., Nazarov, I., Burnaev, E.: Steganographic generative adversarial networks. In: Osten, W., Nikolaev, D.P. (eds.) Twelfth International Conference on Machine Vision, ICMV 2019, SPIE Proceedings, Amsterdam, The Netherlands, 16–18 November 2019, vol. 11433, p. 114333M. SPIE (2019)
14. Yedroudj, M., Comby, F., Chaumont, M.: Steganography using a 3-player game. J. Vis. Commun. Image Represent. **72**, 102910 (2020)
15. Zhang, R., Dong, S., Liu, J.: Invisible steganography via generative adversarial networks. Multim. Tools Appl. **78**(7), 8559–8575 (2019)
16. Hayes, J., Danezis, G.: Generating steganographic images via adversarial training, pp. 1954–1963 (2017)

17. Zhang, K.A., Cuesta-Infante, A., Xu, L., Veeramachaneni, K.: SteganoGAN: high capacity image steganography with GANs. CoRR abs/1901.03892 (2019)
18. Zhu, J., Kaplan, R., Johnson, J., Fei-Fei, L.: HiDDeN: hiding data with deep networks. In: Ferrari, V., Hebert, M., Sminchisescu, C., Weiss, Y. (eds.) ECCV 2018, Part XV. LNCS, vol. 11219, pp. 682–697. Springer, Cham (2018). https:// doi.org/10.1007/978-3-030-01267-0_40
19. Baluja, S.: Hiding images in plain sight: deep steganography. In: Guyon, I., et al. (eds.) Advances in Neural Information Processing Systems 30: Annual Conference on Neural Information Processing Systems 2017, 4–9 December 2017, Long Beach, CA, USA, pp. 2069–2079 (2017)
20. Fridrich, J.J., Kodovský, J.: Rich models for steganalysis of digital images. IEEE Trans. Inf. Forensics Secur. 7(3), 868–882 (2012)
21. Silver, D., Lever, G., Heess, N., Degris, T., Wierstra, D., Riedmiller, M.A.: Deterministic policy gradient algorithms. In: Proceedings of the 31th International Conference on Machine Learning, JMLR Workshop and Conference Proceedings, ICML 2014, Beijing, China, 21–26 June 2014, vol. 32, pp. 387–395. JMLR.org (2014)

An Assurance Case Practice of AI-Enabled Systems on Maritime Inspection

Yongjian Xue[✉], Qian Wei, Xiaoliang Gong, Fan Wu, Yunqi Luo,
and Zhongning Chen

Group Research and Development, DNV, Shanghai, China
`yong.jian.xue@dnv.com`

Abstract. Assuring of AI-enabled systems is challenging and beyond current assurance practices, especially in bridging the gap between assurance process and tools. In this paper, an AI-powered corrosion detection system for maritime inspection is presented as an assurance use case. It serves as a decision support tool for surveyors to assess the coating conditions in ballast tanks. Before deployment, it is crucial to establish confidence of this system as the stakeholders seek to understand the potential risk of adopting it compared to existing or alternative solutions.

Different from other works focused on assurance process or framework, this paper conducts both assurance process and testing methods to create a detailed assurance case. A systematic top-down approach is used to derive the assurance requirements from the system level to the machine learning component level, while testing is conducted using a bottom-up approach to collect the required evidences. Furthermore, a risk-based approach is integrated into the corresponding AI assurance lifecycle, providing valuable insights on analyzing the risk that an AI component may bring into AI-enabled systems.

Keywords: Trustworthy AI · Risk analysis · Assurance case · Testing

1 Introduction

With the rapid development of AI technologies, many systems are enabled by AI components to enhance efficiency and reduce cost. These systems have been widely adopted across various industries, from energy, health care, supply chain, transportation to manufacturing. At the same time, concerns have arisen from different aspects, encompassing both technical and ethical considerations. Among them, the requirements of trustworthy AI are highlighted from privacy, fairness, safety, transparency and others.

Various stakeholders, including communities, international or national standard organizations and policy makers are actively working on how to guide or regulate the AI development in order to address trust issues. For example, the International Organization for Standardization (ISO) and the International

J. Vaidya et al. (Eds.): AIS&P 2023, LNCS 14509, pp. 283–299, 2024.
https://doi.org/10.1007/978-981-99-9785-5_20

Electrotechnical Commission (IEC) are working on a series of technical reports to cover AI-related concerns and provide corresponding guidelines. The China Electronics Standardization Institute, in collaboration with all other research institutes and industrial partners, has also published guidelines on the ethics of trustworthy AI in 2023. In 2021, the European Commission proposed the EU AI Act [2], which emphasizes the classification of risks for different AI applications. This legislation is set to come into force soon and become the world's first comprehensive law to regulate AI.

Despite the numerous requirements and guidelines in place, there remains a huge gap in implementing these requirements in practice for the assurance of AI. This is due to the complexity and dynamics inherent in AI systems. Firstly, it is challenging to align these high-level requirements with specific, actionable requirements or suitable solutions. Secondly, proper tools and methods are needed to facilitate the compliance check for these requirements.

In this paper, we use the Assurance of Machine Learning in Autonomous Systems (AMLAS) [3] framework to conduct an assurance use case study on the AI-based corrosion detection system. The key contributions of this paper are as follows: (1) It provides a detailed and systematic assurance example for AI-enabled systems, (2) A risk analysis is conducted throughout the entire assurance process, which is rarely seen in related case studies, (3) Testing tools and methods are used to collect evidence to support the claims of requirements, which provide insights on how to implement assurance for AI-enabled systems.

The structure of this paper is organized as follows. Section 2 provides an overview of the assurance case. Section 3 conducts the assurance scoping. Section 4 presents the assurance requirements. Section 5 discusses the data management assurance. Sections 6, 7 and 8 cover the model learning, model verification and development assurance. At last, Sect. 9 summarizes the conclusions.

2   Assurance Case Description

In this section, we provide a description of the assurance case, including system description, component description and the operational environment description. These details are typically derived from the AI product specification documents.

2.1   System Description

The assurance case in this study is an AI-powered corrosion detection system called Corrosion.ai that is designed to detect rusted areas in ballast tanks [8]. This web-based tool supports surveyors in assessing coating condition, particularly in estimating the percentage of rusted area. The system architecture is shown in Fig. 1. Firstly, images are captured by cameras. The quality and relevance of images are then checked by experts, and only images satisfying pre-defined constraints are sent to the segmentation algorithm. The segmentation algorithm, which is based on deep neural networks, identifies rusted regions in the image and returns a mask indicating both rusted and rustless areas. Based on the output mask, the corrosion percentage value is calculated, and the corresponding coating condition is rated. It serves as a second opinion for surveyors, aiming to help them make reliable decisions, especially in controversial situations.

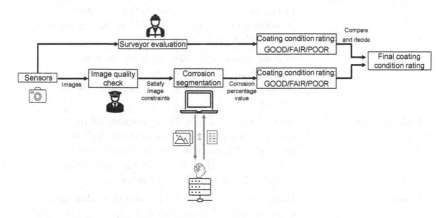

Fig. 1. Corrosion.ai system architecture.

2.2   ML Component Description

The main ML component of Corrosion.ai is a semantic segmentation network with U-Net architecture [7] which identifies corrosion regions in ballast tank images. The encoder part of the U-Net employs SE-ResNeXt-50 [4] with pretrained weights learned on ImageNet. The neural network is implemented using the open-source machine learning framework PyTorch with the 3rd party code based Segmentation Models library. The dice loss between network outputs and corrosion masks produced by human annotators is calculated to generate the supervision signal which optimizes the network's weights. The final weights achieved an IoU score of 0.5042 on the validation dataset and 0.4011 on the test dataset.

2.3   Operational Environment Description

The operational environment description is used to specify the operational design domain (ODD) of the assurance case. For Corrosion.ai, we describe it from the following categories: hull structure types, surfaces, field of view, photo quality, illumination, particulate matter, and supported coating failure types. These categories are shown in Table 1.

3   Assurance Scoping

The product specifications in previous section, including system architecture, operational environment, and supported functions, are utilized as inputs for the assurance scoping.

Table 1. Summary of the operational environment description.

	Description
Hull Structure Types	It supports the majority of hull structure types in ballast tanks belonging to "area under consideration" depicted by [5], including hopper area, tank top, bottom, frame, and bilge.
Surfaces	Surface pre-treatments, such as the removal of mud, oil, and grease on the area, are needed to ensure a dry, clean and light-coloured paint surface.
Field of View	According to [5], a holistic assessment of specified area under consideration shall be provided. Photos shall cover a sufficiently large field of view and be taken at an appropriate distance from the object of interest.
Photo Quality	High-resolution images are recommended, and down-scaling of photos shall be greater than 800 pixels. Motion blur or defocus blur shall be avoided.
Illumination	Sufficient and proper illumination inside ballast tanks is essential for achieving a good corrosion detection result.
Particulate Matter	Impurities in images shall be avoided. Fog, smoke, smog, spatter, dust, and dirt should be limited to a human-recognizable level
Coating Failure Types	i). General corrosion (uniform corrosion), ii). Pitting corrosion (localized corrosion). The detection of coating failures/degradations described by ISO 4628 (e.g., flaking) is not supported.

3.1 System Level Requirements

SYS-REQ1. Corrosion.ai shall achieve human-level performance in the defined Operational Design Domain (ODD).

The rationale of system level requirements includes:

- Usually, a target, e.g., the performance declared in the product specification needs to be established or an existing system, e.g., the AI-enabled system compared to the traditional system needs to be benchmarked. In this case, the performance of Condition Assessment Program(CAP) surveyors is used as the baseline.
- The existing domain-specific rules and standards [5] do not provide an acceptance threshold for corrosion evaluation. However, the qualification of CAP surveyor's results can be used to qualify the Corrosion.ai product. Since the

corrosion estimation and classification (GOOD, FAIR, POOR) are objective, and the surveyors cannot even reach a consensus in boundary situation. But the evaluation variations and misclassification rate of Corrosion.ai shall not be larger than that of CAP surveyors.

- The view of stakeholders is also very important. In this case, from domain experts and users, it is agreed that the Corrosion.ai shall not take over the decision of surveyors.

3.2 Requirements Allocated to ML Component

The ML component of the system is the segmentation algorithm and the corresponding requirements developed from system level can be further broken down to the component level.

COM-REQ1. Corrosion.ai shall give correct segmentation and rating when the input image meets image quality, shooting requirements and environmental requirements.

The final rating is calculated from the ML segmentation component as a percentage value, with the corresponding mappings of: GOOD (3%), FAIR(3–20%), and POOR (>20%). From the final rating results of corrosion estimation, the hazards (consequences) are divided into two parts.

- Underestimation: It rates from POOR to FAIR, POOR to GOOD, and FAIR to GOOD. This may lead to un-suggested repair and maintenance, which can cause damage to the structure and safety issues.
- Overestimation: It rates from GOOD to FAIR, GOOD to POOR, and FAIR to POOR. While the severity is low, but it may still cause reputation loss for the system user, further investigation and additional costs may be needed to re-evaluate the corrosion conditions.

Harzards Analysis. Since Corrosion.ai is a decision assistant tool, human checks and further investigations are still required for real-world applications, especially when it rates a structure as POOR or FAIR. In that case, surveyor's intervention is assumed, and corrosion images are reviewed by them for making claims. As a result, the assurance burden is shared between humans and ML component. Accordingly, we can map the hazards from the system level to the ML component level, as shown in Table 2. The hazard is mitigated from moderate (system level) to minor (component level) if it mis-rates the corrosion percentagefrom GOOD to POOR. For other mis-rating situations, consequences do not change from the system level to the component level.

Considering both the hazards and operational environment, these requirements can be further refined as performance and robustness requirements in the next section.

Table 2. Hazards analysis from system level to component level.

Rating results		GOOD (3%)	FAIR (3–20%)	POOR (>20%)
Ground truth	GOOD (3%)	None	Minor	**Moderate** (System) **Minor** (Component)
	FAIR (3–20%)	Major	None	Minor
	POOR (>20%)	Major	Moderate	None

4   ML Assurance Requirements

From the component requirements, we can further derive the ML specific requirements which include performance requirements and robustness requirements (see Table 3).

Table 3. ML specific assurance requirements.

Performance	
PER-REQ1	The accuracy of "GOOD-FAIR-POOR" rating shall achieve the accuracy of human surveyors
Robustness	
ROB-REQ1	The corrosion segmentation component shall perform as required for all (main) structures of a ballast tank within the defined ODD
ROB-REQ2	The corrosion segmentation component shall identify corrosion irrespective of the background appearance (painting color, typical drawing, mild dirty surface) with respect to the camera
ROB-REQ3	The corrosion segmentation component shall identify corrosion for a distance ranging from 1.5 m to 10 m under good illumination
ROB-REQ4	The corrosion segmentation component shall identify corrosion for a human-recognizable degraded image (lighting, blur, noise)
ROB-REQ5	The corrosion segmentation component shall identify corrosion irrespective of their pose with respect to the camera

4.1   Rationale for ML Specific Requirements

The accuracy of human surveyors or the "human-level performance (HLP)" on assessing GOOD/FAIR/POOR coating conditions is derived from the following activities:

- Seven senior surveyors with experience of 8–14 years are selected by the operator.
- 108 images are selected from different inspection stations, and are reviewed by domain experts. Those images constitute the dataset referred to as pilot dataset.
- For each image, among all seven surveyors' assessments, we remove the two most "outlying" ones and take the median values of the rest as the ground truth assessment.
- Surveyors' accuracy of assessment is then evaluated against the ground truth obtained from the previous step, results are shown in Fig. 2a.

To translate from human-level performance to IoU of image segmentation, the algorithm was tested on several augmented image datasets with degradation. It is shown that a human-level accuracy in Fig. 2a corresponds to 0.345 of IoU. IoU is a common metric for image segmentation, which is more directly to verify the algorithm. In this report, the actual IoU is verified for PER-REQ1 when pixel-level labelled ground truth is available, otherwise, a confusion matrix is evaluated.

4.2   Risk Analysis

For the risk analysis of human-level performance, both the consequence (Table 2) and likelihood (Fig. 2a) shall be considered. To quantify the risk, we replace the confusion matrix as following: corrected rating = 0, minor=1, moderate=2, major=3, then we calculate the risk as:

$$Risk = Consequence \times Likelihood \tag{1}$$

Accordingly, the risk matrix of human-level performance is obtained as shown in Fig. 2b.

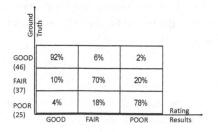
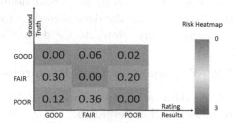

(a) Human-level performance.   (b) Risk of human-level performance.

Fig. 2. Human-level performance and risk analysis.

4.3 ML Assurance Requirements Validation Results

As it is challenging to list all dimensions of variation or decide a quantitative boundary of variation for robustness requirements, the authors invited reviewers, including AI researchers and ship surveyors, to an inspection meeting for the validation of these requirements.

Before the meeting, a dataset of 4,000 images was prepared from a dataset of around 12,000. This dataset was generated by filtering out low quality images and clustering by image hidden features. During the meeting, reviewers went through the image dataset to decide if the image can be used for the CAP survey. This decision generally covers all aspects listed above in robustness requirements.

Finally, image samples were selected to represent the typical variation in the robustness requirements, for instance, typical painting color, drawings, lighting conditions. All reviewers agreed that Corrosion.ai shall at least work well on these scenarios. If more images are to be added to verify robustness requirements, they shall represent the same range of variation as the currently selected image samples.

5 Data Management Assurance

In this section, we firstly describe the data requirements and how these requirements are linked to the previous ML performance and robustness requirements. Then we will give the data verification methods and results as evidences to support the arguments of data requirements.

5.1 Data Requirements

According to [3], the data requirements are specified as relevance, completeness, balance, and accuracy which are listed in Table 4. The relevance requirements describe the data must match the operational domain, from structure, corrosion type, condition, environment, and image quality while the completeness requirements specify the coverage of that domain. The balance requirements describe the distribution of the data according to different dimensions, and the accuracy requirements are related to the data labelling quality.

5.2 Rationale for Data Requirements

All the data requirements are derived from the operational domain and are the results of further development from ML specific component requirements. Generally, the relevance and completeness requirements are derived as the corrosion detection system shall operate in the relevant operational domain, and cover the interested area without expectation for all situations. The balance requirements are also derived from these dimensions and requirements relating to accuracy are derived from labelling accuracy for image segmentation. All these data requirements have been discussed and reviewed by AI experts and ship surveyors in regular project meetings.

5.3   Data Generation

In total, three categories of datasets are created: development data, internal test data, and verification data. The development data and internal test data are ballast tank photos taken from 13 vessels, which are also categorized as training,

Table 4. ML data requirements.

Relevance	
REL-REQ1	All data samples shall represent images of a hull structure that corresponds to the ODD
REL-REQ2	All data samples shall represent images of a corrosion type that corresponds to the ODD
REL-REQ3	All data samples shall represent a condition (including FOV, distance, illumination, pose) that images are captured that corresponds to the ODD
REL-REQ4	All data samples shall represent an environment (including surface, particulate matter, background) that images are captured that corresponds to the ODD
REL-REQ5	All data samples shall represent images of the quality that corresponds to the ODD
Completeness	
COM-REQ1	The data samples shall include images representing all types of hull structures according to the ODD
COM-REQ2	The data samples shall include images representing all corrosion types according to the ODD
COM-REQ3	The data samples shall include images representing all coating conditions (GOOF, FAIR, POOR) according to the ODD
COM-REQ4	The data samples shall include images representing all possible conditions (including FOV, distance, illumination, pose) that corrosion areas may be shot by cameras
COM-REQ5	The data samples shall include images representing all common environments that images are captured that corresponds to the ODD
Balance	
BAL-REQ1	The data set shall contain both positive and negative examples
BAL-REQ2	The percentage of samples of each coating condition (GOOD, FAIR, POOR) in the data set shall be balanced
BAL-REQ3	The percentage of samples of each supported hull structure in the data set shall be balanced
BAL-REQ4	The percentage of samples of each supported failure type in the data set shall be in balanced
Accuracy	
ACC-REQ1	Only rusted areas in data samples shall be labelled
ACC-REQ1	All rusted areas in the data samples shall be correctly labelled

validation, and test subset. They are labeled in pixel-wise level. Since photos taken from different sections of the same vessel usually share similar properties (e.g., the lighting condition and coating color), to guarantee a fair evaluation, the subset split is such that all images of a certain vessel can only be present in the same subset (either training set, validation set or test set).

5.4 Data Validation

The data validation activity is to check whether the generated data sets are sufficient to meet the defined ML data requirements. And the output shall be documented as ML data validation results. Here we only take the training dataset as an example to validate its corresponding data requirements.

Relevance. As shown in Fig. 1, the inputs of Corrosion.ai are controlled by surveyors which rules out the irrelevant images from structure, conditions to environment. As a result, all the relevance requirements REL-REQ1,2,3,4 and 5 are validated through human check. In practice, this part can also be tested by out of distribution (OOD) detection if systems are designed without human supervision.

Completeness. The assurance requirements related to data completeness can be validated with the help of testing tool. Here one tool named as Data Representativeness Testing (DRT) is presented. It uses out-of-distribution detection-based methods [6] to measure how well a scenario (data slice) is represented by a certain dataset, or in other words, how representative of the dataset is for that slice of data generating from the scenario. To conduct DRT, one needs to prepare a reference dataset containing samples from each of the testing scenarios. Then a R-score will be calculated as:

$$R\text{-score} = 100 - \text{Ratio}_{OOD}, \tag{2}$$

where Ratio_{OOD} is the percentage ratio of points classified as out of distribution samples.

For example, to test training set's representativeness for different hull structural elements, a reference dataset containing 103 images of six structures is employed. The R-scores, which measure the training set's representativeness for the six hull structural elements, are all above the acceptance threshold of 90%. This indicates all six hull elements are well covered by Corrosion.ai' s training dataset. To test the three coating conditions, a reference of 54 images are used, results show that only POOR category has a lower R-score than 90%.

Balance. For each sample in the training set, since the coating condition is not given during the annotation process, only a "surrogate" coating condition is derived using the percentage of corrosion pixels in that image. Results show that the POOR coating condition only takes up an extremely small proportion with

0.83% while GOOD accounts for 75.37% and FAIR accounts for 23.80%. This imbalance of distribution could cause the model to under-perform on samples of that coating condition.

Accuracy. All pixel-wise annotations ("masks") have been double-checked by another trained person and surveyors to ensure the requirements of annotation accuracy ACC-REQ1 and ACC-REQ2 are satisfied.

6 Model Learning Assurance

During model learning stage, the model development log and internal test results need to be documented which provide evidence for the assurance argument.

6.1 Internal Test Results

For our use case, the Intersection over Union (IoU) (also referred to as the Jaccard index) is used as target metric for reporting model performance. The internal test evaluated the model performance on both validation set and test set (during practice, a detailed data generation log is required). The IoU scores on both datasets, together with some other metrics of the model are reported, based on that the final confusion matrix are reported in Fig. 3.

6.2 Risk Analysis

Considering both the consequence (Table 2) and likelihood (Fig. 3a and Fig. 3b), similar to Fig. 2b, we can get the risk analysis results of Corrosion.ai on validation and test dataset (illustrated in Fig. 4).

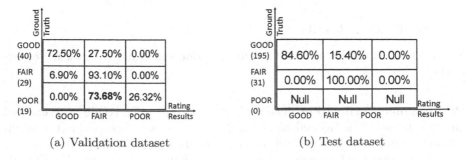

(a) Validation dataset   (b) Test dataset

Fig. 3. Confusion matrix of internal test results.

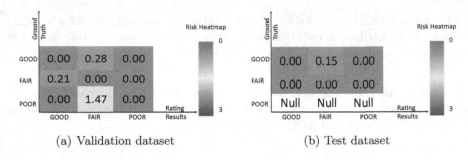

(a) Validation dataset   (b) Test dataset

Fig. 4. Risk analysis of internal test results.

7   Model Verification Assurance

In this section, the model verification activities and corresponding results are reported to support the derived performance and robustness requirements.

7.1   Verification Data

For model verification, ideally we should cover test scenarios or samples as much as possible, but in practice the AI systems are running in very complex environment and the data collection are not endless to cover all corner cases. As a result the preparation of verification dataset are very important, apart from the derived data requirements in previous section, more attention should also be assigned to the high risk scenarios or hard samples. For this case, we prepared one pilot dataset for routine testing, and one high-risk dataset from a large open unlabeled dataset which is selected by model uncertainty quantification method [9], one perturbation dataset which is produced according to the required operational environment.

7.2   Verification Results

The verification results are reported from performance and robustness respectively.

Performance Verification. The performance verification results are reported on pilot dataset (Fig. 5a) and high-risk dataset (Fig. 5b). Comparing the verification results and the internal test results with the human-level results (Fig. 2a), we can find that the Corrosion.ai surpasses the human-level except the POOR-to-FAIR item.

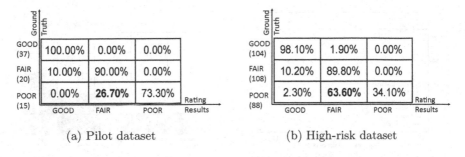

(a) Pilot dataset   (b) High-risk dataset

Fig. 5. Confusion matrix of verification results.

Risk Analysis. Similar to the risk analysis on human-level performance, we can get the risk matrix in Fig. 6. The results show risk reduction for the majority of categories but a slightly increase for POOR-to-FAIR item compared to human-level risk.

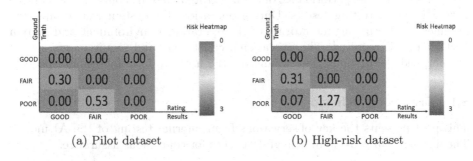

(a) Pilot dataset   (b) High-risk dataset

Fig. 6. Risk analysis of verification results.

Robustness Verification. Robustness requirements are verified on the perturbation data gathered with aim of testing the models to the breaking point but still within the operational domain. Here we give an example of evidence to test the ROB-REQ4, as shown in Fig. 7. The acceptance threshold was set to 86%, which is the ratio of required mIoU of 0.345 to the baseline mIoU of 0.4. Perturbation sensitivity analysis facilitates the comparison of effects of different perturbation types and levels under a unified metric space [1] with consideration of the rationality for their existence. Perturbation score matrix illustrates the model degraded performance on the testing dataset under various perturbation types and levels. As shown by the perturbation results, the Corrosion.ai passes the test for Gaussian noise, spatter, and fog on level 1, but fails for darkness, motion blur, and fog on level 2 and 3.

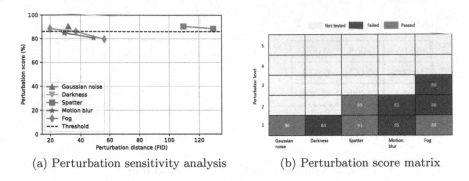

(a) Perturbation sensitivity analysis   (b) Perturbation score matrix

Fig. 7. Image perturbation results.

8 Model Deployment Assurance

After model verification, the AI model will be deployed and integrated into a broader system. The performance of the integrated system should be comparable with ML model testing results. During operation, the system can be updated at times, so continuous monitoring of the operational environment and system performance is required. The operational constraints and requirements should be complied all the time. Here we discuss the use case from following aspects.

8.1 Erroneous Behavior

This part presents the key observations from internal testing of the AI model. The findings will provide meaningful insights for continuous assurance.

Images Out of ODD. The performance of Corrosion.ai will degrade if images do not satisfy the pre-defined ODD. For example, if images are from decks, the coating condition rating results of Corrosion.ai may be significantly different from surveyors. To avoid potential risks, the usage of Corrosion.ai should be strictly restricted to its ODD. Images not from ballast tanks will be rejected in image quality check phase, as shown in Fig. 1.

Poor Quality Images. During system operation, noises or perturbations during sensing or transmission are unavoidable, and sensors may be broken due to extreme weather or improper usage. To mitigate this kind of risk, image pre-processing or rejection are needed before sending these samples to Corrosion.ai.

Disagreements on Results. During integration testing, it is found that Corrosion.ai has a certain probability to disagree with surveyors in evaluating images from ballast tanks of POOR or FAIR coating conditions. Surveyors base their rating grades (GOOD, FAIR and POOR) on their experience and gut-feeling, while Corrosion.ai is hard-coded to calculate corrosion percentage first and assign rating grade based on calculated percentage according to rule table (no "personal" judgment included). In fact, the differences between POOR and FAIR conditions are small, therefore, disagreements are common not only between Corrosion.ai and surveyors, but also among different surveyors. To resolve this dilemma, expert meetings and discussions are required to achieve final decisions.

8.2   Integration Testing

The aim of integration testing is to check whether the integrated system satisfies SYS-REQ1 under Operational scenarios. The integrated system is tested using validation, testing and verification dataset. It is expected that the testing performance will not degrade significantly comparing with ML component testing results using the same dataset.

Integration testing does not focus on the performance of one single module, which should be tested before. Instead, it emphasizes the normal operation of the whole system. For instance, the interface between Corrosion.ai and hardware or other software works well, image acquisition and transmission over the integrated system are correct, timing of different components are correct, etc.

In this Corrosion.ai case, the model is directly deployed on a cloud platform. In many IoT applications, the model is compressed by quantization for a light and fast deployment. This may cause a performance gap, which is required to be verified in this stage.

9   Conclusions

Above all, the main results of the assurance case are summarized as follows:

– To ensure data quality, the training dataset is used as a test case to verify the requirements of DAT-COM-REQ1 and DAT-COM-REQ3. The other requirements can be checked by statistical methods and human inspection.
– For model learning, the internal test results show that Corrosion.ai has better performance than human experts in all categories except for the POOR-to-FAIR category.
– For model verification, the performance results on the pilot dataset and the hard cases selected by the uncertainty quantification method indicate that Corrosion.ai outperforms human experts in all categories except for the POOR-to-FAIR category. For the robustness requirement, the image perturbations test reveals that Corrosion.ai needs to improve its robustness against darkness, motion blur and fog.
– For model deployment and continuous assurance, we provide the possible erroneous behaviors that may occur, and the integration testing requirements that need to be met.

This paper provides a detailed and systematic assurance case on AI-powered corrosion detection system. To derive the assurance requirements, we use a top-down approach that starts from the system level and then goes down to the component level. Furthermore, we specify the machine learning specific requirements, which include the model requirements and the data requirements. To verify these requirements, we use a bottom-up approach that goes from the data requirements to the model learning, the model verification, and the model deployment.

The risk analysis is conducted throughout the entire AI assurance case study. It compares the risk of human-level performance for decision making and the risk of using Corrosion.ai as a decision support tool on different internal testing and verification datasets. It also gives insights on how to use a risk-based approach to conduct an AI assurance case.

During the case study, verification methods are explained and testing tools are also used to collect evidence to support related claims for assurance requirements. For example, the data representativeness testing calculates the R-score for data representativeness measurement; the image perturbation testing is used for robustness verification on possible operational variations and the uncertainty-based sampling is employed to identify high risk scenarios for performance testing.

In the future, to achieve trustworthy AI and assure the AI-enabled systems, it is necessary to explore more practical approaches to link high-level requirements to specific actionable requirements. Due to the complexity and dynamic nature of AI system, the research and development of corresponding verification methods and testing tools are very important to fill this gap.

References

1. Ding, K., Ma, K., Wang, S., Simoncelli, E.P.: Image quality assessment: Unifying structure and texture similarity (2020). CoRR abs/2004.07728, https://arxiv.org/abs/2004.07728
2. European Commission: Artificial intelligence act. Online (2021). https://www.europarl.europa.eu/RegData/etudes/BRIE/2021/698792/EPRS_BRI(2021)698792_EN.pdf
3. Hawkins, R., Paterson, C., Picardi, C., Jia, Y., Calinescu, R., Habli, I.: Guidance on the assurance of machine learning in autonomous systems (amlas) (2021). ArXiv abs/2102.01564
4. Hu, J., Shen, L., Sun, G.: Squeeze-and-excitation networks. In: Proceedings of the IEEE Conference on Computer Vision and Pattern Recognition, pp. 7132–7141 (2018)
5. IACS Recommendation 87: Guidelines for coating maintenance & repairs for ballast tanks and combined cargo/ballast tanks on oil tankers (2015). https://www.iacs.org.uk/publications/recommendations/81-100/rec-87-rev2-cln/
6. Lee, K., Lee, K., Lee, H., Shin, J.: A simple unified framework for detecting out-of-distribution samples and adversarial attacks (2018)
7. Ronneberger, O., Fischer, P., Brox, T.: U-Net: convolutional networks for biomedical image segmentation. In: Navab, N., Hornegger, J., Wells, W.M., Frangi, A.F. (eds.) MICCAI 2015. LNCS, vol. 9351, pp. 234–241. Springer, Cham (2015). https://doi.org/10.1007/978-3-319-24574-4_28

8. Wei, Q., Chen, Y.: An AI-powered corrosion detection solution for maritime inspection activates. In: 20th International Conference on Computer and IT Applications in the Maritime Industries, pp. 97–112 (2021)
9. Wu, F., Wei, Q., Fu, Y.: Adversarial active testing for risk-based AI assurance. In: ESREL 2022 (2022)

Research and Implementation of EXFAT File System Reconstruction Algorithm Based on Cluster Size Assumption and Computational Verification

Enming Lu[1]([✉])[iD] and Fei Peng[2][iD]

[1] Hunan Biological and Electromechanical Polytechnic, Longping High Tech Park, Furong District, Changsha, Hunan, China
14760774@qq.com
[2] Guangzhou University, No. 230, Waihuan West Road, University Town, Guangzhou, Guangdong, China

Abstract. Aim to repair EXFAT file system, a file system reconstruction algorithm based on cluster size assumption and computational verification is proposed. Firstly, an experimental verification study is conducted on the key BPB parameters such as cluster size and first cluster start sector number in Windows EXFAT. After that, the algorithm for calculating and verifying the cluster size is proposed. Finally, the EXFAT file reconstruction system is designed and implemented. Experiments and comparative analysis are carried out with existing algorithms and popular software. The results show that the proposed algorithm is superior in terms of the success rate, temporal attribute, file content, directory structure, as well as the efficiency of its execution. It has great potential in the applications of reconstruction of EXFAT File System formatted by Windows system.

Keywords: EXFAT · file system · reconstruction · clusters · BPB · hypothesis · computational · verification

1 Introduction

With the continuous development of information technology, computers are increasingly involved in people's work and life, playing an increasingly important role in society and life. The use of computers has permeated all aspects of government, military, culture and education and everyday life. They all use computers to access information and process it, while saving their most important information in the form of data files in computers. The high reliance on information technology also poses huge security risks for mankind. The growing prevalence of e-crime has led to an increasing focus on e-discovery techniques. E-discovery treats the document system as a crime scene.

The cluster size is a key BPB parameter in the EXFAT file system, which determines the structure of the entire file system, as well as the location of metafiles and files in

Funded by: 2022 Natural Science General Project of Hunan Biological and Electromechanical Polytechnic "Research and implementation of the Windows EXFAT electronic forensics system based on BPB key parameter calculation and hypothesis verification" (Item No. 22YZK02).

the file system, once the EXFAT file system has been maliciously formatted, completely deleted or otherwise damaged by criminals, the BPB parameters such as cluster size, FAT start sector number and first cluster start sector number will be lost, making it extremely difficult to reconstruct the file system from the remaining information.

Based on this, this paper experimentally verification study the key BPB parameters such as EXFAT cluster size under Windows, and proposes and implements an EXFAT file system reconstruction algorithm based on the assumption of cluster size and computational verification. The algorithm firstly assumes the cluster size and FAT start sector number, then uses the calculation algorithm to get the first cluster start sector number FCS and root directory start cluster number RFS, etc. as verification parameters, and then uses the verification algorithm to verify the cluster size and FAT start sector number by the directory block, data block start sector number, and file and folder start cluster number. File system reconstruction of file names, time attributes, data content, etc., it can maximize the accuracy of system reconstruction. This algorithm solves the defects of the existing EXFAT file system reconstruction, greatly improves the effect of reconstruction, and is a significant addition to the file reconstruction and thus e-discovery after the destruction of the EXFAT file system structure.

2 Related Work

In the past few years, research on file system reconstruction algorithms has been increasing, mainly divided into file reconstruction based on file carving, file reconstruction based on file system metadata, and a combination of the two.

2.1 File Reconstruction Algorithm Based on File Carving

Yoo B et al. describes the development process of file carving [1]. 0, Pal A et al. proposes a method to sculpt AVI, WAV, MP3 multimedia compressed files in NTFS file system by using multimedia compressed file characteristics and by file signature and file footer verification [2]. Thing VLL et al. proposes a method to transform the fragment sculpting problem into a graph theory problem and build a model to achieve reconstruction [3]. Reference [1–3] File carving technique is based on file structure and content only, using information such as file signature values, it can recover file content without matching any file system metadata. However, as there is no metafile directory information, the size of the carved-out files is inconsistent and file names, file times, etc. are difficult to obtain accurately.

2.2 File Reconstruction Algorithm Based on File Carving

Oh J proposed a method to achieve file reconstruction by analyzing and tracking the revision history of file data in $LogFile by simulating MFT, obtaining the location of the file data and the data residing in each instance of modification [4]. Karresand M et al. proposed a method to derive the file reconstruction time in FAT32 using the correlation between the storage location of a file, the location of its directory entries, and the creation time of the file in its nearby location [5]. Literature [4, 5] used $LogFile and time attributes in NTFS respectively, which are key evidence in e-discovery. Karresand M et al.

also proposed a method to improve the efficiency of file carving by comparing the file allocation behavior of block writes and stream writes to determine the location and order in which files are stored [6]. Literature [6, 7] utilized the research carried out into the storage location of files which is determined by metafiles.Alhussein M et al. proposed to reconstruct files by embedding special identifiers in clusters and using the identifiers to keep track of the clusters of individual files [8, 9]. Fellows G researched NTFS volume mounts, directory joins, and $Reparse in forensic research. Cho G-S presented a computer forensic method for detecting timestamp forgeries in the Windows NTFS file system [10]. Cho G S proposed a computer forensic method using $logfile to detect timestamp forgeries in the Windows NTFS file system [11]. BO Dong et al. proposed an anti-anti-forensic approach based on NTFS transaction features and machine learning algorithms [12]. Dp A et al. proposed a new forensic method using four existing Windows artefacts, namely $USNjrnl, linked files, prefetched files, and Windows event logs [13]. Lee W Y etc. proposed a novel method that constructs a creation time bound of files recovered without time information. The method exploits a relationship between the creation order of files and their locations on a storage device managed with the Linux FAT32 file system [14]. Minnaard W studied allocation algorithms and file creation order Reconstruction for Linux FAT32 file system drivers [15]. Karresand M conducted an empirical study of allocation algorithms in NTFS from different perspectives and over time [16]. Nordvik R et al. proposed an approach to forensics in NTFS file systems by using the $ObjId index to record user activity and correlating this index with the corresponding record in the MFT table [17].

2.3 File Reconstruction Algorithm Integrating File Carving and File System Metadata

Ma Guo-fu et al. proposed a comprehensive algorithm for file location, file feature characters and file fragmentation reorganization [18], in which the correlative coefficient proposed by Xie Juan-ying et al. was used in the SVM-based fragmentation classification algorithm [19], but the degree of synthesis was not high, and it failed to take full advantage of the size of the existing or potential cluster.Vandermeer Y et al. proposed a forensic analysis algorithm for mapping directory entries by combining FAT and bitmap [20], which mentions the need to calculate the conversion from sectors to clusters by using the EXFAT partition size and the cluster size, and proposes a correlation coefficient with the "number of HEAPs of sectors retained by the system". The algorithm is related to the "number of HEAPs of sectors retained by the system" for the root directory, but no specific algorithm for HEAPs is proposed [20]. Sitompul O S et al. proposed an Aho-Corasick algorithm that reads the file attributes from the Master File Table (MFT) to check the status of the file, and parses the file attributes, damage status, and extensions through the MFT and the file contents, and thus reconstructs the file [21, 22]. Sahib H I et al. conducted a comparative analysis for NTFS logical file reconstruction technique and Aho-Corasick file reconstruction technique [23].

In conclusion, file sculpting technique based on file signature and file reconstruction technique based on file system metadata are current hotspots, which can solve most of the file system reconstruction problems in NTFS and FAT32 file systems, while at present, there are few studies specifically for EXFAT file system and e-forensic using metadata and file signature, thus this paper presents a study of the reconstruction algorithm for Windows EXFAT file systems based on the assumption of cluster size and computational verification.

3 EXFAT File System Reconfiguration Algorithm

The overall idea of EXFAT file system reconstruction algorithm is shown in Fig. 1:

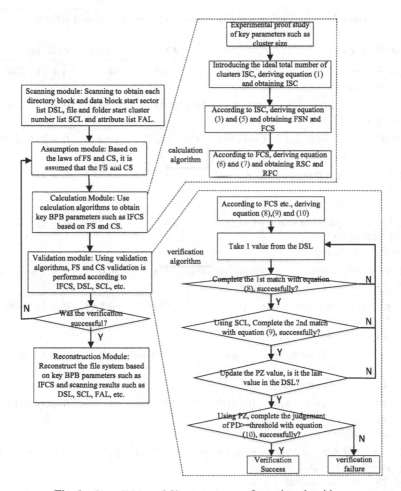

Fig. 1. Overall idea of file system reconfiguration algorithm

3.1 Experimental Verification Study Based on Cluster Size etc.

Windows system is a closed source operating system, it is further partitioned for EXFAT, "the first cluster start sector number" is not immediately after the FAT table, there will be some reserved sectors in the middle, once the number of FAT sectors can determine the "first cluster start sector number" FAT sector number must follow a certain pattern, in order to find the law, the experimental verification study of partitioning test, select the default cluster size, the test area is: $<= 67108846$ MB (≈ 64 TB). Empirical studies as shown in Table 1, $1G = 1024 * 1024 * 1024$, $1 M = 1024 * 1024$, $1 s = 512$ B, *FS* on behalf of the FAT starting sector number, *CS* on behalf of the cluster size, *FCS* on behalf of the first cluster starting sector number.

Table 1. The experimental verification study based on cluster size etc.

Partition Size (B)	*FS* (s)	*CS* (s)	*FCS* (s)
i * 8G <= partition size < (i + 1) * 8G (int i = 0;i <= 3;i + +)	2048	64s	(i + 2) * 2048
i * (4 * 8G) <= partition size < (i + 1) * 3(4 * 8G) (int i = 1;i <= 15;i + +)	2048	4 * 64	(i + 2) * 2048
i * (8 * 8G) <= partition size < (i + 1) * (8 * 8G) (int i = 8;i <= 15;i + +)	2048	8 * 64	(i + 2) * 2048
i * (16 * 8G) <= partition size < (i + 1) * (16 * 8G) (int i = 8;i <= 15;i++)	2048	16 * 64	(i + 2) * 2048
i * (32 * 8G) <= partition size < ((i + 1) * (32 * 8G)-2048 s (int i = 8;i <= 15;i++)	2048	32 * 64	(i + 2) * 2048
(i + 1) * (32 * 8G)-2048 s < = partition size < (i + 1) * (32 * 8G)(int i = 8;i <= 15;i++) special	2048	2048	(i + 1 + 2) * 2048
i * (2 * 64 * 8G) <= partition size < (i + 1) * (2 * 64 * 8G)-4096s (int i = 4;i <= 7;i++)	2*2048	64 * 64	(i + 2) * 4096
(i + 1) * (2 * 64 * 8G)-4096s <= partition size < (i + 1) * (2 * 64 * 8G) (int i = 4;i <= 7;i ++) special	2 * 2048	4096	(i + 1 + 2) * 4096
i* (4*128*8G) <= partition size < (i + 1)* (4*128*8G)-8192s (int i = 2;i <= 3;i++)	4 * 2048	128 * 64	(i + 2) * 8192
(i + 1) * (4 * 128 * 8G)-8192s <= partition size < (i + 1) * (4 * 128 * 8G) (int i = 2;i <= 3;i++) special	4 * 2048	8192	(i + 1) + 2 * 8192
i * (8 * 256 * 8G) <= partition size < (i + 1) * (8 * 256 * 8G)-16384s (int i = 1)	8 * 2048	256 * 64	(i + 2) * 16384
(i + 1) * (8 * 256 * 8G)-16384s <= partition size < (i + 1) * (8 * 256 * 8G) (int i = 1) special	8 * 2048	16384	(i + 1 + 2) * 16384
32T <= partition size < 67108846M	16 * 2048	512 * 64	65536

Analysis of the special cases listed in Table 1 reveals that when the $CS \geq 2048s$, , let the first cluster start sector number when $(i + 1) * (m * n * 8G)$-"1 cluster size" $<=$ partition size $< (i + 1)* (m * n * 8G)$ be $S1$ and the first cluster start sector number when $i* (m * n * 8G) <=$ partition size $< (i + 1)* (m * n * 8G)$-"1 cluster size" be $S2$ $(m = FS/2048, n = CS/64)$, then $S1$-$S2 = FS$. When the $CS < 2048s$, there is no special case at this point since the minimum allocation unit for partitions is 2048s. This leads to the conclusion that EXFAT's FSN and FCS starts to increase the space allocated for one cluster at the location of the $(i + 1) * (m * n 8G)$-"1 cluster size".

3.2 Calculation Algorithm Based on Cluster Size etc.

Through the analysis of EXFAT, the FAT sector number FSN is related to the number of FAT entries, and the number of FAT entries is related to the actual total number of clusters in the partition RSC, so the first cluster start sector number FCS is related to RSC; at the same time, RSC is related to the total sector number TSP, the cluster size CS and the first cluster start sector number FCS, TSP and CS are already randomly determined by the user, so the final RSC is related to FCS again.

At this point an oxymoron is formed. To resolve this oxymoron, the author tries to introduce the concept of ideal total cluster number ISC, which ISC assumes is the number of clusters that the operating system allocates all the space of the partition into clusters. ISC starts from sector 0, RSC starts from FCS, the number of sectors from DBR to the starting sector number of FAT is an integer multiple of the cluster size, and FSN also takes up space, so $ISC > RSC$. Therefore, if the number of FAT entries in an EXFAT partition meets ISC, then it must also meet RSC, so *The number of FAT entries* $= ISC$, according to 0 and 1 are special FAT entries, the FAT entries in the data area start from 2, each FAT entry occupies 4 bytes, and there are 128 FAT entries in a sector of the FAT table. The FSN is an integer multiple of the CS and the FCS is an integer multiple of the FS.

According to the conclusion of 3.1, The formulae for calculating the FCS and FSN are based on the downward rounding feature as follows:

From formula (1):

$$ISC = \lfloor TSP/CS \rfloor \tag{1}$$

In formula (1), ISC represents the ideal total number of clusters, TSP represents the total number of sectors, CS represents the cluster size, and $\lfloor \rfloor$ represents rounding down.

ISC starts from sector 0, and RSC starts from FCS, the number of sectors from DBR to FAT is an integer multiple of the cluster size and FSN also takes up space, so ISC $>$ RSC. It can be concluded that if the number of FAT allocation items can be satisfied ISC during EXFAT partition, it must also be satisfied RSC. Therefore, the number of *FAT allocation items* $= ISC$ is enough. According to the fact that item 0 and item 1 are special FAT items, the FAT items in the data area start from item 2, and each FAT item occupies 4 bytes. There are 128 FAT items in a sector of the FAT items, FSN are integer multiples of CS and FCS are integer multiples of FS. Then, according to the characteristics of rounding down, the calculation formulas for the FSN and FCS are obtained as follows:

From formula (2):

$$FSN = (\lfloor (ISC + 2 - 1) * 4/512/CS \rfloor + 1) * CS \tag{2}$$

Further obtained:

$$FSN = (\lfloor (ISC + 1)/128/CS \rfloor + 1) * CS \tag{3}$$

In formula (3), FSN represents the number of FAT sectors, ISC represents the ideal total number of clusters, CS represents the cluster size, and $\lfloor \rfloor$ represents rounding down.

Further obtained:

$$FCS = (\lfloor (ISC + 2 - 1) * 4/512/FS \rfloor + 2) * FS \tag{4}$$

Further obtained:

$$FCS = (\lfloor (ISC + 1)/128/FS \rfloor + 2) * FS \tag{5}$$

In formula (5), FCS represents the starting sector number of the first cluster, ISC represents the number of ideal total clusters, FS represents the starting sector number of FAT, and $\lfloor \rfloor$ represents rounding down.

It can be obtained from formula (3) and formula (5). $FS + FSN$ is not necessarily followed by FCS, there may be reserved sectors in the middle, mainly because the values of FS and CS may be inconsistent, because FSN are integer multiples of CS and FCS are integer multiples of FS.

When $CS < 2048s$, $FS! = CS$, there may be reserved sectors, FCS is obtained from formula (5).

When $CS >= 2048s$, $FS == CS$, there is no reserved sector, FCS can be obtained directly from $FS + FSN$ or formula (5).

The number of the first cluster in the root directory is related to the initial sector number of the first cluster and the size of \$bitmap. While \$bitmap is determined by the actual total number of clusters, RSC is obtained by using formula (6):

$$RSC = (\lfloor (TSP - FCS)/CS \rfloor \tag{6}$$

In formula (6), RSC represents the actual total number of clusters, TSP represents the total number of sectors, FCS represents the starting sector number of the first cluster, CS represents the cluster size, and $\lfloor \rfloor$ represents rounding down.

Use formula (7) to get the first cluster number of the root directory. The first cluster number of the root directory is:

$$RFC = 4 + \lfloor (RSC - 1)/4096/CS \rfloor + \lfloor 11/CS \rfloor \tag{7}$$

In formula (7), RFC represents the number of the first cluster in the root directory, RFC represents the actual total number of clusters, CS represents the cluster size, and $\lfloor \rfloor$ represents rounding down.

3.3 Verification Algorithm Based on Key Parameters Such as Cluster Size etc.

Since the initial sector number of the first cluster and the size of the cluster are very important in the EXFAT file system, when the EXFAT file system structure is damaged, how to quickly and accurately determine the key parameters of BPB is particularly important in electronic forensics. We can implement it according to the method of hypothesis verification.

The study found that the starting sector of each directory block or data block of different types is actually the starting sector SS of the cluster where the directory block or data block of different types is located, and the starting sector of the cluster is determined by CS, cluster number $DBClu$ and FCS. Therefore, the first hypothesis judgment rule, the second hypothesis judgment rule and the third judgment rule can be obtained. If the first judgment rule is not satisfied, the second judgment rule does not need to be verified. The first assumption judgment rule formula is:

$$(SS - IFCS)\%CS == 0 \tag{8}$$

In formula (8), SS represents the starting sector of the directory block or data blocks of different types, $IFCS$ represents the starting sector number of the temporary first cluster, and CS represents the cluster size.

The second assumption judgment rule formula is:

$$DBClu = (SS - IFCS)/CS + 2$$

$$If\ DBClu \in SCL,\ then\ PZ = PZ + + \tag{9}$$

In formula (9), $DBClu$ represents the cluster number of the directory block or data blocks of different types, SS represents the starting sector of the directory block or data blocks of different types, $IFCS$ represents the starting sector number of the temporary first cluster, CS represents the cluster size, SCL represents the list of folder or file starting cluster numbers, and PZ represents the matching value.

The third judgment rule formula is:

$$PD = (float)PZ/(float)n1 * 100\%$$

$$If\ PD >= Set\ threshold,\ is\ is\ assumed\ that\ the\ verification\ is\ successful \tag{10}$$

Formula (10) indicates that for each FS and CS, the matching degree of all directory blocks and data blocks of different types is calculated. If the set threshold is met, the matching is successful, the verification of FS and CS is completed, and the cycle exits. Otherwise, the verification fails, and the next cycle continues. PD represents the matching degree of all directory blocks and data blocks, PZ represents the matching value, and the initial value is 0, $n1$ represents the total number of directory blocks and data blocks.

3.4 Assumptions and Computational Validation Algorithms Based on Cluster Size etc.

The Windows EXFAT file system reconstruction algorithm based on cluster size assumptions and computational validation is implemented in the following steps:

Step 1: According to the characteristics of the directory block, scan to get the start sector number list of each directory block DSL_d, according to the characteristics of the data block of different types of files, get the start sector number list of each data block DSL - f[i][q] ($i = 0,1,2,3......$, representing doc, xls, ppt, jpg, bmp, etc.), through the information of the directory entries in the directory block, get the folder start cluster number list SCL_d, the list of the start cluster number of the file SCL_fi ($i = 1,2,3......$, representing doc, xls, ppt, jpg, bmp, etc.), through the information of the directory entries in the directory block, get the list of the folder start cluster number, the list of the file start cluster number SCL_f_i ($i = 1,2,3......$, representing doc, xls, ppt, jpg, bmp, etc.) and the list of the file and folder attributes FAL.

FS is a multiple of 2048. The maximum value is 65536. CS is one of 17 data [1, 2, 4, 8, 16, 32, 64, 128, 256, 512, 1024, 2048, 4096, 8192, 16384, 32768, 65536]. It is assumed that the initial sector code of FAT is FS, and the initial value FS = 2048, as the external loop. In the process of assuming the FAT start sector, it is assumed that the cluster size is CS. According to the rule FS and CS, and range of CS, assuming FS and CS, $IFCS$ is calculated. Complete the initial verification of the directory block and the starting sector number of the data block through the data such as, $IFCS$, DSL, and then complete the secondary verification through the calculated starting cluster number of the directory block, the starting cluster number of the data block and the data in SCL, and count the matching values of all the directory blocks and data blocks. Divide the matching values by the total number of the directory blocks and data blocks to obtain the matching degree. Compare the matching degree with the target matching threshold, and finally verify FS and CS.

Step 3: through FS and CS, FSN is obtained according to formula (3), FCS is obtained according to formula (5) and RSC is obtained according to formula (6). Then RFC is obtained according to formula (7) and reused FCL to finally complete electronic forensics.

Step 1 consists of:

Step 1.1: obtain the total number of sectors TSP according to the damaged partition.

Step 1.2: traverse the partition and obtain the starting sector number of each directory block DSL - d[p], through the characteristics of each directory block. The unit is sector, where, $p = 1, 2, 3, \ldots, N_1, N_1$ is the total number of directory blocks. The starting sector number of the data block DSL - f[i][q] is obtained according to the characteristics of different types of files, and the unit is sector. Among them, i, q representing the number of q data block represents the type of file i, i = 1, 2, 3, 14, 5, 6, 7, 8, 8, 9, 10, 11, 12 . . . representing doc, docx, xls, xlsx, ppt, pptx, jpg, gif, bmp, pdf, zip, png (this article only takes the 12 file types as an example), $q = 1, 2, 3, \ldots, M_i, M_i$ representing the total number of files of the type of file i. The starting cluster number of each folder SCL - d[r] is obtained from the directory entries in the directory block. The unit is cluster, $r = 1, 2, 3, \ldots, N_2, N_2$ is the total number of folder starting clusters. The starting cluster number of each file SCL - f[i][s] is obtained from the directory entries in the directory block. The unit is cluster, i, s represents the file s of the file type i where, $i = 1, 2, 3, 14, 5, 6, 7, 8, 8, 9, 10, 11, 12 \ldots$ respectively represents *doc, docx, xls, xlsx, ppt, pptx, jpg,gif, bmp, pdf, zip, png* and other file types, and $s = 1, 2, 3, \ldots, K_i, K_i$ represents the total number of starting clusters of the file types i. Get the attributes of each file and folder $FAL[t]$ through the directory entry in the directory block, $t = 1, 2, 3, \ldots, N_3, N_3$ is the total number of files and folders obtained through the directory entry, $N_3 = N_2 + \sum_{i=1}^{12} K_i$.

The characteristics of the directory block are: 0 bytes of the root directory block are 0x83 or 0x03 (when deleted), 32 bytes must be 0x81, 64 bytes 0x82, 0 bytes of other directory blocks are 0x85 or 0x05 (when deleted), 32 bytes are 0xc0 or 0x40 (when deleted), and 64 bytes are 0xc1 or 0x41 (when deleted). The directory block consists of multiple directory entries.85 attribute contains the time attribute and file attribute of the file. If the fourth byte &0x10 is true, it is a folder, otherwise it is a file. The C0 attribute contains the starting cluster number, size, fragment flag, etc. of the file or folder, and the C1 attribute contains the file name.

The characteristics of different types of files are: *doc, xls* and *ppt* files start with 0xd0cf11e0a1b11ae1, *docx, xlsx* and *pptx* files start with 0x504b0304, *jpg* files start with 0xffd8ffe1, *gif* files start with 0x47494638, *bmp* files start with 0x424d, *pdf* files start with 0x2550442d, *zip* files start with 0x504b03041400, and *png* files start with 0X89504E470D0A1A0A.

Step 2 consists of:

Step 2.1: assume that the starting sector number of external circulation FAT is *FS*, the initial value $FS = 2048$, Then each external cycle is completed $FS = FS*2$.

Step 2.2: assuming that the size of the inner loop cluster is *CS*, the initial value $CS = 65536$, and the matching degree is *PZ*, the initial value $PZ = 0$, each inner loop is completed $CS = CS/2$.

Step 2.3 consists of:

Step 2.3.1: obtain *ISC* by using formula (1). Use formula (3) to get the initial sector number of the temporary starting cluster *IFCS*. Set the initial value, $PZ = 0$, $PZ.d=0, PZ.fi=0$, PZ as the matching value, which is used to count the number of directory blocks and data blocks that conform to the cluster size and the FAT starting sector code. $PZ.d$ represents the matching value of the directory block, $PZ.fi$ represents the matching value of the data block of the type file i.

Step 2.3.2: set the initial value $p = 1$, p representing the directory block number, representing that the matching value is calculated from the first directory block, N1 representing the total number of directory blocks.

Step2.3.3: take out the starting sector number of the directory block p is DSL - $d[p]$. Use formula (8), if DSL - $d[p] - IFCS)\%CS == 0$ is not valid, then p++, otherwise,using formula (9) to obtain the starting cluster number of the directory block $DBClu = (DSL$ - $d[p] - IFCS)/CS + 2$, and judge whether $DBClu$ exists in SCL - $d[r]$. If it exists, that is,$DBClu \in SCL$ - $d[r]$,the matching value $PZ.d = PZ.d + +, p$++; Otherwise p++.

Step 2.3.4: judge whether $p > N1$ is satisfied? If not, skip to step 2.3.3; If yes, the directory block hypothesis verification is completed, and the process jumps to step 2.3.5.

Step 2.3.5: set the initial value $i = 1$, i represents the first file type and $i = 1$ represents the matching value for the first file type.

Step 2.3.6, set $q = 1$, q represents the data block number q, the initial value is 1, representing the data block number 1, fetching M_i, M_i representing the total number of files of the file type i.

Step 2.3.7: take out the starting sector number of the data block No q of the file type i, DSL - f[i][q],using formula (8), If DSL - f[i][q] $- IFCS)\%CS == 0$ is not valid, then $q++$. Otherwise, use formula (9) to obtain the starting cluster number of the directory block $DBClu = (DSL$ - f[i][q] $- IFCS)/CS + 2$, and judge whether $DBClu$ exists in the SCL - f[i][s]. If it exists, that is, $DBClu \in SCL$ - f[i][s],the matching value $PZ.fi = PZ.fi++$, $q++$; Otherwise, $q++$.

Step 2.3.8: judge whether $q > Mi$ is satisfied? If not, skip to step 2.3.6; If yes, the directory block hypothesis verification of i file types is completed, and skip to step 2.3.9.

Step 2.3.9: judge whether $i > 12$ is satisfied? If not, then $i++$,skip to step 2.3.6; If yes, the directory block hypothesis verification of all file types is completed, and the process jumps to step 2.3.10.

Step 2.3.10: calculate the matching value $PZ = PZ.d + \sum_{i=1}^{12} PZ.fi$ and the total matching amount $PL = Min(N1,N2) + \sum_{i=1}^{12} Min(Mi,Ki)$. If $(float)PZ/(float)PL*100\%$ $>= 90\%$, (90% represents the threshold, which is determined according to the user's needs), then *the actual FAT start sector number $= FS$* and *the actual number of sectors per cluster $= CS$*. If the verification is successful, exit the cycle; otherwise, skip to step 2.3.11.

Step2.3.11: calculate $CS = CS/2$, assuming that the value of the sector code of each cluster is halved; If $CS >= 1$, go to step S230 and continue to verify the correctness of CS and FS according to the determination rule. Otherwise, go to step 2.3.12.

Step 2.3.12, calculate $FS = FS *2$. If $FS <= 65536$, go to step 2.2 and continue to verify the correctness of CS and FS according to the decision rule of formula (10). Otherwise, it means that the whole hypothesis verification fails. According to the partition size and the default value in Table 1, the actual FAT start sector number equals the default value, and the actual number of sectors per cluster equals the default value.

Step 3: FS and CS is obtained in step 3. FSN is obtained according to formula (3). FCS is obtained according to formula (5). RSC is obtained according to formula (6). Then RFC is obtained according to formula (7) and FCL reused to finally complete electronic forensics.

4 Experimental Analysis

The author designed and implemented FormatRecovery, a tool for EXFAT file system reconstruction, to simulate and emulate the implementation of the Aho-Corasick algorithm [21], the Forensic Analysis algorithm [20], and the Integrated Approach [18] in EXFAT, using the FormatRecovery with the three algorithms and the current market popular R-Studio_v8.8.171971_Network_Edition, Super-disk-recovery (Chinese), EasyRecovery_Professional_14 three data recovery software to do the test comparison.

Test environment: Intel(R) Core(TM) i5-8400 CPU @2.80 GHz @2.81 GHz; RAM: 16 GB; OS: Windows 10 Professional; Virtual USB flash drive, partition size: 256 GB (262143 MB); File system: EXFAT; Cluster size: 1MB; Number of directories: 16, of which First-level directory: 13, second-level directory: 3; Number of files: 60021; Occupied space: 4367873678B; Destruction: Format the partition into NTFS format and simulate the virus damage to the root directory area, the original data shown in Table 2, and algorithm experiment results data shown in Table 3, the software experiment results data shown in Table 4, the experimental analysis of the comparative data shown in Table 5, FormatRecovery software scanning results shown in Fig. 2:

Table 2. Original data

File. type	quantity	capacity /B	File. type	quantity	capacity /B
ppt	5000	454313472	bmp	000	940811200
pptx	5000	187087801	gif	5000	465086208
doc	5000	134048768	jpg	5000	89463246
docx	5000	65090121	png	5000	1314466769
xls	5000	125440000	zip	21	411056316
xlsx	5000	44764434	pdf	5000	136096225
txt	5000	149118	Total	60021	4367873678

Table 3. Algorithm experimental results data of recovery

File. Type	FormatRecovery		Aho-Corasick		Forensic Analysis		Integrated Approach	
	quantity	capacity /B	quantity	capacity /B	quantity	capacity /B	quantity	capacity /B
ppt	5000	454313472	5000	454313472	4992	453596672	0	0
pptx	5000	187087801	5000	187087801	4996	186940450	4997	186977253
doc	5000	134048768	5000	134048768	4990	133782528	4995	133915648
docx	5000	65090121	5000	65090121	4993	64999097	4996	65038112
xls	5000	125440000	5000	125440000	4993	125264384	4997	125364736
xlsx	5000	44764434	5000	44764434	4496	44728620	4998	44746535
txt	5000	149118	0	0	0	0	0	0
bmp	4979	936860302	4937	928958506	4900	921997400	4885	919175330
gif	5000	465086208	5000	465086208	4998	465065235	4997	465020074
jpg	5000	89463246	5000	89463246	4995	89160993	4994	89121761
png	5000	1314466769	5000	1314466769	4993	1314177475	4997	1313217210
zip	21	411056316	21	411056316	21	411056316	21	411056316
pdf	5000	136096225	5000	136096225	4998	136040743	4999	136066532
Total	60000	4363922780	54958	4355871866	54365	4346809913	49876	3889699507

Table 4. Software experiment result data of recovery

File. Type	FormatRecovery		R-Studio		EasyRecovery		Super- disk-recovery	
	quantity	capacity /B	quantity	capacity /B	quantity	capacity /B	quantity	capacity /B
ppt	5000	454313472	5000	454313472	0	0	0	0
pptx	5000	187087801	5000	187087801	5000	187087801	5000	187087801
doc	5000	134048768	5000	134048768	5000	134048768	4995	132986880
docx	5000	65090121	4998	64987032	5000	65090121	5000	65090121
xls	5000	125440000	5000	125440000	5000	125440000	5000	125440000
xlsx	5000	44764434	5000	44764434	5000	44764434	5000	44764434
txt	5000	149118	5000	149118	0	0	0	0
bmp	4979	936860302	4906	923126228	4888	919739744	4906	923126228
gif	5000	465086208	5000	465079088	5000	465086208	4992	241454961
jpg	5000	89463246	4998	89199862	4994	89251230	4991	88218497
png	5000	1314466769	5000	1314466769	5000	1314466769	4970	742485882
zip	21	411056316	21	411056316	21	411056316	21	411056316
pdf	5000	136096225	5000	136096225	0	0	5000	136096225
Total	60000	4363922780	59923	4349815113	44903	3756031391	49875	3097807345

Table 5. Experimental comparison of EXFAT file system reconstruction

item compared	FormatRecovery	Aho-Corasick	Forensic Analysis	Integrated Approach	R-Studio	EasyRecovery	Super-disk-recovery
Directory number	3	0	0	0	0	0	0
Directory Reconstruction rate	20%	0	0	0	0	0	0
File Name number	60000	0	0	0	5000	0	0
File Name Reconstruction rate	0	99.74%	0	0	8.33%	0	0
File number	60000	54958	54365	49876	59923	44903	49875
File Reconstrution rate (B/sec)	99.97%	91.56%	90.58%	83.10%	99.84%	74.81%	83.10%
Size/B	4363922780	4355871866	4346809913	3889699507	4349815113	3756031391	3097807345
Size Reconstruction rate (B/sec)	99.91%	99.73%	99.52%	89.05%	99.59%	85.99%	70.92%
Scan Time/sec	3m49s	4m32s	4m21s	4m11s	4m42s	12m18s	8m13S
Scan rate (B/sec)	19056431	16043286	16654444	15496811	15424876	5089473	6283585

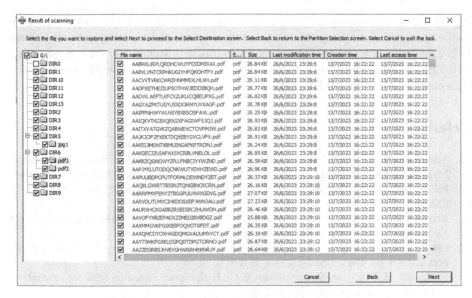

Fig. 2. Scanning results of FileReconstruction software

The experimental results show that the EXFAT file system reconstruction algorithm proposed in this paper is significantly superior in terms of efficiency and success rate compared with algorithm such as Aho-Corasick, Forensic Analysis and Integrated Approach as well as software of R-STUDIO, EasyRecovery and SuperDrive Data Recovery. EXFAT file system formatting, virus damage to the root directory, resulting in the file system structure of the cluster allocation confusion, file and file contents, folder directory and folder contents and other inconsistent relationship, the reconstruction of the file system has become extremely difficult. This method can quickly and accurately calculate important BPB parameters such as cluster size, first cluster start sector number and other important BPB parameters based on the information obtained from scanning, can effectively reconstruct EXFAT file system structure, and completely recover secondary directory names, uncovered file names, files and other data under the root directory based on the information of parameters, directory entries and other information, whereas the other algorithms and software, which are not able to calculate the BPB key parameters. The EXFAT file system structure cannot be reconstructed either and can only perform RAW recovery through special file header and file tail information, which results in the loss of directory names and file names, and certain errors in the size and content of files.

5 Conclusions

In this paper, an EXFAT file system reconstruction algorithm based on cluster size assumption and computational validation is proposed. It makes full use of EXFAT metadata and various types of file signatures, and uses the starting sector number, starting cluster number of directory block, data block, and the list of starting cluster number of files and folders for verification. It combines the advantages of carving technique and logical data reconstruction technique, and it can maximally obtain the key information of file name, time, directory structure, and file content, and the complexity of CS and FCS calculation is low. Experiments show that it is very suitable for the reconstruction of the EXFAT file system formatted by Windows. Our future work will be concentrated on reconstruction of the EXFAT file system formatted by Linux.

References

1. Yoo, B., Park, J., Lim, S., Bang, J., Lee, S.: A study on multimedia file carving method. Multimedia Tools Appl. **61**(1), 243–261 (2011). https://doi.org/10.1007/s11042-010-0704-y
2. Pal, A.: Memon, N.: The evolution of file carving. Sig. Process. Mag. IEEE **26**(2), 59–71 (2009)
3. Vrizlynn, L.L., Ying, T., et al.: Design and analysis of inequality based fragmented file carving algorithms. China Commun. **7**(6), 1–9 (2010)
4. Oh, J., Lee, S., Hwang, H.: NTFS data tracker: tracking file data history based on $LogFile. Forensic Sci. Int. Digit. Investig. (39-), 39 (2021)
5. Karresand, M., Axelsson, S., Dyrkolbotn, G.O.: Disk cluster allocation behavior in windows and NTFS. Mob. Netw. Appl. **25**(3) (2020)
6. Karresand, M., Axelsson, S., Dyrkolbotn, G.O.: Using NTFS cluster allocation behavior to find the location of user data. Digit. Investig. **29**, S51–S60 (2019)
7. Wan, Y.L., Kim, K.H., Lee, H.: Extraction of creation-time for recovered files on windows FAT32 file system. Appl. Sci. **9**(24), 5522 (2019)
8. Alhussein, M., Srinivasan, A., Wijesekera, D.: Forensics filesystem with cluster-level identifiers for efficient data recovery. In: International Conference for Internet Technology and Secured Transactions. IEEE (2012)
9. Alhussein, M., Wijesekera, D.: Multi-version data recovery for cluster identifier forensics filesystem with identifier integrity. Int. J. Intell. Comput. Res. **4**(3), 348–353 (2013)
10. Fellows, G.: NTFS volume mounts, directory junctions and $Reparse. Digit. Investig. **4**(3–4), 116–118 (2007)
11. Cho, G.-S.: A computer forensic method for detecting timestamp forgery in NTFS. Comput. Secur. **34**, 36–46 (2013)
12. Bo, D., Park, K.H., Kim, H.K.: De-wipimization: detection of data wiping traces for investigating NTFS file system. Comput. Secur. **99** (2020)
13. Dp, A., Fba, B.: Artifacts for detecting timestamp manipulation in NTFS on windows and their reliability. Forensic Sci. Int. Digit. Investig. **32** (2020)
14. Lee, W.Y., Kwon, H., Lee, H.: Comments on the Linux FAT32 allocator and file creation order reconstruction. Digit. Investig. **11**(4), 224–233 (2015). 15(DEC.), 119–123
15. Minnaard, W.: The Linux FAT32 allocator and file creation order reconstruction. Digit. Investig. **11**(3), 224–233 (2014)
16. Karresand, M., Dyrkolbotn, G.O., Axelsson, S.: An empirical study of the NTFS cluster allocation behavior over time. Forensic Sci. Int. Digit. Investig. (Jul.), 33S (2020)
17. Nordvik, R., Toolan, F., Axelsson, S.: Using the object ID index as an investigative approach for NTFS file systems. Digit. Investig. **28S**(APR.), S30–S39 (2019)
18. Ma, G., Wang, Z., Cheng, Y.: Recovery of evidence and the judicial identification of electronic data based on EXFAT. In: International Conference on Cyber-Enabled Distributed Computing and Knowledge Discovery, pp. 66–71, October 2015
19. Xie, J., Gao, H.: Statistical correlation and K-means based distinguishable gene subset selection algorithms. J. Softw. **25**(9), 2050–2075 (2014). (in Chinese)

20. Vandermeer, Y., Le-Khac, N.A., Carthy, J., Kechadi, T.: Forensic analysis of the EXFAT artefacts. In: Proceedings of the Conference on Digital Forensics, Security and Law, pp. 83–96, 14p (2018)
21. Sitompul, O.S., Handoko, A., Rahmat, R.F.: File reconstruction in digital forensic. TELKOM-NIKA Indonesian J. Electr. Eng. **16**. https://doi.org/10.12928/TELKOMNIKA.v16i2.8230
22. Sitompul, O.S., Handoko, A., Rahmat, R.F.: IEEE 2016 International Conference on Informatics and Computing (ICIC) - Mataram, Indonesia (2016.10.28–2016.10.29). 2016 International Conference on Informatics and Computing (ICIC) - A File Undelete with Aho-Corasick Algorithm in File Recovery, 427–431 (2016)
23. Sahib, H.I., Rahman, N., Alqasi, A.K., et al. Comparison of Data Recovery Techniques on (MFT) Between Aho-crosick and Logical Data Recovery Based on Efficiency. TELKOM-NIKA (Telecommunication Computing Electronics and Control), **19**(1)(February 2021), 73–78 (2021)

A Verifiable Dynamic Multi-secret Sharing Obfuscation Scheme Applied to Data LakcHouse

Shuai Tang[1][✉][iD], Tianshi Mu[2], Jun Zheng[1], Yurong Fu[1], Quanxin Zhang[1], and Jie Yang[1]

[1] Beijing Institute of Technology, Beijing 100081, China
{TangShuai,zhangqx}@bit.edu.cn
[2] China Southern Power Grid Digital Grid Group Co., Ltd., Guangzhou, Guangdong, China
muts@csg.cn

Abstract. In the context of the evolving Data LakeHouse distributed architecture and the inescapable challenges posed by DM-Crypt, a verifiable dynamic multi-secret sharing obfuscation scheme applied to the Data LakeHouse is proposed. In the proposed scheme, participants select their shadows using a secure one-way function and hide their true identities for self-protection. This scheme can conceal the actual key within any dimension of the homogeneous linear equation system. It can verify whether the distributor, participant, or key restorer has committed fraud by comparing the hash information published by the previous operator on the public bulletin board with the hash information calculated by the current operator. Enables dynamic addition or deletion of participants, dynamic key modification, and periodic key updates. Among these dynamic operations, it is fully dynamic only when participants are added or deleted, as long as the remaining participants meet the minimum decryption threshold. In other cases, the process is semi-dynamic, requiring modifications to information related to other participants. The security of the scheme is based on the Shamir threshold scheme, the asymmetric key encryption system (RSA), a secure and tamper-resistant hash function, and a secure one-way computation function.

Keywords: Data LakeHouse · Key Management · Verifiable Dynamic Multi-Secret Sharing · Threshold scheme

1 Introduction

Given the limitations of Data Warehouses, including closedness and high coupling, the Data Lake emerged. It is defined as a raw data repository that can store various formats. While the Data Lake is suitable for data storage, it lacks some key functions, such as no support for transactions, a lack of consistency and isolation, and no guarantee of data execution quality. These shortcomings

J. Vaidya et al. (Eds.): AIS&P 2023, LNCS 14509, pp. 316–327, 2024.
https://doi.org/10.1007/978-981-99-9785-5_22

make it unrealistic for the Data Lake to support read and write access, batch processing, and streaming jobs. Consequently, the original goal of establishing a Data Lake has not been achieved, and in many cases, the original advantages of traditional Data Warehouses have been forfeited. Especially now, multiple systems typically coexist within enterprises, including a Data Lake alongside multiple Data Warehouses and other dedicated systems. This not only increases the difficulty of operation and maintenance, but more importantly, the transfer of data between different systems will add a lot of delay and cannot improve the timeliness of the data. This not only increases the complexity of operation and maintenance, but, more significantly, data transfer between different systems will introduce significant delays and fail to enhance data timeliness.

Data LakeHouse [1] is a new type of open architecture that combines the advantages of Data Warehouse and Data Lake. It is built on the low-cost and highly flexible data storage architecture of the Data Lake and inherits the data processing and management functions of the Data Warehouse, as Fig. 1 shows. Therefore, Data LakeHouse has the advantages of strong reliability, an open data format, high scalability, transaction support, and high performance, and is widely used in diverse scenarios such as data science, machine learning, SQL queries, and analysis [2,18].

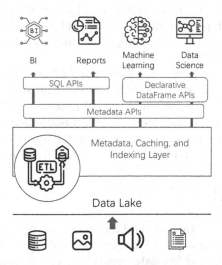

Fig. 1. Example Lakehouse system design, with key components shown in green. The system centers around a metadata layer that may add transactions, versioning, and auxiliary data structures over files in an open format, and can be queried with diverse APIs and engines. (Color figure online)

However, with the rise of LakeHouse architecture, data security and privacy protection issues have gradually become prominent. The data stored in the Lake-House often contains a large amount of sensitive information, such as personal privacy data, corporate operating data, customer confidential data, etc. Therefore, effective measures must be taken to protect these data from unauthorized access and leakage. Some scholars protect data security from the perspectives of attributes [33,34], channels [16,28,29,35], data models [26], encryption schemes [19], etc., and some scholars have proposed a key agreement protocol [30,32,34] to protect system data security in a cloud computing environment similar to a LakeHouse integration.

Data encryption is a common measure to protect the data security of the LakeHouse system. Among them, DM-Crypt, a powerful disk encryption tool, has been widely used in the field of disk encryption, which can effectively prevent unauthorized access, data leakage, etc. However, the key management of DM-Crypt is often criticized. Traditionally, the key of DM-Crypt is manually entered by the user or stored on the local computer, which has many potential security risks. For example, users may choose weak passwords to reduce encryption security. Secondly, if the user decides to store the key in the local computer, then when the computer is attacked or damaged, the key may be leaked or lost, making the data inaccessible or stolen by the attacker.

This paper presents a dynamic multi-secret sharing obfuscation scheme as a forward-looking technology to address the key management problem of the LakeHouse structure. The main process is to expand and obfuscate the secret that needs to be protected. After obtaining the obfuscated multi-secrets, distribute them in shards to each participant and allow runtime modifications to the obfuscation method and shares, i.e., adding, deleting, updating existing shares, or updating shares periodically.

The structure of the paper is organized as follows. In the next section, we introduce some related work. In Sect. 3, we briefly introduce the development of the Data LakeHouse architecture and the pain points of DM-Crypt. In Sect. 4, a verifiable dynamic multi-secret sharing obfuscation scheme applied to the Data LakeHouse is proposed. In Sect. 5, we analyze the verifiability and dynamics of our scheme. Finally, in Sect. 6, the conclusion of this paper is presented.

2   Related Work

Secret sharing is an important method for protecting privacy and ensuring information security, and it has played a vital role in fields such as finance, medical care, cloud computing, and big data analysis. Secret sharing was first proposed in 1979 by Shamir [23] and Blakley [4]. Many researchers have improved this scheme to have dynamic, verifiable, multiple secrets, etc.

To overcome the limitation that communication groups can only share a single secret and each secret can only be used once, multi-secret sharing (MSS) [6,11,12,31] was proposed. Although it solves the problem that only one secret can be shared at a time, there is no way to verify the honesty of distributors and participants, which means that dishonest distributors/participants cannot be dealt with. A paper [24] points out that it is possible to introduce discrete logarithms into the scheme. Some other papers [8,9] propose to use discrete logarithm and the intractability of the RSA cryptosystem to improve the YCH scheme. Still, it cannot solve the problem of resource consumption, resulting in slow operation.

Chor et al. [7] proposed the first verifiable secret sharing (VSS) in 1985. It allows each participant to verify their share. If the participant recognizes that he has received an invalid share, he can ask the distributor to regenerate the share for distribution. Generally speaking, invalid shares may be caused by communication channel noise, deception, or malicious attacks. Feldman [10] and Pedersen [20] developed non-interactive VSSs based on cryptographic commitment schemes, but they all can only verify their share. Stadler [25] proposed the first publicly verifiable secret sharing (PVSS) scheme, in this scheme, each participant can mutually verify that the others' shares are valid.

Cachin [5] and Pinch [21] propose dynamic secret sharing schemes respectively, but each participant's shadow should not change after dynamic entry or exit. In the scheme of the paper [14], participants can dynamically join or exit their own shadow, which increases the dynamics. Paper [15,22,27] proposes a similar scheme based on Lagrange interpolation or Hermitian interpolation. In 2012 Hu et al. [13] and in 2015 Mashhadi et al. [17] respectively proposed verifiable multi-secret sharing schemes. The latter paper thinks that participants can dynamically add or delete shared secrets, and participants can dynamically join or exit. In 2017 A et al. [3]. Proposed a verifiable general structure based on the ECC(Elliptic Curve Cryptography) algorithm. This scheme has high efficiency, but cannot generate multiple keys at the same time.

3   Preliminaries

3.1   LakeHouse Architecture

The general architecture of the Data LakeHouse is as follows: The top layer is the service interface, responsible for retrieving processed data from the computing engine. The layer above, functioning as the computing engine, is responsible for extracting data from the Data LakeHouse and providing it to the service interface layer in near real-time, utilizing batch or stream processing. The middle layer is the metadata layer, primarily tasked with managing the metadata of both the Data Lake and the Data Warehouse, and implementing specific optimizations. The bottom layer, serving as the original data layer, is responsible for storing the raw data. The primary storage options consist of locally deployed Hadoop Distributed File Systems and remotely hosted cloud storage solutions like AWS S3, Google Cloud Storage, and Object Storage Service, among others.

Managed cloud services typically offer the advantages of ease of management, high flexibility, global performance, and online accessibility. Nevertheless, prolonged usage can lead to high costs, particularly in the case of extensive and frequent data access. Moreover, it cannot be compared to HDFS in terms of data privacy and security. In some near-real-time scenarios, enterprises may opt to utilize local HDFS as the ultimate storage solution. Hence, the security and key management of local storage data have emerged as challenges that require attention.

3.2 DM-Crypt

DM-Crypt is a transparent block device encryption subsystem, a component of the device mapper infrastructure, and utilizes encryption routines from the kernel's encryption API. It can encrypt entire disk partitions, software RAID volumes, logical volumes, as well as files. It appears as a block device and can be used for backing up filesystems, swaps, or as an LVM physical volume. It is highly suitable as a foundational storage encryption tool for Data LakeHouse. DM-crypt employs the initrd to request a password from the user at the console or to insert a smart card before the regular boot process. Security issues in the encryption process often arise in key management, including weak passwords, social engineering attacks, password leaks, forgetfulness, and so on. In distributed systems, secret sharding is commonly employed for key management.

4 The Proposed Scheme

The basic idea of the proposed scheme is illustrated as follows. The secret sharing system consists of a distributor (D), a secret restorer (C) and multiple participants $(P = \{P_1, P_2, \cdots, P_n\})$. The distributor will directly obtain the password $(originP)$ from the DM-Crypt encrypted disk. The RSA asymmetric encryption algorithm is initially used to establish a secure communication tunnel with other participants. The distributor D will then expand and obfuscate the original password, resulting in multiple secret sequences $(S = \{S_1, S_2, \cdots, S_k\})$. Designate the last item, S_k as the true secret, and the other items as false secrets. The purpose of this operation is to conceal the actual original password. Finally, the obfuscated multi-secrets are divided using the dynamic multi-secret sharing algorithm. The dynamism of this scheme is achieved through modifying the access structure, adding or removing participants, and obfuscating secrets.

4.1 Initialization Phase

The initialization phase aims to exchange public information, such as publishing the list of supported cipher suites and computing the public key of P_j. Through the following steps, D can securely communicate with participants.

(1) D chooses a secure asymmetric encryption algorithm (such as RSA, ECC), a safe and collision-resistant hash function $h(\cdot)$, a large prime number p, and a generator g of F_p.
(2) D choose a safe one-way function $f(\cdot, \cdot)$ and selects an integer u. It is easy to compute the value of $f(u, v)$ when given two parameters u and v. But it is tough to calculate in the following cases:
 a. Given u and $f(u, v)$, it is extremely difficult to compute v.
 b. Given v and $f(u, v)$, it is extremely difficult to compute u.
 c. Given one of v and u, but without knowledge about $f(\cdot, \cdot)$, it is extremely difficult to compute $f(u, v)$.
 d. For $u' \neq u$, given u and $f(\cdot, \cdot)$, it is difficult to compute $f(u, v)$.

    e. Given v , it is difficult to find two different values u_1, u_2, such that
$f(u_1, v) = f(u_2, v)$.

(3) D publishes $\{p, g, F_p, u, h(\cdot), f(\cdot, \cdot)\}$.

(4) Each participant select choose its own identifier$(P_j, j = 1, 2, \cdots, n)$ and compute the identity of themselves

$$I_j = f(u, P_j), j = 1, 2, \cdots, n \tag{1}$$

And then constructs a pair of asymmetric encrypted public keys and private keys. Take the RSA algorithm as an example:

  a) P_j firstly selects two large prime numbers p_j and q_j. Then calculate $m_j = p_j \cdot q_j$ and $\varphi(m_j) = (p_j - 1)(q_j - 1)$, where φ is the Euler function.

  b) P_j selects one random e_j, $1 \le e_j \le \varphi(n_j)$, satisfied $gcd(e_j, \varphi(n_j)) = 1$.

  c) computes the inverse d_j of e_j over the domain modulo φ, where $1 \le d_j \le \varphi(m_j)$, satisfied $e_j \cdot d_j \bmod \varphi = 1$.

(5) P_j publishes $< I_j, e_j, m_j >$. I_j is used as an index, which only corresponds to pair $< e_j, m_j >$ and cannot correspond to P_j, effectively protecting the identity information of the participants.

(6) D should ensure the uniqueness of each set of data pairs in the set $\{< I_j, e_j, m_j >\}$ released by the participants. Otherwise, the participants should re-select e again.

4.2   Construction Phase

The main purpose of this stage is to construct a shared secret among n participants by operating on the original secret and using a one-way function to verify whether the participants and the distributor have engaged in fraudulent behavior.

(1) D expands and obfuscates the $originP$ to create $S = \{S_1, S_2, \cdots, S_k\}$. Among them, S_k is the actual key, and the others are fake keys intended to obfuscate. Then calculate the hash value of S_i:

$$h_{S_i} = h(S_i) \tag{2}$$

(2) Randomly select distinct numbers $x_j \in F_p^*, j = 1, 2, \cdots, n$. Then compute the following equation from each participant's public key:

$$y_j = x_j^{e_j} \bmod m_j \tag{3}$$

(3) For each secret S_i, construct a polynomial of degree $i - 1$, which is given by:

$$F_i = (S_i + d_1 x + d_2 x^2 + \cdots + d_{i-1} x^{i-1}) \bmod p, \quad i = 1, 2, \cdots, k \tag{4}$$

In this equation, $F_i(0) = S_i$.

(4) For $i = 1, 2, \cdots, k$ and $j = 1, 2, \cdots, n$, compute the following:

$$V_{ij} = F_i(I_j) \tag{5}$$

$$c_{ij} - h^i(x_j) \oplus x_j \tag{6}$$

$$h_{c_{ij}} = h(c_{ij}) \tag{7}$$

$$R_{ij} = Vij - c_{ij} \bmod p \tag{8}$$

In the above formula, V_{ij} is a set of equations with $[k, n]$ dimensions. c_{ij} represents a set of pseudo-secret shares, signifying a protected intermediate result. $h_{c_{ij}}$ is later used to verify whether any participant cheated when publishing the hash value in the recovery phase. R_{ij} is a set of data that must be published and is used by each participant to compute their shard key.

(5) D publishes all $\{\{y_j\}, \{h_{S_i}\}, \{h_{c_{ij}}\}, \{R_{ij}\}\}$.

(6) P_j uses their private key d_j to calculate compute their shard key:

$$x_j = y_j^{d_j} \tag{9}$$

4.3   Recovery Phase

In a general analysis, suppose we wish to recover the index r secret S_r. In this case, r or more participants are required to cooperate in performing the following steps:

(1) Each participant computes their own pseudo-secret share and the corresponding hash value as follows:

$$c_{rj} = h^r(x_j) \oplus x_j \tag{10}$$

$$h^*_{c_{rj}} = h(c_{rj}) \tag{11}$$

Participant P_j first checks whether $h^*_{c_{rj}}$ is as same as $h_{c_{rj}}$ published by D. If they are different, it indicates that D has cheated; otherwise, P_j send c_{rj} to the secret restorer C.

(2) After C receives the c_{rj} sent by P_j, C will also calculate the corresponding hash value $h^{**}_{c_{rj}}$ by Eq. (12). If it is different from the published $h_{c_{rj}}$, this indicates that at least one participant has cheated. C will request all participants to resend data until everything matches. If C received all the correct data, $C_{rj}(j = 1, 2, \cdots, r)$, the index r secret can be reconstructed as follows:

$$h^{**}_{c_{rj}} = h(c_{rj}) \tag{12}$$

$$S_r = \sum_{j=1}^{r} (c_{rj} + R_{rj}) \prod_{t=1, t \neq j}^{r} \frac{-I_t}{I_j - I_t} \bmod p \tag{13}$$

After obtaining the secret S_r, C will calculate its hash value $h^*_{S_r}$ using Eq. (14). Then, if the hash value matches $h(S_r)$ published by D, the correct secret is obtained; otherwise, it suggests that D is dishonest.

$$h^*_{S_r} = h(S_r) \tag{14}$$

5   Verifiability and Dynamics Analysis

5.1   Verifiability Analysis

This scheme achieves verifiability by comparing the hash value read from the public bulletin board with the hash value calculated from the corresponding data.

During the recovery phase, participant P_j reads the values of y_j and $h_{c_{rj}}$ from the public bulletin board. P_j will compute its pseudo-secret c_{rj} and then compute $h^*_{c_{rj}}$ through eqs. (10) and (11). If $h^*_{c_{rj}} \neq h_{c_{rj}}$, then D is considered to have cheated in publishing y_j; otherwise, the verification is passed.

When C restores the key, P_j should actively transfer c_{rj} to C securely. Simultaneously, C should read $h_{c_{rj}}$ from the public bulletin board. C will first verify whether the c_{rj} passed by P_j is valid, by calculating through Eq. (12). If there exists $h^{**}_{c_{rj}} \neq h_{c_{rj}}$ $(j = 1, 2, \cdots, n)$, it can be concluded that at least one participant has cheated.

If there is no problem with the previous process, then C will calculate the secret S_r through Eq. (13) and calculate its hash value $h^*_{S_r}$ through Eq. (14). If $h^*_{S_r}$ matches h_{S_r} as read from the public bulletin board, then the verification passes; otherwise, it indicates that D is dishonest.

5.2   Dynamics Analysis

In actual operation, the following situations may occur that require reconsideration:

  i. Need to add or delete participants.
 ii. Need to modify the threshold.
iii. Need to modify the shards periodically.
 iv. Need to modify the origin key.

We will analyze these situations one by one to verify whether the scheme proposed in this paper satisfies the dynamic characteristics.

(1) **Addition of New Participants**
When new participants $\{P_{n+1}, P_{n+2}, \cdots, P_{n+l}\}$ want to become participants, they first retrieve $f(\cdot, \cdot)$ and u from public bulletin board to compute their identifiers, as shown in Eq. (1). Each P_{n+t} should choose two large prime numbers p_{n+t}, q_{n+t} and e_{n+t} $(1 \leq e_{n+t} \leq \varphi(m_{n+t}))$. Then, they compute m_{n+t}, $\varphi(m_{n+t})$ and $d_{n+t}(1 \leq d_{n+t} \leq \varphi(m_{n+t}))$. After these calculations, the uniqueness of $< I_{n+t}, e_{n+t}, m_{n+t} >$ should be ensured. Otherwise, e_{n+t} should be re-selected and recomputed.
In the construction phase, D selects x_{n+t} and calculates y_{n+t} using Eq. (3), where $x_{n+t} \in F_p^*$, and $x_{n+t} \neq x_j(j = 1, 2, \cdots, n)$. Then D calculates $V_{i,n+t}$, $c_{i,n+t}$, $h_{i,n+t}$, $R_{i,n+t}$ using eqs. (5) to (8).

The content to be published during the addition participant phase is: $\{\{<I_{n+t}, e_{n+t}, m_{n+t} >\}, \{y_{n+t}\}, \{V_{i,n+t}\}, \{c_{i,n+t}\}, \{h_{c_{i,n+t}}\}, \{R_{i,n+t}\}\}$
This allows new participants to calculate their own shards x_{n+t} using Eq. (9) without modifying the shards of the original participants. The shards of the new participants are also valid for recovery keys.

(2) **Deletion of Participants**

When deleting participants, they can be safely removed as long as the number of remaining participants reaches the threshold required for key recovery. The method for deleting participant P_j is to remove the $h_{c_{ij}}$ of P_j from the public bulletin board. If P_j wishes to recover the key after deletion but his hash authentication has expired and cannot pass the authentication, the security of the shared key will not be compromised.

(3) **Modification of the Threshold**

If someone wants to modify the threshold for secret recovery, they need to consider that our actual key is always stored in the highest dimension. They need to reselect the value of k and start calculating Eq. (4) again. Then they must modify the values of $h_{c_{ij}}$ and R_{ij} on the public bulletin board. Afterward, every P_j should recalculate their own new shard x_j. This will modify the data of other participants, but there's no need to reinitialize and recalculate y_i. Therefore, in this case, this scheme is not fully dynamic.

(4) **Periodic Modification of Shards**

This situation is more complex than modifying the threshold. Since the threshold has not changed, the secret shard needs to be regenerated, which means D needs to reselect $x_j (x_j \in F_p^*, j = 1, 2, \cdots, n)$ for every P_j and then compute Eq. (3). In this case, it is necessary to republish new value of $\{y_j\}$, $\{h_{c_{ij}}\}$, and $\{R_{ij}\}$. Afterward, every P_j should recalculate their own new shard x_j. This fulfills the objective of periodically changing shards. However, this scheme is not fully dynamic in this case either.

(5) **Modification of the Origin Key**

In this case, you need to start over from the beginning of the construction phase. This is because the first step in the construction phase involves the use of the origin key. If the origin key is modified, the key group must also be adjusted. In this case, it is necessary to republish new values of $\{h_{S_i}\}$, $\{y_j\}$, $\{h_{c_{ij}}\}$ and $\{R_{ij}\}$. Afterward, every P_j should recalculate their own new shard x_j. This completes the purpose of periodically changing shards. However, this scheme is not fully dynamic in this case either.

6   Conclusion

Based on the Data LakeHouse distributed system, we have proposed this verifiable dynamic multi-secret sharing obfuscation scheme in this paper. This scheme also incorporates the advantages of previous ones, but we have expanded and obfuscated a single key into a sequence of keys, making it more difficult to crack. Analysis shows that our scheme is computationally verifiable and dynamic. This scheme achieves verifiability mainly by comparing the hash value read from the public bulletin board with the hash value calculated from the corresponding data. This scheme supports the following dynamics: removing or adding participants dynamically, modifying the threshold dynamically, changing the shards periodically, and modifying the origin key dynamically, but only the first of these satisfies full dynamism. Moreover, it is easy to implement and is applicable in practical scenarios.

References

1. Armbrust, M., Ghodsi, A., Xin, R., Zaharia, M.: Lakehouse: a new generation of open platforms that unify data warehousing and advanced analytics. In: Proceedings of CIDR, vol. 8 (2021)
2. Begoli, E., Goethert, I., Knight, K.: A lakehouse architecture for the management and analysis of heterogeneous data for biomedical research and mega-biobanks. In: 2021 IEEE International Conference on Big Data (Big Data), pp. 4643–4651. IEEE (2021)
3. Binu, V.P., Sreekumar, A.: Secure and efficient secret sharing scheme with general access structures based on elliptic curve and pairing. Wireless Pers. Commun. **92**, 1531–1543 (2017)
4. Blakley, G.R.: Safeguarding cryptographic keys. In: Managing Requirements Knowledge, International Workshop on, pp. 313–313. IEEE Computer Society (1979)
5. Cachin, C.: On-line secret sharing. In: Boyd, C. (ed.) Cryptography and Coding 1995. LNCS, vol. 1025, pp. 190–198. Springer, Heidelberg (1995). https://doi.org/10.1007/3-540-60693-9_22
6. Chien, H.-Y., Jan, J.-K., Tseng, Y.-M.: A practical (t, n) multi-secret sharing scheme. IEICE Trans. Fundam. Electron. Commun. Comput. Sci. **83**(12), 2762–2765 (2000)
7. Chor, B., Goldwasser, S., Micali, S., Awerbuch, B.: Verifiable secret sharing and achieving simultaneity in the presence of faults. In: 26th Annual Symposium on Foundations of Computer Science (SFCS 1985), pp. 383–395. IEEE (1985)
8. Massoud Hadian Dehkordi and Samaneh Mashhadi: An efficient threshold verifiable multi-secret sharing. Comput. Stand. Interfaces **30**(3), 187–190 (2008)
9. Massoud Hadian Dehkordi and Samaneh Mashhadi: New efficient and practical verifiable multi-secret sharing schemes. Inf. Sci. **178**(9), 2262–2274 (2008)
10. Feldman, P.: A practical scheme for non-interactive verifiable secret sharing. In: 28th Annual Symposium on Foundations of Computer Science (SFCS 1987), pp. 427–438. IEEE (1987)
11. Harn, L.: Efficient sharing (broadcasting) of multiple secrets. IEE Proc.-Comput. Digital Tech. **142**(3), 237 (1995)
12. He, J., Dawson, E.: Multistage secret sharing based on one-way function. Electron. Lett. **30**(19), 1591–1592 (1994)
13. Chunqiang, H., Liao, X., Cheng, X.: Verifiable multi-secret sharing based on LFSR sequences. Theoret. Comput. Sci. **445**, 52–62 (2012)
14. Huang, Y., Yang, G.: Pairing-based dynamic threshold secret sharing scheme. In: 2010 6th International Conference on Wireless Communications Networking and Mobile Computing (WiCOM), pp. 1–4. IEEE (2010)
15. Hwang, R.-J., Chang, C.-C.: An on-line secret sharing scheme for multi-secrets. Comput. Commun. **21**(13), 1170–1176 (1998)

16. Liang, C., Qiu, K., Zhang, Z., Yang, J., Li, Y., Jingjing, H.: Towards robust and stealthy communication for wireless intelligent terminals. Int. J. Intell. Syst. **37**(12), 11791–11814 (2022)
17. Mashhadi, S., Dehkordi, M.H.: Two verifiable multi secret sharing schemes based on nonhomogeneous linear recursion and LFSR public-key cryptosystem. Inf. Sci. **294**, 31–40 (2015)
18. Oreščanin, D., Hlupić, T.: Data lakehouse-a novel step in analytics architecture. In: 2021 44th International Convention on Information, Communication and Electronic Technology (MIPRO), pp. 1242–1246. IEEE (2021)
19. Pang, P., Aourra, K., Xue, Y., Li, Y.Z., Zhang, Q.X.: A transparent encryption scheme of video data for android devices. In: 2017 IEEE International Conference on Computational Science and Engineering (CSE) and IEEE International Conference on Embedded and Ubiquitous Computing (EUC), vol. 1, pp. 817–822. IEEE (2017)
20. Pedersen, T.P.: Non-interactive and information-theoretic secure verifiable secret sharing. In: Feigenbaum, J. (ed.) CRYPTO 1991. LNCS, vol. 576, pp. 129–140. Springer, Heidelberg (1992). https://doi.org/10.1007/3-540-46766-1_9
21. Pinch, R.G.E.: On-line multiple secret sharing. Electron. Lett. **32**(12), 1087–1088 (1996)
22. Qu, J., Zou, L., Zhang, J.: A practical dynamic multi-secret sharing scheme. In: 2010 IEEE International Conference on Information Theory and Information Security, pp. 629–631. IEEE (2010)
23. Shamir, A.: How to share a secret. Commun. ACM **22**(11), 612–613 (1979)
24. Shao, J., Cao, Z.: A new efficient (t, n) verifiable multi-secret sharing (VMSS) based on YCH scheme. Appl. Math. Comput. **168**(1), 135–140 (2005)
25. Stadler, M.: Publicly verifiable secret sharing. In: Maurer, U. (ed.) EUROCRYPT 1996. LNCS, vol. 1070, pp. 190–199. Springer, Heidelberg (1996). https://doi.org/10.1007/3-540-68339-9_17
26. Sun, H., Tan, Y., Zhu, L., Zhang, Q., Li, Y., Shangbo, W.: A fine-grained and traceable multidomain secure data-sharing model for intelligent terminals in edge-cloud collaboration scenarios. Int. J. Intell. Syst. **37**(3), 2543–2566 (2022)
27. Tadayon, M.H., Khanmohammadi, H., Haghighi, M.S.: Dynamic and verifiable multi-secret sharing scheme based on hermite interpolation and bilinear maps. IET Inf. Secur. **9**(4), 234–239 (2015)
28. Tan, Y., Xinting, X., Liang, C., Zhang, X., Zhang, Q., Li, Y.: An end-to-end covert channel via packet dropout for mobile networks. Int. J. Distrib. Sens. Netw. **14**(5), 1550147718779568 (2018)
29. Tan, Y., Zhang, X., Sharif, K., Liang, C., Zhang, Q., Li, Y.: Covert timing channels for IoT over mobile networks. IEEE Wirel. Commun. **25**(6), 38–44 (2018)
30. Tan, Y., Zheng, J., Zhang, Q., Zhang, X., Li, Y., Zhang, Q.: A specific-targeting asymmetric group key agreement for cloud computing. Chin. J. Electron. **27**(4), 866–872 (2018)
31. Yang, C.-C., Chang, T.-Y., Hwang, M.-S.: A (t, n) multi-secret sharing scheme. Appl. Math. Comput. **151**(2), 483–490 (2004)
32. Zhang, Q., Li, Y., Song, D., Tan, Y.: Alliance-authentication protocol in clouds computing environment. China Commun. **9**(7), 42–54 (2012)
33. Zhang, Q., Yong Gan, L., Liu, X.W., Luo, X., Li, Y.: An authenticated asymmetric group key agreement based on attribute encryption. J. Netw. Comput. Appl. **123**, 1–10 (2018)

34. Zhang, Q., Wang, X., Junling Yuan, L., Liu, R.W., Huang, H., Li, Y.: A hierarchical group key agreement protocol using orientable attributes for cloud computing. Inf. Sci. **480**, 55–69 (2019)
35. Zhang, X., Liang, C., Zhang, Q., Li, Y., Zheng, J., Tan, Y.: Building covert timing channels by packet rearrangement over mobile networks. Inf. Sci. **445**, 66–78 (2018)

DZIP: A Data Deduplication-Compatible Enhanced Version of Gzip

Hengying Xiao[✉] and Yangyang Liu[✉]

University of Electronic Science and Technology of China, Chengdu, China
2692645221@qq.com, sagelyy@uestc.edu.cn

Abstract. Data deduplication is a common method for reducing storage space in backup storage systems. Despite extensive research aimed at improving the efficiency of data deduplication, we have observed poor compatibility between compressed data and deduplication. Specifically, two files with significant duplicate content cannot be deduplicated once they are compressed. In this paper, we delve into the internals of gzip and investigate the primary cause of this issue: the default compression-ratio-based heuristic blocking algorithm within deflate introduces a boundary offset issue. We introduce DZIP, which incorporates a content-defined chunking algorithm into gzip to maintain the redundancy of similar files after compression. The dataset-driven evaluation demonstrates that data compressed by DZIP can achieve a deduplication ratio of up to 86.2% compared to uncompressed data, with the compression ratio remaining largely unchanged compared to gzip, while achieving a throughput of up to 96% of gzip.

1 Introduction

Backup data storage constitutes a crucial component within modern data storage systems, finding widespread application in scenarios such as network hard drives and data centers(i.e. Google Drive, Dropbox). Within backup data storage system, users periodically upload local files backup to cloud servers for persistent storage. Studies [21] indicates a substantial amount of redundancy among the uploaded backups, which has inspired researchers to integrate data deduplication into backup data storage system [16,24] to improve storage efficiency. Data deduplication partitions uploaded backups into chunks and ensures that identical physical chunk is stored only once, thereby enhancing storage efficiency.

Nevertheless, the formats of user-uploaded files are diverse and unpredictable, with certain file formats not inherently compatible with data deduplication. For instance, in a cost-saving effort, users might compress different versions of files before uploading them to the cloud server. We found that even if the two versions of the files share significant duplicate content before compression, this duplicate content is substantially lost after compression.

J. Vaidya et al. (Eds.): AIS&P 2023, LNCS 14509, pp. 328–341, 2024.
https://doi.org/10.1007/978-981-99-9785-5_23

Addressing the aforementioned challenges, this paper takes the widely used compression software, *gzip* [9], as a case study to investigate the incompatibility between compressed data and deduplication. Our insight is that gzip did not take into account the potential for deduplication of compressed data. Consequently, its internal use of a compression-ratio-based heuristic blocking algorithm led to boundary offset issues.

We present DZIP, in which we replace the built-in blocking algorithm with a content-defined chunking algorithm. The contributions of this paper are as follows:

1. This paper employs a combination of data-driven experiments and theoretical analysis to investigate the issue of duplicated content loss in compressed data.
2. This paper introduces DZIP, which is an enhanced version of gzip. It improves compatibility between compressed data and deduplication by modifying the internal blocking algorithm of gzip.
3. This paper conducted evaluations of DZIP, and the results indicate that data compressed through DZIP achieve a deduplication ratio of 82.6% of the original data's deduplication ratio with only a slight reduction in throughput and without compromising the compression ratio.

The rest of this paper is organized as follows: In Sect. 2, we provide a detailed overview of the backup storage system and two common data reduction techniques, deduplication and compression. In Sect. 3, we leverage data-driven experiments and conduct theoretical analysis to highlight the limitations of gzip in deduplication, which serve as the motivation for the design of DZIP. In Sect. 4 and Sect. 5, we delve into the design details of DZIP and present the corresponding evaluation results, respectively. Finally, in Sect. 6 and Sect. 7, we discuss related research and conclude this paper.

2   Background

Backup Storage System. Figure 1 illustrates the basic architecture of the reduction-based storage system: the client periodically backs up local files and uploads the backup to the server. Upon receiving the client's backup, the server first performs data reduction and then writes the reduced data to the underlying physical storage. The physical storage is built on top of storage media, responsible for persisting data to devices such as hard disks and solid-state drives (SSDs). Common data reduction techniques include data deduplication, lossless compression, delta compression, etc. This paper primarily focuses on the first two. Details about these techniques will be elaborated in the following paragraphs.

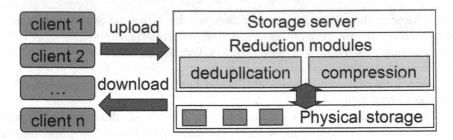

Fig. 1. The fundamental architecture of a data deduplication-based backup storage system. The client sends upload and download commands, while the server is responsible for data reduction and persistent storage.

Deduplication. Deduplication is employed to eliminate data redundancy in both backup and primary workloads [6,12,20,26,27]. It begins by using chunking algorithm(e.g., fixed-size chunking, RabinCDC [2,19]) to partition files into multiple consecutive chunks. The size of a chunk generally varies from 4 KiB to 8 KiB depending on the chunking algorithm and the scenario. Subsequently, it employs hashing algorithms like MD5 to calculate the content hash of each chunk as its *fingerprint*. Chunks with the same fingerprint are considered to have identical content, and vice versa [1]. Finally, in physical storage, data deduplication ensures that chunks with the same fingerprint are stored only once, while maintaining two data structures: a fingerprint index and a file recipe index. The former records the physical storage addresses of each unique chunk, while the latter records the fingerprint sequence of chunks contained in each file.

Compression. In backup storage system, compression is usually applied after data deduplication to further reduce the size of unique chunks. This paper focuses on the GNU software *gzip* [9] and the compression algorithm deflate [5] it uses. Deflate is a stream-based lossless compression algorithm that comprises two main components: LZ77 encoding and Huffman encoding.

LZ77 encoding [25], achieves data compression by identifying and eliminating recurring data patterns. To accomplish this, it scans the data stream for repeated patterns and substitutes them with references pointing to earlier occurrences. The search process is constrained to a predefined length forward from the current position, meaning patterns located beyond this searching range are not considered for reference. Whether or not a pattern is located, LZ77 generates a short code (l, d) to represent the outcome of each search, in which l represents the length of the pattern, and d represents the distance between the current pattern and the found pattern. If no patterns are found during the search, LZ77 directly outputs the current byte, which is stored in l.

After the LZ77 encoding, gzip employs a *compression-ratio-based heuristic blocking algorithm* to divide the short code sequence output by Lz77 into multiple blocks, then applying Huffman encoding [11] to each block individually. Specifically, it scans the LZ77 short code sequence, accumulating each code it encounters into the ongoing block. Subsequently, gzip estimates the compression ratio that the current block would achieve post Huffman encoding. If this estimated compression ratio meets the predefined criteria, the current block is terminated, and a new block begins. At this phase, gzip employs Huffman encoding to encode the terminated block by building a dynamic Huffman tree and saves the encoded binary data into the compressed file.

3   Observation

We then study the limitations of applying deduplication on compressed data through data-driven evaluation, and motivate the design of DZIP through theoretical analysis.

3.1   Limitations

We selected 5 different versions of the gcc software packages, ranging in size from 508.7 MiB to 512.8 MiB (see Sect. 5 for details), we then measured the deduplication ratio (the ratio of the total size of logical chunks to the total size of unique chunks) before and after compression. Specifically, The experiments were divided into two groups: in the first group, the 5 versions were left unprocessed, while in the second group, the same 5 versions were compressed using gzip with default configuration. For each group, we applied the RabinCDC [2, 19] algorithm to partition each version of the release into chunks with an average chunk size of 4 KiB and conducted deduplication. Finally, for each group, we calculated the deduplication ratio after processing each version. The results for both groups are depicted in Fig. 2.

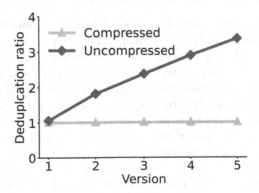

Fig. 2. The deduplication ratios of two groups of backups, one compressed and the other uncompressed.

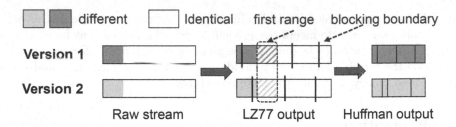

Fig. 3. The two steps of gzip compression, LZ77 encoding, and Huffman coding, both result in the loss of duplicated data.

From the figure, it can be observed that for the compressed group, as the number of versions participating in deduplication increases, the deduplication ratio remains around 1. However, for the group where deduplication is performed directly without compression, the deduplication ratio steadily increases as more versions are involved. When all 5 versions of data are processed, the deduplication ratio reaches 3.4. This indicates that there is a significant amount of duplicate content among the uncompressed versions of these software packages, which disappears after compression. Therefore, the duplicate data among multiple versions is challenging to preserve after they are compressed.

3.2   Motivation

Analysis. Next, we delve into the reasons behind the above evaluation result. Gzip's compression of the original data stream involves two main phases: LZ77 encoding and Huffman encoding. We contend that both of these phases can impact the deduplication ratio, as outlined in detail below:

1. **The LZ77 encoding amplifies the differences between two version of files that share duplicate content.** The output data stream from LZ77 (i.e., the short code sequence) is not solely determined by the current processing data, it also relies on data within a preceding searching range (refer to Sect. 2 for details). As illustrated in Fig. 3, for a segment of deduplicated content in two versions of a file, the LZ77 short code output for data within the first searching range depends on its preceding content, resulting in it being encoded differently. However, the content following the first searching range remains unaffected and, therefore, still remains the same. Therefore, LZ77 encoding reduces the duplicate data between the two versions of the files.
2. **The compression-ratio-based heuristic blocking algorithm employed by gzip introduces boundary shifts.** As depicted in Fig. 3, for the output data stream from LZ77 with identical content, their blocking positions are likely to differ (due to potentially different starting blocking positions). Consequently, the frequency distribution of short codes within the current block diverges, and Huffman coding encodes each l and d in every LZ77 short code into a bit sequence based on their frequencies. This triggers a butterfly effect, resulting in completely different compressed data.

Our Approach. The analysis above highlights that both stages of gzip, namely LZ77 and Huffman encoding, introduce adverse effects on data deduplication. To address this concern, we are motivated to enhance gzip's compatibility with data deduplication by focusing on two dimensions: LZ77 searching range size and blocking algorithm. Our design focus is centered on addressing boundary shifts. Our insight is to integrate a content-defined chunking algorithm into gzip, replacing the previous heuristic blocking algorithm to mitigate boundary offset issues. When it comes to the LZ77 searching range, although a smaller searching range reduces the amplification of differences, it also significantly lowers the compression ratio. Therefore, we are considering modifying the default searching range size to strike a balance between compression ratio and deduplication ratio, which will be discussed in our future work.

4   Dzip Design

4.1   Design

We introduce DZIP, a solution that utilizes a content-defined chunking algorithm to deal with the LZ77 short code sequence instead of gzip's default heuristic blocking algorithm. The content-defined chunking algorithm, a well-established technique, involves analyzing a data stream using a sliding window, calculating hash values for each window, and partitioning the file into data chunks based on patterns in these hash values. This approach effectively addresses the boundary shift problem. Consequently, our modification aims to resolve the compatibility issue between compressed data and data deduplication.

Specifically, DZIP maintains a fixed-size sliding window (e.g., 48). Whenever new LZ77 short code is generated, DZIP computes the rabin fingerprint [19] within the window and checks whether the current fingerprint adheres to a specific pattern (e.g., congruent to 0 modulo 2^{10}). If satisfied, dzip immediately concludes the current block and initiates a new block. The terminated block is subsequently Huffman-encoded and written into the file as compressed data stream. Although we replaced the previous default heuristic blocking algorithm, this modification had minimal impact on the compression ratio of the data, which will be detailed in the evaluation section.

As the deflate RFC [5] does not impose any strict requirements on the blocking algorithm, our design is fully compatible with existing decompression software (e.g., zlib [7]). Furthermore, although our research primarily focuses on gzip, it's worth noting that many compression softwares or algorithms, such as zstd [4], also internally employ the method of blocking before compression based on Huffman coding. Therefore, our work may potentially be applicable to other software as well.

4.2   Implementation

We made modifications to the source code of gzip version 1.2.6 by integrating an open-source implementation [18] of RabinCDC and altering the blocking algorithm within the deflate implementation.

Due to the fact that the open source implementation of RabinCDC operates at a byte-level granularity, and since LZ77 output short codes exceed 256, without the possibility of splitting individual LZ77 short codes, we customized the open-source implementation to handle a wider input range.

Table 1. The detailed information about the dataset we used. The deduplication ratio is computed using the RabinCDC algorithm with an average chunk size of 4 KiB, while the compression ratio is measured using gzip with default configuration.

Versions	5
Total Size	2552.7 MiB
Deduplication ratio	3.39
Compression ratio	5.08

5 Evaluation

We conduct dataset-driven experiments to evaluate the performance of DZIP. Our evaluation shows the following key findings:

- In terms of reduction ratio, DZIP retains a significant amount of duplicate content compared to gzip with almost no loss in compression ratio. The deduplication ratio of data compressed by DZIP can reach 82.6% of the deduplication ratio of the original data (Exp#1).
- In the evaluation of storage efficiency in the backup storage system, DZIP employed a pre-compression followed by deduplication approach, resulting in a final physical storage size only 36.6% of that achieved by gzip using the same approach. Furthermore, compared to the approach of deduplication followed by compression, DZIP significantly reducing both server CPU and network overhead (Exp#2).
- In terms of throughput, DZIP achieves compression speeds reaching up to 96% of gzip (Exp#3).

5.1 Setup

We obtained five consecutive backup versions of GCC releases from the GNU gcc website [8]. These releases were distributed as *tar.gz* format archive files. To prevent incompatibility issues between the *tar* format and data deduplication [17] that could affect our evaluation results, we initially decompressed these compressed archive files. Subsequently, we preprocessed these extracted archive files using mtar [17] to create a *tar.fm* format archive file that is compatible with data deduplication. The only difference between this format and the traditional *tar* format is the rearrangement of internal metadata to support deduplication. Table 1 provides some details about this dataset.

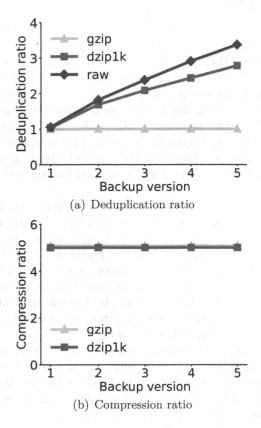

(a) Deduplication ratio

(b) Compression ratio

Fig. 4. Exp#1 (Reduction ratios): The average deduplication ratio and compression ratio after processing each backup version

We conduct experiments on a machine running Debian 11 with Linux kernel 5.10. It equips with a 16-core Intel Xeon E5-2683v4 CPU, 128 GB DDR4 RAM, quad 4 TiB SATA HDDs in a RAID5 array.

In our evaluation, in addition to comparing DZIP with gzip under default configuration, we also configured multiple dzip instances with various blocking granularity, ranging from 128, 256, 512, and all the way up to 8k (i.e., 8192), and so on. The blocking granularity refers to the average number of LZ77 short codes each time dzip preserves before performing Huffman coding.

We primarily focus on the following two aspects: (i) Reduction ratios: This encompasses both compression and deduplication, quantified as the ratio between the original size of the backup data before reduction and its size after reduction. (ii) Throughput: The total volume of data that can be compressed within a unit of time.

5.2 Results

Exp#1 (Reduction Ratios). To evaluate the performance of DZIP, we initially used DZIP (with a blocking granularity of 1024,labeled with "dzip1k") and gzip to individually compress each of the backup version processed by mtar.Subsequently, we conducted deduplication on each backup (with an average block size of 4 KiB). After processing each backup file, we measured the average deduplication ratio and compression ratio. The results are shown in Fig. 4. Additionally, to evaluate the impact of DZIP on deduplication ratio, we also plotted the step deduplication ratio of uncompressed backup versions in the figure (labeled as "raw").

For deduplication ratio, we observed from Fig. 4(a) that the deduplication ratio under the DZIP approach increases with the number of backups. After processing all 5 backups, DZIP achieved a deduplication ratio of 2.80, which is 82.6% of the deduplication ratio without compression, which is 3.39. In contrast, for gzip, the deduplication ratio of its compressed backups remained around 1.0. This difference arises because the default deflate blocking algorithm used by gzip introduces boundary shifts that result in the loss of duplicate content, while DZIP, utilizing a content defined chunking approach, preserves the redundancy.

As for compression ratio, which is shown in Fig. 4(b), the compression ratio of DZIP and gzip are nearly indistinguishable. After processing all 5 backups, DZIP achieved a compression ratio of 5.01, which is 98.8% compared to gzip, which is 5.08. This indicates that when the blocking granularity of DZIP is 1024, while DZIP abandoned the heuristic blocking algorithm, its actual compression ratio is almost unaffected.

Exp#2 (Storage Efficiency). We evaluate the storage efficiency of DZIP in a basic backup storage system. Specifically, we compare three different data reduction approaches:

1. **dzip:** First, we compress each backup file on the client side using DZIP with different blocking granularity. Then, we upload these backups to the server for deduplication.
2. **gzip:** This approach is the same as dzip, but we compress backups using gzip instead of DZIP.
3. **raw:** We directly upload the original backup versions to the server, where the server deduplicates these backups first and then applies gzip compression to each unique chunk separately.

All three approaches have a deduplication granularity of average 4 KiB using RabinCDC. Figure 5(a) shows the physical sizes after processing all backup versions, while Fig. 5(b) displays the deduplication ratio and compression ratio for each approach after processing all backups.

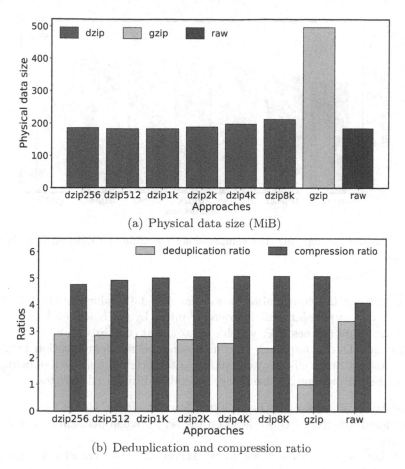

(a) Physical data size (MiB)

(b) Deduplication and compression ratio

Fig. 5. Exp#2 (Storage efficiency): The final physical data size and reduction ratios of different approaches.

From Fig. 5(a), it can be observed that for different instances of DZIP, the final physical data size of the backup files decreases first and then increases. When the block granularity is 8192, the physical data reaches its maximum size, at 212.81 MiB, which is 115.3% of the physical data size of the raw approach. However, when the block granularity is 1024, the physical data reaches its minimum size, at 182.25 MiB, only 98.7% of the raw scheme's size. In contrast, for gzip, the backups' physical size in this approach is 497.27 MiB, which is 269.3% of the size in the raw approach.

The results in Fig. 5(b) illustrate the reasons for the above phenomenon. For DZIP, as its blocking granularity increases, the deduplication ratio gradually decreases, but the compression ratio decreases gradually. Furthermore, in the gzip approach, there is almost no deduplication among different backups (The deduplication ratio is only 1.011, which is very close to 1), resulting in a significantly larger final physical size. Lastly, for the raw scheme, although it has a

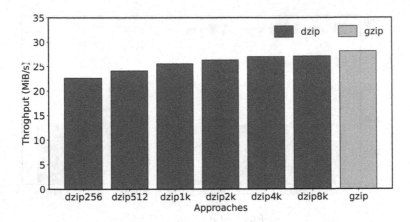

Fig. 6. Exp#3 (Throughput): The compression throughput for different DZIP instance and gzip

clear advantage in deduplication ratio, reaching 3.38, which is 112.1% that of dzip1024, its compression ratio is severely limited by the block-level compression scheme and only reaches 4.08, which is 81.6% that of dzip1024.

Although DZIP's approach of compressing before deduplication results in slightly lower storage efficiency compared to the original approach of deduplication before compression, it also brings several additional advantages:

1. The data compression has been shifted from the server side to the client side, which reduces the CPU overhead on the server when performing online data storage.
2. The data sent from the client to the server has been transformed from the original data into smaller-sized compressed data, which saves network overhead.

Therefore, we argue that the marginal storage space waste introduced by DZIP is justified.

Exp#3 (Throughput). Due to DZIP's design, which replaces the original heuristic blocking algorithm to RabinCDC, we evaluated its impact on throughput. Specifically, we opted for a specific version of the file and conducted 10 repeated measurements for each instance of DZIP and gzip to determine the time required for compressing the entire file. We then calculated the average compression time and subsequently derived the throughput. The results are depicted in Fig. 6.

From the figure, it can be observed that the throughput of DZIP increases with the blocking granularity. For instance, at a blocking granularity of 128, the throughput reaches 22.64 MiB/s, while it reaches 27.10 MiB/s at a blocking granularity of 8192. This is because coarser blocking granularity reduces the number of LZ77 short codes that participate in Huffman coding during each iteration of the deflate algorithm. Consequently, it increases the frequency of

Huffman coding, which in turn raises CPU overhead and lowers throughput. On the other hand, gzip has the highest throughput, reaching 28.11 MiB/s. At an 8k blocking granularity, DZIP's throughput is only 96.4% that of gzip. This indicates that content-defined chunking algorithms have higher overhead compared to compression-ratio-based heuristic blocking algorithms.

6 Related Work

Most existing research in the optimization of data deduplication backup systems has predominantly focused on improving I/O locality (e.g., MFdedup [27], ALACC [3]), optimizing indexing (e.g., DDFS [24], sparse indexing [16]), and enhancing block algorithms (e.g., FastCDC [23], bimodal CDC [13]). In the realm of combining data compression and deduplication, DEC [10] leverages the locality introduced by data deduplication to cluster similar chunks, thereby enhancing compression ratio. Nitro [14], Cache Dedup [15] and Austere Cache [22] apply both data compression and deduplication to flash cache devices to improve cache efficiency.

In addressing the incompatibility between file formats and deduplication, the research most closely related to dzip is mtar [17]. Using the tar format as an example, mtar explores the reasons why tar archive files are challenging to deduplicate and enhances compatibility between tar format archive files and data deduplication through metadata rearrangement.

7 Conclusion

We took gzip as an example to investigate the reasons why different versions of files with a significant amount of duplicate content cannot be deduplicated after compression. We then proposed DZIP, which is an enhanced version of gzip. In the design of DZIP, we replaced gzip's original compression-ratio-based heuristic blocking algorithm with a content-based chunking algorithm to improve its compatibility with deduplication. Our dataset-driven experiments show that DZIP, compared to gzip, retains a significant amount of redundancy in compressed data, achieving a deduplication ratio of 82.6% of the original data's deduplication ratio. Additionally, DZIP's throughput can reach up to 96% of gzip's with minimal impact on compression ratio.

References

1. Black, J.: Compare-by-hash: a reasoned analysis. In: Proceedings of USENIX ATC, pp. 85–90 (2006)
2. Broder, A.Z.: Some applications of Rabin's fingerprinting method. In: Capocelli, R., De Santis, A., Vaccaro, U. (eds.) Sequences II, pp. 143–152. Springer, New York (1993). https://doi.org/10.1007/978-1-4613-9323-8_11

3. Cao, Z., Wen, H., Wu, F., Du, D.H.: ALACC: accelerating restore performance of data deduplication systems using adaptive Look-Ahead window assisted chunk caching. In: Proceedings of USENIX FAST (2018)
4. Collet, Y.: Zstandard compression and the 'application/zstd' media type. https://datatracker.ietf.org/doc/html/rfc8878
5. Deutsch, L.P.: Deflate compressed data format specification version 1.3. https://www.ietf.org/rfc/rfc1951.txt
6. El-Shimi, A., Kalach, R., Kumar, A., Oltean, A., Li, J., Sengupta, S.: Primary data deduplication-large scale study and system design. In: Proceedings of USENIX ATC (2012)
7. loup Gailly, J., Adler, M.: zlib. https://www.zlib.net/
8. GNU: GCC. https://gcc.gnu.org/
9. GNU: GZIP. https://www.gnu.org/software/gzip/
10. Han, Z., et al.: DEC: an efficient deduplication-enhanced compression approach. In: Proceedings of IEEE ICPADS (2016)
11. Huffman, D.A.: A method for the construction of minimum-redundancy codes. Proc. IRE **40**(9), 1098–1101 (1952)
12. Kotlarska, I., Jackowski, A., Lichota, K., Welnicki, M., Dubnicki, C., Iwanicki, K.: InftyDedup: scalable and cost-effective cloud tiering with deduplication. In: Proceedings of USENIX FAST (2023)
13. Kruus, E., Ungureanu, C.: Bimodal content defined chunking for backup streams. In: Proceedings of USENIX FAST (2010)
14. Li, C., Shilane, P., Douglis, F., Shim, H., Smaldone, S., Wallace, G.: Nitro: a capacity-optimized SSD cache for primary storage. In: Proceedings of USENIX ATC (2014)
15. Li, W., Jean-Baptise, G., Riveros, J., Narasimhan, G., Zhang, T., Zhao, M.: CacheDedup: in-line deduplication for flash caching. In: Proceedings of USENIX FAST (2016)
16. Lillibridge, M., Eshghi, K., Bhagwat, D., Deolalikar, V., Trezise, G., Camble, P.: Sparse indexing: large scale, inline deduplication using sampling and locality. In: Proceedings of USENIX FAST (2009)
17. Lin, X., et al.: Metadata considered Harmful... to deduplication. In: Proceedings of USENIX HotStorage (2015)
18. Neumann, A.: rabin-cdc. https://github.com/fd0/rabin-cdc
19. Rabin, M.O.: Fingerprinting by random polynomials. Technical report (1981)
20. Srinivasan, K., Bisson, T., Goodson, G., Voruganti, K.: iDedup: latency-aware, inline data deduplication for primary storage. In: Proceedings of USENIX FAST (2012)
21. Wallace, G., et al.: Characteristics of backup workloads in production systems. In: Proceedings of USENIX FAST (2012)
22. Wang, Q., Li, J., Xia, W., Kruus, E., Debnath, B., Lee, P.P.: Austere flash caching with deduplication and compression. In: Proceedings of USENIX ATC (2020)
23. Xia, W., et al.: FastCDC: a fast and efficient content-defined chunking approach for data deduplication. In: Proceedings of USENIX ATC (2016)
24. Zhu, B., Li, K., Patterson, H.: Avoiding the disk bottleneck in the data domain deduplication file system. In: Proceedings of USENIX FAST (2008)
25. Ziv, J., Lempel, A.: A universal algorithm for sequential data compression. IEEE Trans. Inf. Theory **23**(3), 337–343 (1977)

26. Zou, X., Xia, W., Shilane, P., Zhang, H., Wang, X.: Building a high-performance fine-grained deduplication framework for backup storage with high deduplication ratio. In: Proceedings of USENIX ATC (2022)
27. Zou, X., Yuan, J., Shilane, P., Xia, W., Zhang, H., Wang, X.: The dilemma between deduplication and locality: can both be achieved? In: Proceedings of USENIX FAST (2021)

Efficient Wildcard Searchable Symmetric Encryption with Forward and Backward Security

Xi Zhang[1,2,4], Ye Su[3], Zhongkai Wei[4], Wenting Shen[5], and Jing Qin[4(✉)]

[1] College of Computer and Cyber Security, Hebei Normal University,
Shijiazhuang 050024, China
[2] Hebei Provincial Key Laboratory of Network and Information Security,
Hebei Normal University, Shijiazhuang 050024, China
[3] School of Information Science and Engineering, Shandong Normal University,
Jinan 250358, China
[4] School of Mathematics, Shandong University,
Jinan 250100, China
qinjing@sdu.edu.cn
[5] College of Computer Science and Technology, Qingdao University,
Qingdao 266071, China

Abstract. Wildcard Searchable Symmetric Encryption can achieve flexibility and pattern matching while protecting data privacy. There is a promising future for combining secure wildcard search with emerging technologies such as AI. However, there are challenges in reducing the communication cost and improving security. In this paper, we propose an efficient wildcard searchable symmetric encryption with forward and backward security. The complexity of communication costs in the search protocol is $O(1)$, independent of the number of characters of wildcard keywords, or files in the search result. It is achieved by a double-compressed index in which the character set and file identifiers are encoded simultaneously. Then, the double-compressed index provides a possibility to achieve oblivious keyword query and update, so that the proposed scheme only reveals the query type (search or update) without anything else. It achieves forward and backward security using the distributed multi-point function and an additively homomorphic symmetric encryption scheme. Detailed security proof and the theoretical comparison show the improvement in security and efficiency. The acceptable overheads of the proposed scheme are presented by the extensive performance evaluation.

Keywords: Searchable Symmetric Encryption · Wildcard Keyword Search · Data Update · Forward and Backward Security

1 Introduction

As a technique with flexibility and pattern matching, wildcard search allows users to represent one or more unknown letters using a wildcard, so that users

J. Vaidya et al. (Eds.): AIS&P 2023, LNCS 14509, pp. 342–357, 2024.
https://doi.org/10.1007/978-981-99-9785-5_24

can query words or phrases matched to a specified pattern. In particular, there are two kinds of wildcards % and _. The multiple-character wildcard % is used to indicate letters with any length, and the single-character wildcard _ only replaces one letter. Due to the flexible expression, wildcard search is widely used in various fields and has been incorporated into search engines (e.g., Google, Bing), databases (e.g., SQL Databases), text editors (e.g., Sublime Text), and so on.

Wildcard search is also a powerful tool for AI programs. For example, there is an AI project to train a ball recognizer so the trainer needs to search for photos of all balls as training data. Assuming there are "Baseball, Football, Basketball, Volleyball, Soccerball, Softball, Beachball, Handball, Fireball, Eightball, ..." in the data set, users can use the wildcard keyword "%ball" to capture all poten-tial variations rather than searching for each possible word separately. Formally speaking, the wildcard keyword W represents all variant keywords $w_1, w_2, \ldots$ in a dataset. Overall, this flexible and pattern-based information retrieval can enhance the user's ability to find relevant content even when they do not have precise details about what they're looking for. So the efficiency of AI training could be further improved and developed using wildcard search.

Recently, frequent data leakage incidents have captured users' attention to data security. As valuable assets, datasets, even without sensitive information, are to be protected. Encrypting the entire dataset can prevent information leak-age but simultaneously destroy the search functionality in the process. The method to download, decrypt, and search is not economical and efficient. There-fore, searchable encryption was proposed and provided an alternative way to achieve privacy-preserving search [16]. For users who upload datasets and then query data by themselves, protecting the dataset with a symmetric system in the context of searchable encryption is preferred, i.e., Searchable Symmetric Encryption (SSE). Therefore, we mainly focus on SSE in this paper.

Flexible queries in SSE have always attracted attention, and the wildcard keyword search is a potential and powerful assistance for emerging technologies such as AI. In addition to performing flexible queries and then data aggregation without compromising data privacy, it is also promising to be applied in data regulations, especially for industries subjected to strict data privacy regulations (e.g., GDPR, HIPAA). For example, it can help them comply with these reg-ulations by ensuring that data remains encrypted while enabling analysis and research.

However, there still exist some challenges in SSE supporting wildcard search thus far. On the one hand, the computation or communication complexity often is linear with the number of keywords or the size of search results. It is eager to improve the efficiency of wildcard search over encrypted data for more promising applications. On the other hand, in terms of data updates, file-injection attacks utilize only a small amount of information to compromise the privacy of data [24]. Forward and backward security is the necessary defense for any SSE supporting data update [12]. The same holds true for SSE with wildcard search. In this paper, we solve the above challenges simultaneously, and the contributions are summarized in three folds:

- We design a double-compressed index to simultaneously describe the keywords and file sets. Keywords are presented for wildcard search by character sets using a modified extraction rule and then compressed in bloom filters. The file sets containing each keyword are encoded into a ciphertext and thus reduces communication costs.
- We propose an SSE scheme with the double-compressed index to achieve a forward and backward secure wildcard search. Specifically, users query wildcard keywords in an oblivious way and cloud servers learn nothing except the middle computation result. Whereas the method of no differential update is adopted to resist file-injection attacks.
- We formally prove that the proposed SSE scheme supporting wildcard search is forward and backward secure under a simulation-based paradigm. Then the comparison with existing works is presented theoretically. In addition, the extensive performance evaluation shows that the proposed scheme is efficient in terms of computational and communication overheads.

The rest of the paper is organized as follows. The related works review is presented in Sect. 2, followed by descriptions of the system model, adversarial model, security model, and design goals. Then, Sects. 4 and 5 show the preliminaries and proposed construction. Its security proof is in Sect. 6. Section 7 shows the performance evaluation and comparison. Finally, we conclude this paper in Sect. 8.

2   Related Work

To search for a wildcard keyword W, the straightforward approach is to query each keyword $w \in W$. Indeed, wildcard searchable encryption was first considered in the seminal paper on symmetric searchable encryption (SSE) by Song et al. [16]. This work achieved wildcard search in a limited form by enumerating all possible keywords. For example, to query $ab[a - z]$, users would need to generate and query 26 keywords $\{aba, abb, \ldots, abz\}$. However, it's evident that both the leakage and the number of queries increase rapidly as the constraints on keywords decrease. To enhance the efficiency and security of simultaneously querying multiple keywords, multi-keyword search has emerged as an alternative for achieving wildcard keyword search [1]. It encompasses conjunctive/conjunctive keyword search [6,22] and boolean search [11,20]. In terms of security, the design of forward and backward secure schemes for multi-keyword search has become essential [13,19]. However, users often face the challenge of complex keyword enumeration for W and the computation of trapdoors, especially when dealing with a relatively large keyword space. Improving computation and communication efficiency has become a prominent area of research. Additionally, existing works have explored techniques such as hidden vector encryption [15], homomorphic encryption [21], proxy middleware [10], and inner product [14] to achieve wildcard search within the context of public encryption.

Another popular approach to realize wildcard keyword search is by describing a keyword using characters, allowing cloud servers to determine whether a keyword $w \in W$ by assessing the subset relationship between the keyword and the trapdoor. Suga et al. [17] proposed a position-specific keyword search scheme, where the positions of letters in a keyword need to be known. The rule to extract characters is represented as $\{s \parallel 1, e \parallel 2, c \parallel 3, \ldots, y \parallel 8, \text{null} \parallel 9\}$ for the word "security". To further query keywords with letters in unknown positions, Hu and Han [8] improved the extraction rule by incorporating 2-gram information, reverse order, and the existence of letters, resulting in $\{s \parallel 1, e \parallel 2, c \parallel 3, \ldots, y \parallel 8; y \parallel -1, t \parallel -2, \ldots, s \parallel -8; se \parallel 0, ec \parallel 0, \ldots, ty \parallel 0; s \parallel 0, e \parallel 0, c \parallel 0, \ldots, y \parallel 0\}$. This enhancement allows users to query, for example, $s\%oo\%$ using the character set $\{s \parallel 1, oo \parallel 0, o \parallel 0\}$. Similar work includes schemes proposed by Zhao and Nishide [25], and Hu et al. [9]. However, the search cost in these works is linear with the number of keywords, as keywords are judged in the index one by one. To enhance search efficiency, Zhang et al. [23] introduced a three-step search method with a tree-based index. They expanded the functionalities of single-character wildcard search by introducing AB$\parallel 1$. They classified characters into three types: A-characters to describe the existence of letters, AB-characters to indicate the relative position of letters, and BF-characters to provide further keyword details. The search process, from A to AB and then BF characters, narrows down the range of keywords step by step, avoiding the linear comparison of character sets.

From the perspective of security, many schemes supporting wildcard search use the security model of searchable symmetric encryption. For example, Hu et al. [9,14] follow the IND-CKA or IND-CKA1 model. The scheme in Bösch et al. [1] is adaptive secure. Suga et al. [17]proposed a security model: Indistinguishability against Chosen Position-Specific Keyword Attack (IND-CPSKA). Zhang et al. [23] put forth a nonadaptive Indistinguishability Against Chosen Character Set Attacks (IND-CCSA). To date, there are fewer schemes to achieve forward and backward security [18].

3   Models and Design Goals

3.1   System Model

As shown in Fig. 1, our system model consists of users and cloud servers.

- *Users*: Users are entities with resource-limited devices such as phones and computers. They purchase storage space from the cloud to store their dataset, and then require keyword search and data update services. Specifically, They have a file universe $\mathbb{F} = \{F_1, \ldots, F_n\}$, and users' file sets are subsets of this universe. Each file F_i has a unique identifier id_i and is associated with some keywords. The keyword universe is denoted as $\mathbb{W} = \{w_1, \ldots, w_m\}$. Therefore, the dataset can be described by an index $\mathbb{D}$ with m pairs $(w, \mathbb{D}(w))$ where $\mathbb{D}(w)$ is the set of identifiers of files related to keyword w. Users search for specific keywords w and wildcard keywords W to retrieve $\mathbb{D}(w)$ and $\mathbb{D}(W)$, where $\mathbb{D}(W)$ is the union of $\mathbb{D}(w)$ for all w in W. They also update the index $\mathbb{D}$ with pairs (w, id, op) where $op =$"Add" or "Delete". To achieve data security and protect users' privacy, the index $\mathbb{D}$ and files set $\mathbb{F}$ are encrypted to $\mathbb{ED}$ and $\mathbb{EF}$ respectively, before uploading to cloud servers. Then the keyword search and data update services are run securely between users and cloud servers. Users are honest and do not engage in any behaviors that could endanger system security.

- *Cloud Servers* (CSs): Cloud servers have enough storage space and powerful computational capabilities, so they are service providers for resource-limited users. They store $\mathbb{ED}$ and $\mathbb{EF}$ uploaded from users and run algorithms or protocols to provide users with services. Specifically, they receive trapdoors tr_W to target the search algorithm, and (w, id, op) in a secure format to invoke the update protocol. CSs are honest-but-curious entities, meaning that, they faithfully follow the algorithms or protocols but may passively analyze users' information due to curiosity and commercial interests. Additionally, communications among CSs are safe and collusion is not allowed.

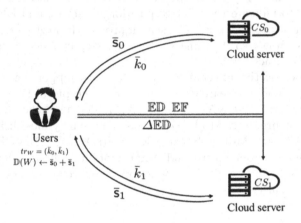

Fig. 1. The system model

Similar to existing works [5,7], we mainly focus on how users query the encrypted index using keywords in a privacy-preserving manner. The process of retrieving files $\{F_1, \ldots, F_n\}$ securely with identifiers is an independent work.

3.2 Security Model

The syntax of wildcard searchable symmetric encryption consists of one algorithm and two protocols:

- **Setup**$(\lambda, \mathbb{K}, \mathbb{D}) \rightarrow \mathbb{ED}$: This algorithm is invoked by users to encrypt the index $\mathbb{D}$. Users input a security parameter λ, the inverted index $\mathbb{D}$ and a key set $\mathbb{K}$, and output an encrypted index $\mathbb{ED}$. Finally, $\mathbb{ED}$ is outsourced to CSs.

- **Search**$(W, \mathbb{K}; \mathbb{ED}) \rightarrow \mathbb{D}(W)$: This protocol is executed between users and CSs. Users input the key set $\mathbb{K}$ and the queried wildcard keyword W to compute a trapdoor tr_W. CSs input the encrypted index $\mathbb{ED}$, and output the search result $\mathbb{D}(W)$ using the trapdoor.
- **Update**$((w, id, op), \mathbb{K}; \mathbb{ED}) \rightarrow \mathbb{ED}'$: This protocol is run between users and CSs. Users input the update information (w, id, op) and the key set $\mathbb{K}$ to protect data privacy. CSs input the encrypted index $\mathbb{ED}$ and update the $\mathbb{D}(w)$ in $\mathbb{D}$. That is, users send the encrypted update information $\Delta\mathbb{ED}$ to CSs. Finally CSs output the latest version of the encrypted index $\mathbb{ED}'$.

Security Model: The forward and backward security of scheme Π is captured by an ideal/real world simulation paradigm. The information leaked to CSs is defined as a leakage function $\mathcal{L} = (\mathcal{L}^{Setup}, \mathcal{L}^{Search}, \mathcal{L}^{Update})$. Then, a simulator $\mathcal{S}$ who has the leakage function is interactive with an adversary $\mathcal{A}$ in the ideal world $\mathsf{Ideal}_{\mathcal{A},\mathcal{S}}^{\Pi}(\lambda)$. If the adversary $\mathcal{A}$ cannot distinguish $\mathsf{Ideal}_{\mathcal{A},\mathcal{S}}^{\Pi}(\lambda)$ from the real world $\mathsf{Real}_{\mathcal{A}}^{\Pi}(\lambda)$ where a challenger $\mathcal{C}$ responds to $\mathcal{A}$, we say the scheme Π is $\mathcal{L}$-adaptive-secure. Additionally, if the leakage function $\mathcal{L}^{Search}$ only contains the wildcard keyword query operation s, the $\mathcal{L}^{Update}$ only shows the update operation u, we say the scheme Π further achieves forward and backward security. Formally,

Definition 1. ($\mathcal{L}$-adaptive-secure). *For each PPT adversary $\mathcal{A}$, if there exists a simulator $\mathcal{S}$ who invokes the leakage function $\mathcal{L}$ to simulate the ideal world $\mathsf{Ideal}_{\mathcal{A},\mathcal{S}}^{\Pi}(\lambda)$ such that*

$$|Pr[\mathsf{Real}_{\mathcal{A}}^{\Pi}(\lambda) = 1] - Pr[\mathsf{Ideal}_{\mathcal{A},\mathcal{S},\mathcal{L}}^{\Pi}(\lambda) = 1]| \leq \epsilon,$$

where λ is the security parameter and ϵ is a negligible function of λ. We say the scheme Π is $\mathcal{L}$-adaptive-secure.

Definition 2 (Forward and Backward Security). *If the leakage function of the $\mathcal{L}$-adaptive-secure scheme Σ is*

$$\mathcal{L}^{Search}(w) = \mathcal{L}'(\mathsf{OP}),$$

$$\mathcal{L}^{Update}((w, id, op)) = \mathcal{L}''(\mathsf{OP}),$$

where $\mathcal{L}'$ and $\mathcal{L}''$ are stateless functions, and $\mathsf{OP} \in \{\mathsf{s}, \mathsf{u}\}$, we say the $\mathcal{L}$-adaptive-secure scheme Σ is forward and backward secure.

3.3 Design Goals

Based on the system and security models, the following design goals should be achieved.

- Efficient wildcard searchable symmetric encryption. The proposed scheme can achieve wildcard keyword search efficiently without limitations on wildcard types. The user's communication or computation costs are independent of the size of the character sets or search results.

- Keyword and file update. Users can add(delete) keywords or files without running the setup algorithm again. Specifically, they achieve this goal by changing the $\mathbb{D}(w)$ in index $\mathbb{D}$ so that the encrypted index $\mathbb{ED}$ can be updated by cloud servers. The search result is based on the latest version of $\mathbb{ED}$.
- Forward and backward security. The goal is to ensure the data privacy of the proposed scheme in the search and update protocol.

4   Preliminaries

4.1   Character Set and Presentation

Based on existing works, wildcard keywords can be described by a character set according to some extraction rules. Here, the rule of character extraction is to extract the keyword itself, normal order, reverse order, 2-gram, 3-gram, AB$\|$1, and AB-character [23]. Take the keyword "security" as an example, its character set $\mathbb{CS}$ is "$security; s \parallel 1, e \parallel 2, c \parallel 3, \ldots, y \parallel 8; y \parallel -1, t \parallel -2, \ldots, s \parallel -8; se \parallel 0, ec \parallel 0, \ldots, ty \parallel 0; sec \parallel 0, ecu \parallel 0, \ldots, ity \parallel 0; sc \parallel 1, eu \parallel 1, cr \parallel 1, ui \parallel 1, rt \parallel 1, iy \parallel 1; se, sc, su, \ldots, sy, ec, eu, \ldots, ey, cu, cr, \ldots, cy, \ldots, ty$".

To represent the character set $\mathbb{CS}$ for searching wildcard keywords, we use the bloom filter, a popular space-saving tool, to achieve an approximate set membership test. Generally, it consists of three polynomial-time algorithms.

- BF.Gen$(\mathsf{p}, \mathsf{b}_1) \rightarrow (\{h_i\}_{i=1}^{\mathsf{b}_2}, B = 0_{\mathsf{b}_3})$: Users input the false positive rate p and the number of characters b_1 in $\mathbb{CS}$, output a hash functions set $\{h_1, \ldots, h_{\mathsf{b}_2}\}$, and the initialized bloom filter $B = 0_{\mathsf{b}_3}$, a b_3 bits zero string, where the number of hash functions $\mathsf{b}_2 = -\log_2^{\mathsf{p}}$ and the size of bloom filter $\mathsf{b}_3 = \frac{\mathsf{b}_1 \times \mathsf{b}_2}{\ln 2}$.
- BF.Upd$(\mathbb{CS}, B, T) \rightarrow B'$: For each character $e \in \mathbb{CS}$, users compute $\{h_i(e)\}_{i=1}^{\mathsf{b}_2}$, obtaining a position set $T = \{i_1, \ldots, i_t\}$. Then users set $1 := B[i_t]$ to obtain the updated bloom filter B'.
- BF.Chk$(e, B', T') \rightarrow b$: CSs input a position set T' evaluated from $\mathbb{CS}'$ and the updated bloom filter B', then output $b = \bigwedge B'[i_j]_{j=1}^{t}$, where $b = 0$ means the character set $\mathbb{CS}'$ is not a subset of $\mathbb{CS}$, otherwise, $\mathbb{CS}' \subseteq \mathbb{CS}$.

4.2   Distributed Multi-point Function

Distributed Multi-point Function (DMPF) is a secure tool to achieve efficient PIR [2,4]. A multi-point function $\bar{f}$ outputs $\bar{y}_i = \bar{f}(\bar{x}_i)$ at some particular points $\bar{x}_i$ and 0 else.

A two-party distributed multi-point function $\bar{f}$ computes the sum of $\Sigma_i \bar{y}_i$. It divides $\Sigma_i \bar{y}_i$ into two shares where each part can be evaluated by keys $\bar{k}_0$ and $\bar{k}_1$ respectively. Therefore, each party evaluates its share using its key $\bar{k}_j$, then users can recover $\Sigma_i \bar{y}_i$. Formally,

- DMPF.Gen$(\lambda, \bar{f}) \rightarrow (\bar{k}_0, \bar{k}_1)$: Users input a security parameter λ and a multi-point function $\bar{f}$ with particular points $\{(\bar{x}_i, \bar{y}_i)\}$, then output a pair of key $(\bar{k}_0, \bar{k}_1)$.

– DMPF.Eval($\bar{k}_j, x$) → $\bar{y}_j$: Each party P_j, $j \in \{0,1\}$ inputs the key $\bar{k}_j$ and x, then outputs $\bar{y}_j$ where if $x = \bar{x}_i$, $\bar{y}_i = \bar{f}(\bar{x}_i) = \bar{y}_0 + \bar{y}_1$; otherwise, $\bar{y}_i = \bar{y}_0 + \bar{y}_1 = 0$.

4.3 Additively Homomorphic Symmetric Encryption Scheme

The Additively Homomorphic Symmetric Encryption(AHSE) has three algorithms [3].

– AHSE.Gen(λ, k, f, l) → sk_l: Users input four parameters to initialize the additively homomorphic symmetric encryption scheme, that is, the security parameter λ, the symmetric key k, the pseudorandom function f and a random number l. At last, users output a symmetric key $sk_l = f_k(l)$.
– AHSE.Enc(sk_l, m, h, r) → Enc(m): Users input the key sk_l, a length-matching hash function $h : \{0,1\}^\lambda \rightarrow \{0,1\}^{\lambda_1}$, and a nonce r to encrypt a plaintext m. The output of this algorithm is $(l, \mathsf{Enc(m)})$ where $\mathsf{Enc(m)} = \mathsf{m} + h(f_{sk_l}(r))$ mod M and $|M| = \lambda_1$.
– AHSE.Dec($sk_l, r, \mathsf{Enc(m)}$) → m: Users input the key sk_l and a ciphertext Enc(m), they can compute the corresponding plaintext m = Enc(m) − $h(f_{sk_l}(r))$ mod M.

Homomorphic Addition: For two ciphertexts $(l_1, \mathsf{Enc(m_1)})$ and $(l_2, \mathsf{Enc(m_2)})$, the sum of two ciphertexts is $((l_1, l_2), \mathsf{Enc(m_1)} + \mathsf{Enc(m_2)})$, and its plaintext is $\mathsf{m_1} + \mathsf{m_2} = \mathsf{Enc(m_1)} + \mathsf{Enc(m_2)} - h(f_{sk_{l_1}}(r)) - h(f_{sk_{l_2}}(r))$ mod M.

4.4 Super-Increasing Sequence

If there is $\alpha_i > \mathsf{t} \cdot \Sigma_{j<i}\alpha_j, i, j \in [1, n]$ in a sequence $\mathbb{S} = \{\alpha_1, \ldots \alpha_n\}$, we say it is a super-increasing sequence (SIS). One can encode a set $S = \{(a_i, b_i)\}$ into a number s. That is, $a_i \in [1, \mathsf{t}], b_i \in [1, n]$ and each (a_i, b_i) means $a_i \cdot \alpha_{b_i}$. SIS has two algorithms: Encode and Decode.

– SIS.Ecd($\mathbb{S}, S$) → s: Users input a SIS $\mathbb{S}$ and a set S, then output an encoded number $\mathsf{s} = \Sigma_{(a_i, b_i) \in S}(a_i \cdot \alpha_{b_i})$.
– SIS.Dcd($\mathbb{S}, \mathsf{s}$) → S: Users input the sequence $\mathbb{S}$ and the encoded number s and then output the set S using modular operation.

5 Detailed Construction

In this section, we present the details of the proposed wildcard searchable symmetric encryption scheme. To query a wildcard keyword, users first interact with cloud servers to determine the keyword set W and its $\mathbb{D}(W)$. Then, the encrypted files in the search result are obtained with an additional round of interaction. Generally, we consider the first round merely, and its communication complexity is linear with $|W|$, even with $\mathbb{D}(W)$. The proposed scheme focuses on reducing the

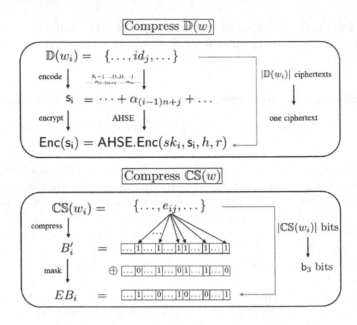

Fig. 2. Procedures to compress $\mathbb{CS}$ and $\mathbb{D}(w)$

communication complexity and then improving the security in update protocol, i.e., forward and backward security.

On the whole, we design a double-compressed index so that users can search wildcard keywords W in an oblivious way. The index $\mathbb{D}$ consists of the character set $\mathbb{CS}$ and $\mathbb{D}(w)$. The $\mathbb{CS}$ is compressed by bloom filters, whereas the $\mathbb{D}(w)$ is encoded using SIS.Ecd. When users query the W, the keywords are filtered by the bloom filters, then the search result $\mathbb{D}(w)$ is recovered from SIS.Dcd. To achieve low communication cost, each file in $\mathbb{F}$ is mapped into an element of $\mathbb{S}$ so that the search result $\mathbb{D}(W)$ can be compressed. By using the above index, the search and update protocols can be achieved obliviously by using DMPF and AHSE. Additionally, it needs only one round of interaction. The algorithms and protocols of our construction are detailed below.

In **Setup** algorithm, users choose the security parameter λ and define the key set $\mathbb{K} = \{\mathsf{p}, k, \mathsf{b}_1, \mathbb{S}, \mathbb{S}_1, \hat{k}, \tilde{k}, r'\}$ where $\mathbb{S} = \{\alpha_1, \ldots \alpha_{mn}\}$, $\mathbb{S}_1 = \{\beta_1, \ldots \beta_t\}$, $\hat{k}, \tilde{k}$ are keys of hash functions $\hat{h}, \tilde{h}$. Then, users input λ and b_1 to initialize the bloom filter and the hash function set $\{h_i\}_{i=1}^{\mathsf{b}_2}$ by running BF.Gen. Then for each keyword $w_i \in \mathbb{W}$, users extract the character set $\mathbb{CS}$, and generate its bloom filter B' by invoking BF.Upd, and then the bloom filter B'_i is encrypted into EB_i by masking it with the fragment of $\tilde{h}(\tilde{k}, r')$ from $[(i-1)\mathsf{b}_3 + 1]$-th bit to $(i\mathsf{b}_3)$-th bit. Additionally, $\mathbb{D}(w_i) \subseteq \{\alpha_{(i-1)n}, \ldots, \alpha_{in}\}$ is encoded to s_i by SIS.Ecd$(\mathbb{S}, S)$ where $S = \{\ldots, (1, \alpha_j), \ldots\}$ for each $\alpha_j \in \mathbb{D}(w_i)$. Users also invoke AHSE.Gen to obtain the symmetric key $sk_i = f_k(\hat{h}(\hat{k}, i))$ so that the encoded number s_i can be encrypted as Enc(s_i) homomorphically. The compression process is shown in

Fig. 2. Finally, the encrypted index $\mathbb{ED} = \{\ldots, (p(i), \mathsf{Enc}(\mathsf{s_i}), EB_i), \ldots\}$ is sent to cloud servers where p is a pseudorandom permutation.

When users want to query a wildcard keyword W, they first extract the character set $\mathbb{CS}$. Secondly, they compute the position set $T = \{i_1, i_2, \ldots, i_t\}$ from $\{h_i(e)\}_{i=1}^{b_2}$ where $e \in \mathbb{CS}$. Then, users can get a multi-point function $\bar{f}$ with particular points $\{(i_l, \beta_l)\}$ where $l \in [1, t]$. Thirdly, users invoke the DMPF.Gen$(\lambda, \bar{f})$ to compute the trapdoor tr_W: the key pair $(\bar{k}_0, \bar{k}_1)$. After receiving $\bar{k}_0$ and $\bar{k}_1$ from users, CS_j runs the $\bar{y}_j = \mathsf{DMPF.Eval}(\bar{k}_j, x)$ to obtain $\bar{s}_{j,p(i)} = (\Sigma_{l=1}^{t}(\bar{y}_j \cdot EB_i'[i_l]) + 1) \cdot \mathsf{Enc}(\mathsf{s_i})$ for each item i in $\mathbb{ED}$. Finally, CS_j sends $\bar{s}_j = \Sigma_{p(i)}\bar{s}_{j,p(i)}$. According to the property of DMPF, there is $\mathsf{Enc}(\bar{s}) = \bar{s}_0 + \bar{s}_1 = \Sigma_{p(i)}(\bar{s}_{0,p(i)} + \bar{s}_{1,p(i)}) = \Sigma_{p(i)}((\Sigma_{l=1}^{t}(\beta_l \cdot EB_i'[i_l]) + 2) \cdot \mathsf{Enc}(\mathsf{s_i}))$. Thus, users can decrypt $\bar{s} = \mathsf{AHSE.Dec}(\Sigma sk_i, r, \mathsf{Enc}(\bar{s}))$ invoke the decode algorithm $\mathsf{SIS.Dcd}(\mathbb{S}, \bar{s})$ to obtain the search result.

Specifically, there are three situations. Take $t = 4$ as an example, (1) If the string of $B_i'[i_1 : i_4]$ and $\tilde{h}(\tilde{k}, r')[(i-1) \cdot b_3 + i_1 : (i-1) \cdot b_3 + i_4]$ are 1111 and 1100, EB_i equals to 0011. Thus, users can learn that $\beta_3 + \beta_4 + 2$ is the coefficient of $\mathsf{s_i}$ from the $\bar{s}$. Then, users could recover that $B_i'[i_1 : i_4] = 1100 \oplus 0011 = 1111$, meaning that the keyword $w_i \in W$, and that $\mathbb{D}(w_i)$ evaluated from $\mathsf{s_i}$ belongs to the search result. (2) If the string of $B_i'[i_1 : i_4]$ and $\tilde{h}(\tilde{k}, r')[(i-1) \cdot b_3 + i_1 : (i-1) \cdot b_3 + i_4]$ are 1010 and 1100, EB_i equals to 0110. Thus, users can obtain that $\beta_2 + \beta_3 + 2$ is the coefficient of $\mathsf{s_i}$ from the $\bar{s}$. Then, users could recover that $B_i'[i_1 : i_4] = 1100 \oplus 0110 = 1010$, meaning that the keyword $w_i \notin W$. (3) If the string of $B_i'[i_1 : i_4]$ and $\tilde{h}(\tilde{k}, r')[(i-1) \cdot b_3 + i_1 : (i-1) \cdot b_3 + i_4]$ are 1010 and 1010, EB_i equals to 0000. Thus, users can learn that 2 is the coefficient of $\mathsf{s_i}$ from the $\bar{s}$. Then, users could recover that $B_i'[i_1 : i_4] = 1010 \oplus 0000 = 1010$, meaning that the keyword $w_i \notin W$.

In the **Update** protocol, the updates $\{(w_i, id_j, op)\}$ is achieved by adding or deleting elements $\alpha_{(i-1)n+j}$ from $\mathbb{S}$. More specifically, each operation (w_i, id_j, op) can be regarded as the update of $\mathbb{D}(w)$ so that users only renovate the $\mathsf{Enc}(\mathsf{s_i})$. The correctness is ensured by the homomorphic addition property of AHSE.

6 Security Analysis

Theorem 1. *The proposed scheme can achieve forward and backward security if AHSE is IND-CPA secure, DMPF is function private, and pseudorandom functions are secure.*

Proof. We employ a series of indistinguishability games to demonstrate that the proposed scheme achieves both forward and backward security. We start with the first game, denoted as G_0, which represents the real world. Subsequently, we introduce modifications to G_0 in order to create additional games. Ultimately, we construct a game representing the ideal world, which is simulated using the leakage function.

Game G_0: We define G_0 to be equivalent to $\mathsf{Real}_{\mathcal{A}}^{\Pi}(\lambda)$, so $Pr[\mathsf{Real}_{\mathcal{A}}^{\Pi}(\lambda) = 1] = Pr[G_0 = 1]$.

Game G_1: It is obtained by replacing all instances of EB_i in $(p(i), \mathsf{Enc}(\mathsf{s_i}), EB_i)$. Specifically, when users want to encrypt B_i' using $\tilde{h}$, they replace $\tilde{h}(\tilde{k}, r')$ with a long random bit string. If an adversary $\mathcal{B}_1$ can distinguish G_0 and G_1 with non-negligible probability, it means $\mathcal{B}_1$ can determine which string is the output of $\tilde{h}(\tilde{k}, r')$. This would imply a breach of pseudorandom security, contradicting its definition. Therefore, we have $|Pr[G_0 = 1] - Pr[G_1 = 1]| \leq \mathsf{Adv}_{\mathsf{PRF}, \mathcal{B}_1}(\lambda)$.

Game G_2: It differs from G_1 only in how $\mathsf{Enc}(\mathsf{s_i})$ is generated. In the real world and G_1, $\mathsf{s_i}$ is encrypted using AHSE.Enc. According to the IND-CPA security of AHSE, the simulator $\mathcal{S}$ chooses random numbers and encrypts them to replace all ciphertexts in $(p(i), \mathsf{Enc}(\mathsf{s_i}), EB_i)$ in G_2. This replacement maintains indistinguishability between G_1 and G_2 unless the probability of distinguishing them is negligible. In formal terms, $|Pr[G_1 = 1] - Pr[G_2 = 1]| \leq \mathsf{Adv}_{\mathsf{AHSE}, \mathcal{B}_2}(\lambda)$. Where $\mathcal{B}_2$ is an adversary capable of breaking the IND-CPA security of AHSE.

Game G_3: G_3 builds upon G_2 but differs in the **Search** protocol. There are two servers that do not communicate with each other. Consequently, in response to a query W, each server CS_j receives only a key $\bar{k}_j$ and its evaluation result, gaining no information about the search result. Furthermore, the key pair is changed by users through the selection of a new random number, ensuring that cloud servers learn nothing about the queried keywords. The simulator in G_3 can arbitrarily select a wildcard keyword and send the output key pair as the trapdoor. Leveraging the security of DMPF, an adversary $\mathcal{B}_3$ cannot ascertain the computed function, therefore, cannot distinguish between the two games. This leads to the inequality $|Pr[G_2 = 1] - Pr[G_3 = 1]| \leq \mathsf{Adv}_{\mathsf{DMPF}, \mathcal{B}_3}(\lambda)$.

Game G_4: It demonstrates how to simulate the **Update** protocol in G_3.

To update $\mathbb{ED}$, a $\varDelta\mathbb{ED}$ is generated using $\{(w_i, id_j, op)\}$ by encrypting elements in $\mathbb{S}$ or using zeros. Due to the IND-CPA security of AHSE, the cloud servers gain no knowledge about the updated keywords and files. In G_4, the simulator $\mathcal{S}$ has the flexibility to generate a random $\varDelta\mathbb{ED}$ to replace the one in G_3. These two games are indistinguishable, or else an adversary $\mathcal{B}_2$ would be capable of breaking the IND-CPA security. The relationship can be expressed as follows: $|Pr[G_3 = 1] - Pr[G_4 = 1]| \leq \mathsf{Adv}_{\mathsf{AHSE}, \mathcal{B}_2}(\lambda)$.

In the ideal world, where the simulator $\mathcal{S}$ communicates with the adversary $\mathcal{A}$ following the same rules as in G_4, we have $Pr[\mathsf{Ideal}_{\mathcal{A}, \mathcal{S}}^{\varPi}(\lambda) = 1] = Pr[G_4 = 1]$. This equality reflects the indistinguishability of the ideal world and the game G_4.

Finally, we can design a simulator $\mathcal{S}$ for each adversary $\mathcal{A}$, and $\mathcal{S}$ only uses the leakage function to ensure that the difference between the real world and ideal world satisfies the following inequality: $|Pr[\mathsf{Real}_{\mathcal{A}}^{\varPi}(\lambda) = 1] - Pr[\mathsf{Ideal}_{\mathcal{A}, \mathcal{S}}^{\varPi}(\lambda) = 1]| \leq \mathsf{Adv}_{\mathsf{PRF}, \mathcal{B}_1}(\lambda) + 2\mathsf{Adv}_{\mathsf{AHSE}, \mathcal{B}_2}(\lambda) + \mathsf{Adv}_{\mathsf{DMPF}, \mathcal{B}_3}(\lambda)$

Thus, the proposed scheme achieves both forward and backward security.

7 Comparison and Performance Evaluation

In this section, we compare the proposed scheme with existing works and then present the performance evaluation of the proposed scheme.

7.1   Theoretical Comparison

To present a comprehensive comparison, we compare the proposed scheme with existing works using multi-keyword search or character set extraction to achieve wildcard search. For the latter, there are some keywords not matched with the wildcard keyword, which is defined as the false positive. "Y/N" means that a scheme has/has not the false positive. Then, the number of cloud servers required in schemes is listed and CSs are honest-but-curious (semi-honest). In terms of functionalities, we evaluate whether a scheme can support data update, single-character wildcard search, and multiple-character wildcard search. Finally, the security of these schemes is shown. The notation $\checkmark$ represents that a scheme achieves the responding function or security, otherwise, using ⊚. The comparison result is shown in Table 1. Specifically, the authors of Libertas [18] proposed a backward construction from an encryption scheme with *which key concealing* and an SSE scheme supporting add operations and wildcard queries. They use Zhao and Nishide's work [25] as the SSE scheme which can support update, single and multiple character search. Although both Libertas [18] and the proposed scheme can achieve the functionalities and security mentioned in Table 1, Libertas leaks more information than our scheme such as search pattern. And that information is protected by the oblivious query with the help of two servers.

Then, we further analyze the complexity and storage cost. Here, n_Δ is the number of update pairs (w, id); $n_{\Delta w}$ or $n_{\Delta F}$ is the number of update keywords or files in $\{(w, id, op)\}$; In existing works, the trapdoor is linear in the number of characters extracted from wildcard keywords W since each character is mapped into bloom filters in encrypted indexes. We consider the users' communication and computation complexity of the search and update protocol respectively. The comparison result is shown in Table 2, and the proposed scheme achieves efficient communication in wildcard search.

Table 1. The comparison of schemes

Scheme	False Positive	Server Required	Security Model	Functionalities			Security		
				Update	Single Character	Multiple Character	Adaptive Secure	Forward Security	Backward Security
BEIS-I [6]	N	1	Honest-but-curious	$\checkmark$	$\checkmark^1$	$\checkmark^1$	$\checkmark$	$\checkmark$	⊚
Hu et al. [8]	Y	1		-	⊚	$\checkmark$	⊚	⊚	⊚
TBIS [23]	Y	1		$\checkmark$	$\checkmark$	$\checkmark$	⊚	⊚	⊚
Libertas [18]	Y	1		$\checkmark$	$\checkmark$	$\checkmark$	$\checkmark$	$\checkmark$	$\checkmark$
Our scheme	Y	2		$\checkmark$	$\checkmark$	$\checkmark$	$\checkmark$	$\checkmark$	$\checkmark$

$\checkmark^1$: it achieves wildcard search by listing and searching all possible keywords.

7.2   Performance Comparison

Experimental Setup. Our evaluations are written and tested in Java. They are implemented on a Windows 10 PC with an Intel Core i7-8550U CPU running at 2.00 GHz and 16 GB of memory. The instantiations of DMPF and AHSE are based on the schemes presented in Castro et al. [4] and Castelluccia et al. [3]

Table 2. The computation and communication complexity comparison

Scheme	Wildcard search		Update		Client Storage	Index Storage
	Communication	Computation	Communication	Computation		
BEIS-I [6]	$O(\|\mathbb{W}\|)$	$O(\|\mathbb{W}\| + \|\mathbb{F}\|)$	$O(n_{\Delta F})$	$O(n_{\Delta F})$	$O(1)$	$O(\|\mathbb{W}\|)$
Hu et al. [8]	$O(\|\mathbb{CS}\| + \|W\|)$	$O(\|\mathbb{CS}\| + \|W\|)$	-	-	$O(1)$	$O(\|\mathbb{W}\|)$
TBIS [23]	$O(\|\mathbb{CS}\| + \|W\|)$	$O(\|\mathbb{CS}\| + \|W\|)$	$O(n_{\Delta w})$	$O(n_{\Delta w})$	$O(1)$	$O(\|\mathbb{W}\| + S(T))$
Libertas [18]	$O(\|\mathbb{CS}\| + \Sigma_{w \in W}\|\mathbb{D}(w)\|)$	$O(\|\mathbb{CS}\| + \Sigma_{w \in W}\|\mathbb{D}(w)\|)$	$O(n_{\Delta})$	$O(n_{\Delta})$	$O(1)$	$O(\Sigma_{w \in W}\|\mathbb{D}(w)\|)$
Our scheme	$O(1)$	$O(\|\mathbb{CS}\| + \|\mathbb{W}\|)$	$O(\|\mathbb{W}\|)$	$O(\|\mathbb{W}\|)$	$O(1)$	$O(\|\mathbb{W}\|)$

respectively. In our experiments, the false positive rate of the bloom filters p is set to 0.01 and the length of keyword is set to 5 or 10 respectively. Thus, b_1 equals to 31 or 91. According to the BF.Gen, there are $b_2 = 7$, $b_3 = 314$ when $|w| = 5$, and $b_3 = 919$ when $|w| = 10$.

Evaluation. The experiments are designed to measure the effects of changes in index size, wildcard queries, and updates under varying keyword lengths.

Table 3. The time cost of the setup algorithm

the number of keyword length/files	1000	5000	10000
$|w| = 5$(e.g.:index)	440.74622 ms	6240.05638 ms	42604.45749 ms
$|w| = 10$(e.g.:university)	468.109867 ms	6682.04146 ms	42998.27587 ms

Table 4. The impact on t by varying wildcard keywords

wildcard keywords	abc%d	%ab	%abc	%abc%	a_bcd	a_b_c	ab%d
$|\mathbb{CS}|$	18	4	9	6	23	11	10
t^*	126	28	63	42	161	77	70

*: We use the maximum of t evaluated by $b_2 \cdot |\mathbb{CS}|$ where $b_2 = 7$.

Varying Index Size. The number of keywords and $|\mathbb{D}(w)|$ is set at 100 by default. To change the index size, we set the number of files as 1000, 5000, and 10000, respectively. As shown in Table 3, the time cost of the setup algorithm grows rapidly with the number of files but changes very little with the keyword length. This reflects that generating EB_i has little impact on the setup time with changes in the keyword length. The increase in time cost with varying index size may be caused by the rapidly increasing value in SIS.

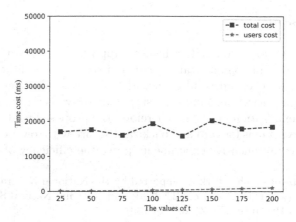

Fig. 3. The time cost of the search protocol

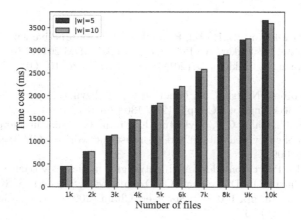

Fig. 4. The time cost of the update protocol

Varying Wildcard Query. In this experiment, we set the keyword length to be 5, and the number of files to be 1000. To demonstrate the time cost of the search protocol under different wildcard keywords, we first observe that the number of positions t in T is the key factor influenced by $|\mathbb{CS}|$, as shown in Table 4. Then, we vary t from 25 to 200 to simulate different wildcards. The time cost of the search algorithm is depicted in Fig. 3. To showcase the efficiency of the proposed scheme, we also provide the time cost from the users' perspective, illustrating its efficiency for users. *Varying updates.* To evaluate the time cost of the update protocol, we vary the keyword length and the number of files. As shown in Fig. 4, the time cost of the update protocol grows steadily with the number of files but changes very little with the keyword length. Considering that users compute a $\mathsf{Enc}(\alpha_{(i-1)n+j})$ for each (w_i, f_j, op), it is reasonable to obtain this result. The growth of time cost with varying the number of files is caused by the increasing value in SIS.

8   Conclusion

In this paper, we focus on SSE schemes supporting wildcard search. Faced with the challenges in communication cost and security, we designed a double-compressed index to encode and reduce the size of search results. It further achieves oblivious search and update using additively secret sharing in DMPF and additively encryption in AHSE. Therefore, the proposed wildcard searchable symmetric encryption is forward and backward secure with lower communication overhead. The comprehensive experiments prove the efficiency of our scheme.

Acknowledgements. This work is supported by the National Natural Science Foundation of China (No. 62072276, No. 62302280) and National Natural Science Foundation of Shandong Province (No. ZR2023QF133).

References

1. Bösch, C., Brinkman, R., Hartel, P., Jonker, W.: Conjunctive wildcard search over encrypted data. In: Jonker, W., Petković, M. (eds.) SDM 2011. LNCS, vol. 6933, pp. 114–127. Springer, Heidelberg (2011). https://doi.org/10.1007/978-3-642-23556-6_8
2. Boyle, E., Gilboa, N., Ishai, Y.: Function secret sharing: improvements and extensions. In: Proceedings of CCS, pp. 1292–1303 (2016)
3. Castelluccia, C., Chan, C.F., Mykletun, E., Sudik, G.: Efficient and provably secure aggregation of encrypted data in wireless sensor networks. ACM Trans. Sens. Netw. **5**(3), 1–36 (2009)
4. de Castro, L., Polychroniadou, A.: Lightweight, maliciously secure verifiable function secret sharing. In: Dunkelman, O., Dziembowski, S. (eds.) EUROCRYPT 2022. LNCS, vol. 13275, pp. 150–179. Springer, Cham (2022). https://doi.org/10.1007/978-3-031-06944-4_6
5. Dauterman, E., Feng, E., Luo, E., Popa, R.A., Stoica, I.: {DORY}: an encrypted search system with distributed trust. In: Proceedings of NSDI, pp. 1101–1119 (2020)
6. Du, M., Wang, Q., He, M., Weng, J.: Privacy-preserving indexing and query processing for secure dynamic cloud storage. IEEE Trans. Inf. Forensics Secur. **13**(9), 2320–2332 (2018)
7. Gui, Z., Paterson, K.G., Patranabis, S.: Rethinking searchable symmetric encryption. In: 44th IEEE Symposium on Security and Privacy (SP 2023) (2023)
8. Hu, C., Han, L.: Efficient wildcard search over encrypted data. Int. J. Inf. Secur. **15**, 539–547 (2016)
9. Hu, C., Han, L., Yiu, S.: Efficient and secure multi-functional searchable symmetric encryption schemes. Secur. Commun. Netw. **9**(1), 34–42 (2016)
10. Hua, J., Liu, Y., Chen, H., Tian, X., Jin, C.: An enhanced wildcard-based fuzzy searching scheme in encrypted databases. World Wide Web **23**(3), 2185–2214 (2020)
11. Kamara, S., Moataz, T.: Boolean searchable symmetric encryption with worst-case sub-linear complexity. In: Coron, J.-S., Nielsen, J.B. (eds.) EUROCRYPT 2017, Part III. LNCS, vol. 10212, pp. 94–124. Springer, Cham (2017). https://doi.org/10.1007/978-3-319-56617-7_4

12. Li, J., et al.: Searchable symmetric encryption with forward search privacy. IEEE Trans. Depend. Secure Comput. **18**(1), 460–474 (2019)
13. Li, M., Jia, C., Du, R., Shao, W.: Forward and backward secure searchable encryption scheme supporting conjunctive queries over bipartite graphs. IEEE Trans. Cloud Comput. **11**(1), 1091–1102 (2023)
14. Li, Y., Ning, J., Chen, J.: Secure and practical wildcard searchable encryption system based on inner product. IEEE Trans. Serv. Comput. **16**(3), 2178–2190 (2023)
15. Sedghi, S., van Liesdonk, P., Nikova, S., Hartel, P., Jonker, W.: Searching keywords with wildcards on encrypted data. In: Garay, J.A., De Prisco, R. (eds.) SCN 2010. LNCS, vol. 6280, pp. 138–153. Springer, Heidelberg (2010). https://doi.org/10.1007/978-3-642-15317-4_10
16. Song, D.X., Wagner, D., Perrig, A.: Practical techniques for searches on encrypted data. In: Proceedings of IEEE S&P, pp. 44–55 (2000)
17. Suga, T., Nishide, T., Sakurai, K.: Secure keyword search using bloom filter with specified character positions. In: Takagi, T., Wang, G., Qin, Z., Jiang, S., Yu, Y. (eds.) ProvSec 2012. LNCS, vol. 7496, pp. 235–252. Springer, Heidelberg (2012). https://doi.org/10.1007/978-3-642-33272-2_15
18. Weener, J., Hahn, F., Peter, A.: Libertas: backward private dynamic searchable symmetric encryption supporting wildcards. In: Sural, S., Lu, H. (eds.) DBSec 2022. LNCS, vol. 13383, pp. 215–235. Springer, Cham (2022). https://doi.org/10.1007/978-3-031-10684-2_13
19. Wu, Z., Li, R.: OBI: a multi-path oblivious RAM for forward-and-backward-secure searchable encryption. In: 30th Annual Network and Distributed System Security Symposium, NDSS 2023, San Diego, California, USA, 27 February–3 March 2023. The Internet Society (2023)
20. Xu, W., Zhang, J., Yuan, Y.: Enabling dynamic multi-client and Boolean query in searchable symmetric encryption scheme for cloud storage system. KSII Trans. Internet Inf. Syst. **16**(4), 1286–1306 (2022)
21. Yang, Y., Liu, X., Deng, R.H., Weng, J.: Flexible wildcard searchable encryption system. IEEE Trans. Serv. Comput. **13**(3), 464–477 (2020)
22. Yuan, D., Zuo, C., Cui, S., Russello, G.: Result-pattern-hiding conjunctive searchable symmetric encryption with forward and backward privacy. Proc. Priv. Enhanc. Technol. **2023**(2), 40–58 (2023)
23. Zhang, X., Zhao, B., Qin, J., Hou, W., Su, Y., Yang, H.: Practical wildcard searchable encryption with tree-based index. Int. J. Intell. Syst. **36**(12), 7475–7499 (2021)
24. Zhang, Y., Katz, J., Papamanthou, C.: All your queries are belong to us: the power of {File-Injection} attacks on searchable encryption. In: 25th USENIX Security Symposium (USENIX Security 16), pp. 707–720 (2016)
25. Zhao, F., Nishide, T.: Searchable symmetric encryption supporting queries with multiple-character wildcards. In: Chen, J., Piuri, V., Su, C., Yung, M. (eds.) NSS 2016. LNCS, vol. 9955, pp. 266–282. Springer, Cham (2016). https://doi.org/10.1007/978-3-319-46298-1_18

Adversarial Attacks Against Object Detection in Remote Sensing Images

Rong Huang[1], Li Chen[2], Jun Zheng[1], Quanxin Zhang[2], and Xiao Yu[3(✉)]

[1] School of Cyberspace Science and Technology, Beijing Institute of Technology,
Beijing 100081, China
[2] School of Computer Science, Beijing Institute of Technology, Beijing 100081, China
[3] Department of Computer Science and Technology,
Shandong University of Technology, Zibo 255022, China
yuxiao8907118@163.com

Abstract. With the continuous development of artificial intelligence technology and the increasing richness of remote sensing data, deep convolutional neural networks(DNNs) have been widely used in the field of remote sensing images. Object detection in remote sensing images has achieved considerable progress due to DNNs. However, DNNs have shown their vulnerability to adversarial attacks. The object detection models in remote sensing images also have this vulnerability. The complexity of remote sensing object detection models makes it difficult to implement adversarial attacks. In this work, we propose an adversarial attack method against the remote sensing object detection model based on the L_∞ norm, which can make the detector blind–that is, the detector misses a large number of objects in the image. Because some remote sensing images are too large, we also designed a pre-processing method to segment and pre-process the huge images, which is combined with the attack method. Our proposed attack method can effectively perform adversarial attacks on remote sensing object detection models.

Keywords: Adversarial Attack · Remote Sensing Images · Object Detection models

1 Introduction

Remote sensing images have always been important data sources for natural resource surveys, military monitoring, and urban planning management [1]. With the rapid development of deep learning in the field of computer vision, especially the extensive application of DNNs in image recognition and object detection tasks [2], the analysis and processing of remote sensing images have also ushered in new opportunities and challenges.

In recent years, deep learning has made remarkable progress in the field of object detection [3–5]. Many deep learning object detection models have been proposed and applied to remote sensing image object detection [6–9], such as

J. Vaidya et al. (Eds.): AIS&P 2023, LNCS 14509, pp. 358–367, 2024.
https://doi.org/10.1007/978-981-99-9785-5_25

ROI-Transformer [7], R3Det [8], and Gliding Vertex [9]. The excellent performance of these deep learning-based object detectors further promotes the application of remote sensing technology in real life, such as natural resource monitoring, urban planning, traffic control, etc. These application fields often rely on the highly reliable prediction results of the model to make decisions. Wrong prediction results may cause serious consequences. Therefore, the safety issue of the model cannot be ignored.

However, the fragility of DNNs has been exposed in recent years [10]. The models are vulnerable to adversarial attacks. Adversarial attack refers to the small perturbation of the input samples to make the model produce wrong results when performing tasks such as classification or detection. These perturbations are invisible to the naked eye but can significantly change the model's predictions. In the field of remote sensing image tasks, the remote sensing classification model has been proven to be insecure [11], and the remote sensing object detection model also has such fragility theoretically. However, due to the fact that the adversarial attack on the target detection model is more difficult than the adversarial attack on the image classification model, and the size of the remote sensing image is huge and there are many small objects in any direction distributed in the picture, which to a certain extent caused the attack difficulty. Therefore, there is still a lack of systematic research on adversarial attacks on target detection in remote sensing images.

We propose an effective adversarial attack algorithm for object detection models in remote sensing images. Remote sensing images are mainly divided into remote sensing optical images and remote sensing Synthetic Aperture Radar(SAR) images. Optical images are images obtained by optical photography systems using photosensitive film as the medium. They are large in size and contain grayscale information in multiple bands [7]. SAR sensors only record echo information in one band, which is a big difference from optical images [12]. Therefore, in this study, we designed different algorithms for two different image data.

Our main contributions are as follows:

For remote sensing SAR images, we designed an adaptation of the adversarial attack algorithm to the existing remote sensing object detection model. This attack can significantly reduce the accuracy of the target detection model and cause the detector to miss a large number of targets in the image. The adversarial attack is successfully completed and the vulnerability of the remote sensing target detection model is confirmed (Fig. 1).

For remote sensing optical images, due to the large size of the image, we designed a specific cutting pre-processing algorithm–Split, and combined with the confrontation algorithm MIM, we also successfully completed the adversarial attack, and the object detection model could not detect the target.

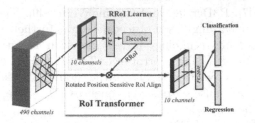

Fig. 1. RoI Transformer model network structure

2   Related Work

2.1   Remote Sensing Image Dataset

Remote sensing image dataset is one of the important resources in the field of remote sensing technology and computer vision. It contains images of the earth's surface obtained from spacecraft, drones or other remote sensing platforms. The objects in these images are small and dense, and the background is complex. These data sets It has a wide range of applications in various application fields, such as environmental monitoring, urban planning and national defense. Remote sensing image datasets can be divided into remote sensing optical images and remote sensing SAR images.

Remote sensing optical images are usually high-altitude and long-distance acquisition images, which are shot by artificial satellites, space shuttles, etc., equipped with space cameras. The working band of the imaging sensor is the visible light band from ultraviolet to infrared, which is the most commonly used working band in traditional aerial photographic reconnaissance and aerial photographic mapping. Well-known remote sensing optical target detection datasets include DOTA, HRSC2016, etc. Among them, the original size range of the image in the DOTA dataset is about 800 * 800–4000 * 4000, far exceeding the 1000*1000 size of MSCOCO in the common dataset.

Synthetic Aperture Radar is an active Earth observation system that can be installed in spacecraft such as aircraft, satellites, etc. SAR sensors basically belong to the microwave frequency band, and the wavelength is usually at the centimeter level. Since the imaging geometry of SAR is of the slant range projection type, while optical images are of the center projection type, SAR images are quite different from optical images in terms of imaging mechanism, geometric characteristics, radiation characteristics, etc.

2.2   Remote Sensing Image Object Detection

Because of the special nature of remote sensing images, object detection models for remote sensing images naturally have some different characteristics compared with general object detection models. Due to the large size of the remote sensing image for target detection, most GPUs cannot support directly sending the entire high-resolution image directly into the calculation during the training or prediction process. Therefore, in order to improve the detection accuracy when processing large-scale images, the current remote sensing image object detection model cuts a single large image into multiple small images that can be processed through certain rules. Another property of object detection models in remote sensing images is that they are more suitable for rotating object detectors. Due to the directionality and dense distribution of targets in the image, the commonly used horizontal target detection in the past may cause multiple targets to be squeezed together by a horizontal detection frame, resulting in misalignment between the bounding box and the object, and it is difficult to train the detector to extract target features and ensures precise localization of the identified target. Rotating target detection is more suitable for remote sensing image target detection tasks, and can draw more accurate frames.

2.3 Adversarial Attack

Adversarial attacks refer to adding small perturbations directly to the pixels of remote sensing images through optimization algorithms to maximize the loss function of the target detection model, thereby causing the modified sample results to deviate from the true values to achieve the effect of the attack. Good fellow et al. first proposed the Fast Gradient Symbol Method (FGSM) in 2016 to create adversarial samples [13]. A multi-step optimization algorithm was subsequently proposed. Madry et al. proposed PGD [14] in 2018. This multi-step optimization algorithm can better find the global optimal point compared with the single-step algorithm. Dong et al. proposed the momentum iterative fast gradient symbolic method MI-FGSM [15], which achieves higher transferability to black-box models by incorporating momentum into the attack process.

3 Methodology

For remote sensing SAR images, only one band of information is included, and the image is black and white, so we use common adversarial attack algorithms to attack the model. Let x be an clean input image and D be a detector. Send the picture x directly to the detector D, obtain a series of target detection classification outputs $Y = \{Y_1, Y_2, \ldots, Y_n\}$. Our attack goal is to make the detector D miss or fail to detect as many target objects in the picture as possible, Therefore, we set the attack target class $y' = 0$. Donate $Y' = \{y_1', y_2', \ldots, y_n'\}$ represents the set of adversarial attack target categories for the entire image. The adversarial method is based on MI-FGSM, which is an iterative fast gradient sign method based on momentum. The initialization is a small random noise, and x^* is updated by the following gradient backpropagation:

$$x_{t+1}^* = x_t^* - \alpha sign(m_{t+1}) \tag{1}$$

where α is the step size of each update, We use m_t to represent the gradient information of the every iteration. Updated m_t via the formula:

$$m_{t+1} = \mu \cdot m_t + \frac{\nabla_x J\left(Y_t^*, Y_t^{'}\right)}{||\nabla_x J\left(Y_t^*, Y_t^{'}\right)||_\infty}, \tag{2}$$

The μ represents the attenuation factor. J represents the cross entropy loss function:

$$J(p_n, Y_n^{'}) = -\log_p(Y_n^{'}|p_n) \tag{3}$$

In each iteration, we update x_{t+1}^* by $\nabla_x J\left(Y_t^*, Y_t^{'}\right)$.

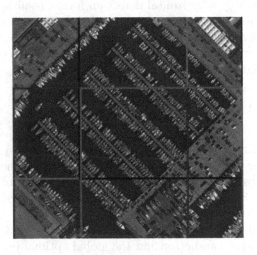

Fig. 2. Split demos

For remote sensing optical images, due to the special nature of the image, large size and various types of objects, we cut the images, then split them and send them to the model to obtain the prediction results. The cutting process is shown in Fig. 2. We use $Split(x)$ to describe the preprocessing operation of cropping. After crop we can get a patches set $P = \{p_1, p_2, \ldots, p_n\}$ on x. If we send all p in the P to D, we will get the set of all proposal regions on P marked as $B = \{B_1, B_2, \ldots, B_n\}$ and the detected label set $Y = \{Y_1, Y_2, \ldots, Y_n\}$. Notice that B_n is also a small set, $B_n = \{b_1, b_2, \ldots, b_m\}$, b_m stands for the mth proposal region on p_n. Same like B_n, $Y_n = \{y_1, y_2, \ldots, y_m\}$ corresponds to the class of b_m and $y_m \in \{0, 1, 2, \ldots, C\}$, where C is the number of classes, Our goal is making the D ignores the objects in the x as much as possible, so we set up an adversarial target class $y_m^{'} = 0$ for each proposal region on p_n, Donate $Y_n^{'} = \{y_1^{'}, y_2^{'}, \ldots, y_m^{'}\}$. And the adversarial target label is $Y' = \{Y_1^{'}, Y_2^{'}, \ldots, Y_n^{'}\}$, Then we update x^* by:

$$x_{t+1}^* = Clip_{(x,\epsilon)}\{x_t^* - \alpha sign\left(m_{t+1}\right)\}. \tag{4}$$

We use m_t to represent the gathered gradients and t is iterations. We update m_t by:

$$m_{t+1} = \mu \cdot m_t + \frac{\nabla_x J\left(p_{t\_n}^*, Y_{t\_n}^{'}\right)}{||\nabla_x J\left(p_{t\_n}^*, Y_{t\_n}^{'}\right)||_\infty}, \tag{5}$$

In each iteration we update the adversarial example x_{t+1}^* by backpropagating the gradient $\nabla_x J\left(p_{t\_n}^*, Y_{t\_n}'\right)$

$$J(p_\text{n}, Y_\text{n}') = -\log_\text{p}(Y_\text{n}'|p_\text{n}) \tag{6}$$

4 Experiments

Datasets. We used the HRSID and SSDD from the remote sensing SAR image dataset, and the HRSC and DOTA from the remote sensing optics dataset. Among them, HRSID, SSDD and HRSC are conventional remote sensing images, including ships, DOTA is special. It is a large-scale remote sensing image datasets. It selects 15 categories for annotations of target objects contained in the image, including airplanes, ships, piggy banks, baseball infields, tennis courts, basketball courts, track and field fields, seaports, Bridges, large vehicles, small vehicles, helicopters, soccer pitches, circular courses, swimming pools.

Metrics. In object detection tasks, mean Average Precision (mAP) is often used to evaluate model accuracy. AP can be regarded as the area under the precision-recall curve divided by the number of categories, that is, the average precision value of recall in the range of (0, 1). mAP is the main evaluation index of the target detection algorithm. The higher the mAP value, the better the detection effect of the target detection model on a given dataset. In our experiment, in order to evaluate the effect of the adversarial attack, we will use mAP drop, which is the difference between the mAP of the target detection model on the original dataset and the mAP on the adversarial sample data set to measure the attack success rate.

Threat Models. In this experiment, our main attack object is RoI Transformer, a mature remote sensing image object detector. It is a three-stage target detection model proposed by Ding et al. of Wuhan University. In order to solve the mismatch between the horizontal detection frame in conventional target detection and the rotating target in remote sensing images, and the high computational complexity and matching efficiency faced by rotating target detection low question. Before the model is attacked, the detection performance in the DOTA verification set, HRSC test set, HRSIC test set, and SSDD test set is good. For the DOTA dataset, the mAP is 75.03% mAP, and for the HRSC dataset, it can reach 98.91%. For HRSIC, 84.92% can be achieved, and for SSDD is 84.31%.

Attack Parameters. For the experiment of remote sensing SAR images, we set the number of iterations t to 50, and the modified maximum pixel value α is 0.2, L_∞ Norm max perturbation ε is 1, For the experiment of remote sensing optics images, we set the number of iterations t to 30, and the modified maximum pixel value α is 0.1, L_∞ Norm max perturbation ε is 1. This is because the pixel values of remote sensing SAR images vary greatly, so differentiated processing is required.

5 Results

In experiments on remote sensing SAR images, we achieved significant results. As shown in Fig. 3, in both datasets, the object detection model was unable to detect the target in the image and we successfully implemented the hidden attack. In the HRSID dataset, the mAP value dropped from 84.92% to 6.32%. Most of the ships in the post attack images could not be successfully identified by the detector. In the SSDD dataset, the mAP value dropped from 84.31% to 28% (Table 1).

Table 1. mAP drop for different datasets

	DOTA	HRSC	HRSID	SSDD
mapDrop	4.99%	94.95%	78.6%	56.31%

Our effect is equally pronounced in attacks against remotely sensed optical images. For the 15 categories of images in the DOTA dataset, the AP value of each category has dropped significantly, and the mAP value after the attack has dropped below 5%, successfully hiding the target (Fig. 4).

In the HRSC dataset, the AP value is reduced from 98.91% to 3.96% when attacking images of the category of ships, achieving excellent attack performance (Fig. 5).

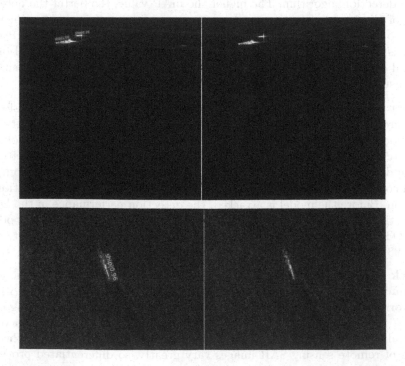

Fig. 3. Visualization results after detection of adversarial examples attacking SAR images (upper left: original HRSID, upper right: adversarial examples of HRSID, lower left: original image of SSDD, lower right: adversarial examples of SSDD)

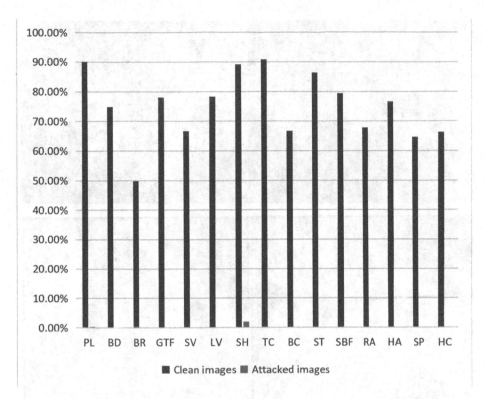

Fig. 4. Various AP detection results on DOTA confrontation samples

Fig. 5. Visualized results after detection of adversarial samples attacking remote sensing optical images (upper left: original image of HRSC, upper right: adversarial sample of HRSC, lower left: original image of DOTA dataset, lower right: adversarial sample of DOTA dataset)

6 Conclusions

In this work, We proposed corresponding adversarial attack methods for different scenarios of remote sensing images. For remote sensing SAR images, we migrated the MIM method to the classic target detection model and obtained good attack results; for remote sensing optical images, according to the image Due to the huge size, we designed a cutting preprocessing algorithm corresponding to it. Combined with the classic attack algorithm MIM, it also achieved a significant attack effect on remote sensing optical images, and the image changes cannot be seen by the naked eye. In future research, we will consider developing more stealthy and effective attack algorithms for remote sensing images. In addition, we also plan to study the adversarial defense strategy of the object detection model to improve the robustness of the model.

References

1. Cheng, G., Yang, C., Yao, X., Guo, L., Han, J.: When deep learning meets metric learning: remote sensing image scene classification via learning discriminative CNNs. IEEE Trans. Geosci. Remote Sens. **56**(5), 2811–2821 (2018)
2. Duan, K., et al.: Keypoint triplets for object detection. In: Proceedings of the IEEE International Conference on Computer Vision, Seoul, Republic of Korea, pp. 27–32 (2019)
3. Jiao, L., et al.: A survey of deep learning-based object detection. IEEE Access **7**, 128837–128868 (2019)
4. Redmon, J., Farhadi, A.: Yolo9000: better, faster, stronger. In: Proceedings of the IEEE Conference on Computer Vision and Pattern Recognition, pp. 7263–7271 (2017)
5. Liu, L., Zixuan, X., He, D., Yang, D., Guo, H.: Local pixel attack based on sensitive pixel location for remote sensing images. Electronics **12**(9), 1987 (2023)
6. Yi, J., Wu, P., Liu, B., Huang, Q., Qu, H., Metaxas, D.: Oriented object detection in aerial images with box boundary-aware vectors. In: Proceedings of the IEEE/CVF Winter Conference on Applications of Computer Vision, pp. 2150–2159 (2021)
7. Ding, J., Xue, N., Long, Y., Xia, G.-S., Lu, Q.: Learning ROI transformer for oriented object detection in aerial images. In: Proceedings of the IEEE/CVF Conference on Computer Vision and Pattern Recognition, pp. 2849–2858 (2019)
8. Yang, X., Yan, J., Feng, Z., He, T.: R3det: refined single-stage detector with feature refinement for rotating object. In: Proceedings of the AAAI Conference on Artificial Intelligence, vol. 35, pp. 3163–3171 (2021)
9. Yongchao, X., et al.: Gliding vertex on the horizontal bounding box for multi-oriented object detection. IEEE Trans. Pattern Anal. Mach. Intell. **43**(4), 1452–1459 (2020)
10. Wang, Y., et al.: Towards a physical-world adversarial patch for blinding object detection models. Inf. Sci. **556**, 459–471 (2021)
11. Czaja, W., Fendley, N., Pekala, M., Ratto, C., Wang, I.-J.: Adversarial examples in remote sensing. In: Proceedings of the 26th ACM SIGSPATIAL International Conference on Advances in Geographic Information Systems, pp. 408–411 (2018)
12. Wei, S., Zeng, X., Qizhe, Q., Wang, M., Hao, S., Shi, J.: HRSID: a high-resolution SAR images dataset for ship detection and instance segmentation. IEEE Access **8**, 120234–120254 (2020)
13. Kurakin, A., Goodfellow, I., Bengio, S.: Adversarial machine learning at scale. arXiv preprint arXiv:1611.01236 (2016)
14. Makdry, A., Makelov, A., Schmidt, L., Tsipras, D., Vladu, A.: Towards deep learning models resistant to adversarial attacks. Stat **1050**, 9 (2017)
15. Dong, Y., et al.: Boosting adversarial attacks with momentum. In: Proceedings of the IEEE Conference on Computer Vision and Pattern Recognition, pp. 9185–9193 (2018)

Hardware Implementation and Optimization of Critical Modules of SM9 Digital Signature Algorithm

Yujie Shao, Tian Chen, Ke Li, and Lu Liu[✉]

Beijing Institute of Technology, Beijing 100081, China
liulu@bit.edu.cn

Abstract. SM9 is an identity-based cryptographic algorithm based on elliptic curves, which has high security and low management costs. However, its computational complexity restricts its development and application. This paper implements and optimizes the critical modules of SM9 digital signature algorithm based on FPGA. We simplify modular addition and subtraction, avoiding the use of large number comparators and saving approximately 50% of LUTs compared to traditional methods. The modular multiplication adopts the Montgomery modular multiplication algorithm, which only takes 0.24 μ s to realize modular multiplication operation on F_p. For complex modules, this paper analyzes the dependency relationship between calculations and parallelizes irrelevant operations to improve the parallelism within and between modules at different levels, greatly reducing the number of computation cycles required. In addition, this paper utilizes multiplexers to achieve resource reuse while ensuring computational performance. This research is not only of great significance for the high-performance implementation of SM9, but also has reference value for the implementation of other cryptographic algorithms based on elliptic curves.

Keywords: SM9 · FPGA · Montgomery modular multiplication · Miller loop · R-ate bilinear pairing

1 Introduction

SM9 algorithm is an identification cryptographic algorithm based on elliptic curves. The algorithm standard was released in 2016 [1], which includes digital signature algorithm, key exchange protocol, key encapsulation mechanism and public key encryption algorithm. The implementation of traditional public key cryptosystems mainly relies on Public Key Infrastructure (PKI), where the Certificate Authority (CA) ensures the legitimacy of the user identity and public key. The identity-based cryptographic algorithm eliminates the process of issuing digital certificates by CA, and the public key is calculated by the user's unique identity [2]. This greatly reduces the cost of operation and maintenance,

J. Vaidya et al. (Eds.): AIS&P 2023, LNCS 14509, pp. 368–381, 2024.
https://doi.org/10.1007/978-981-99-9785-5_26

and is receiving increasing attention from scholars [3]. However, the high computational complexity of identity cipher algorithms limits their development and
application.

The underlying operations of SM9 are modular operations, and the upper
operations include point addition, point doubling, point multiplication, bilinear
pairing, etc. SM9 cryptographic algorithm is a typical computationally intensive
task that consumes a large amount of CPU and memory resources, and the powerful logical processing power of the CPU cannot be fully utilized. Compared to
CPU, FPGA supports hardware programming without the need for instruction
decoding and data strobes. It can be optimized at the hardware level for specific
tasks to achieve low latency and high controllability. The parallelism of FPGA
is also much higher than that of CPU. With sufficient resources, functional units
can be replicated in large numbers to improve computational efficiency. Therefore, using FPGA to implement some or all modules of SM9 to improve its computational efficiency has become the choice of many researchers. [4] designed an
ultra-high radix interleaved modular multiplication algorithm based on Virtex-7
FPGA, which can complete 256-bit modular multiplication operation in $0.56\,\mu s$.
[5] proved that the computational efficiency of Miller loop under the Jacobian
coordinate system is 5% higher than that under the projection coordinate system. [6] improved the parallelism at different levels of the pairing algorithm,
which can complete pairing computation within 3.4ms.

This paper implements the SM9 digital signature algorithm based on FPGA,
and optimizes the performance and resources of critical parts, mainly including
the following aspects:

1. We optimize modular addition, modular subtraction, modular multiplication,
   and modular inversion on F_p to simplify logic and reduce resources.
2. We analyze the correlation of different steps within the module, parallelize
   unrelated calculations, improve the degree of parallelism at different levels,
   and use multiplexers to achieve resource reuse while ensuring computing performance.
3. We re-plan Miller loop to hide modular multiplication calculations, and output partial results of point doubling early to initiate point addition operation
   in advance to reduce the number of cycles.

2   SM9 Digital Signature Algorithm

2.1   Data Representation

The SM9 algorithm involves the prime field F_p and extension fields $F_{p^2}, F_{p^4}, F_{p^{12}}$.
For extension fields, this paper adopts the following tower extension scheme:

$$F_{p^2}[u] = F_p[u]/(u^2 - \alpha), \alpha = -2 \tag{1}$$

$$F_{p^4}[v] = F_{p^2}[v]/(v^2 - \xi), \xi = u \tag{2}$$

$$F_{p^{12}}[w] = F_{p^4}[w]/(w^3 - v), v^2 = \xi \tag{3}$$

where $u^2 = \alpha, v^2 = u, w^3 = v$.

During storage and operation, the element on F_p is a 256-bit number with a range of $[0, p - 1]$. The element on F_{p^2} is represented by two elements on F_p, as shown in Eq. (4). The element on Γ_{p^4} is represented by four elements on F_p, as shown in Eq. (5). The element on $F_{p^{12}}$ is represented by twelve elements on F_p, as shown in Eq. (6).

$$(a_1, a_0) = a_1 u + a_0 \tag{4}$$

$$(a_3, a_2, a_1, a_0) = (a_3 u + a_2)v + (a_1 u + a_0) \tag{5}$$

$$(a_{11}, a_{10}, a_9, a_8, a_7, a_6, a_5, a_4, a_3, a_2, a_1, a_0) = [(a_{11}u + a_{10})v + (a_9 u + a_8)]w^2$$
$$+ [(a_7 u + a_6)v + (a_5 u + a_4)]w + [(a_3 u + a_2)v + (a_1 u + a_0)] \tag{6}$$

where $a_i \in F_p$.

2.2   Algorithm Flow

The SM9 digital signature algorithm is shown in Algorithm 1.

Algorithm 1: SM9 digital signature algorithm

Input: system parameters, message M, signature private key ds_A
Output: (h, s)
1: $g = e(P_1, P_{pub-s})$
2: generate random number r
3: $w = g^r$
4: $h = H_2(M\|w, N)$
5: $l = (r - h) \bmod N$, if $(l == 0)$ then *goto* 2
6: $S = [l]ds_A$
7: return (h, s)

P_1 and N are system parameters. P_1 is the generator of the additive cyclic group G_1, in the form of (x_{P_1}, y_{P_1}), where $x_{P_1}, y_{P_1} \in F_p$. P_{pub-s} is the signature master public key, in the form of $(x_{P_{pub-s}}, y_{P_{pub-s}})$, where $x_{P_{pub-s}}, y_{P_{pub-s}} \in F_{p^2}$.

3   Implementation

Basic operations on F_p include modular addition, modular subtraction, modular multiplication and modular inversion, and the operands are 256-bit numbers. When performing modular addition and modular subtraction operations, considering the characteristics of unsigned addition and subtraction, we make full use of carry information to simplify the operation logic and reduce resources. The modular multiplication operation adopts an efficient calculation method — the Montgomery modular multiplication algorithm [7], which effectively avoids the division operation. When performing modular inversion operations, use displacement instead of division to improve efficiency.

3.1 F_p Units

Modular Addition on F_p. The modular addition operation is shown in Algorithm 2.

Algorithm 2: Modular addition on F_p

Input: x, y, p
Output: $z = add(x, y) = x + y \bmod p$
1: $\{c_0, z_0\} = x + y$
2: $\{c_1, z_1\} = z_0 + (-p)$
3: if $(c_0 == 0$ and $c_1 == 0)$
4:   return z_0
5: else
6:   return z_1

The modular addition module uses two 256-bit adders with carry. One adder calculates $x + y$ and outputs 256-bit z_0 and 1-bit c_0. If c_0 equals 1, which means $x + y$ is greater than p, then $z = \{c_0, z_0\} - p = \{1, z_0\} - p = z_0 - p$. If c_0 equals 0, then

$$z = \begin{cases} z_0 - p, & \& z_0 \geq p \\ z_0, & \& z_0 < p \end{cases} \tag{7}$$

We use c_1 to determine the size relationship between z_0 and p to avoid using a comparator. The efficiency of an adder is generally higher than that of a subtracter, so subtracting p is changed to adding $-p$, and $-p$ can be calculated in advance without any delay. **Modulo Subtraction on F_p** The modular subtraction operation is shown in Algorithm 3.

Algorithm 3: Modular subtraction on F_p

Input: x, y, p
Output: $z = sub(x, y) = x - y \bmod p$
1: $\{c_0, z_0\} = x - y$
2: $z_1 = z_0 + p$
3: if $(c_0 == 1)$
4:   return z_0
5: else
6:   return z_1

Step 1 uses a subtractor with carry. When x is not less than y, c_0 equals 1, otherwise, c_0 equals 0. Therefore, if c_0 equals 1, output z_0 directly, otherwise, output z_1. The calculation of z_1 is show in Eq. (8).

$$\begin{cases} z_0 = x + 2^k - y \\ z_1 = x + p - y = x + 2^k - y + p = z_0 + p \end{cases} \tag{8}$$

Modular Multiplication on F_p. The modular multiplication on F_p is the most core operation unit in SM9 algorithm and the fundamental component of complex operations. The efficiency of modular multiplication determines that of the SM9 digital signature algorithm. This paper adopts the Montgomery modular multiplication algorithm, which is an efficient algorithm for implementing modular multiplication of large integers. The idea is to convert the modular operation on large prime numbers p into the modular operation on R (2^k) to avoid complex division operations. Wang [6] parallelized the high-radix Montgomery algorithm [8], as shown in Algorithm 4.

Algorithm 4: Parallel high-radix Montgomery multiplication on F_p [6]

Input: x, y, p, w, where $w = -p^{-1} \bmod r$, $r = 2^n$
Output: $z = mul(x, y) = xyR^{-1} \bmod p$, where $R = r^m$
1: $z = 0, v = 0$
2: for i=0 to m-1 do
  2.1: $\{c_a, z_0\} = z_0 + x_i y_0$
  2.2: $t_i = z_0 w \bmod r$
  2.3: $\{c_b, z_0\} = z_0 + t_i p_0$
  2.4: $\{c_a, z_1\} = z_1 + x_i y_1 + c_a$
  2.5: for $j = 1$ to $m - 2$ do
    2.5.1: $\{c_b, z_{j-1}\} = z_j + t_i p_j + c_b$
    2.5.2: $\{c_a, z_{j+1}\} = z_{j+1} + x_i y_{j+1} + c_a$
  2.6: $\{c_b, z_{m-2}\} = z_{m-1} + t_i p_{m-1} + c_b$
  2.7: $\{v, z_{m-1}\} = c_a + c_b + v$
3: if $z \geq p$ then $z = z - p$
4: return z

Steps 2.1, 2.2, 2.3, 2.4, 2.5.1, 2.5.2, 2.6, and 2.7 in Algorithm 4 can be summarized in the form of $(A + B) + (C * D)$ and packaged as an *addmul* module. The hardware architecture is shown in Fig. 1. Steps 2.3 and 2.4, 2.5.1 and 2.5.2 have no data dependencies and can be calculated in parallel, so instantiate two *addmul* modules and use them cyclically. Use Algorithm 3 to improve step 3, then the final output result is determined by carry. The implementation of modular multiplication is shown in Algorithm 5.

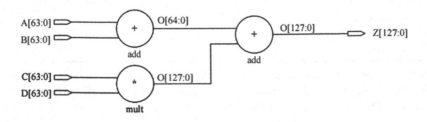

Fig. 1. *addmul* module.

Algorithm 5: Modified parallel high-radix Montgomery multiplication

---

Input: x, y, p, w, where $w = -p^{-1} \ mod \ r$, $r = 2^{64}$
Output: $z = mul(x, y) = xyR^{-1} \ mod \ p$, where $R = 2^{256}$
1: $z = 0$, $v = 0$, $i = 0$, $a = addmul_0(0, 0, x_0, y_0)$, goto 3
2: $v = b_1$, $z_3 = b_0$, $a = addmul_0(z_0, 0, x_i, y_0)$
3: $z_0 = a_0$, $b = addmul_1(0, 0, a_0, w)$
4: $t_i = b_0$, $a = addmul_0(a_0, 0, b_0, p_0)$, $b = addmul_1(z_1, a_1, x_i, y_1)$
5: $a = addmul_0(b_0, a_1, t_i, p_1)$, $b = addmul_1(z_2, b_1, x_i, y_2)$
6: $z_0 = a_0$, $z_2 = b_0$, $a = addmul_0(b_0, a_1, t_i, p_2)$, $b = addmul_1(z_3, b_1, x_i, y_3)$
7: $z_3 = b_0$, $z_1 = a_0$, $a = addmul_0(b_0, a_1, t_i, p_3)$
8: $z_2 = a_0$, $b = addmul_1(b_1, a_1, v, 1)$, if $(i < 3)$ then begin $i = i + 1$, goto 2 end
9: $\{c, z\} = \{b_0, z_2, z_1, z_0\} + (-p)$
10:if $(c == 0$ and $b[64] == 0)$ then $z = \{b_0, z_2, z_1, z_0\}$
11:return z

The result of $mul(x, y)$ is $xyR^{-1} \ mod \ p$. It is necessary to perform $mul(z, R^2)$ calculation to obtain $xy \ mod \ p$. In this paper, the operands are converted to the Montgomery field first. Then, all operations are performed on the Montgomery field. Finally, the result is transferred from the Montgomery field to F_p, as shown in Algorithm 6.

Algorithm 6: Field conversion

Input: x, y, p, where $x, y \in F_p$
Output: $z = xy \ mod \ p$
1: $x_m = mul(x, R^2)$, $y_m = mul(y, R^2)$
2: $z_m = mul(x_m, y_m)$
3: $z = mul(z_m, 1)$
4: return z

Modular Inversion on F_p. Modular inversion is implemented by the extended Euclidean algorithm, specifically using addition, subtraction and displacement operations, as shown in Algorithm 7.

Algorithm 7: Modular inversion on F_p

Input: a, p where $a \in F_p$
Output: $z = inv(a) = a^{-1} \ mod \ p$
1: $u = a$, $v = p$, $m = 1$, $n = 0$
2: if $(u \neq 1$ and $v \neq 1)$ repeat
   2.1: if $(u[0] == 0)$ repeat $u = u \gg 1, m = (m[0] == 0) \ ? \ m \gg 1 : (m + p) \gg 1$
   2.2: if $(v[0] == 0)$ repeat $v = v \gg 1, n = (n[0] == 0) \ ? \ n \gg 1 : (n + p) \gg 1$
   2.3: if $(u \geq v)$ then $u = u - v, m = m - n$
   2.4: else $v = v - u, n = n - m$
3: if $(u == 1)$ return m
4: else return n

Steps 2.1 and 2.2 are data independent, so they are executed in parallel. Step 2.3 is merged into step 2.1, and step 2.4 is merged into step 2.2 to significantly reduce the number of cycles required. In addition, the modular inversion operations of SM9 are mostly connected to multiplication, in the form of ba^{-1}. Initializing m with b can convert $inv(a)$ into $inv\_mul(a, b)$.

3.2 F_{p^n} Units

Unlike above work that focuses on algorithm selection and optimization, the operations on the extension field are a combination of lower level computing units. Operations on F_{p^2}, F_{p^4}, $F_{p^{12}}$ are respectively composed of operations on F_p, F_{p^2}, F_{p^4}. So, the focus of optimization work is on scheduling. We analyze whether there is a dependency relationship between each step, execute irrelevant operations in parallel, set an appropriate number of basic units to reuse, and achieve a balance between resources and efficiency.

Modular Addition and Subtraction on F_{p^n}. The modular addition and subtraction operations on F_{p^n} are to perform modular addition and subtraction on the corresponding elements on F_p. For $A = a_1 u + a_0$, $B = b_1 u + b_0$, where $A, B \in F_{p^2}$, $a_i, b_i \in F_p$:

$$A \pm B = (a_1 u + a_0) \pm (b_1 u + b_0) = (a_1 \pm b_1)u + (a_0 \pm b_0) \tag{9}$$

Modular Multiplication on F_{p^n}. The principle of modular multiplication on F_{p^n} is similar to polynomial multiplication.

For modular multiplication on F_{p^2}, where $A, B \in F_{p^2}$, $a_i, b_i \in F_p$:

$$\begin{aligned} A \cdot B &= (a_1 u + a_0)(b_1 u + b_0) = a_1 b_1 u^2 + (a_1 b_0 + a_0 b_1)u + a_0 b_0 \\ &= (a_1 b_0 + a_0 b_1)u + (-2 a_1 b_1 + a_0 b_0) \end{aligned} \tag{10}$$

when instantiating four modular multiplication modules on F_p, the number of cycles required for calculation is minimal.

For modular multiplication on F_{p^4}, where $A, B \in F_{p^4}$, $A_i, B_i \in F_{p^2}$, $a_i, b_i \in F_p$:

$$\begin{cases} A = A_1 v + A_0 = (a_3 u + a_2)v + (a_1 u + a_0) \\ B = B_1 v + B_0 = (b_3 u + b_2)v + (b_1 u + b_0) \\ A \cdot B = (A_1 v + A_0)(B_1 v + B_0) = A_1 B_1 v^2 + (A_1 B_0 + A_0 B_1)v + A_0 B_0 \\ \qquad = (A_1 B_0 + A_0 B_1)v + (A_1 B_1 u + A_0 B_0) \end{cases} \tag{11}$$

Similar to Eq. (10), there are four modular multiplication operations on F_{p^2} in Eq. (11). In this paper, the method of Karatsuba [9] is used to reduce the number of multiplications, as shown in Eq. (12):

$$\begin{aligned} A \cdot B &= (A_1 B_0 + A_0 B_1)v + (A_1 B_1 u + A_0 B_0) \\ &= [(A_1 + A_0)(B_1 + B_0) - (A_1 B_1 + A_0 B_0)]v + (A_1 B_1 u + A_0 B_0) \end{aligned} \tag{12}$$

For Eq. (12), instantiating three modular multiplication modules on F_{p^2} can achieve the highest efficiency. Comparing Eq. (11) with Eq. (12), the number of modular multiplications in Eq. (11) is greater than that in Eq. (12), but the implementation logic of Eq. (11) is simpler. When the saved resources cannot offset the negative effects of increasing circuit complexity and decreasing clock frequency, optimization fails [10]. This is why resource optimization is not performed on Eq. (10). The final implementation of modular multiplication on F_{p^4} is shown in Algorithm 8.

Algorithm 8: Modular multiplication on F_{p^4}

Input: $A = (a_3, a_2, a_1, a_0)$, $B = (b_3, b_2, b_1, b_0)$, p, where $A, B \in F_{p^4}$
Output: $Z = mul\_4(A, B)$
1: $a_{02} = add_0(a_0, a_2)$, $a_{13} = add_1(a_1, a_3)$
    $(x_1, x_0) = mul\_2_0((a_1, a_0), (b_1, b_0))$, $(t_1, t_0) = mul\_2_1((a_3, a_2), (b_3, b_2))$
2: $b_{02} = add_0(b_0, b_2)$, $b_{13} = add_1(b_1, b_3)$
3: $(y_1, y_0) = mul\_2_2((a_{13}, a_{02}), (b_{13}, b_{02}))$
4: $t = add_0(t_1, t_1)$
5: $z_0 = sub_0(p, t)$
6: $(y_1, y_0) = (sub_0(y_1, x_1), sub_1(y_0, x_0))$
7: $(y_1, y_0) = (sub_0(y_1, t_1), sub_1(y_0, t_0))$, $(x_1, x_0) = (add_0(x_1, t_0), add_1(x_0, z_0))$
8: return $((y_1, y_0), (x_1, x_0))$

In step 1, the time consumption of modular multiplication operation is much greater than that of modular addition and subtraction operation. So, steps 2 and 3 do not need to wait for the completion of modular multiplication operation in step 1.

The implementation and optimization for modular multiplication on $F_{p^{12}}$ is similar to that on F_{p^4}.

Modular Inversion on F_{p^n}. The implementation idea of modular inversion is similar to the modular multiplication on F_{p^n}. However, due to the limited use of the modular inversion operation, the implementation of the modular inversion adopts a resource-saving architecture without excessive parallelization.

3.3   R-Ate Bilinear Pairing

Both in terms of resources and time, bilinear pairing is the core of the SM9 digital signature algorithm, greatly affecting its performance. This paper adopts R-ate bilinear pairing and optimizes it, including reconstructing Miller loop and simplifying Frobenius mapping operation. Reconstructing Miller loop can hide modular multiplication operation and reduce the number of cycles by starting point doubling early. Simplifying the implementation of Frobenius mapping is beneficial for the calculation of π_p, π_{p^2} and fixed exponential power. The implementation of R-ate bilinear pairing is shown in Algorithm 9.

Algorithm 9: R-ate bilinear pairing

Input: $P \in E(F_p)[r]$, $Q \in E'(F_{p^2})[r]$, $a = 6t + 2$
Output: $y = e(P, Q)$
1: $a = \sum_{i=0}^{L-1} a_i 2^i$, $a_{L-1} = 1$
2: $T = Q$, $f = 1$
3: for $i = L - 2$ to 0 do
  3.1: $f = f^2$, $l_{TT} = l_{T,T}(P)$, $T = [2]T$
  3.2: $f = f \cdot l_{TT}$
  3.3: if $(a_i == 1)$ then
    3.3.1: $l_{TQ} = l_{T,Q}(P)$, $T = T + Q$
    3.3.2: $f = f \cdot l_{TQ}$
4: $Q_1 = \pi_p(Q)$, $Q_2 = \pi_{p^2}(Q)$
5: $f = f \cdot l_{T,Q_1}(P)$, $T = T + Q_1$
6: $f = f \cdot l_{T,-Q_2}(P)$, $T = T - Q_2$
7: $f = f^{(q^{12}-1)/r}$
8: return f

Miller Loop. Miller loop (step 3 in Algorithm 9) involves modular multiplication on $F_{p^{12}}$, line operation, and point operation. To avoid duplicate calculations [6,11], $l_{TT} = l_{T,T}(P)$ and $T = [2]T$ are realized in *ltt_double* module as a whole, and $l_{TQ} = l_{T,Q}(P)$ and $T = T + Q$ are implemented as a whole in *ltq_add* module.

Analyzing step 3 in Algorithm 9, when a_i is equal to 1, steps 3.2 and 3.3.1 have no dependency on each other. And step 3.3.2 is independent of the point and line operations in step 3.1 of the next cycle. When a_i is equal to 0, step 3.2 is independent of the point and line operations in step 3.1 of the next cycle. Therefore, we schedule it as shown in Fig. 2.

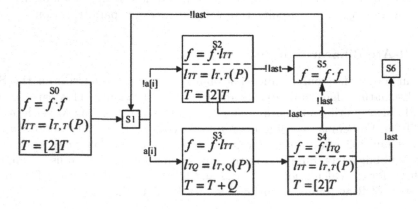

Fig. 2. Reconstructed Miller loop state machine.

S3 is a combination of steps 3.2 and 3.3.1. Since the number of cycles required to perform a calculation of *ltt_double* or *ltq_add* is much greater than that to perform a calculation of modular multiplication on $F_{p^{12}}$, S5 is performed after the modular multiplication in S2 or S4 so as not to occupy additional time cycles. Our scheduling scheme achieves the concealment of all modular multiplication operations on $F_{p^{12}}$ during the loop process, as shown in Algorithm 10.

Algorithm 10: Modified Miller loop

Input: $P \in E(F_p)[r]$, $Q \in E^{'}(F_{p^2})[r]$, $a = \sum_{i=0}^{L-1} a_i 2^i$, $a_{L-1} = 1$
Output: f, T
1: $i = L - 2$, $last = 0$
2: $f = f \cdot f$, $l_{TT} = l_{T,T}(P)$, $T = [2]T$
3: if $(i > 0)$ then $i = i - 1$ else $last = 1$
   if $(a_i == 1)$ then *goto* 5
4: $f = f \cdot l_{TT}$
   if $(last == 0)$ then begin $l_{TT} = l_{T,T}(P)$, $T = [2]T$, *goto* 7 end else *goto* 8
5: $f = f \cdot l_{TT}$, $l_{TQ} = l_{T,Q}(P)$, $T = T + Q$
6: $f = f \cdot l_{TQ}$, $l_{TT} = l_{T,T}(P)$, $T = [2]T$, if $(last == 1)$ then *goto* 8
7: $f = f \cdot f$, *goto* 3
8: return f, T

When performing point addition or point doubling operations, Jacobian coordinate system is used to avoid complex inversion operations [12]. Analyzing Algorithm 9, the point addition operation always follows the point doubling operation. And the point doubling operation will first calculate the Z coordinate of $[2]T$, while the point addition operation will first use the Z coordinate for calculation. Therefore, after *ltt_double* module calculates the Z coordinate, it outputs a valid signal for the Z coordinate, enabling *ltq_add* module to initiate calculations related to the Z coordinate in advance, thereby reducing the number of cycles.

Frobenius Mapping. In step 4 of Algorithm 9, π_p and π_{p^2} are Frobenius once and twice automorphism, which are point to point mappings defined as follows:

$$\begin{cases} \pi_p : E \to E, \ \pi_p(x,y) = (x^p, y^p) \\ \pi_{p^2} : E \to E, \ \pi_{p^2}(x,y) = (x^{p^2}, y^{p^2}) \end{cases} \tag{13}$$

The definition of Frobenius mapping $\varphi(x)$ is as follows:

$$\varphi(x) = x^p \tag{14}$$

and

$$\begin{cases} \varphi(x + y) = x^p + y^p = \varphi(x) + \varphi(y) \\ \varphi(x \cdot y) = x^p \cdot y^p = \varphi(x) \cdot \varphi(y) \end{cases} \tag{15}$$

According to the above definitions and properties, for $f \in F_{p^{12}}$:

$$\varphi(f) = f^p = (f_{11}uvw^2)^p + (f_{10}vw^2)^p + (f_9uw^2)^p + (f_8w^2)^p + (f_7uvw)^p \\ + (f_6vw)^p + (f_5uw)^p + (f_4w)^p + (f_3uv)^p + (f_2v)^p + (f_1u)^p + (f_0)^p \tag{16}$$

According to Fermat's theorem, when p is a prime number and integer f_i is less than p:

$$f_i^p = f_i \bmod p \tag{17}$$

The value of $(u^i v^j w^k)^p$ can be calculated in advance, Therefore, f^p can be obtained by multiplying f_i in each dimension by a fixed value.

In Algorithm 9, the calculation results of π_p and π_{p^2} need to be calculated together with the results of the Miller loop, and the number of cycles required by Miller loop is much larger than that of π_p and π_{p^2}. Therefore, using a resource-saving architecture to implement π_p and π_{p^2}, only instantiating one modular multiplication module on F_p can meet the needs. For the modular exponentiation operation in step 7 of Algorithm 9, the exponent is a polynomial related to p, and it is decomposed as follows [13]:

$$\frac{p^{12} - 1}{r} = (p^6 - 1)(p^2 + 1) = \frac{p^4 - p^2 + 1}{r} \tag{18}$$

We use f^{p^2} and f^{p^6} to simplify the calculation of modular exponentiation.

4   Result Analysis

We use Verilog to implement the SM9 digital signature algorithm, verify its correctness through simulation, synthesize and implement it based on the Xilinx Virtex UltraScale+ platform.

Modular addition, modular subtraction, modular multiplication and modular inversion are the most fundamental operational units in the SM9 digital signature algorithm, which determine the performance of the upper modules. We have optimized and implemented them.

Table 1 and Table 2 compare the improved implementation methods of modular addition and modular subtraction in this paper with the traditional implementation methods. Due to the avoidance of comparison between large numbers, approximately 50% of LUTs are saved compared to the original methods, and the maximum clock frequency is increased by 15.80% and 34.87% respectively.

Table 1. Comparison of 256-bit modular addition.

Plan	Frequency(MHz)	LUT
Improved	257.33	389
Original	222.22	820

Table 2. Comparison of 256-bit modular subtraction.

Plan	Frequency(MHz)	LUT
Improved	275.86	374
Original	204.54	719

Table 3 compares the different implementation methods of modular multiplication. The modular multiplication module in this paper has the lowest latency, but uses the most DSPs.

Table 4 shows the number of cycles used by the parallel and serial modular inversion modules to perform inversion operations on randomly selected numbers. Compared with the serial method, the parallel method reduces the number of clock cycles by approximately 27.55%. From the experimental results and theoretical analysis, it can be seen that the parallel modular inversion method significantly saves the number of cycles used while the resources are almost unchanged.

Table 3. Comparison of 256-bit modular multiplication.

Plan	Platform	Clock Cycles	Frequency(MHz)	Latency(μ s)	LUT	DSP
This paper	Virtex UltraScale+	29	121.57	0.24	1400	32
[14]	Virtex-7	112	225	0.50	1917	9
[4]	Virtex-7	151	268.1	0.56	9100	16
[15]	Virtex-7	193	207.1	0.93	1151	/

Table 4. Comparison of 256-bit modulo inversion.

Plan	Num0	Num1	Num2	Num3
Parallel	404	398	383	378
Serial	541	546	542	528

Table 5. The number of basic modules and cycles used in complex modules.

Field	Module	mul	add	sub	inv	Clock Cycles
F_p	point doubling	3	2	1	/	99
	point addition	3	/	2	/	120
	point multiplication	7	2	3	2	41,406*
	$\pi_p + \pi_{p^2}$	1	/	/	/	249
F_{p^2}	modular multiplication	4	1	1	/	33
	modular inversion	2	/	/	2	468*
	point doubling + line	20	11	9	/	109
	point addition + line	16	4	8	/	108
F_{p^4}	modular multiplication	12	5	5	/	39
	modular inversion	10	2	2	2	543*
$F_{p^{12}}$	modular multiplication	108	59	45	/	44
	modular inversion	82	36	44	2	619*
	modular exponentiation	216	118	90	/	2,836*
	Miller loop	144	74	62	/	8,830

*: The number of clock cycles is not fixed and depends on the input.

We start from the prime field and realize calculations on extension fields in sequence. Operation modules on F_p are the basis of other modules, and the upper complex modules are composed of lower modules combined with logic circuits. This paper adopts a performance-based design idea, and sets the minimum number of instances on the basis of ensuring performance to achieve resource saving. Table 5 shows the resource usage and number of clock cycles required of complex modules.

5   Conclusion

This paper implementes SM9 digital signature algorithm based on FPGA, simplifies the implementation logic of the underlying operation modules and applies fine-grained parallelism. For basic modules on F_p, comparators in modular addition and subtraction are omitted, and the Montgomery modular multiplication algorithm is improved to achieve circuit simplification and resource conservation. For complex modules, high parallelism within and between modules is achieved, greatly reducing the number of computation cycles. However, in order to improve the computational speed, more resources are used, which needs to be optimized in future work.

References

1. State Cryptography Administration. SM9 Identification Cryptography Algorithm (2016:3). GM/T 0044.2016
2. Zhu, H., Tan, Y., Yu, X., et al.: An identity-based proxy signature on NTRU lattice. Chin. J. Electron. **27**(2), 297–303 (2018)
3. Zhu, H., Tan, Y., Zhu, L., et al.: An efficient identity-based proxy blind signature for semioffline services. Wirel. Commun. Mob. Comput. (2018)
4. Xiao, H., Liu, Y., Li, Z., Liu, G.: Algorithm-hardware co-design of ultra-high radix based high throughput modular multiplier. IEICE Electron. Express **18**(10), 1–6 (2021)
5. Zhen, P., Tu, Y., Xia, B., et al.: Research on the miller loop optimization of SM9 bilinear pairings. In: 2017 IEEE 17th International Conference on Communication Technology (ICCT), pp. 138–144. IEEE (2017)
6. Wang, A.T., Guo, B.W., Wei, C.J.: Highly-parallel hardware implementation of optimal ate pairing over Barreto-Naehrig curves. Integration **64**, 13–21 (2019)
7. Montgomery, P.L.: Modular multiplication without trial division. Math. Comput. **44**(170), 519–521 (1985)
8. Miyamoto, A., Homma, N., Aoki, T., et al.: Systematic design of RSA processors based on high-radix Montgomery multipliers. IEEE Trans. Very Large Scale Integration (VLSI) Syst. **19**(7), 1136–1146 (2010)
9. Karatsuba, A.A.: The complexity of computations. In: Proceedings of the Steklov Institute of Mathematics-Interperiodica Translation, vol. 211, pp. 169–183 (1995)
10. Sun, W., Wirthlin, M.J., Neuendorffer, S.: FPGA pipeline synthesis design exploration using module selection and resource sharing. IEEE Trans. Comput. Aided Des. Integr. Circuits Syst. **26**(2), 254–265 (2007)
11. Hao, Z., Guo, W., Wei, J., et al.: Dual processing engine architecture to speed up optimal ate pairing on FPGA platform. In: 2016 IEEE Trustcom/BigDataSE/ISPA, pp. 584–589. IEEE (2016)
12. Chatterjee, S., Sarkar, P., Barua, R.: Efficient computation of Tate pairing in projective coordinate over general characteristic fields. In: Park, C., Chee, S. (eds.) ICISC 2004. LNCS, vol. 3506, pp. 168–181. Springer, Heidelberg (2005). https://doi.org/10.1007/11496618_13
13. Scott, M., Benger, N., Charlemagne, M., Dominguez Perez, L.J., Kachisa, E.J.: On the final exponentiation for calculating pairings on ordinary elliptic curves. In: Shacham, H., Waters, B. (eds.) Pairing 2009. LNCS, vol. 5671, pp. 78–88. Springer, Heidelberg (2009). https://doi.org/10.1007/978-3-642-03298-1_6
14. Amiet D, Curiger A, Zbinden P. Flexible FPGA-based architectures for curve point multiplication over GF(p), pp. 107–114. IEEE (2016)
15. Islam, M.M., Hossain, M.S., Shahjalal, M., Hasan, M.K., Jang, Y.M.: Area-time efficient hardware implementation of modular multiplication for elliptic curve cryptography. IEEE Access **8**, 73898–73906 (2020)

Post-quantum Dropout-Resilient Aggregation for Federated Learning via Lattice-Based PRF

Ruozhou Zuo[1,2], Haibo Tian[1,2], and Fangguo Zhang[1,2(✉)]

[1] School of Computer Science and Engineering, Sun Yat-sen University, Guangzhou 510006, China

[2] Guangdong Key Laboratory of Information Security Technology, Guangzhou 510006, China

isszhfg@mail.sysu.edu.cn

Abstract. Machine learning has greatly improved the convenience of modern life. As the deployment scale of machine learning grows larger, the corresponding data scale also increases, leading to a large number of small and medium-sized organizations wishing to use their respective data to train models together, even though this may bring risks of violating data privacy regulations and privacy leakage. To meet this demand, federated learning was proposed, which can satisfy the needs of various organizations to expand the training data scale without directly sharing data, while avoiding violations of data privacy regulations and privacy leakage. General federated learning usually allows clients to train local models independently, and then aggregate them on a central server to build a global model in a privacy-preserving manner. There are various ways to protect privacy, such as homomorphic encryption, differential privacy, etc. Among these methods, one type of federated learning scheme is based on homomorphic pseudorandom functions. This type of scheme is relatively simple to construct, has a smaller communication scale, is more resilient to disconnections, and has high scalability. However, the security aggregation with cryptographic primitives based on classic assumptions such as DDH cannot resist quantum attacks, and since the protected gradient vectors are usually tens of thousands of dimensions, obtaining the aggregation results requires solving tens of thousands of discrete logarithms, which leads to some loss of efficiency. In this paper, we proposed a secure aggregation scheme based on HPRG over lattice, which has practical efficiency and resilience to dropout and can resist quantum attacks due to the hardness of the RLWE assumption. Moreover, our scheme only requires polynomial multiplication and addition (usually treated as vectors in implements), thus significantly improving computational efficiency.

Keywords: Privacy-Preserving · Secure Aggregation · Post-quantum Cryptography · Federated Learning · Lattice Cryptography

© The Author(s), under exclusive license to Springer Nature Singapore Pte Ltd. 2024
J. Vaidya et al. (Eds.): AIS&P 2023, LNCS 14509, pp. 382–399, 2024.
https://doi.org/10.1007/978-981-99-9785-5_27

1   Introduction

With the rapid development of machine learning (ML), the demand for data scale is increasing, and ML has aroused great interest in different industries, including the Internet, finance, IoT, academia, government, etc. [1–3]. As such, federated learning (FL) [4] has emerged, giving rise to horizontal federated learning and vertical federated learning, providing cross-silo or cross-device machine learning services for different scenarios and institutions. Federated learning allows multiple participants to train models together without privacy leaks. Unlike traditional centralized machine learning, in federated learning, participants generally train models independently on their local datasets, computing gradients, and then upload the gradients to a central server for aggregation, thus constructing a global model. However, even if only these gradients are leaked, there is a risk of revealing the local data of individual participants [5]. In this connection, some cryptographic mechanisms can be proposed to solve the issue, such as secure multi-party computation (MPC) [6], homomorphic encryption (HE) [7], differential privacy (DP) [8], and other mechanisms, to preserve privacy while aggregating the local models in federated learning.

Generally, we can complete PPML by blocking or encrypting the uploaded gradient and ensuring that the server can aggregate the ciphertext, which can be achieved through methods such as DP [9], MPC [10], or HE [11]. For example, applying DP to gradients before uploading to the server can protect gradient information, while the server can obtain approximate aggregation results with acceptable errors, although this can result in a trade-off between security and model performance. Whereas, HE-based PPML directly encrypts gradients. Due to the homomorphism of HE, the server can directly calculate the ciphertext to obtain encrypted aggregation results and continue to train on it. This makes the structure of HE-based PPML simple and easy to implement. However, the practicality of HE-based PPML depends on the performance of the HE used, and the selection of HE may be limited for specific ML models.

On the other hand, due to the rapid popularization of mobile internet, a large number of users involved in federated learning in practical application scenarios may be using mobile devices. This means that there could be significant differences in the computing resources and communication environments of different clients. Therefore, federated learning needs to consider not only privacy protection but also resilience to disconnections. Some dropout-resilient FL applications have been deployed, such as Gboard (Google Keyboard) on Android [12], which has deployed FL on mobile devices to enhance the relevance of contextual suggestion inputs. In this situation, FL deployed on a large number of mobile devices inevitably involves disconnection issues, and some devices may exit the system due to environmental changes.

First, we revisit the basic method of addressing dropout issues based on pairwise masks proposed in the pioneer work SecAgg [13]. Suppose there is a group of clients $\mathcal{U}$, where each client $u_i \in \mathcal{U}$'s private input is ω_i, we need to compute $\sum \omega_i$ without revealing any ω_i to any participant. In SecAgg, a set of masks is generated to encrypt ω_i as follows:

$$y_i = \omega_i + \sum_{i<j} \mathrm{PRG}\left(s_{i,j}\right) - \sum_{i>j} \mathrm{PRG}\left(s_{j,i}\right)$$

where PRG represents a pseudorandom generator that can generate a set of pseudorandom sequences $\mathrm{PRG}(s)$ based on the seed s. In the above equation, s_j^i and s_i^j are pair-wise keys generated by key agreement, i.e., $s_j^i = s_i^j$, so when all y_i are added, all pseudorandom sequences serving as masks will be revealed, thus obtaining $\sum \omega_i$:

$$\sum_{u_i \in \mathcal{U}} y_i = \sum_{u_i \in \mathcal{U}} \left(\omega_i + \sum_{i<j} \mathrm{PRG}\left(s_{i,j}\right) - \sum_{i>j} \mathrm{PRG}\left(s_{j,i}\right) \right) = \sum_{u_i \in \mathcal{U}} \omega_i$$

For dropout users, SecAgg solves this by secret sharing the keys used to generate the masks. However, every two users have to execute a pair-wise key agreement in SecAgg, resulting in $O(n^2)$ communication and computational overhead. In the follow-up work of SecAgg [14,15], researchers also pointed out its inefficiency. Therefore, using a method to generate masks that doesn't rely on interactive communication can significantly improve the protocol's performance. A method that has recently attracted attention is based on Homomorphic Pseudorandom Function (HPRF) [16]. In this regard, some representative works have been proposed [17] [18]. For instance, in HomoAgg [17], clients encrypt gradients through HPRF, and by utilizing the homomorphism of Shamir secret sharing [6], the demasking of the model is completed through seed aggregation, thereby greatly reducing communication and computational overhead. However, since its HPRG is based on discrete logarithms, although the impact on performance is acceptable under the assumption that the values of gradient vectors are not large [19], it cannot resist quantum attacks. Our scheme refers to the construction of classic lattice-based encryption like BGV [20], BFV [21] and CKKS [22], and constructs an efficient PRF based on RLWE, making our PPML protocol resistant to quantum attacks and practically efficient.

In this paper, we propose an efficient, quantum-attack-resistant, and dropout-resistant federated learning secure aggregation protocol using an RLWE-based PRF and a distributed Shamir multi-secret sharing scheme. Overall, the improved efficiency and resistance to quantum attacks mainly come from the additive homomorphic pseudorandom function based on RLWE, and the dropout resilience comes from Shamir secret sharing. Specifically, we replace the PRF based on discrete logarithms in the original scheme with an RLWE-based PRF. In this case, clients no longer use multiplication and exponentiation to construct masks but use addition and multiplication of polynomials, which will further reduce the computational overhead compared to previous work.

Our contributions: Compared to existing works, our contributions are as follows:

- The proposed scheme uses a lattice-based PRF, making it resistant to quantum attacks.

- The proposed scheme only involves the addition and multiplication of vectors, thus it is more efficient and significantly reduces computational overhead.
- The proposed scheme involves fewer cryptographic primitives, making it easier to apply or further improve.

Organization: Organization of the paper: The rest of the paper is organized as follows. In Sect. 2, we review the underlying cryptographic primitives and define the notations used by our proposed scheme. Then we proceed to our proposed protocol in Sect. 3, followed by the security analysis in Sect. 4, performance analysis, and discussions in Sect. 5. Finally, we give the conclusions in Sect. 6.

2 Cryptographic Primitives

This section mainly describes the preliminaries of the Shamir multi-secret sharing scheme, homomorphic pseudorandom functions (HPRF), and the signature scheme.

2.1 Secret Sharing

The Shamir secret sharing scheme [23] is a cryptographic technique that divides confidential information, known as the secret S, into n shares of data.

In a standard (t, n) Shamir secret sharing scheme, the secret S and its corresponding shares $S_1, ..., S_n$ are represented as elements in a finite field $\mathbb{Z}_P$, where P is a prime number and $0 < t \leq n < P$. The scheme involves one secret holder, u_s, and a group of n participants $u_1, ..., u_n$.

It is worth noting that Shamir secret sharing is additive homomorphic [6].

For example, assume there are n users while each user has d secrets to be shared. Let's focus on u_1 first, assume the secrets are $s_1^1, s_1^2, ..., s_1^d$ (denoted as a vector $S_1 = (s_1^1, s_1^2, ..., s_1^d)$). The secrets S_1 are shared among n users $u_1, u_2, ..., u_n$. Then u_1 selects a polynomial of degree $t - 1$ as:

$$f_1(x) = a_0 + a_1 x + ... + a_{t-1} x^{t-1} \bmod P$$

where $a_0 = s_1^1, a_1 = s_1^2, ..., a_{t-1} = s_1^d$, and assume $t = d$. Then, the shares of each group of $\{s_1^i | 1 \leq i \leq d\}$ to be sent to u_i are computed as $\{x_i, f_1(x_i)\}$, where x_i is usually sampled from $\{1, 2, ..., n\}$ (denoted as X) in Shamir-based applications [13,24,25]. Similarly, another secret holder u_2 with secrets $s_2^1, \{s_2^2, ..., s_2^d\}$ selects a polynomial of degree $t - 1$ as:

$$f_2(x) = b_0 + b_1 x + ... + b_{t-1} x^{t-1} \bmod P$$

where $b_0 = s_2^1, b_1 = s_2^2, ..., b_{t-1} = s_2^d$. The n shares of S_2 are computed as $\{f_2(x_i)\}$, where $x_i \in \{1, 2, ..., n\}$, and they are sent to u_i as $\{x_i, f_2(x_i)\}$. Note that each u_i can locally compute $\{x_i, f_1(x_i) + f_2(x_i)\}$, then at least t users can reconstruct the sum $S_1 + S_2$ of secrets S_1 and S_2 by Lagrange interpolation over $\{x_i, f_1(x_i) + f_2(x_i)\}$ without learning S_1 or S_2, where $x_i \in X$.

Furthermore, we can reconstruct $\sum_{i=1}^{n} S_i$. More specifically, in our protocol, each user u_i only provides $\sum_{j=1}^{n} f_j(x_i)$ to the server, and the server can reconstruct $\sum_{i=1}^{n} S_i$ to demask with $\{x_i, \sum_{j=1}^{n} f_j(x_i)\}$ and obtain final output without learning anything about the secrets as follows:

$$\begin{cases} \sum_{j=1}^{n} f_j(x_1) = \sum_{j=1}^{n} s_j^1 + \sum_{j=1}^{n} s_j^2 x_1^1 + \sum_{j=1}^{n} s_j^3 x_1^2 + ... + \sum_{j=1}^{n} s_j^t x_1^{t-1} \bmod P, \\ \sum_{j=1}^{n} f_j(x_2) = \sum_{j=1}^{n} s_j^1 + \sum_{j=1}^{n} s_j^2 x_2^1 + \sum_{j=1}^{n} s_j^3 x_2^2 + ... + \sum_{j=1}^{n} s_j^t x_2^{t-1} \bmod P, \\ ... \\ \sum_{j=1}^{n} f_j(x_n) = \sum_{j=1}^{n} s_j^1 + \sum_{j=1}^{n} s_j^2 x_n^1 + \sum_{j=1}^{n} s_j^3 x_n^2 + ... + \sum_{j=1}^{n} s_j^t x_n^{t-1} \bmod P. \end{cases}$$

Through the equations mentioned above, the central server can learn each entry of the sum of secret vectors, i.e., $\{\sum_{j=1}^{n} s_j^i\}$, and obtain the sum of the secrets $\sum_{i=1}^{n} S_i$, where $i \in \{1, 2, ..., n\}$.

Furthermore, we can reconstruct $\sum_{i=1}^{n} S_i$. More specifically, in our protocol, each user u_i only provides $\sum_{j=1}^{n} f_j(x_i)$ to the server, and the server can reconstruct $\sum_{i=1}^{n} S_i$ to demask with $x_i, \sum_{j=1}^{n} f_j(x_i)$ and obtain final output without learning anything about the secrets.

2.2 Homomorphic Pseudorandom Generator Based on Lattice

A pseudorandom function (PRF) is an essential tool in cryptography. It refers to a deterministic function $F : \{0,1\}^{\lambda} \times \{0,1\}^{in} \rightarrow \{0,1\}^{out}$, where the input is from the key space $\times$ input space. A PRF is considered secure if, for each input, its output appears to be uniformly sampled from $\{0,1\}^{out}$, resembling a random selection, i.e., for a uniform $k \in \{0,1\}^{\lambda}$, an oracle for $F(k, \cdot)$ is computationally indistinguishable from an oracle for a uniform function $f(\cdot)$ [16], where $f : \{0,1\}^{in} \rightarrow \{0,1\}^{out}$. In a similar vein to the definition of PRF, a pseudorandom generator (PRG) is an efficient algorithm capable of generating a sequence of pseudo-random numbers.

To be more precise, a PRG is an efficiently computable function $G : \mathcal{S} \rightarrow \mathcal{Y}$, where S represents the input space and Y denotes the output space. The PRG function ensures that for a uniformly chosen input $s \in \mathcal{S}$ and a uniformly chosen output $y \in \mathcal{Y}$, the distribution $\{G(s)\}$ is computationally indistinguishable from the distribution of $\{y\}$.

Assume a PRF $F : \mathcal{K} \times \mathcal{X} \rightarrow \mathcal{Y}$ and two groups $(\mathcal{K}, \oplus)$, $(\mathcal{Y}, \otimes)$. If for any $F(k_1, x)$ and $F(k_2, x)$, there exists an efficient algorithm to compute $F(k_1 \oplus k_2, x) = F(k_1, x) \otimes F(k_2, x)$, then the pseudorandom function F is said to be key-homomorphic, i.e., a HPRF is key-homomorphic with respect to its key. Similarly, for a PRG $G : \mathcal{S} \rightarrow \mathcal{Y}$, any $G(s_1, x)$ and $G(s_2, x)$, if $G(s_1 \oplus s_2, x) = G(s_1, x) \otimes G(s_2, x)$, then the PRG G is said to be seed-homomorphic, and such a PRG is referred to as a homomorphic PRG (HPRG).

Similar to the HPRF based on hash functions and discrete logarithms described in [17], we have constructed an HPRF based on RLWE according to the symmetric encryption scheme of CKKS [22]. Let $\mathbb{R}_q = \mathbb{Z}_q[x]/(X^N + 1)$ be a polynomial ring modulo q, $\mathcal{DG}(\sigma^2)$ be a discrete Gaussian distribution, and $\mathcal{S}_h$ be a uniform distribution over the set of signed binary vectors in $\{\pm 1\}^N$ whose Hamming weight is exactly h, and the function $F : (\mathcal{S}_h) \times \mathbb{R}_q \rightarrow \mathbb{R}_q$ be

$$F(s, x) = xs + e$$

where $e \in \mathcal{DG}^N(\sigma^2)$. We can observe that F is additive homomorphic:

$$F(s_1 + s_2, x) \approx F(s_1, x) + F(s_2, x)$$

According to the hardness of the RLWE problem [26], if s is randomly elected in $\mathcal{S}_h$, then $F(s, \cdot)$ is indistinguishable from random samples in $\mathbb{R}_q$. Therefore, an HPRG can be constructed based on $F(s, \cdot)$ and a set of x [27]. Specifically, assuming $a_0 \in \{0, 1\}^N$ is a system parameter and $H : \{0, 1\}^* \rightarrow \{0, 1\}^N$ is a secure hash function, with $a_i = H(a_{i-1})$, the HPRG $G(s)$ can generate a pseudorandom sequence, i.e., $F(s, a_0), F(s, a_1), F(s, a_2)\ldots$. The length of the sequence is $\lceil d/N \rceil$, where d is the dimension of the vector to be masked.

2.3 Digital Signature

A signature system serves to verify the source of a message.

When a message is signed using the signer's secret key, it can be confirmed that the message originated from the signer. A signature system consists of a set of algorithms $(\mathcal{F}_{gen}, \mathcal{F}_{sig}, \mathcal{F}_{ver})$ such that:

- $\mathcal{F}_{gen}$ is a randomized algorithm that generates a secret key sk and a public key pk.
- $\mathcal{F}_{sig}$ is a randomized algorithm that takes the secret key sk and a message m as inputs, and produces a signature σ.
- $\mathcal{F}_{ver}$ is a deterministic algorithm that take the public key pk, the message m, and the signature σ as inputs, and returns 1 if σ is a valid signature for m, and 0 otherwise.

The signature scheme provides protection against universal forgery under chosen message attack (UF-CMA). This means that an individual without access to the secret key is unable to generate a valid signature for a message they have not previously seen signed. In essence, the likelihood of an adversary successfully creating a pair (m^*, σ^*) without knowing the secret key sk, where σ^* is a valid signature for m^* and m^* has been unknown for the adversary so far, is extremely low, i.e., the probability is negligible.

3   Proposed Scheme

According to the SecAgg scheme, the participants mainly consist of a central server S and n clients. Each client u_i has a locally trained gradient ω_i, and the central server needs to collect these gradients to update the global model and distribute it to the users. The goal of our scheme is for the central server to compute the sum of the gradients without knowing any information about ω_i. Additionally, the scheme should be able to aggregate correctly even if some users drop out during the process.

Threat model: In many federated learning applications, participants may come from different companies or institutions, some of which may even have competing interests. This creates a motivation to obtain data from other participants. Due to these reasons, we have considered two threat models, namely the semi-honest model and the malicious model. In the semi-honest model, the adversary will execute the protocol normally but will attempt to infer the honest participants' data. In the malicious model, the adversary not only tries to infer information from other clients but may also collude and send false messages.

3.1 Masking Gradients

To protect the local model from leakage, it is necessary to hide the gradients before uploading them to the central server. The most straightforward approach is to add a mask. In our scheme, we use a one-time pad to mask the gradients. Let's assume the gradient of client u_i to be ω_i, the scale for eliminating errors to be Δ, and the corresponding mask to be r_i. The masked gradient of client u_i is then calculated as:

$$y_i = \Delta\omega_i + r_i$$

Subsequently, the client sends y_i to the central server. Assuming the server can somehow obtain the sum of all participants' masks without errors, denoted as $R = \sum_{u_i \in \mathcal{U}} r_i - e$, then it can obtain:

$$z = \left\lfloor (\sum_{u_i \in \mathcal{U}} y_i - R)/\Delta \right\rceil = \left\lfloor (\sum_{u_i \in \mathcal{U}} y_i - \sum_{u_i \in \mathcal{U}} r_i + e)/\Delta \right\rceil = \left\lfloor \sum_{u_i \in \mathcal{U}} \omega_i + e/\Delta \right\rceil = \sum_{u_i \in \mathcal{U}} \omega_i$$

where e denotes the sum of all errors after aggregation.

3.2 Handling Dropping Out

According to Sect. 2.2, for simplicity, we assume that for each client u_j, $\mathcal{F}_{share}$ is a share generation function, $\mathcal{F}_{rec}$ is a secret reconstruction function, and the mask of client u_j is $r_j = G(s_j)$, where G is a homomorphic pseudorandom generator, s_j is the seed of client u_j, and $\mathcal{F}_{share}(s_j) \to \left\{i, s_j^i\right\}_{u_i \in \mathcal{U}}$ are the shares generated from s_j. Therefore, we can reconstruct s_j by $\mathcal{F}_{rec}(\left\{i, s_j^i\right\}_{u_i \in \mathcal{U}'}) \to s_j$ where $|\mathcal{U}'| \geq t$. Then each client u_i calculates the sum of the shares sent to itself, noted as:

$$S^i = \sum_{u_j \in \mathcal{U}} s_j^i \bmod P$$

According to the additive homomorphism mentioned in Sect. 2.2, the server can obtain the sum of seeds S without learning any information about any seed of clients, i.e., $\mathcal{F}_{rec}(\{i, S^i\}_{u_i \in \mathcal{U}'}) \to S$ where $|\mathcal{U}'| \geq t$. Then, the server can demask with the HPRG as follows:

$$R = \sum_{u_i \in \mathcal{U}} r_i = \sum_{u_i \in \mathcal{U}} G(s_i) = G(\sum_{u_i \in \mathcal{U}} s_i) = G(S)$$

By the method mentioned above, it's able to handle dropout clients and make the protocol dropout-resilient with a (t, n) threshold.

3.3   High-Level View

Once the server obtains R, it can proceed with demasking. Let's summarize the above steps into the following protocol:

1. Each client u_i selects a random seed s_i from $\mathcal{S}_h$.
2. Each client u_i computes n shares of the seed s_i using the Shamir secret sharing scheme as $\mathcal{F}_{share}(s_i) \to \{j, s_i^j\}_{u_j \in \mathcal{U}}$, and then sends shares to the corresponding client u_j.
3. Each client u_i computes the mask $r_i = G(s_i)$ and the masked gradient $y_i = \Delta\omega_i + r_i$, and sends y_i to the server. Let $\mathcal{U}'$ be the set of connected clients that sent y_i. The server receives all masked gradients y_i from $\mathcal{U}'$, and computes the sum $\sum_{u_i \in \mathcal{U}'} y_i = \Delta\sum_{u_i \in \mathcal{U}'} \omega_i + \sum_{u_i \in \mathcal{U}'} r_i = \Delta\sum_{u_i \in \mathcal{U}'} \omega_i + R$.
4. Each client u_i computes the sum of the shares received from $\mathcal{U}'$, i.e., $S^i = \sum_{u_j \in \mathcal{U}'} s_j^i$, and sends S^i to the server.
5. The server reconstructs the sum S of the seeds using the Shamir secret sharing scheme as $\mathcal{F}_{rec}(\{i, S^i\}_{u_i \in \mathcal{U}'}) \to S$, and then computes the sum R of the masks as $R = G(S)$.

With the above protocol, we utilize the additive homomorphism of Shamir secret sharing to avoid the need for pair-wise masks and additional secret sharing, as seen in SecAgg. This significantly reduces the communication overhead and enhances resilience against client disconnections [17].

3.4   Proposed PPFL Protocol

We call our scheme LHAgg (Lattice-based Homomorphic Aggregation). Our federated learning protocol involves a central server and n clients. During the protocol execution, the gradients sent by each client u_i to the server are denoted as an input vector ω_i, where d elements belong to $\mathbb{Z}_q$.

Similar to SecAgg, assuming there are time constraints on the communication between the central server and clients, if a client's communication times out, the server considers it as a dropout. For convenience, we assume that each client has a set of keys based on the same public-key infrastructure for encrypting messages, denoted as $Enc(msg, pk)$ and $Dec(msg, sk)$. Clients may drop out at any round of the protocol, and as long as the server receives masked gradients from t non-dropout users, it can output the correct aggregation result. Our scheme has resilience to dropouts. Specifically, client u_i has already shared its seed s_i through secret sharing in the first round. Therefore, if client u_i drops out after uploading the masked gradient y_i in the second round, y_i can still be correctly used for calculating the aggregation result.

We can observe that compared to other PPML schemes, our scheme significantly reduces the complexity of the computation in the protocol. This is because we utilize a PRF based on RLWE, which transforms the aggregation computation into vector addition. The detailed description is as follows:

Lattice-based Homomorphic Aggregation for Federated Learning (LHAgg)

Participants: A server S and a set of clients U.

Private inputs: Each client u_i has a gradient noted as a vector ω_i, a secret key for encryption c_i^{SK}, and a secret key for signature d_i^{SK}. The server has a secret key for encryption c_s^{SK} and a secret key for signature d_s^{SK}.

Public inputs: The set of clients U, the threshold $t < n = |U|$, the field $\mathbb{Z}_P$ for Shamir secret sharing scheme with function $\mathcal{F}_{share}$ and $\mathcal{F}_{rec}$, the algorithm $(\mathbb{R}, N, q)$ for HPRG which samples two polynomial rings $\mathbb{R}=\mathbb{Z}[x]/(X^N+1)$ and $\mathbb{R}_q=\mathbb{Z}_q[x]/(X^N+1)$, the discrete Gaussian distribution $\mathcal{DG}(\sigma^2)$ of variance σ^2, a uniform distribution $\mathcal{S}_h$ for sampling, the scale Δ, and the security parameter κ. Each client u_i's and server's public key for encryption c_i^{PK}, c_s^{PK}, and their public keys for signature d_i^{PK}, d_s^{PK}. Note that $P > q > n * max(\omega_i)$.

Outputs: The aggregation result $\sum_{u_i \in U_2} x_i$, where $U_2 \subseteq U$. $(U_3 \subseteq U)$

Round 1:

Client u_i:

1) Randomly picks $s_i \in \mathcal{S}_h$. Generates (t, n) Shamir secret shares of s_i, i.e.,

$$\mathcal{F}_{share}(s_i) \rightarrow \left(\left\{ j, s_i^j \right\}_{u_j \in U} \right)$$

2) Sends $\left(\text{Enc}\left(s_i^j, c_j^{PK} \right), \sigma_{i,j}^1 \right)$ to client $u_j \in U$, where the signature $\mathcal{F}_{sig}\left(\text{Enc}\left(s_i^j, c_j^{PK} \right), d_i^{SK} \right) \rightarrow \sigma_{i,j}^1$ (Denote U_1 as the set of clients $u_i \in U$ that at least t shares of s_i have been received by other clients).

3) Receives $\left(Enc\left(s_j^i, c_i^{PK} \right), \sigma_{j,i}^1 \right)$ from all the clients $u_j \in U$, and then computes $s_j^i = \text{Dec}\left(Enc\left(s_j^i, c_i^{PK} \right), c_i^{SK} \right)$. If $\mathcal{F}_{ver}\left(\text{Enc}\left(s_j^i, c_i^{PK} \right), d_j^{PK}, \sigma_{j,i}^1 \right) = 0$, aborts.

Server S:

1) Collects messages from at least t clients.

2) Sends messages to corresponding clients.

Round 2:

Client u_i:

1) Generates the mask vector r_i using HPRG with the seed s_i, i.e., $r_i = \text{G}(s_i)$.

2) Computes the input masked by r_i, i.e., $y_i = \Delta \omega_i + r_i$.

3) Sends y_i with the signature $\mathcal{F}_{sig}\left(y_i, d_i^{SK} \right) \rightarrow \sigma_i^2$ to the server S.

Server $\mathcal{S}$:

1) Receives all $\boldsymbol{y}_i$ with the signature σ_i^2 from clients (Denoted as $\mathcal{U}_2 \subseteq \mathcal{U}_1$). If $\mathcal{F}_{ver}\left(\boldsymbol{y}_i, d_i^{PK}, \sigma_i^2\right) = 0$, remove client u_i from $\mathcal{U}_2$.

Round 3:

Client u_i :

1) Fetches the list of $\mathcal{U}_2$ from the server $\mathcal{S}$ with the server's signature $\mathcal{F}_{sig}\left(\mathcal{U}_2, d_s^{SK}\right) \rightarrow \sigma_s^3$. If $\mathcal{F}_{ver}\left(\mathcal{U}_2 , d_s^{PK}, \sigma_s^3\right) = 0$, aborts.

2) Sends $\mathcal{F}_{sig}\left(\mathcal{U}_2, d_i^{SK}\right) \rightarrow \sigma_i^4$ to the server $\mathcal{S}$.

Server $\mathcal{S}$:

1) Receives σ_i^4 from at least t clients (Denoted as $\mathcal{U}_3 \subseteq \mathcal{U}_2$) and forwards to the clients in $\mathcal{U}_3$.

Round 4:

Client u_i :

1) If the protocol does not consist of step 3 for consistency checking, fetches the list of $\mathcal{U}_2$ from the server $\mathcal{S}$. Otherwise, if $|\mathcal{U}_3| < t$ or for any $u_j \in \mathcal{U}_3$, $\mathcal{F}_{ver}\left(\mathcal{U}_2, d_j^{PK}, \sigma_j^4\right) = 0$, aborts.

2) Computes the sum of shares of s_j^i from all the clients $u_j \in \mathcal{U}_2$, i.e., $S^i = \sum_{u_j \in \mathcal{U}_2} s_j^i \bmod P$. ($\mathcal{U}_3$ under malicious threat model)

3) Sends S^i with the signature $\mathcal{F}_{sig}\left(Enc\left(S^i, c_s^{PK}\right), d_i^{SK}\right) \rightarrow \sigma_j^5$ to the server.

Server $\mathcal{S}$:

1) Receives S^j from all $u_j \in \mathcal{U}_4 \subseteq \mathcal{U}_3$. If $\mathcal{F}_{ver}\left(Enc\left(S^j, c_s^{PK}\right), d_j^{PK}, \sigma_j^5\right) = 0$, remove u_j from $\mathcal{U}_4$. Proceed until $|\mathcal{U}_4| > t$.

2) Reconstructs the seed for demasking, i.e., $\mathcal{F}_{rec}\left(\{j, S^j\}_{u_j \in \mathcal{U}_4}\right) \rightarrow S \bmod P$.

3) Generates the mask $\boldsymbol{R}$ using HPRG with the seed S , i.e., $\boldsymbol{R} = G(S)$.

4) Computes $z = (\sum_{u \in \mathcal{U}_2} \boldsymbol{y}_i - \boldsymbol{R})/\Delta = (\sum_{u_i \in \mathcal{U}_2} \boldsymbol{y}_i - \sum_{u_i \in \mathcal{U}_2} \boldsymbol{r}_i)/\Delta = \sum_{u_i \in \mathcal{U}_2} \boldsymbol{\omega}_i$. ($\mathcal{U}_3$ under malicious threat model)

5) Outputs $z = \sum_{u_i \in \mathcal{U}_2} \boldsymbol{\omega}_i$. ($\mathcal{U}_3$ under malicious threat model)

Protocol 1. The description of LHAgg for one round of training. The underlined parts are only for malicious threat model.

4 Security Analysis

In this section, we analyze the security of Protocol 1, denoted as π.

We assume the set of all clients as $\mathcal{U}$, the central server as $\mathcal{S}$, and the adversary as a subset of clients denoted as $\mathcal{C}$, where $|\mathcal{C}| < t$. Depending on different threat models, the adversary may include the central server or not, and all adversaries can be either semi-honest or malicious.

We claim that our protocol is secure if no attacker has any knowledge of the inputs of honest clients. More specifically, given a set of adversaries $\mathcal{C}$ belonging to $\mathcal{U}$, any values sent by honest clients $\mathcal{U} \setminus \mathcal{C}$ during the protocol execution should be uniformly random.

Let ω_i and x_i represent the input and view of clients respectively, $\mathsf{REAL}_{\pi,\mathcal{C}}^{\mathcal{U}} = \{x_i | i \in \mathcal{C}\}$ denote the view of adversaries in the actual execution of the protocol. Let $\mathsf{SIM}_{\mathcal{C}}^{\mathcal{U}} = \{x_i | i \in \mathcal{C}\}$ denote another view of adversaries, where the inputs of honest participants are randomly chosen by a simulator and represented as x_c. According to the security requirements of the protocol, our protocol is secure iff the distribution of REAL and SIM is indistinguishable. In addition, due to the lattice-based property of the PRF in this paper, its security can be reduced to the hardness of decisional RLWE problem [26] in the single-user scenario, and to the n-decision RLWE problem in the multi-user scenario, demonstrating semantic security. We refer interested readers to [28] for details.

Since malicious servers may actively delete some honest clients, we use a random oracle model similar to that in SecAgg [13] for the security proof of malicious models with malicious servers. Specifically, it is assumed that SIM can access a random oracle IDEAL such that:

$$\mathsf{IDEAL}(t,\mathcal{U},\mathcal{C}) = \begin{cases} \sum_{u_i \in \mathcal{U} \backslash \mathcal{C}} \omega_i & |\mathcal{U} \backslash \mathcal{C}| > t \\ \bot & |\mathcal{U} \backslash \mathcal{C}| \leq t \end{cases}$$

by which SIM can ignore client disconnections and obtain the correct sum of gradients of honest clients.

Simply, we consider the malicious model with a malicious server. Suppose the simulator has access to the oracle IDEAL that can ignore client disconnections and provide the correct sum. For convenience, we assume all communication in the protocol execution is performed through a trusted third party $\mathcal{T}$.

Theorem 1 (Security against malicious clients and malicious server). *There exists a PPT simulator* SIM *such that for all* $t, \mathcal{U}, \mathcal{C} \subseteq \mathcal{U} \cup \{\mathcal{S}\}$

$$\mathsf{REAL}_{\pi,\mathcal{C}}^{\mathcal{U}}(n,t;\omega_{\mathcal{U}}) \equiv \mathsf{SIM}_{\mathcal{C}}^{\mathcal{U}}(n,t;x_{\mathcal{C}})$$

where $\equiv$ *denotes that the distributions are indistinguishable.*

Proof Similarly, this is also proven through the hybrid argument.

*H_0: In this hybrid, the joint view of $\mathcal{C}$ in SIM is exactly distributed like that in REAL.

*H_1: In this hybrid, for all honest clients $u_i \in \mathcal{U} \backslash \mathcal{C}$, we select a uniformly random element X_i^j from the field and replace u_i's share s_i^j with X_i^j. Due to the security of the Shamir secret sharing scheme, this hybrid is indistinguishable from the previous hybrid.

*H_2: In this hybrid, if any party provides an invalid signature $\sigma_{i,j}^1$ in Round 1, SIM aborts. The security of the signature scheme ensures that this hybrid is indistinguishable from the previous one.

*H_3: In this hybrid, the mask r_i of honest client u_i is replaced with a uniformly random polynomial in the corresponding ring $\mathbb{R}_q$, and the masks of adversaries are set to 0. Based on the hardness of the decisional RLWE problem, this hybrid is indistinguishable from the previous one.

*H_4: In this hybrid, if any party provides an invalid signature σ_i^2 in Round 2, SIM aborts. Since this violates the security of the signature scheme, this hybrid is indistinguishable from the previous one.

*H_5: In this hybrid, if the server provides an invalid signature σ_s^3 in Round 3, SIM aborts. The unforgeability of the signature scheme ensures that this hybrid is indistinguishable from the previous one.

*H_6: In this hybrid, if any party provides an invalid signature σ_i^4 in Round 4, SIM aborts. The unforgeability of the signature scheme ensures that this hybrid is indistinguishable from the previous one.

*H_7: In this hybrid, we denote the set $\mathcal{U}_2$ sent by the server to clients in Round 3 as $\mathcal{U}_2'$. If SIM observes any two distinct signed $\mathcal{U}_2'$, the protocol aborts. The security of the signature scheme ensures that the adversary cannot forge a signature with non-negligible probability, which means this hybrid is indistinguishable from the previous one.

*H_8: In this hybrid, if any party provides an invalid signature σ_i^4 in Round 4, SIM aborts. The unforgeability of the signature scheme ensures that this hybrid is indistinguishable from the previous one.

*H_9: In this hybrid, the simulator SIM obtains the sum of $\{\omega_i | u_i \in \mathcal{U}_2' \setminus \mathcal{C}\}$ by querying IDEAL and selects x_i such that $\sum_{u_i \in \mathcal{U}_2 \setminus \mathcal{C}} x_i = \sum_{u_i \in \mathcal{U}_2 \setminus \mathcal{C}} \omega_i \mod q$. Note that querying IDEAL does not affect the protocol simulation process. Therefore, this hybrid is indistinguishable from the previous one.

Thus, SIM can simulate Protocol 1 without requiring any input from honest clients, resulting in SIM $\equiv$ REAL, completing the proof. $\qquad\square$

5   Performance Analysis

In this section, we will sequentially analyze the computational and communication complexities of our protocol and present our experimental results from various aspects.

5.1   Complexity Analysis

Computation Overheads. Our protocol structurally resembles HomoAgg, thus the computational cost for each client largely depends on Shamir secret sharing and mask generation. Given the adoption of multi-secret sharing, the complexity of (t, n) secret sharing is approximately $O(\frac{n^2 N}{t})$. For an input vector of dimension d, each client needs to generate $\frac{d}{N}$ masks. Therefore, the total computational complexity for each client is $O(\frac{n^2 N}{t} + \frac{d}{N})$. The server needs to reconstruct the vector sum of N/t seed fragments from N/t groups of n Shamir secret shares. We employ a pre-computation approach similar to SecAgg for secret reconstruction, hence the cost for a single reconstruction is $O(n)$, with the total computational overhead being $O(\frac{nN}{t})$.

Communication Overheads. Each client's communication includes $\frac{nN}{t}$ ciphertext shares and a set of d/N N-dimensional mask vectors, hence the complexity is $O(\frac{nN}{t} + d)$. The server's communication mainly involves forwarding $O(n^2)$ encrypted $\frac{nN}{t}$ shares and receiving d/N N-dimensional mask vectors from n clients, hence the total communication complexity is $O(\frac{n^2N}{t} + nd)$.

5.2   Experiment

Protocol Performance. Our experiments are conducted on a single-threaded Python implementation, running on a BKunYun C-16–2 instance with Intel Xeon Cascade Lake 8255C (2.5GHz), 16 cores, and 32GB memory. For cryptographic primitives, we use AES-GCM with a 128-bit key for authenticated encryption, a specialized (t, n) Shamir multi-secret sharing scheme to generate seed shares, and cryptographic schemes based on RLWE, such as CKKS [22], from the Tenseal library to construct HPRG for mask generation. Additionally, performance can be improved with faster interpolation algorithms [29]. As our scheme is cryptography-based, it can achieve precise decryption, thereby introducing no additional errors to model training. We implement and analyze our experiments from two aspects: (i) Test whether the performance of the LHAgg secure aggregation protocol is practical; (ii) Train and test model effects using our protocol based on some classic datasets and machine learning models. We will test and compare performance from aspects such as the average runtime of clients per training round, server runtime, and communication volume.

To simplify non-critical parts of the experiment, we directly apply our LHAgg protocol to model training aggregation by simulating it on separate machines. The average client runtime is estimated by dividing the total runtime of the client simulation part by the number of clients. Similarly, the server runtime is represented by the corresponding part. The communication volume of a single client with the server is represented by the sum of the average total communication volume per round. Our experiments will train a classic CNN model based on the MINIST dataset to verify the usability of LHAgg.

As previously mentioned, our protocol is oriented towards PPFL systems, where clients are typically mobile devices of different specifications in diverse environments. Therefore, from the perspective of a single client, the protocol's communication volume has the most direct impact, and the client runtime is shorter than the server runtime, since clients can be considered to run in parallel, we only consider client communication cost. For the server, the server runtime constitutes the majority of the protocol's runtime, hence it is the primary indicator in our experimental data. As shown in Fig. 1 and Fig. 2, we can observe that, under the settings of 500 clients, 100K vector size, and RLWE polynomial degree N=4096, the coefficient modulus size of 24 bits, the communication cost for each client in one training round is approximately 0.55MB, and one round of aggregation can be completed within about 90 s, indicating that LHAgg is practical in terms of performance.

In addition to runtime and communication volume costs, concerning dropout resistance, as our protocol structurally resembles HomoAgg, its dropout resis-

a suitable encoding, any decision problem can be viewed as such a member-ship problem. The membership problem is decidable iff the following **characteristic function** g_L of L is recursive:

$$g_L(w) = \begin{cases} 1 & \text{if } w \text{ is in } L, \\ 0 & \text{if } w \text{ is not in } L. \end{cases}$$

Indeed, g_L being recursive gives an immediate algorithm for solving the membership problem: Given an arbitrary w, we just compute the value $g_L(w)$. Conversely, if the membership problem for L is decidable, Church's thesis guarantees that g_L is recursive.

The following notation will be used frequently in the sequel. Let f be a partial recursive function of n variables. We denote by

$$f(x_1, \ldots, x_n)\!\downarrow \qquad (\text{resp. } f(x_1, \ldots, x_n)\!\uparrow)$$

the fact that f converges (resp. diverges) for the argument $(x_1, \ldots, x_n)$— that is, that the value $f(x_1, \ldots, x_n)$ is (resp. is not) defined. The same notation is used also in connection with Turing machines:

$$TM(x_1, \ldots, x_n)\!\downarrow \qquad (\text{resp. } TM(x_1, \ldots, x_n)\!\uparrow)$$

means that TM halts (resp. loops) with the input $(x_1, \ldots, x_n)$.

Pairing functions, defined later, provide a convenient transition from tuples of integers to integers. Consider the function φ of two variables ranging over positive integers, defined by

$$\varphi(x, y) = \frac{1}{2}(x + y - 1)(x + y - 2) + x.$$

It is easy to see that $\varphi(x, y)$ assumes its values according to Table 4.1. Hence, φ assumes every positive integer z as its value exactly once; there are unique integers $\psi_1(z)$ and $\psi_2(z)$ with the property

$$\varphi(\psi_1(z), \psi_2(z)) = z.$$

In fact, $\psi_1(z)$ (resp. $\psi_2(z)$) is the index for the row (resp. column) containing the number z. Clearly the functions φ, ψ_1, and ψ_2 are recursive.

TABLE 4.1

x \ y	1	2	3	4	5	...
1	1	2	4	7	11	
2	3	5	8	12		
3	6	9	13			
4	10	14				
5	15					
⋮						

Inductively define functions $\varphi_2, \varphi_3, \ldots$ as follows:

$$\varphi_2(x_1, x_2) = \varphi(x_1, x_2),$$

$$\varphi_i(x_1, \ldots, x_i) = \varphi(\varphi_{i-1}(x_1, \ldots, x_{i-1}), x_i) \qquad \text{for } i \geq 3.$$

Clearly, every φ_i assumes every positive integer as its value exactly once: For every $z \geq 1$, there are unique integers $\psi_1^{(i)}(z), \ldots, \psi_i^{(i)}(z)$ such that

$$\varphi_i(\psi_1^{(i)}(z), \ldots, \psi_i^{(i)}(z)) = z.$$

We are now ready for some basic undecidability results. We first consider the **halting problem** for Turing machines: to decide whether $TM_i(x)\downarrow$ holds for a given pair (i, x) of positive integers. The next theorem shows that this problem is undecidable. The way of representing x is irrelevant.

Theorem 4.4.   *The function g defined by*

$$g(\varphi(i, x)) = \begin{cases} 1 & \text{if } TM_i(x)\downarrow \\ 0 & \text{if } TM_i(x)\uparrow \end{cases}$$

is not recursive.

Proof.   Observe first that by the properties of the pairing function φ, the function g is well defined, total, and maps the set of positive integers into $\{0, 1\}$.

We proceed indirectly, assuming that g is recursive. Then the function $g_1(i, x) = g(\varphi(i, x))$ is also recursive. Let b be an index for g_1—that is, $g_1 = f_b$. We deduce a contradiction.

Consider the partial function $\tau(x, i)$, defined as follows. If $f_i(x, x) = 0$ then $\tau(x, i) = 1$. Otherwise (if $f_i(x, x) \neq 0$ or if $f_i(x, x)$ is undefined), $\tau(x, i)$ is undefined. The following constitutes an intuitive algorithm for computing the values of $\tau(x, i)$. Given a pair (x, i), start the ith Turing machine TM_i with the input (x, x) and see what happens. If TM_i halts with (x, x) and the final output is 0, then $\tau(x, i) = 1$. In other cases (that is, the final output is not 0 or TM_i loops with $(x, x,))$, $\tau(x,i)$ is undefined.

By Church's thesis, $\tau(x, i)$ is partial recursive and, consequently, has an index a:

$$\tau(x, i) = f_a(x, i).$$

By Theorem 4.1 with $m = n = 1$, there is a recursive function s such that

$$f_a(x, i) = f_{s(a, i)}(x). \tag{13}$$

Considering the recursive function $s_1(i) = s(a, i)$, we may write (13) in the form

$$\tau(x, i) = f_{s_1(i)}(x). \tag{14}$$

Substituting $x = s_1(b)$, $i = b$, we infer that

$$f_{s_1(b)}(s_1(b))\!\downarrow \quad \text{iff} \quad \tau(s_1(b),b)\!\downarrow \quad \text{iff} \quad f_b(s_1(b),s_1(b)) = 0, \qquad (15)$$

the latter equivalence being due to the definition of τ.

On the other hand, by the definition of g_1,

$$f_{s_1(b)}(s_1(b))\!\downarrow \quad \text{iff} \quad g_1(s_1(b), s_1(b)) = 1. \qquad (16)$$

By (15) and (16),

$$g_1(s_1(b), s_1(b)) = 1 \quad \text{iff} \quad f_b(s_1(b), s_1(b)) = 0,$$

a contradiction because $g_1 = f_b$. □

The above argument has been formulated keeping in mind those readers who still might want to replace Church's thesis by "machine-language programming." Therefore, Church's thesis appears only in a very immediate context. The following argument is simpler but uses Church's thesis in a more involved fashion.

Assuming g and, hence, g_1 to be recursive, we conclude that the function $\tau(x)$ defined as follows is partial recursive: If $g_1(x, x) = 0$, then $\tau(x) = 1$. Otherwise, $\tau(x)$ is undefined. Let a be an index for τ. Hence, by the definition of τ, $f_a(a)\!\downarrow$ iff $g_1(a, a) = 0$. On the other hand, by the definition of g_1, $f_a(a)\!\uparrow$ iff $g_1(a, a) = 0$, a contradiction!

Theorem 4.4 is the very basic undecidability result. The proof can be modified in a number of ways but is always diagonal in flavor. The reader might wonder why binary functions $f_i(x, x)$ were considered in the definition of τ, rather than unary functions $f_i(x)$. Indeed, τ can be defined in terms of the unary functions, but then the final part of the argument becomes more complicated because the "inverses" ψ_i of the pairing function φ have to be used.

The proof given above for Theorem 4.4 immediately yields the following corollary.

Theorem 4.5.   *The function*

$$g(x) = \begin{cases} 1 & \text{if } TM_x(x)\!\downarrow \\ 0 & \text{if } TM_x(x)\!\uparrow \end{cases}$$

is not recursive.

Proof.   The contradiction in the proof of Theorem 4.4 was based on the consideration of values $f_{s_1(b)}(s_1(b))$—that is, on what happens when the Turing machine $TM_{s_1(b)}$ is started with the input $s_1(b)$. □

Theorem 4.5 is usually referred to as the *undecidability of the self-applicability problem for Turing machines*. A Turing machine TM_x is started with its own index x (that is, with an input completely describing the Turing machine in question). What happens is undecidable!

Let us repeat the brief argument showing the undecidability. (In fact, this is the most straightforward undecidability proof.) Assume that the self-applicability problem is decidable, which means that the function g in Theorem 4.5 is recursive. Then the function g_1 such that $g_1(x) = g(x)$ if $g(x) = 0$ (whereas $g_1(x)$ is undefined if $g(x) = 1$) is partial recursive. (Indeed, a Turing machine computing g can immediately be converted to a Turing machine computing g_1 by starting a loop from the output 1.) Let b be an index of g_1. Then $g_1(b) = f_b(b)$ is defined iff $g(b) = 0$ iff $TM_b(b)\uparrow$ iff $f_b(b)$ is undefined. Consequently, $f_b(b)$ is defined iff $f_b(b)$ is undefined, a contradiction.

In general, proofs about undecidability are either *direct* or *indirect*. Direct proofs (such as that of Theorem 4.4) make use of some diagonalization argument. Indirect proofs have the following format. We know that a problem P is undecidable. To show the undecidability of a problem P', we prove that if P' were, in fact, decidable, then P would also be decidable, which is a contradiction. Thus, the indirect argument uses a **reduction:** An algorithm for solving P' yields an algorithm for solving P. In this fashion we use problems already known to be undecidable to generate new undecidable problems. Of course, it is sufficient to exhibit an intuitive algorithm to show the decidability of a problem. In most cases, it is still very difficult to establish the borderline between decidability and undecidability. For instance, the equivalence problem is decidable for regular languages and undecidable for context-free languages, but the exact borderline is unknown.

We still want to emphasize, with forthcoming proofs in mind, that an equality between two functions means that the functions are defined for the same arguments, and that the values are the same whenever they are defined. In all cases considered (such as equations (13) and (14)), it should be clear what the intended variables of the functions are. Therefore, we have not introduced any special notation to indicate the variables (such as the λ-notation [Ro]).

The next theorem, customarily called the **recursion theorem,** is a powerful tool in constructions involving some fixed-point requirement. In the statement of the theorem, we apply the *convention* that f_i is the partial recursive function of *one* variable computed by the Turing machine TM_i. Whenever functions of several variables are considered, the number of variables is indicated in the notation.

Theorem 4.6.   *For every recursive function g, there is a natural number n (called a fixed point of g) such that*

$$f_n = f_{g(n)}. \tag{17}$$

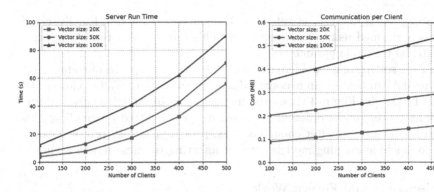

Fig. 1. Server Run Time   **Fig. 2.** Communication Cost

tance is essentially close to it, and the performance comparison under different dropout rates with SecAgg and SecAgg+ can refer to the data in [20].

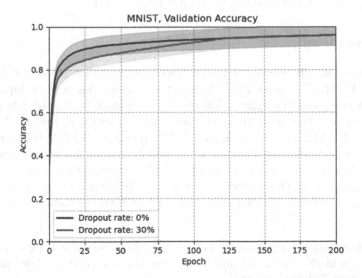

Fig. 3. Validation accuracy progression over training runs on MNIST. Batch size is fixed to 128.

Model Accuracy. The MNIST database is a frequently used benchmark for image recognition, composed of 60,000 training samples and 10,000 test samples. Each sample is a 28×28 grayscale image of a handwritten digit. We train a classifier that includes two convolutional layers activated by ReLU, each followed by max pooling, and a densely activated ReLU layer with 32 nodes. Finally, classification is performed using a softmax layer. This model has a total of approximately 26,000 trainable parameters.

Under the setting of a fixed batch size of 128, after 200 epochs of training, our MNIST model can achieve an average validation accuracy of 95.9%, close to the results of FLDP [9]. Figure 3 shows the improvement in model accuracy during the training process under different dropout rates, even with dropout clients, the model will still converge after enough epochs, proving that our protocol can be applied to machine learning models without affecting the final model accuracy.

5.3   Discussion and Future Works

The focus of LHAgg is to protect the data uploaded by clients, without considering the model distributed to clients during aggregation. Therefore, LHAgg is still susceptible to the membership inference attack [30]. A natural solution is to use schemes based on fully homomorphic encryption [31]. Furthermore, for other known attacks against distributed machine learning, such as backdoor attacks [32] and the Eclipse attack [33], our scheme can also be combined with corresponding defense methods to counter these attacks.

6   Conclusion

We proposed an efficient post-quantum secure aggregation protocol LHAgg constructed by RLWE-based PRF. Without revealing the private input of the clients, it computes the aggregation result of the encrypted gradients uploaded to the central server by a group of clients participating in federated learning. Moreover, LHAgg is a post-quantum PPFL protocol. Additionally, our protocol has two versions, each resistant to attacks from semi-honest adversaries and active malicious adversaries, respectively. With appropriate parameter settings, the performance of our protocol is proven to be practical. As post-quantum cryptographic schemes develop, our protocol can further enhance post-quantum security.

Acknowledgement. This work is supported by the National Natural Science Foundation of China (No. 62272491) and Guangdong Major Project of Basic and Applied Basic Research(2019B030302008).

References

1. Chandnani, N., Khairnar, C.N.: A reliable protocol for data aggregation and optimized routing in IoT WSNs based on machine learning. Wirel. Pers. Commun. **130**(4), 2589–2622 (2023)
2. Long, Guodong, Tan, Yue, Jiang, Jing, Zhang, Chengqi: Federated Learning for Open Banking. In: Yang, Qiang, Fan, Lixin, Yu, Han (eds.) Federated Learning. LNCS (LNAI), vol. 12500, pp. 240–254. Springer, Cham (2020). https://doi.org/10.1007/978-3-030-63076-8_17

3. Nguyen, D.C., Ding, M., Pathirana, P.N., Seneviratne, A., Li, J., Poor, H.V.: Federated learning for internet of things: a comprehensive survey. IEEE Commun. Surv. Tutorials **23**(3), 1622–1658 (2021)

4. Yang, Q., Liu, Y., Chen, T., Tong, Y.: Federated machine learning: Concept and applications. ACM Trans. Intell. Syst. Technol. 10(2), 12:1–12:19 (2019). https://doi.org/10.1145/3298981

5. Zhu, L., Liu, Z., Han, S.: Deep leakage from gradients. In: Wallach, H.M., Larochelle, H., Beygelzimer, A., d'Alché-Buc, F., Fox, E.B., Garnett, R. (eds.) Advances in Neural Information Processing Systems 32: Annual Conference on Neural Information Processing Systems 2019, NeurIPS 2019, December 8–14, 2019, Vancouver, BC, Canada, pp. 14747–14756 (2019). https://proceedings.neurips.cc/paper/2019/hash/60a6c4002cc7b29142def8871531281a-Abstract.html

6. Oded, G.: Secure multi-party computation. manuscript. preliminary version 78(110) (1998)

7. Gentry, C.: Fully homomorphic encryption using ideal lattices. In: Mitzenmacher, M. (ed.) Proceedings of the 41st Annual ACM Symposium on Theory of Computing, STOC 2009, Bethesda, MD, USA, May 31 - June 2, 2009, pp. 169–178. ACM (2009). https://doi.org/10.1145/1536414.1536440

8. Dwork, C., Roth, A.: The algorithmic foundations of differential privacy. Found. Trends Theor. Comput. Sci. **9**(3–4), 211–407 (2014)

9. Stevens, T., Skalka, C., Vincent, C., Ring, J., Clark, S., Near, J.P.: Efficient differentially private secure aggregation for federated learning via hardness of learning with errors. In: Butler, K.R.B., Thomas, K. (eds.) 31st USENIX Security Symposium, USENIX Security 2022, Boston, MA, USA, August 10–12, 2022, pp. 1379–1395. USENIX Association (2022), https://www.usenix.org/conference/usenixsecurity22/presentation/stevens

10. Gehlhar, T., Marx, F., Schneider, T., Suresh, A., Wehrle, T., Yalame, H.: SAFEFL: MPC-friendly framework for private and robust federated learning. In: 2023 IEEE Security and Privacy Workshops (SPW), San Francisco, CA, USA, May 25, 2023, pp. 69–76. IEEE (2023). https://doi.org/10.1109/SPW59333.2023.00012

11. Jaehyoung, P., Hyuk, L.: Privacy-preserving federated learning using homomorphic encryption. Appl. Sci. 12(2) (2022). https://doi.org/10.3390/app12020734,https://www.mdpi.com/2076-3417/12/2/734

12. Yang, T., et al.: Applied federated learning: Improving google keyboard query suggestions. CoRR abs/1812.02903 (2018). https://arxiv.org/abs/1812.02903

13. Bonawitz, K.A., et al.: Practical secure aggregation for privacy-preserving machine learning. In: Thuraisingham, B., Evans, D., Malkin, T., Xu, D. (eds.) Proceedings of the 2017 ACM SIGSAC Conference on Computer and Communications Security, CCS 2017, Dallas, TX, USA, October 30 - November 03, 2017, pp. 1175–1191. ACM (2017). https://doi.org/10.1145/3133956.3133982

14. Kalikinkar, M., Guang, G.: PrivFL: Practical privacy-preserving federated regressions on high-dimensional data over mobile networks. In: Proceedings of the 2019 ACM SIGSAC Conference on Cloud Computing Security Workshop, pp. 57–68 (2019)

15. Guo, Jiale, Liu, Ziyao, Lam, Kwok-Yan., Zhao, Jun, Chen, Yiqiang: Privacy-Enhanced Federated Learning with Weighted Aggregation. In: Lin, Limei, Liu, Yuhong, Lee, Chia-Wei. (eds.) SocialSec 2021. CCIS, vol. 1495, pp. 93–109. Springer, Singapore (2021). https://doi.org/10.1007/978-981-16-7913-1_7

16. Goldreich, O., Goldwasser, S., Micali, S.: How to construct random functions. In: Goldreich, O. (ed.) providing sound foundations for cryptography: On the Work of Shafi Goldwasser and Silvio Micali, pp. 241–264. ACM (2019). https://doi.org/10.1145/3335741.3335752

17. Liu, Z., Guo, J., Lam, K., Zhao, J.: Efficient dropout-resilient aggregation for privacy-preserving machine learning. IEEE Trans. Inf. Forensics Secur. **18**, 1839–1854 (2023)

18. Yang, S., Chen, Y., Tu, S., Yang, Z.: A post-quantum secure aggregation for federated learning. In: Proceedings of the 12th International Conference on Communication and Network Security, ICCNS 2022, Beijing, China, December 1–3, 2022, pp. 117–124. ACM (2022). https://doi.org/10.1145/3586102.3586120

19. Elaine, S., T-H. Hubert, C., Eleanor, R., Richard, C., Dawn, S.: Privacy-preserving aggregation of time-series data. ACM Trans. Sen. Netw **5**(3), 1–36 (2009)

20. Brakerski, Z., Gentry, C., Vaikuntanathan, V.: Fully homomorphic encryption without bootstrapping. Electron. Colloquium Comput. Complex. TR11-111 (2011). https://eccc.weizmann.ac.il/report/2011/111

21. Brakerski, Zvika: Fully Homomorphic Encryption without Modulus Switching from Classical GapSVP. In: Safavi-Naini, Reihaneh, Canetti, Ran (eds.) CRYPTO 2012. LNCS, vol. 7417, pp. 868–886. Springer, Heidelberg (2012). https://doi.org/10.1007/978-3-642-32009-5_50

22. Cheon, Jung Hee, Han, Kyoohyung, Kim, Andrey, Kim, Miran, Song, Yongsoo: Bootstrapping for Approximate Homomorphic Encryption. In: Nielsen, Jesper Buus, Rijmen, Vincent (eds.) EUROCRYPT 2018. LNCS, vol. 10820, pp. 360–384. Springer, Cham (2018). https://doi.org/10.1007/978-3-319-78381-9_14

23. Shamir, A.: How to share a secret. Commun. ACM **22**(11), 612–613 (1979)

24. So, J., Güler, B., Avestimehr, A.S.: Turbo-aggregate: Breaking the quadratic aggregation barrier in secure federated learning. IEEE J. Sel. Areas Inf. Theory **2**(1), 479–489 (2021)

25. Bell, J.H., Bonawitz, K.A., Gascón, A., Lepoint, T., Raykova, M.: Secure single-server aggregation with (poly)logarithmic overhead. In: Ligatti, J., Ou, X., Katz, J., Vigna, G. (eds.) CCS '20: 2020 ACM SIGSAC Conference on Computer and Communications Security, Virtual Event, USA, November 9–13, 2020, pp. 1253–1269. ACM (2020). https://doi.org/10.1145/3372297.3417885

26. Lyubashevsky, Vadim, Peikert, Chris, Regev, Oded: On Ideal Lattices and Learning with Errors over Rings. In: Gilbert, Henri (ed.) EUROCRYPT 2010. LNCS, vol. 6110, pp. 1–23. Springer, Heidelberg (2010). https://doi.org/10.1007/978-3-642-13190-5_1

27. Banerjee, Abhishek, Fuchsbauer, Georg, Peikert, Chris, Pietrzak, Krzysztof, Stevens, Sophie: Key-Homomorphic Constrained Pseudorandom Functions. In: Dodis, Yevgeniy, Nielsen, Jesper Buus (eds.) TCC 2015. LNCS, vol. 9015, pp. 31–60. Springer, Heidelberg (2015). https://doi.org/10.1007/978-3-662-46497-7_2

28. Tian, H., Wen, Y., Zhang, F., Shao, Y., Li, B.: A distributed threshold additive homomorphic encryption for federated learning with dropout resiliency based on lattice. In: Chen, X., Shen, J., Susilo, W. (eds.) Cyberspace Safety and Security - 14th International Symposium, CSS 2022, Xi'an, China, October 16–18, 2022, Proceedings. Lecture Notes in Computer Science, vol. 13547, pp. 277–292. Springer (2022). https://doi.org/10.1007/978-3-031-18067-5_20

29. von zur Gathen, J., Gerhard, J.: Modern Computer Algebra (3. ed.). Cambridge University Press (2013)

30. Shokri, R., Stronati, M., Song, C., Shmatikov, V.: Membership inference attacks against machine learning models. In: 2017 IEEE Symp. on Secur. and Priv., SP 2017, San Jose, CA, USA, May 22–26, 2017. pp. 3–18. IEEE Comput. Soc. (2017). https://doi.org/10.1109/SP.2017.41
31. Froelicher, D., et al.: Scalable privacy-preserving distributed learning. Proc. Priv. Enhancing Technol. **2021**(2), 323–347 (2021)
32. Bagdasaryan, E., Veit, A., Hua, Y., Estrin, D., Shmatikov, V.: How to backdoor federated learning. In: Chiappa, S., Calandra, R. (eds.) The 23rd International Conference on Artificial Intelligence and Statistics, AISTATS 2020, 26–28 August 2020, Online [Palermo, Sicily, Italy]. Proceedings of Machine Learning Research, vol. 108, pp. 2938–2948. PMLR (2020). https://proceedings.mlr.press/v108/bagdasaryan20a.html
33. Tian, H., Li, M., Ren, S.: ESE: Efficient security enhancement method for the secure aggregation protocol in federated learning. Chinese J. Electron. 32(3), 542–555 (2023). 10.23919/CJE.2021.00.370

Practical and Privacy-Preserving Decision Tree Evaluation with One Round Communication

Liang Xue[1]([✉])[iD], Xiaodong Lin[1][iD], and Pulei Xiong[2][iD]

[1] University of Guelph, Guelph N1G 2W1, Canada
{xliang07,xlin08}@uoguelph.ca
[2] National Research Council of Canada, Mississauga L5K, Canada
Pulei.Xiong@nrc-cnrc.gc.ca

Abstract. Machine learning enables organizations and individuals to improve efficiency and productivity. With an abundance of data and computational resources, large companies can build complex machine learning models and provide prediction services to clients. One example is decision tree evaluation, where a client can access the trained decision tree model with its input and obtain the classification result. However, the privacy issues on model parameters and clients' inputs and results need to be addressed. In this paper, we propose a privacy-preserving decision tree evaluation scheme, where we first design an improved interval encoding method that can hide parameters representing an interval. Then, based on the interval encoding method, hash functions, and the Diffie-Hellman key agreement technique, a model owner can generate a set of encodings for the decision tree model and send them to a client, who can determine the classification result based on its input and the encodings. The proposed scheme conceals the model parameters from clients and preserves the data privacy of clients, and only one round of communication between the two entities is needed. We provide a formal security proof that demonstrates the privacy preservation property of our scheme. Performance evaluation shows the practicability of the proposed scheme.

Keywords: Machine learning · Decision tree evaluation · Privacy preservation

1 Introduction

With the high demand for machine learning in many industries such as finance [1], manufacturing [2], and healthcare [3], machine learning as a service (MLaaS) has spouted in recent years. It can provide out-of-the-box predictive analysis for various use cases and offer tools that can accelerate the model building and deployment. Small companies and individuals can use machine learning services for their business or specific needs without investing in any specialized infrastructure. A popular predictive service for MLaaS is decision tree evaluation, which

J. Vaidya et al. (Eds.): AIS&P 2023, LNCS 14509, pp. 400–414, 2024.
https://doi.org/10.1007/978-981-99-9785-5_28

can be used for the diagnosis of diseases, detection of fraud, and investment analysis. A decision tree is a machine learning algorithm that can be utilized for classification and regression tasks. A decision tree model uses a tree structure to make decisions based on input data. For the decision tree evaluation service, high-tech companies can collect extensive training datasets and leverage their strong computational and storage resources to train a large decision tree model with high accuracy. Then, they can provide decision tree evaluation services for clients, who have their inputs and wish to obtain the classification results for the inputs, and gain benefits from clients.

Using the decision tree evaluation service, clients can obtain high-quality classification results from a model owner at a low cost. However, there are some privacy issues during the service. A client's input may contain its sensitive information such as health conditions and financial status, thus, it is reluctant to send its private input to the model owner to obtain the classification result. Meanwhile, the result can also contain private information of the client that it prefers not to disclose to the model owner. On the other hand, the owner cannot just send the model parameters to clients due to the following reasons: First, the model is acquired after significant resource consumption and belongs to the owner's intellectual property; Second, through the model's parameters, one could potentially deduce sensitive information of the training dataset by launching model inversion attacks [4]; Third, once the model is leaked or public, the owner cannot consistently gain benefit from the model. Therefore, it is desired to attain the classification result without leaking a client's input and output to the model owner and the owner does not need to share the original model parameters with clients.

In recent years, there has been a few research work on privacy-preserving decision tree evaluation. Most of the work utilizes homomorphic encryption and the garbled circuit [5–7]. The basic idea is that considering that there are many comparisons in a decision tree, the authors first design a secure integer-comparison protocol based on homomorphic encryption, and with the comparison protocol and public key encryption, a client can interact with the model owner to obliviously obtain the classification result for its input without knowing the model parameters. For methods using homomorphic encryption, there are usually multiple rounds of communication between the model owner and a client to ensure that the client can decrypt the result on the correct path. Different from the schemes that adopt public-key cryptographic techniques, Banerjee et al. [8] proposed a privacy-preserving decision tree evaluation scheme, where the authors first design an encoder that can hide parameters in an interval membership function, and then they utilize the encoder to construct an obfuscator for a decision tree. The model owner publishes the obfuscator to enable privacy-preserving decision tree evaluation by clients. However, there is a false positive problem in their scheme. That is, in their encoding scheme for an interval, a value that is not in the interval can be judged to be in this interval, resulting in the accuracy problem for the decision tree evaluation. Moreover, once the obfuscated decision tree model is public, the model owner cannot gain benefit from each client that accesses the model.

In this paper, we improve the scheme [8] by solving the accuracy problem in it, and a model owner does not publish its model parameters. Instead, for each client, it can generate a set of encodings for the decision tree model and gain benefit from the client. An eavesdropping attacker cannot utilize the decision tree evaluation service. To be specific, we consider a full binary decision tree, i.e., the classification label can be 1 or 0, and there are two phases in our decision tree evaluation scheme. In the pre-processing phase, based on our modified encoding method for an interval, a model owner can generate pre-encodings for each leaf node with label 1. In the evaluation phase, a client sends its public key to the model owner, who generates a secret key based on the Diffie-Hellman key agreement protocol [9]. Then, the owner generates encodings for the decision tree based on the results in the first phase and the secret key, and sends the encodings of the tree to the client. The client also generates the secret key and generates encodings for its input based on the key and our improved encoding scheme. Then, the client can check whether there is any intersection between the two encoding sets to attain the final classification result. The main contribution of this work can be summarized as follows:

- We propose a privacy-preserving decision tree evaluation scheme where there is only one round of communication between a model owner and a client, and no complicated cryptographic operations such as homomorphic encryption is involved.
- We provide a formal proof for the proposed scheme in the random oracle mode. We demonstrate that an adversary cannot distinguish whether it interacts with a model owner or with a simulator. Thus, model privacy is preserved. Moreover, the model owner has no knowledge of clients' inputs and outputs.
- We analyze the computational and communication overhead for our scheme, and compare its performance with the scheme in [8] and PDTC [10]. The results show that our scheme is more practical.

The rest of the paper is organized as follows: We review the related work in Sect. 2, and we define the system model, the threat model, and design goals in Sect. 3. In Sect. 4, we introduce the necessary preliminaries for designing our scheme. In Sect. 5, we present the overview of our scheme and the detailed construction. In Sect. 6, we provide a formal security proof for the proposed scheme, followed by the performance evaluation in Sect. 7. We give conclusions in Sect. 8.

2   Related Work

In this section, we briefly review the existing literature pertaining to privacy-preserving decision tree evaluation.

Bost et al. [5] first proposed three major classification schemes for hyperplane decision, Naive Bayes, and decision trees respectively. For a decision tree, they assign a boolean variable for each node of the tree and construct a polynomial for the tree. The polynomial is a sum of terms, where each is a multiplication of the decision value on a leaf node and boolean values related to the passing nodes that can lead to the leaf node. Moreover, fully homomorphic encryption (FHE) is involved to preserve the privacy of clients. To avoid expressing a decision tree as a high-degree polynomial, Tai et al. [6] proposed a privacy-preserving decision tree evaluation scheme where each path of a decision tree is represented as a linear function. They exploit the structure of decision trees and put forward a concept called path cost. The idea is as follows: Let $b \in \{0,1\}$ be the comparison result as an internal node. The cost of the left (right) edge of the node is set to be b $(1-b)$ such that the sum of the edge cost along the correct path for an input should be 0. The decision value of each leaf node is added to its randomized path cost and the results are sent to a client. Wu et al. [7] first constructed a decision tree evaluation protocol under a semi-honest model, and then they extended the protocol to provide robustness such that it can defend against malicious attackers. In their scheme, a model owner first permutes the decision tree, and then the owner and a client engage in a secure comparison protocol such that the client learns each comparison result in the permuted tree. In the end, the client can obtain the classification result by executing an oblivious transfer protocol with the owner. Liu et al. [11] proposed a privacy-preserving and practical decision tree training and evaluation scheme in a twin cloud architecture. They improve a secure comparison scheme proposed by Damgard et al. [12], and express a decision tree as a polynomial. A classification result can be derived by calculating the polynomial. Xue. et al. [10] proposed a secure and privacy-preserving scheme based on additively homomorphic encryption and the secret sharing technique. In their secure two-party comparison protocol, the inputs of a client and a model owner can be compared efficiently instead of in a bit-by-bit manner. Banerjee et al. [8] proposed a non-interactive and privacy-preserving prediction scheme for decision trees. They first designed a new technique for encoding interval membership functions, and applied the technique to the privacy-preserving evasive decision tree evaluation. Hao et al. [13] proposed a privacy-preserving decision tree evaluation protocol based on a lattice-based fully homomorphic encryption scheme. The designed integer-comparison algorithm can resist quantum attacks.

3   Models and Goals

In this section, we first define our system model and the threat model, then we give the design goals of this work.

3.1   System Model

As shown in Fig. 1, in our system, there are two types of entities: a decision tree model owner (MO) and clients. MO collects training data and trains a decision tree model for a specific application, for example, diagnosis of diseases. MO can provide the decision tree evaluation service to different clients. Clients can use the service provided by MO and pay to MO. However, MO is reluctant to share its model with clients, and clients would like to preserve their data privacy for both inputs and returned results. Thus, it is desired that at the end

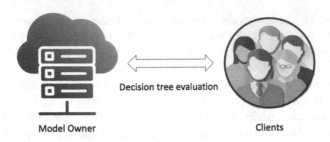

Fig. 1. System model

of the service, a client can learn the classification result without leaking both the input and the result. Considering that clients may have limited computation and communication resources, the overhead for clients should be small.

3.2 Threat Model

We assume both MO and clients are honest but curious, which means both parties will correctly execute the protocol, but try to learn as much information as possible during the protocol execution. The server wants to protect the decision tree model, which includes the comparison function at each internal node, the structure of the model, and the decision label at each leaf node. The data that a client wants to protect are its input and the corresponding model output.

To avoid the model extraction attacks, we consider the evasive decision tree model [8]. A decision tree model C is evasive if for every $\lambda \in \mathbb{N}$ and an input $x = (x_1, \ldots, , x_n)$, $Pr[C(x) = 1] \leq \rho(\lambda)$, where ρ is a negligible function for λ. For the evasive decision tree models, an adversary cannot extract the model except with a negligible probability. In our system model, adversaries can be clients, MO, or an external attacker, who can eavesdrop on the communication channel.

3.3 Design Goals

There are two design goals for privacy-preserving decision tree evaluation, which are security and efficiency.

- Security: A secure decision tree evaluation scheme should ensure the confidentiality of decision tree model parameters and the privacy of clients, i.e., the input vectors and model outputs are concealed from MO. Meanwhile, clients can obtain accurate classification results.
- Efficiency: Clients may be resource-constrained and MO also wants to reduce the overall protocol run time. Thus, the interaction between MO and clients as well as the computational cost of them should be small.

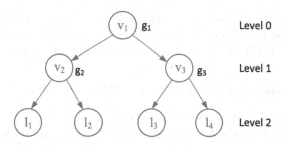

Fig. 2. A decision tree

4   Preliminaries

4.1   Decision Tree

A decision tree is a popular machine learning algorithm and is usually used for classification tasks. Given an input and a decision tree model, one can get the corresponding classification result for the input after the decision tree evaluation. A decision tree has an inverted tree structure, where each internal node is related to a comparison function. For simplification, we consider a full binary tree T of depth d as shown in Fig. 2. The root node is at level 0, and leaf nodes are at level d. There are 2^d leaf nodes $(l_1, \ldots, l_{2^d})$ and $2^d - 1$ internal nodes $(v_1, \ldots, v_{2^d-1})$, where v_1 is the root node. Let $\{n, l\} \in N$ and $[n]$ denote $\{1, \ldots, n\}$. An input is of the form $X = (x_1, \ldots, x_n)$, where $x_i \in [0, 2^l)$. Each internal node v_j is related with an index i of x_i $(i \in [n])$ and an integer t_j, where $t_j \in [0, 2^l)$ and is an integer. For each internal node v_j, where $j \in [2^d - 1]$, comparison function g_j outputs 1 if $x_i \leq t_j$, and the left branch of v_j is taken. Otherwise, it outputs 0, and the right branch of v_j is taken. Each leaf node $\{l_k\}_{k \in [2^d]}$ is related to a specific class or a decision label, which is denoted as z_k.

For the evaluation of an input X, the decision tree algorithm traverses the decision tree from the root node and follows the appropriate path based on the comparison results at the internal nodes. The classification result of X is the decision label of the reached leaf node. In this work, we only focus on the prediction phase of a decision tree model and how the model is trained is out of the scope of this paper.

4.2   Obfuscator

An obfuscator is a probabilistic polynomial-time algorithm that makes a program difficult to understand or reverse engineer while preserving its functionality. Barak et al. [14] provide the definition of an obfuscator $\mathcal{O}$, which is an efficient "compiler" that takes as input a program P and outputs a new program $\mathcal{O}(P)$ that satisfies the subsequent two properties.

– Functionality: $\mathcal{O}(P)$ performs the identical function as P.

– "Virtual black box" property: Anything that can be efficiently derived from $\mathcal{O}(P)$ can also be efficiently derived when provided with oracle access to P, i.e., one can only have input-output access to the program.

4.3   Computational Diffie-Hellman (CDH) Problem

For a multiplicative cyclic group G of order q and g is a generator of the group. Given (g, g^a, g^b) for some $a, b \in Z_q$, a probabilistic polynomial-time (PPT) adversary $\mathcal{A}$ is required to output g^{ab}. The success probability of $\mathcal{A}$ in solving the CDH problem is defined to be:

$$Suc_{A,G}^{CDH} = Pr[\mathcal{A}(g, g^a, g^b, g^{ab}) = 1 : a, b \in_R Z_q]$$

The CDH assumption [15] is that for every PPT $\mathcal{A}$, $Suc_{A,G}^{CDH}$ is negligible.

5   Our Constructions

5.1   Overview

Let the input of a client ($\mathcal{C}$) be a vector $X = \{x_1, \ldots, x_n\}$, where $x_i \in [0, 2^l)$. At an internal node v_j, an $\{x_i\}_{i \in [n]}$ is compared with t_j, where $t_j \in [0, 2^l)$ and is related to v_j. If $x_i \le t_j$, the left branch of v_j would be taken. Otherwise, the right branch would be taken. We assume that each x_i is compared at most twice during the decision tree evaluation. Thus, after going through a path, each leaf node corresponds to n intervals $\{I_i\}_{i \in [n]}$ for each $\{x_i\}_{i \in [n]}$, and the intersection of these intervals represents a leaf node, which has a decision label. Without loss of generality, we consider a full binary decision tree, i.e., the decision label can be 0 or 1. To avoid the model extraction attacks, we focus on the evasive decision tree model [8].

There are two phases in our scheme: the preprocessing phase and the evaluation phase. To reduce the interactions between MO and $\mathcal{C}$, in the preprocessing phase, for each leaf node with label 1, MO first encodes each interval $\{I_i\}_{i \in [n]}$, and then it encodes the intersection of the intervals. In the second phase, $\mathcal{C}$ requests the decision tree evaluation service with an input X, and $\mathcal{C}$ sends its public key pk_C to MO. Based on pk_C, MO generates the encodings of leaf nodes with label 1, which we call the encodings (E_M) of the decision tree model, and sends them to $\mathcal{C}$. $\mathcal{C}$ also generates the encodings (E_C) of its input. Then $\mathcal{C}$ can check whether there is intersection between E_M and E_C. If an intersection exists, the evaluation result for the input is 1. Otherwise, the result is 0.

5.2   Detailed Scheme

Notation: Let λ be a security parameter. A function $\rho : R \to R$ is negligible in λ if for any $b > 0$, we have $|\rho(\lambda)| < 1/\lambda^b$. In our scheme, a client's input is $X = (x_1, \ldots, x_n)$ where $x_i \in [0, 2^l)$ and are integers, and n, l are polynomials in λ. The depth of the decision tree is d. The internal nodes and the leaf nodes are represented as $V = \{v_1, \ldots, v_{2^d-1}\}$ and $S = \{s_1, \ldots, s_{2^d}\}$. Each internal node is related with a comparison function g_j, whose inputs are $t_j \in [0, 2^l)$ and an $\{x_i\}_{i \in [n]}$, and the output is $b_j \in \{0, 1\}$. $|\mathcal{X}|$ denotes the number of elements in a set $\mathcal{X}$. Table 1 lists some important parameters in our scheme.

Table 1. System parameters

Acronym	Definition
X	Input vector; $X = (x_1, \ldots, x_n)$
I_i	Interval of x_i for a leaf node; $i \in [n]$
u, u'	Hamming weights
H	Hash function; $H : \{0,1\}^* \to \{0,1\}^w$
E_I	A set of encodings of an interval I

In the following, we first introduce our improved interval encoding method, which is a building block for our privacy-preserving decision tree evaluation scheme. Then, we present our detailed scheme, which includes the pre-processing phase and the evaluation phase.

Interval Encoding. In the decision tree evaluation with an input $X = (x_1, \ldots, x_n)$, the comparisons on an x_i along each path such as $x_i \leq t_j$ and $x_i > t_{j'}$ form an interval I_i. Following the idea of [8], an interval can be encoded as some sub-intervals of the form $[a, a + 2^\zeta)$, where $a \in [0, 2^l)$ and $\zeta \in \{0, \ldots, l\}$. To be specific, a comparison $x \leq t$, where $x, t \in [0, 2^l)$, can divide the interval $[0, 2^l)$ into two parts: $\mathcal{X} \in [0, t+1)$ and $\mathcal{X}' \in [t+1, 2^l)$.

For the interval $\mathcal{X} = [0, t+1)$, we can split it into several sub-intervals based on the binary representation of $|\mathcal{X}|$. Specifically, we first calculate the hamming weight of $|\mathcal{X}|$, which is the number of 1 in the binary representation of $|\mathcal{X}|$, and we denote it as $u = hw(|\mathcal{X}|)$. Then, we record the bit positions of 1 in the binary representation of $|\mathcal{X}|$, which are $p_1, \ldots, p_u$, where $p_1 > \cdots > p_u$, and the bit positions can be $\{0, \ldots, l\}$. Interval $\mathcal{X}$ can be divided into u disjoint sub-intervals $[a_j, a_j + 2^{p_j})$ where $j \in [u]$ and $a_1 = 0$. Take $\mathcal{X} = [0, 38)$ and $l = 6$ as an example, where $|\mathcal{X}| = 38$ and the binary representation of $|\mathcal{X}|$ is 100110. $p_1 = 5$, $p_2 = 2$, and $p_3 = 1$. Thus, $a_1 = 0$, $a_2 = a_1 + 2^5 = 32$, $a_3 = a_2 + 2^2 = 36$, and $a_4 = a_3 + 2^1 = 38$. The sub-intervals for $\mathcal{X} = [0, 38)$ are $\mathcal{I}_\mathcal{X} = \{[0, 32), [32, 36), [36, 38)\}$.

For the interval $\mathcal{X}' = [t' + 1, 2^l)$, we can also divide the interval into several intervals based on the binary representation of $|\mathcal{X}'|$. Let u' be the hamming weight of $|\mathcal{X}'|$ and $p_1, \ldots, p_{u'}$ denote the bit positions of 1 in the binary encoding of $|\mathcal{X}'|$, where $p_1 < \cdots < p_{u'}$. Then, the interval $\mathcal{X}'$ can be partitioned into u' sub-intervals $[a_j, a_j + 2^{p_j})$, where $j \in [u']$ and $a_1 = t' + 1$. Take $\mathcal{X}' = [24, 64)$ as an example. For interval $\mathcal{X}'$, $|\mathcal{X}'| = 40$, and its binary representation is 101000.

$p_1 = 3$ and $p_2 = 5$. Thus, $a_1 = 24$, $a_2 = a_1 + 2^3 = 32$, and $a_3 = a_2 + 2^5 = 64$. The sub-intervals for $\mathcal{X}'$ are $\mathcal{I}_{\mathcal{X}'} = \{[24, 32), [32, 64)\}$.

Based on $\mathcal{I}_{\mathcal{X}}$ and $\mathcal{I}_{\mathcal{X}'}$, one can obtain the intersection (I) of the two intervals $\mathcal{X}$ and $\mathcal{X}'$, which is also a set of sub-intervals of the form $[a_j, a_j + 2^t)$ for some $t \in \{0, \ldots, l\}$ and $a_j \in [0, 2^l)$. Let $k = |I|$, which is the number of the sub-intervals in I. To encode these sub-intervals, we first define a family of functions $F = \{f_0, \ldots, f_{l-1}\}$, where $f_i(y) : \{0, 1\}^l \to \{0, 1\}^{l-i}$ is defined as $f_i(y) = \lfloor \frac{y}{2^i} \rfloor$. Let $H : \{0, 1\}^* \to \{0, 1\}^w$ be a hash function such that $l < w$ and w is polynomial in λ. Moreover, H is injective on all the strings of length less or equal to $2l$. With F, H, and I, we can encode the interaction I as several hash values as below:

- I can be represented as a set of $\{[a_j, a_j + 2^t)\}_{j \in [k]}$. For each $j \in [k]$, compute $\eta_j = f_t(a_j)$.
- Calculate $h_j = H(2^t || \eta_j)$, where $2^t \in \{0, 1\}^l$. If the length of 2^t is less than l, prepend some 0s to the string.
- The encodings of I are $\{h_1, \ldots h_k\}$, which we denote as a set E_I.

To check whether an element $x \in [0, 2^l)$ belongs to one of the sub-intervals of the intersection I, one can first calculate $f_i(x)$ and $H(2^i || f_i(x))$, where $i \in \{0, \ldots, l - 1\}$, and check whether $H(2^i || f_i(x)) \in E_I$.

We can see that if x is in I, there exists an $i \in \{0, \ldots, l - 1\}$ and a $j \in [k]$ such that $f_i(x) = \eta_j$ and $h_j = H(2^t || \eta_j) = H(2^i || f_i(x))$. With the input being the concatenation of 2^t and η_j, the output h_j fixes an interval of I. For x that does not belong to I, there is no intersection between $H(2^i || f_i(x))_{i \in \{0, \ldots, l-1\}}$ and the encodings of I.

Pre-processing Phase. In this phase, MO generates pre-encodings for its decition tree model. For decision tree evaluation, since each $\{x_i\}_{i \in [n]}$ is compared at most twice in a path, $x_i > c_i$ and $x_i \leq (c_i + w_i)$ generates an interval $x_i \in (c_i, c_i + w_i]$, where $c_i, c_i + w_i \in \{t_j\}_{j \in [2^d - 1]}$. Thus, each leaf node can be represented by the intersection of n intervals, as the result of a conjunction of comparisons.

MO encodes each leaf node with label 1 as follows:

1. For $i \in [n]$, generate the encodings for the corresponding interval of x_i. That is, MO first generates a set of sub-intervals ($\mathcal{I}_X^i$) for the interval $[0, c_i + w_i + 1)$, and then generates a set of the sub-intervals ($\mathcal{I}_{X'}^i$) for $[c_i + 1, 2^l)$. After that, MO determines the intersections of these two sets, which is denoted as I^i. Then, MO uses functions F and H to obtain the encodings of I^i, which is denoted as a set $E_{I_i} = \{h_1^i, \ldots, h_{\xi_i}^i\}$ and ξ_i is $|E_{I_i}|$, where $\xi_i \leq l$.
2. For $i \in [n]$ and $j \in [\xi_i]$, generate strings with the form $||_{i=1}^n h_j^i$. That is, for all $i \in [n]$, MO chooses an element in E_{I_i} and concatenates n entries in different E_{I_i} in order. The resulting strings are the pre-encodings of the leaf nodes with the decision label 1, which is denoted as a set A. Denote $\gamma = |A|$.

In the pre-processing phase, MO also generates its private-public key pair. Let G be a multiplicative group of integers modulo p, where p is a prime and g is a primitive root modulo p.

MO randomly chooses an $sk_M \in Z_p^*$, and computes $pk_M = g^{sk_M} \bmod p$. The private-public key pair for MO is (sk_M, pk_M). Let $H_d : \{0,1\}^* \to \{0,1\}^\kappa$ be a hash function, where κ is polynomial in λ. MO publishes public parameters $pp = \{G, p, g, pk_M, l, n, F, H_d, \gamma\}$.

Evaluation Phase. In this phase, a client $\mathcal{C}$ requests the decision tree model evaluation service from MO and obtains the classification result for its input.

$\mathcal{C}$ first creates its private key and public key as follows: it chooses a random $sk_C \in Z_p^*$, and computes $pk_C = g^{sk_C} \bmod p$. The private-public key of $\mathcal{C}$ is (sk_C, pk_C). Then, $\mathcal{C}$ with input $Y = \{y_1, \ldots, y_n\}$ generates the encodings for its input as follows:

1. For each $\{y_i\}_{i \in [n]}$, $\mathcal{C}$ computes $E_{y_i} = \{H(2^0 || f_0(y_i)), \ldots, H(2^{l-1} || f_{l-1}(y_i))\}$, which are denoted as $E_{y_i} = \{h_1^i, \ldots, h_l^i\}$.
2. $\mathcal{C}$ calculates $s = pk_M^{sk_C} \bmod p$, which can be seen as a secret shared by MO and $\mathcal{C}$.
3. For each $i \in [n]$ and $j \in [l]$, $\mathcal{C}$ computes $H_d((||_{i=1}^n h_j^i) || s)$, and for each i, the input can only contain one element in E_{y_i}. That is, for each $i \in [n]$, $\mathcal{C}$ chooses one of the hashes in E_{y_i}, concatenate them together in order of i, and add s at the end of the string. Then, $\mathcal{C}$ put the string into hash function H_d. The resulting hash values, which we denote as a set B, are the encodings of $\mathcal{C}'s$ input Y.

To obtain the evaluation result for input Y, $\mathcal{C}$ initiates a connection to MO, and sends pk_C to MO. MO then computes $s' = pk_C^{sk_M} \bmod p$. After that, for each string $\bar{s}_i \in A$, where $i \in [\gamma]$, MO calculates $\bar{h}_i' = H_d(\bar{s}_i || s')$. The set of all $\bar{h}_i'$ is denoted as a set A'. MO sends set A' to $\mathcal{C}$.

We can see that if Y belongs to the class with label 1, there is a hash value α in B such that $\alpha \in A'$, as Y would reach one of the leaf nodes with label 1 in the decision tree. Thus, the client can check whether there is an intersection between set A' and set B. If that is the case, the classification result for input Y is 1. Otherwise, the result is 0.

6 Security Proof

In this section, we prove that the proposed privacy-preserving decision tree evaluation scheme is secure in the random-oracle model.

In our scheme, client $\mathcal{C}$ does not need to send its input to MO. MO sends a set of encodings of its decision tree model to $\mathcal{C}$, who compares the elements in sets A' and B to obtain the model output. Thus, $\mathcal{C}'s$ inputs and outputs are not leaked to MO. We will prove that set A' does not leak the parameters of the model, which include the comparison functions at each internal node and the labels at each leaf node. Set A' can be seen as an obfuscator of the decision tree model. We follow the definition of the security model in [8], where an obfuscator scheme is secure if it achieves the property of virtual block-box, i.e., anything that can be efficiently computed by $\mathcal{O}(C)$ can also be efficiently computed only given oracle access to C. The formal definition of the security model is as follows:

Definition 1. *Let λ be a security parameter. Denote $C = \{C_\lambda\}$ be a family of polynomial-size programs with input of length $n(\lambda)$. Let $D = \{D_\lambda\}$ be a set of distribution ensembles, in which D_λ is a distribution over C_λ. Virtual blackbox obfuscation for the family C and the distribution D satisfies that: For a probabilistic polynomial-time (PPT) adversary $\mathcal{A}$, there exists a polynomial size simulator S with oracle access to C such that*

$$|Pr_{C \leftarrow C_\lambda}[\mathcal{A}(\mathcal{O}(C)) = 1] - Pr_{C \leftarrow C_\lambda}[S^C(1^\lambda) = 1]| \leq \epsilon(\lambda)$$

where $\epsilon(\lambda)$ is a negligible function.

Next, we prove the security of our scheme based on the security model.

Theorem 1. *The proposed scheme securely executes the decision tree evaluation under random oracle model, provided that the two hash functions $H : \{0,1\}^* \rightarrow \{0,1\}^w$ and $H_d : \{0,1\}^* \rightarrow \{0,1\}^\kappa$ are collision-resistant [16] and the CDH problem is hard.*

Informally, we prove that the advantage of an attacker, who can access the obfuscated decision tree model ($\mathcal{O}(M)$), in obtaining the parameters of the model is not larger than a simulator which can access an oracle of the decision tree model, which can be proved by construct a simulator such that the adversary cannot distinguish whether it is interacting with $\mathcal{O}(M)$ or the simulator. The detailed proof is as follows.

Proof. We model the hash function $H : \{0,1\}^* \rightarrow \{0,1\}^w$ and $H_d : \{0,1\}^* \rightarrow \{0,1\}^\kappa$ as random oracles. Let C_λ be a family of full binary decision trees. For every $C \in C_\lambda$, each leaf node with label 1 is parameterized by $(c_1, \ldots, c_n)$ and $(w_1, \ldots, w_n)$, where for $i \in [n]$, $c_i, w_i \in [0, 2^l)$. C evaluates whether an input $x = (x_1, \ldots, x_n)$ reaches one of the leaf nodes with label 1.

Let (pk_S, sk_S) and $(pk_\mathcal{A}, sk_\mathcal{A})$ be public-private key pairs for S and $\mathcal{A}$, respectively. For the encodings of the decision tree model, given γ and $pk_\mathcal{A}$, S randomly chooses γ elements $\{h'_1, \ldots, h'_\gamma\}$ from the co-domain of H_d as the simulated encodings for the model.

A simulator S can simulate the oracles H and H_d as follows: S maintains two tables T_1 and T_2 to record the inputs and outputs for H and H_d. Initially, T_1 and T_2 are empty. For the encodings of $\mathcal{A}'s$ inputs, we assume that $\mathcal{A}$ needs to first access random oracles H and H_d, and $\mathcal{A}$ can make polynomially many queries to the oracles. For oracle H, we denote (u, v) to be a query submitted by $\mathcal{A}$ and the result returned to $\mathcal{A}$ respectively. When oracle H receives a new query u', it first checks whether u' exists in table T_1. If u' is in T_1, S returns the corresponding v'. Otherwise, S randomly choose a value v' from the co-domain of H, adds (u', v') to T_1, and returns v'.

For oracle H_d, in our scheme, $\mathcal{A}$ can access H_d with an input of form $(||_{i=1}^n h_j^i)||s$, where s is a secret key owned by $\mathcal{A}$ and $\mathcal{O}(M)$. When a new H_d query is submitted, S first checks whether the query exists in T_2. If that is the case, S returns the same output to $\mathcal{A}$. Otherwise, an input s^* should be a concatenation of n outputs of oracle H followed by a secret s. S parses the input as two parts. The first part is a sequence of n strings $(s_i)_{i \in [n]}$ and is of length $n\kappa$ bits. The second part should be the binary representation of $pk_{\mathcal{A}}^{sks}$. $\mathcal{S}$ first checks whether the second part (s) is correct based on its private key and $\mathcal{A}'s$ public key. If s is not correct, $\mathcal{S}$ randomly chooses a value val from $\{0,1\}^\kappa$, and adds (s^*, val) to table T_2. Otherwise, for the first part, $\mathcal{S}$ searches table T_1 to check whether all $\{s_i\}_{i \in [n]}$ are in T_1. If not, $\mathcal{S}$ randomly chooses a $val \in \{0,1\}^\kappa$, and assigns val as the output of h_d for s^*. Otherwise, for each s_i which is an output of oracle H, $\mathcal{S}$ finds the input u_i such that (u_i, s_i) is in T_1. Then, based on $\{u_i\}_{i \in [n]}$, which is in the form of $2^{t_i} \| a_i$ and $2^{t_i} \in \{0,1\}^l$, $\mathcal{S}$ recovers x_i as $x_i = 2^t \times a_i$. With the recovered $x^* = (x_1, \ldots, x_n)$, $\mathcal{S}$ submits x^* to an oracle $(O(C))$ of the decision tree model.

If $O(C)$ returns 0, $\mathcal{S}$ randomly chooses a value val from $\{0,1\}^\kappa$, and adds (s^*, val) to table T_2. Otherwise, based on x^* and $O(C)$, $\mathcal{S}$ determines $(c_1, \ldots, c_n)$ and $(w_1, \ldots, w_n)$ using Binary search such that if $x_i \in (c_i, c_i + w_i]$ for all $i \in [n]$, the decision tree output is 1. Based on $(c_i, c_i + w_i]$, $\mathcal{S}$ can calculate the encodings of these intervals, and obtain a set of pairs of (u, v). $\mathcal{S}$ adds the pairs of (u, v) in T_1. If there exists (u', v) in T_1, and $u' \neq u$, $\mathcal{S}$ aborts. Based on the set of inserted v, $\mathcal{S}$ calculates the inputs for H_d, where each input is a concatenation of n hash values followed by the correct s, as described in our scheme. $\mathcal{S}$ also marks these inputs with flag 1. For an input with flag 1, $\mathcal{S}$ assigns a value from the set $\{h_1', \ldots, h_\gamma'\}$, which is chosen initially by $\mathcal{S}$, as its output, and adds this input and its output to table T_2.

We can see that the probability of $\mathcal{S}$ aborts is equal to the conflicts that occur in T_1, which is negligible if H is collision-resistant. Thus, the simulated view provided by $\mathcal{S}$ is identical to the real view, and $\mathcal{A}$ cannot differentiate between them. Therefore, the advantage of $\mathcal{A}$ is no larger than the simulator $\mathcal{S}$ that can only access an oracle of C. Furthermore, since the intractability of the CDH problem, an adversary without the secret s cannot obtain the encodings of the decision tree model.

7 Performance Evaluation

For the performance of our scheme, since MO can calculate pre-encodings for its decision tree model in the pre-processing phase, the computational cost for MO in the evaluation phase is small. To be specific, in the first phase, we omit the computation cost to generate a set of intervals (I_i) for each $x_i \in X$, where $i \in [n]$ and the intervals are fixed for each leaf node with label 1. To generate the encodings (E_{I_i}) for each I_i, at most l hash operations, i.e., H evaluation, are needed. For the pre-encodings of the decision tree model, MO only needs to generate concatenations of n encodings from E_{I_i} in order, where $i \in [n]$. Thus, in the preprocessing phase, the computational overhead for MO is $ln \cdot T(H)$, where $T(H)$ denotes the time cost for a hash evaluation on H.

For the evaluation phase, client C needs to execute an exponentiation operation in group G to obtain the secret s. We denote $T(EXP)$ as the time cost for the exponentiation operation. To compute the encodings (B) of its input $Y = \{y_1, \ldots, y_n\}$, C would execute ln hash operations on H for Y and at most l^n hash operations on H_d. Thus, the computation overhead for C is $ln \cdot T(H)$ $+l^n \cdot T(H_d) +T(EXP)$. For MO, if the number of leaf nodes with label 1 is a, MO needs to execute an exponentiation operation in group G and al^n hash operations on H_d. Let $T(H_d)$ denote the time cost for a hash evaluation on H_d. In this phase, the computation overhead for MO is $al^n \cdot T(H_d) +T(EXP)$. For the communication overhead, there is only one round of communication between MO and C, and the communication overhead is $|p| + \kappa al^n$. Note that since pk_{MO} is public, only C needs to transmit its public key pk_C to MO. When $|p| = 1024$, $\kappa = 256$, and $a = 1$, two example parameters [8] for decision tree models and the communication overhead of our scheme are shown in Table 2. Since we assume that an $\{x_i\}_{i\in[n]}$ can be compared at most twice, d should be less than $2n$.

Table 2. Communication overhead in two examples

d	l	n	λ	Communication cost (bits)
5	64	4	128	$1024 + 64^4$
3	64	2	64	$1024 + 64^2$

To evaluate the computational cost of our scheme, we first test the time cost for the hash operation and the exponentiation operation in group G on a notebook with Intel(R) Core(TM) i5-1135G7 @ 2.40GHz and 16 GB RAM. In the simulation, the Miracl library [17] is invoked. When $|p| = 1024$ and $w = \kappa = 256$, the average time cost for a hash operation and an exponentiation operation is 0.006 ms and 0.56 ms. Compared with the scheme in [8], the overall computational overhead of our scheme is around 1 ms higher. However, there is no false positive problem in our scheme.

Considering that a client may have limited computational resources, we compare the computational cost on the client side with PDTC scheme [10] when the decision nodes in a tree range from 32 to 512, where $a = 1$, $l = 16$, and $n = 5$. The computational results are shown in Fig. 3.

We can see that when n and l are small, the computational cost of the client in our scheme is fixed while for the PDTC scheme, the computational cost of the client increases with the number of decision nodes in the tree. From Fig. 3, when the number of the decision nodes is larger than 350, our scheme has a less computational cost for the client than PDTC [10].

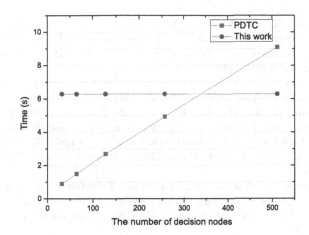

Fig. 3. Computational cost on a client vs. the number of the decision node

8 Conclusions

In this paper, we proposed a new privacy-preserving decision tree evaluation scheme, where based on the improved interval encoding method and a negotiated secret key, a model owner can create an obfuscated decision tree model and send the encodings of the tree to a client, who can also encode its input and compare the two set of encodings to attain the final classification result for its input. No complex cryptographic operations needs to be performed by two entities, and they only require a single round of communication. Formal security proof demonstrates that our scheme can achieve model privacy for the model owner and data privacy for clients. In future work, we will focus on further improving the efficiency of the evaluation for the decision tree models and other machine-learning models.

Acknowledgement. This project was supported in part by collaborative research funding from the National Research Council of Canada's Artificial Intelligence for Logistics Program.

References

1. Cao, L.: AI in finance: challenges, techniques, and opportunities. ACM Comput. Surv. (CSUR) **55**(3), 1–38 (2022). https://doi.org/10.1145/3502289
2. Fahle, S., Prinz, C., Kuhlenkötter, B.: Systematic review on machine learning (ML) methods for manufacturing processes-identifying artificial intelligence (AI) methods for field application. Procedia CIRP **93**, 413–418 (2020). https://doi.org/10.1016/j.procir.2020.04.109
3. Liang, J., Qin, Z., Xue, L., Lin, X., Shen, X.: Efficient and privacy-preserving decision tree classification for health monitoring systems. IEEE Internet Things J. **8**(16), 12528–12539 (2021). https://doi.org/10.1109/JIOT.2021.3066307

4. Zhang, Y., Jia, R., Pei, H., Wang, W., Li, B., Song, D.: The secret revealer: generative model-inversion attacks against deep neural networks. In: Proceedings of the IEEE/CVF Conference on Computer Vision and Pattern Recognition, pp. 253–261. 10.48550/arXiv. 1911.07135

5. Bost, R., Popa, R.A., Tu, S., Goldwasser, S.: Machine learning classification over encrypted data. Cryptology ePrint Archive (2014). 10.14722/ndss.2015.23241

6. Tai, Raymond K. H.., Ma, Jack P. K.., Zhao, Yongjun, Chow, Sherman S. M..: Privacy-Preserving Decision Trees Evaluation via Linear Functions. In: Foley, Simon N.., Gollmann, Dieter, Snekkenes, Einar (eds.) ESORICS 2017. LNCS, vol. 10493, pp. 494–512. Springer, Cham (2017). https://doi.org/10.1007/978-3-319-66399-9_27

7. Wu, D.J., Feng, T., Naehrig, M., Lauter, K.: Privately evaluating decision trees and random forests. Cryptology ePrint Archive (2015). popets-2016-0043

8. Banerjee, S., Galbraith, S.D., Russello, G.: Obfuscating decision trees. Cryptology ePrint Archive

9. Diffie, W., Hellman, M.E.: New directions in cryptography. In: Democratizing Cryptography: The Work of Whitfield Diffie and Martin Hellman, pp. 365–390 (2022). https://doi.org/10.1145/3549993.3550007

10. Xue, L., Liu, D., Huang, C., Lin, X., Shen, X.S.: Secure and privacy-preserving decision tree classification with lower complexity. J. Commun. Inf. Netw. 5(1), 16–25 (2020)

11. Liu, L., Chen, R., Liu, X., Su, J., Qiao, L.: Towards practical privacy-preserving decision tree training and evaluation in the cloud. IEEE Trans. Inf. Forensics Secur. 15, 2914–2929 (2020). https://doi.org/10.1109/TIFS.2020.2980192

12. Damgard, I., Geisler, M., Kroigard, M.: Homomorphic encryption and secure comparison. Int. J. Appl. Crypt. 1(1), 22–31 (2008). https://doi.org/10.1504/IJACT. 2008.017048

13. Hao, Y., Qin, B., Sun, Y.: Privacy-preserving decision-tree evaluation with low complexity for communication. Sensors 23(5), 2624 (2023). https://doi.org/10. 3390/s23052624

14. Barak, B., et al.: On the (IM) possibility of obfuscating programs. J. ACM (JACM) 59(2), 1–48 (2012). https://doi.org/10.1145/2160158.2160159

15. Boneh, D., Shen, E., Waters, B.: Strongly unforgeable signatures based on computational diffie-hellman. In: Public Key Cryptography-PKC 2006: 9th International Conference on Theory and Practice in Public-Key Cryptography, New York, NY, USA, 24–26 April, 2006. Proceedings 9, pp. 229–240 (2006). https://doi.org/10. 1007/11745853_15

16. Ishai, Y., Kushilevitz, E., Ostrovsky, R.: Sufficient conditions for collision-resistant hashing. In: Theory of Cryptography: Second Theory of Cryptography Conference, TCC 2005, Cambridge, MA, USA, 10–12 February , 2005. Proceedings 2, pp. 445–456 (2005). https://doi.org/10.1007/978-3-540-30576-7_24

17. MIRACL Library. Website. https://github.com/miracl/MIRACL

IoT-Inspired Education 4.0 Framework for Higher Education and Industry Needs

Xie Kanqi[✉], Luo Jun, and Liao Bo Xun

Guangzhou College of Technology and Business, Guangzhou 510800, Guangdong, China
{xiekanqi,luojun}@gzgs.edu.cn

Abstract. Education 4.0 provide strong talents for Industry 4.0. To better meet the needs of enterprises for talents, higher education needs to formulate corresponding education measures. Using the Internet of things technology combined with the concept of education 4.0 to provide a guarantee for industry 4.0 is a useful solution, this paper makes relevant research and discussion on this issue, including what are the technical challenges faced by higher education under the requirements of education 4.0, and how to match university education and social needs under the requirements of Education 4.0. According to the proposed problem statement, this paper proposed a based concept of OBE education educational framework 4.0, expecting to set up the cultivation of the enterprise associated indicators, drawing portraits method, assisting the talent demand of college and enterprise and establishing a good educational circle. The designed framework and indicators have been demonstrated and analyzed through research and investigation and have certain pertinence and typical significance.

Keywords: Industry 4.0 · Education 4.0 · Portrait drawing · Talent cultivate · University Innovation

1 Introduction

1.1 A Subsection Sample

A highly industrialized and data-intensive advanced society requires a highly skilled workforce, and the requirement of digital transformation reshaping the way industrial business processes, just as the current Industry 4.0 transformation, also requires a higher quality workforce. The concept of Education 4.0 is to solution of this situation, which focused on securing the future workforce of Industry 4.0-a conception that emerged from the German Industrial Revolution with a scientific, technological, and industrial base [1] and was framed by DIN [2]. Among them, digital education Ecosystem (DEEs) is the result of education digitization and digital technology promotion. DEE refers to the structure connected to each other in educational activities and different geographically distributed e-learning information and communication technology (ICT) infrastructure, information systems which connected through the Internet of Things (IoT) and cyber-physical systems (CPS) [3]. Education 4.0 is a guarantee to serve Industry 4.0. Figure 1

J. Vaidya et al. (Eds.): AIS&P 2023, LNCS 14509, pp. 415–429, 2024.
https://doi.org/10.1007/978-981-99-9785-5_29

illustrates these concepts and illustrates the flow of data and information among the development of the Education 4.0 ecosystem (e.g., students, educators, universities, schools, enterprises). The information exchange of technology and knowledge is to ensure the availability of the high-quality workforce required for Industry 4.0. It is essential to ensure interoperability across the industry 4.0 ecosystem [4]. To solve this future labor demand, the demand for education is constantly increasing, and the concept of Education 4.0 has also emerged. Different from the high attention and maturity of Industry 4.0, the topic of Education 4.0 is still in the early stage, and the research work does not have a very clear direction. Establishing the link between the necessity of Industry 4.0 for workforce skills and capabilities and exploring the cooperation between the emerging digital education ecosystem and Education 4.0 are the topics of high concern at present.

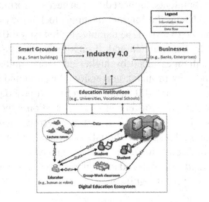

Fig. 1. DEE and Industry 4.0

1.2 Problem Statement

The importance of Education 4.0 has been gradually recognized in recent studies, and it is also a newly emerging hot research field [5]. Education 4.0 combines many emerging technologies. For example, students can use artificial intelligence (AI), cognitive technology, data analysis, Internet of Things (IoT) and other methods to make their learning more suitable for their needs, and teachers can combine traditional classrooms and virtual classrooms for various training activities [6]. The research focusing on education 4.0 is still relatively scattered and has not formed a coherent view for the time being. How IoT technology combined with the concept of education 4.0, and to provide protection for industry 4.0 is a problem statement that worthy of study. This paper makes relevant research and discussion on this issue, including:

(1) What are the technical challenges faced by higher education under the requirements of Education 4.0?
(2) How to match university education and social needs under the requirements of Education 4.0?

This paper introduces the technical challenges in Education 4.0. Then, with the help of OBE concept, the education 4.0 framework based on the IoT is proposed. The designed framework set effective university education goals based on enterprise needs, and effectively assists both schools and enterprises in quality assessment. Finally, the paper concludes with a summary and future prospect.

The organization structure of this paper is as follows; the relevant literature review is carried out on the next chapter. The technical challenges faced by the education model of Education 4.0 are expounds, and the OBE concept is introduced. Then the framework of higher education under Education 4.0 is designed and explained in the scheme. Finally, is the summary and prospection.

2 Recent Research

2.1 Education 4.0

Education 4.0 is a new paradigm in the field of teaching, which aim to help students and new generation learners for the upcoming industrial experiment revolution [7], which requires new skills, including but not limited to advanced robotics, industrial Internet of Things (I-IoT) technology, 3D printing technology and other cutting-edge technologies. In addition, after the impact of the COVID-19 epidemic, the traditional education industry has also found that it needs to accelerate the transition of personalized education, which is also part of the connected digital ecosystem. Traditional education relies on face-to-face teaching with hard copy materials, and all assessments and examinations are paper based, limiting space and time for teaching and training. The connection of digital information brings a new dawn to this situation. Through a variety of digital technologies, many learning materials, such as video lectures, audio books, etc. can be more effectively. With the help of it, electronic handouts can be widely used at anytime and anywhere. It is also easy to obtain any type of information in a short period, such as computer-based examinations can provide results and feedback to students immediately. In addition, Education 4.0 also looks forward to the use of AI to assist self-regulated learning, with smart sensors and wearable devices, AR/VR can assist distance learning facilities to enhance the remote presence and distance learning experience [8]. This AI-based assessment and early progress recognition system, enabling students to learn at their own pace and supporting student success [9]. Furthermore, a flexible and more immersive learning environment is expected to be provided in the future, such as the facilities of smart campus. The description above is the potential development of education 4.0 content, and IoT devices will make a great impacted in education 4.0.

In the self-adaptive framework of Education 4.0, [10] introduced the benefits of information obtained through intelligent devices, including biofeedback information of students' physiological data. AI algorithms can improve self-regulated learning, provide suggestions to improve teaching effects and protect students' health. Using AI algorithms education 4.0 can improve the well-being and health of students [11]. Through big data collection, groups with strategic importance to students' success can be identified, supervision and reminder can be maintained when students are at risk [12]. Also using "intelligent" remote tutors in Education 4.0 to answer simple questions with expertise from lecture manuscripts through Natural Language processing (NLP), text mining, and biofeedback [13]. These are the requirements that traditional education cannot provide, and the advances brought by technology will continue to widen the gap between this two, which lead to an increasing difference between the requirements of academia and industry.

2.2 Opportunities and Challenges

Education 4.0 is the future of the education system. Future education will create ubiquitous, immersive, adaptive, and personalized learning experiences, which imposes requirements for the rapid delivery of information. For example, students and researchers need to connect and control physics university laboratories remotely, which was previously difficult to ensure due to the bandwidth limitations of mainstream communication networks (e.g., remote handling of robots located in university laboratories requires very high communication performance), but with the development of 5G communication technology, these challenges will continue to be solved. Article [14] provides a vision of Education 4.0 by emphasizing the key role of 5G as an enabler. Key enabling technologies and use cases for Education 4.0 are also investigated, especially for remote labor and training cases. In addition, the technical challenges of Education 4.0 are identified and potential 5G solutions are evaluated. Finally, the remote circuit design laboratory prototype is used as a case study to discuss teaching and learning perspectives, and to emphasize the necessity of 5G for education 4.0, which will bring new development opportunities to more traditional industries.

However, the research field of education 4.0 is expanding, there are still facing many challenging. For example, the security link of hardware resources. There is no clear agreement on how ICT infrastructure, e-learning systems, devices, sensors, and data information should be related to each other, and the implementation of interoperable DEE is still not realized. In addition, there is the issue of resource provision and distribution. For educational institutions, educational resources are limited within the management, such as the shortage of teaching staff, the high cost of acquiring and implementing new technologies, the difficulty of curriculum redesign, and some other problems of educational institutions themselves.

The protection of data and information flow in Education 4.0, and the execution of DEE are also a key factor to be considered. For example, to avoid unauthorized access to stored information, security attacks and misuse of data. It is important to have regulations and appropriate tools or mechanisms in place to ensure security and privacy, especially when educational institutions collect or store private identity information. It is also mentioned in [15] that the challenges faced by Education 4.0 include: lack of technology infrastructure and technical support, reluctance of educators to adopt Industry 4.0-compliant technologies in teaching, and lack of legislation and regulations for data protection and privacy. In addition, digital infrastructure, effective financial planning, revised curriculum are all challenges that must be faced. In addition to the above problems, how to improve the problems in traditional education is also a problem in the development of Education 4.0.

An analysis of the relevant literature identified five technical challenges for Education 4.0 and DEE that have not yet been adequately addressed:

a. ICT infrastructure [16, 17]. Realizing the vision of Education 4.0 requires the design of adequate architecture and ICT infrastructure.
b. Interoperability. DEE and digital factories embed heterogeneous ICT infrastructures, (e-learning) systems, sensors, and devices that need to exchange data and interpret data in the same protocol.
c. Developing AI methods and algorithms for education [18, 19].
d. Implement the Education 4.0 vision to use and provide the latest technologies in educational activities.
e. Security, data protection, and privacy [14].

2.3 OBE

Outcomes-based education (OBE), which is an educational concept, should be paid attention to in the research of Education 4.0. This concept originates from the engineering education accreditation (EEA) system, which is an important direction of engineering higher education reform. OBE is output oriented education, which is an advanced concept, followed by the reform and practice of higher education at home and abroad. The OBE concept, first introduced by Spady in 1981, advocates that everyone can succeed and focuses on the achievement, maintenance, and assurance of student learning. In the teaching field, teachers and students must be familiar with their learning outcomes, encourage students to achieve higher standards than before through deep learning, and emphasize that teachers should continuously improve teaching conditions to facilitate students to achieve learning outcomes. This concept proposes a higher standard for improving the quality of higher education, requiring all educational activities and curriculum design around learning outcomes. At the same time, it also shown a certain direction for the reform and practice of higher education, that is we must pay attention to the training mode of students 'post-education ability with special emphasis on ability training and learning output. The OBE concept emphasizes outcome orientation, which requires teachers to incorporate training objectives, work needs and practical ability into the course design, which can improve the teaching effect and the quality of talent training [20].

OBE is a performance-based approach to education for the development of advanced higher education curricula and is becoming the de facto standard in many established and emerging education systems, such as Europe, USA, Australia, Malaysia, India. This approach focuses on developing applied skills rather than emphasizing eloquence in the educational process. The applied learning or skill development of students in OBE is defined by the result, and the process development is defined by implementing the goal, which promotes the definition of the course and its organization, the selection of course content, teaching methods and assessment process. Thus, OBE is an outcome education approach in which decisions about courses are driven by the learning outcomes that students should demonstrate at the end of each course and at the end of graduation. This OBE mechanism is essential to produce knowledgeable, creative, highly skilled, flexible, innovative engineering graduates with critical thinking, problem solving, and entrepreneurial spirit to meet the challenges of the fourth industrial revolution [21]. OBE creates an atmosphere where students are driven by what they can learn and use to solve real life problems. Therefore, OBE is inductive teaching, which many scholars believe it is a better way to motivate students to learn. In inductive teaching, the teacher presents students with a specific challenge at the beginning, such as a complex problem. OBE defines the process and practice of engineering education the real-world problem to be solved, the data to explain a particular research phenomenon, or the case study to be analyzed. The learning process is supported by high quality shared resources, teaching and assessment and follows standards. In a nutshell, OBE is a combination of three capabilities:

a. Practicality: the ability to know how to do things and make decisions.
b. Fundamentals: Understanding what you are doing and why.
c. Reflective: learning and adaptation through self-reflection; Apply knowledge appropriately and responsibly. Learners take responsibility for their own learning and are motivated by feedback and affirmation of their own worth [22].

With the development of education 4.0, the OBE concept can better carry out its results-oriented goals and use digital means to assist in talent training education, which will be a big integration opportunity of education 4.0 and OBE concept.

3   Framework of University Talent Portrait for Industry Demand

The digitalization of the economy and the transition to Industry 4.0 requires creation of an educational system that not only creates professional competencies for future engineers, but also enhances their creativity. At the same time, incentive systems play an important role, as personal interest in your future career allows you to make informed choices, ensuring future career development and job satisfaction in this respect. The implementation of a progressive learning approach will provide synergies that improve the quality of processes in transportation, production, education, and other systems. The accumulated experience shows that the use of modern learning methods in engineering education contributes to the development of engineering competencies, which are necessary for high-tech companies and industries. The engineering education system needs to be improved so that the educational environment motivates students to pursue engineering majors [23].

To ensure it, [24] describes a framework for the deployment of new teaching and learning systems for the industry 4.0 vision, based on three dimensions: technology, teaching, and organization. The authors also introduce two case studies involving collaborative networks and open innovation to demonstrate the use of the proposed framework. Education 4.0 needs to guarantee six dimensions: knowledge, skills, and qualifications; Teaching; Learning; Implement the vision and teaching method of Education 4.0. Electronic assessment and quality assurance. Some researcher discusses different technologies for education 4.0 that can be realized through 5G communications:

The digitization of educational systems cannot be limited to the creation of digital copies of familiar textbooks, the digitization of documents, and digital teacher training. The digital economy requires a comprehensive approach to education systems that will set new goals and change the structure and content of the education process. One such promising area of activity is to enhance the use of resources. The decision-making system based on the data processing results of learners' digital twins will not only help to improve the informatization of the educational process, but also to improve the entire educational institution [25].

Combined with the OBE concept, colleges carry out oriented talent training according to the needs of the industry, aiming to cultivate talents with the needs of the industry. Therefore, it is necessary to keep close to the latest industry trends, collect data in stages, and customize the final data evaluation project.

SDLC is a process to generate high quality and low-cost applications in minimum time. It provides a well-structured step flow that helps businesses easily produce high-quality, well-tested, ready-to-use software products [26]. SDLC's goal is to produce quality software that meets and exceeds all customer expectations and needs. The SDLC defines and outlines a detailed plan containing various stages, each of which contains its own processes and deliverables. Compliance with SDLC increases development speed and minimizes project risks and costs associated with alternative production methods.

As described in article [27], in terms of requirements analysis and management. Waterfall attempts to analyze the requirements from the beginning and not implement them until they are fully understood, documented, and almost frozen. Unlike waterfall methods, agile methods accept the fact that requirements may not be detailed enough at the beginning of the project, but they will evolve over time and stakeholders and will have more insight into what they need. Thus, agile welcomes requirements change but focuses on handling them correctly. It allows the customer to be continuously involved in the project and evolve requirements. In this study, the goal is clear, but in the design process, the feedback collected from the industry is used to adjust the teaching strategy and evaluation index in time, so the researcher adopts the Agile Model based on the six stages of SDLC to carry out the project, to obtain a more perfect and accurate Framework. Therefore, this paper proposes a method to summarize each student's specific college portrait based on daily collection data, as a method for enterprises to observe the performance of students in school. In Requirement analysis, it mainly takes the development of times and the needs of the industry as the starting point. In the framework design, testing is based on the analysis and comparison of the collected data. Deployment says to analyze the data and industry requirements for a certain period. In addition, the education 4.0 framework proposed in this study is based on the OBE concept, which means that in the implementation process, it will be improved and modified according to the needs, so the needs that may be generated in each process and the overall education framework are shown in Fig. 2.

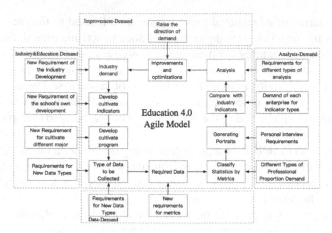

Fig. 2. Framework of Education 4.0

Based on this framework, the process of education 4.0 covered by AI technology described by [28], it can be known in [29, 10] that AI technology and wearable device assistance can cover the process of education 4.0, and the data generated by the education process can be obtained through the following seven aspects:

1) Orientation: an entrance test that assesses students' prior knowledge of the subject (passive adaptation) [30]; An overview of course content, motivation for activity diagrams showing exam scores, and strategic planning with inherent learning objectives; Biometric registration and authentication methods.

2) Digital preparation: Personalized content according to the type of learning (interactive books or videos), self-monitored learning control for students through adaptive quizzes and self-assessments (continuous adaptation) [31] Biofeedback.

3) Interactive presence: teachers discuss case studies and act as coaches; AR/VR experience; Students work in groups; Hands-on experimentation and creating Spaces such as the Education 4.0 Learning Lab (E4LL).

4) Collaboration: "Communities of practice" with student materials for students, students engage in so-called assignments for short-term projects - with increasing difficulty and enhanced problem-based tasks, such as the Digital Technology Learning Lab (LL4DT) [32].

5) Follow-up and performance: Feedback - self-assessment, in which correct solutions and answers are explained and studied; Recognize the level of learning material covered and assess the evolution of knowledge.

6) Reflection and motivation: The "Early Recognition System" applies neural networks for self-monitoring and self-observation to continue the educational process. A future scenario could be an extension of the learning analytics cockpit based on online and physiological activity data [33].

7) Assessment and examination: Electronic assessment - Part of the examination is an automatically assessed competency test. We conducted experiments with latent semantic Analysis (LSA) and Word2Vec [34].

Through these dimensions, data of various indicators can be obtained. Due to the promotion of Industry 4.0, most colleges are equipped with various intelligent devices and constantly transforming to smart campus. The large amount of data generated and collected can be used as the basis for students 'daily life and learning habits. Extensive use of online platforms, introduction of personalized educational trajectories and courses, new spatial opportunities, and formats. With the help of the IoT and AI, teachers, classmates, and other administrative can score through different dimensions, which was used as the basis for the multi-dimensional portrait of the student. [35] shows the modular construction of a ubiquitous monitoring framework for relevant personnel (including students, teachers, and other administrative personnel), and sets up a framework for self-realizing dynamic data collection, monitoring, classification, and prediction in the education 4.0 environment. The proposed framework is very useful for automatic real-time assessment and certification in the education 4.0 environment. Combined with the assimilation of IoT sensors, effective data collection can be carried out in the environment of smart campus, and the information flow contained in the collection in smart campus is shown in Fig. 3.

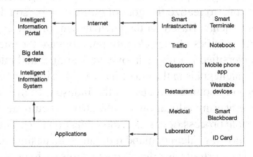

Fig. 3. Framework of Information Collection

Before designing the portrait data of students, according to the OBE concept, we need to master the dynamics of the industry and the required talents to get the key points. With the help of expert system, industry expert team are evaluated to design our framework, set reasonable enterprise indicators according to the characteristics of the industry, set secondary indicators according to the enterprise indicators and the characteristics of the school as the evaluation standard of the school, and then classify each daily indicator as the evaluation standard. The structure is shown in Fig. 4.

Through the results of each layer of indicators by importing the daily data of each student as the evaluation standard, the exclusive portrait data is generated, and the data required for the portrait is generated. Finally, a more authoritative portrait proportion map will be obtained. As mentioned above, Industry 4.0 needs more "innovative talents", and "innovative talents" embody their social value in more fields, avoiding becoming "craftsmen" with narrow knowledge, single skills, and narrow employment. Engineering innovation talents are "architects" who can cope with the challenges of globalization and systematically construct complex social, economic, and political environments. The connotation of engineering innovation capability includes engineering knowledge application, innovation research, problem analysis and solution, team cooperation, etc. CDIO

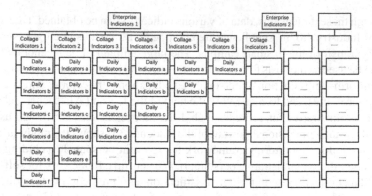

Fig. 4. Framework of the Indicators

(Conception, Design, Implementation, Operation), international engineering education model is the latest achievement of international engineering education reform in recent years. The key to improving the practical ability of engineering talents in China is to cultivate innovative thinking, to realize the goal of cultivating personalized and innovative emerging engineering talents. Based on the environment of new engineering construction in colleges and universities, the article [36] puts forward the talent training mode of "orderly" learning, diversified learning, innovative learning, and international learning. This model meets the talent goals in the era of Industry 4.0, improve engineering practice ability, and shape the core competitiveness of the industry. In the era of artificial intelligence, emerging engineering education should strive to cultivate innovative practice ability through the whole process of engineering education [37]. Therefore, considering various aspects, the following talent demand indicators are designed. This paper designs six indicators for student portraits as reference standards to draw portrait radar charts as it shown in Table 1.

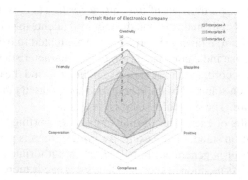

Fig. 5. Portrait Radar of Electronics Company

For each reference index, students' daily activities in school are used as the secondary index to calculate the score, as shown in the table.

Table 1. 6 indicators of electronics company.

Indicators	Creativity	Discipline	Positive	Compliance	Cooperation	Friendly
Score	1–10	1–10	1–10	1–10	1–10	1–10

Table 2. 5 Details of Creative Indicators.

Indicators	Project	Article	Schemes	Performance	Creative Idea
Score	1–10	1–10	1–10	1–10	1–10

Table 3. Details of Discipline Indicators

Indicators	Classes Attendance Rate (AR)	Appointment Activities AR	Campus Activities AR	Social Activities AR	Corporate Activities AR
Score	1–10	1–10	1–10	1–10	1–10

Table 4. Details of Positive Indicators

Indicators	Interaction	Competition	Campus Activities	Social Activities	Enterprise Activities
Score	1–10	1–10	1–10	1–10	1–10

Table 5. Compliance

Indicators	Examination Rules	Classroom Rules	Dormitory Rules	Practical Activities	Enterprise Rules
Score	1–10	1–10	1–10	1–10	1–10

Table 6. Cooperative

Indicators	In Team Projects	Assisting Classmate	Assisting Teachers	Social Activities	Enterprise Practice
Score	1–10	1–10	1–10	1–10	1–10

Each indicators continue to set several indicators, which contain daily data collection, the specific index block diagram is shown in Table 2, 3, 4, 5, 6, 7 and Fig. 6.

Table 7. Friendly

Indicators	Interactivity in Classroom	Participation Rate in Competition	Expressive Force in Campus Activities	Evaluation of Social Practice	Evaluation of Enterprise Practices
Score	1–10	1–10	1–10	1–10	1–10

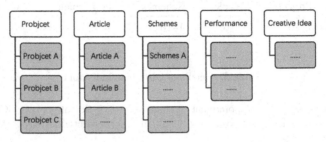

Fig. 6. Specific Index of the Creative Indicator

By integrating these data, the portrait of a student can be portrayed from multiple dimensions as is shown in Fig. 7.

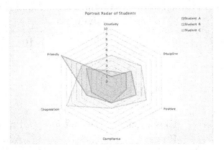

Fig. 7. Portrait Radar of Students

The portrait radar data will be continuously improved with the accumulation of data of students in school. Finally, a portrait of its own four years will be drawn upon graduation. Enterprises can better understand whether it meets their own needs according to such a portrait. In addition, colleges and universities also comprehensively judge their own teaching arrangements according to the characteristics of students 'portraits, and whether the teaching meets the needs of it. In addition, the reference indicators should also be periodically adjusted according to the needs and characteristics of the industry, and timely adapt to the industry update. The adoption of digital measurement standards can make talent training more accurate and efficient, and more in line with the characteristics of education 4.0.

4 Conclusion

Based on the OBE concept, this paper focuses on cultivation of applied talents, which requires colleges and universities to make innovation and reform in education, teaching, scientific research, and social services. An education 4.0 education framework inspired by the Internet of things and industry needs is proposed. The following work has been done:

1. Designed an education framework based on education 4.0 for college students to enroll and graduate.
2. A data collection and application method is proposed, which uses the data generated by the daily behavior of students to draw portraits.
3. The importance of the framework is discussed.

The future work of this paper will track the implementation efficiency of the framework, improve the running efficiency, and improve the accuracy of portrait rendering. In addition, when establishing data samples, how to set more targeted indicators, how to reasonably protect information data collection, and ensure the accuracy, security and privacy of information are also an important direction of this paper's future research.

References

1. Wermann, J., Colombo A.W., Pechmann A., et al.: Using an interdisciplinary demonstration platform for teaching Industry 4.0, Procedia Manufacturing 31, pp. 302—308 (2019)
2. DIN Deutsches Institut für Normung e.V.: DIN SPEC 91345:2016–04: Referenzarchitektur-modell Industrie 4.0 (RAMI4.0) (2016)
3. Chituc, C.-M.: A Framework for Education 4.0 in Digital Education Ecosystems. In: Camarinha-Matos, L.M., Boucher, X., Afsarmanesh, H. (eds.) Smart and Sustainable Collaborative Networks 4.0: 22nd IFIP WG 5.5 Working Conference on Virtual Enterprises, PRO-VE 2021, Saint-Étienne, France, November 22–24, 2021, Proceedings, pp. 702–709. Springer International Publishing, Cham (2021). https://doi.org/10.1007/978-3-030-85969-5_66
4. Chituc, C. -M.:"An Analysis of technical challenges for education 4.0 and digital education ecosystems. In: 2022 IEEE German Education Conference (GeCon), pp. 1–5. Berlin, Germany (2022)
5. Imran, M.A., Sambo, Y.A., Abbasi, Q.H.: Enabling 5G Communi- cation Systems to Support Vertical Industries. Wiley, Hoboken, NJ, USA (2019)
6. Toma, M. -V., Turcu, C. E.:"Towards Education 4.0: Enhancing Traditional Textbooks with Augmented Reality and Quick Response codes. In: 2022 International Conference on Development and Application Systems (DAS), pp. 144–149, Suceava, Romania (2022)
7. Hussin, A.A.: 'Education 4.0 made simple: Ideas for teaching.' Int. J. Educ. Literacy Stud. 6(3), 92–98 (2018)
8. Prasad, P.B., Padmaja, N., Kumar, B.S., Aravind, B.S.: Industry 4.0: Augmented and virtual reality in education. In: Innovating With Augmented Reality: Applications in Education and Industry. CRC Press (2021)
9. Ciolacu, M., Tehrani, A.F., Binder,L., Svasta, P.M.: Education 4.0 - Artificial Intelligence Assisted Higher Education: Early recognition System with Machine Learning to support Students' Success. In: 2018 IEEE 24th International Symposium for Design and Technology in Electronic Packaging (SIITME), pp. 23–30, Iasi, Romania (2018)

10. Ciolacu, M.I., Binder, L., Svasta, P., Tache, I., Stoichescu, D.: Education 4.0 – jump to innovation with IoT in higher education. In: Proceeding of the 2019 IEEE 25th International Symposium for Design and Technology in Electronic Packaging (SIITME), pp. 135–141, Cluj-Napoca, Romania (2019)

11. Awada, A.I., Mocanu, I.: A platform to promote a more active lifestyle between students. In: Procedings of the 16th International eLearning and Software for Education Conference (eLSE), Bucharest, Romania, vol. 3, pp. 355–362 (2020)

12. Ciolacu, M.I., Svasta, P., Hartl, D., Görzen, S.: Education 4.0: smart blended learning assited by Artificial Intelligence, biofeedback and sensors. In: Proceedings of the 14th IEEE International Symposium on Electronics and Telecommunications (ISETC 2020), pp.1–4,Timisoara, Romania(2020)

13. Ciolacu, M.I., Binder, L., Popp, H.: Enabling IoT in Education 4.0 with biosensors from wearables and artificial intelligence. In: Proceedings of the 2019 IEEE 25th International Symposium for Design and Technology in Electronic Packaging (SIITME), Cluj-Napoca, pp. 17–24, Romania (2019)

14. Kyzilkaya, B., Zhao, G., Sambo, Y.A., Li, L., Imran, M.A.: 5G-Enabled Education 4.0: Enabling technologies, challneges, and solutions. IEEE Access, vol. 9, pp. 166962–166969, (2021)

15. Mian, S.H., et al.: Adapting Universities for Sustainability Education in Industry 4.0: Channel of Challenges and Opportunities. Sustainability. 12, 15, 6100 (2020)

16. Costan, E., et al.: Education 4.0 in developing economies: a systematic literature review of implementation barriers and future research agenda. Sustainability. 13, 22, 12763 (2021)

17. Ferreira Costa, A.C., de Mello Santos, V.H., de Oliveira, O.J.: Towards the revolution and democratization of education: a framework to overcome challenges and explore opportunities through Industry 4.0. Inform. Educ. 21(1), 1–32(2022)

18. Ionita Ciolacu, M., Binder, L., Popp, H.: Enabling IoT in Education 4.0 with bio-sensors from wearable and artificial intelligence, IEEE SIITME, pp. 17–24 (2019)

19. Halili, S.H.: Technological advances in Education 4.0. Online J. Distance Educ. e-Learn. 1(7), 63–69 (2019)

20. Bin, Z. H., Ming, W.: Research on the Innovation Mode of Educational Informatization based on OBE Concept. In: 2021 16th International Conference on Computer Science & Education (ICCSE), pp. 839–846, Lancaster, United Kingdom (2021)

21. Hassan, M.M.S.: Outcome-based education: A new dimension in higher education," The Independent, International Engineering Alliance (IEA), Washington Accord Signatories. [Online]. https://www.ieagreements.org/accords/washington/signatories/2020/09/05

22. Syeed, M.M., Shihavuddin, A., Uddin, M.F., Hasan, M., Khan, R.H.: Outcome based education (obe): defining the process and practice for engineering education. IEEE Access 10, 119170–119192 (2022)

23. Makarova, I., Pashkevich, A., Shepelev, V., Mukhametdinov, E., Fatikhova, L., Mavrin, V.: Motivation role in engineering education in the transition to a digital society. In: 2020 21st International Conference on Research and Education in Mechatronics (REM), pp. 1–6 Cracow, Poland (2020)

24. Miranda, J., Ramírez-Montoya, M.S., Molina, A.: Education 4.0 Reference Framework for the Design of Teaching-Learning Systems: Two Case Studies Involving Collaborative Networks and Open Innovation. In: Camarinha-Matos, L.M., Boucher, X., Afsarmanesh, H. (eds.) PRO-VE 2021. IAICT, vol. 629, pp. 692–701. Springer, Cham (2021). https://doi.org/10.1007/978-3-030-85969-5_65

25. Martynov, V., Filosova, F., Egorova, Y.: information architecture to support engineering education in the era of industry 4.0. In: 2022 VI International Conference on Information Technologies in Engineering Education (Inforino), pp. 1–5, Moscow, Russian Federation (2022)

26. Khan, R.A., Khan, S.U., Khan, H.U., Ilyas, M.: Systematic literature review on security risks and its practices in secure software development. In: IEEE Access, vol. 10, pp. 5456–5481 (2022)
27. Demirel, S.T., Das, R.: Software requirement analysis: research challenges and technical approaches. In: 2018 6th International Symposium on Digital Forensic and Security (ISDFS), pp. 1–6, Antalya, Turkey (2018)
28. Ciolacu, M. I., Marcu, A.E., Vladescu, M., Stoichescu, D., Svasta, P.: Education 4.0 Lab for Digital Innovation Units-Collaborative Learning in Time of COVID-19 Pandemic. 2021 IEEE 27th International Symposium for Design and Technology in Electronic Packaging, SIITME 2021 - Conference Proceedings, pp. 268–271 (2021)
29. Ciolacu, M.I., Marcu, A.-E., Vladescu, M., Stoichescu, D., Svasta, P.: Education 4.0 lab for digital innovation units – collaborative learning in time of COVID-19 pandemic. In: 2021 IEEE 27th International Symposium for Design and Technology in Electronic Packaging (SIITME), pp. 268–271, Timisoara, Romania(2021)
30. Faisal, M.H., AlAmeeri, A.W., Alsumait, A.A.: An adaptive e-learning framework: crowd-sourcing approach. In: Proceedings of the 17th International Conference on Information Integration and Web-based Applications & Services (iiWAS '15). Association for Computing Machinery, Article 13, 1–5, New York, USA (2015)
31. Ioniţă Ciolacu, M., Svasta, P., Hartl, D., Görzen, s.: "Education 4.0: smart blended learning assisted by artificial intelligence, biofeedback and sensors. In: 2020 International Symposium on Electronics and Telecommunications (ISETC), pp1–4, Timisoara, Romania, (2020)
32. Humpe, A., Brehm, L.: Problem-based learning for teaching new technologie. In: 2020 IEEE Global Engineering Education (EDUCON), pp. 493–496, Porto, Portugal (2020)
33. Ciolacu, M.I., Tehrani, A.F., Binder, L., Svasta, P.: Education 4.0 – Artificial Intelligence assited higher education – early recognition system to support students' success. In: Proceedings of the 2018 IEEE 24th International Symposium for Design and Technology in Electronic Packaging (SIITME), Iasi, Romania, pp. 23–30, (2018)
34. Ciolacu, M.I., Binder, L., Popp, H.: Enabling IoT in Education 4.0 with biosensors from wearables and artificial intelligence. In: Proceedings of the 2019 IEEE 25th International Symposium for Design and Technology in Electronic Packaging (SIITME), pp. 17–24, Cluj-Napoca, Romania (2019)
35. Verma, A., Singh, A., Anand, D., Aljahdali, H.M., Alsubhi, K., Khan, B.: IoT Inspired Intelligent Monitoring and Reporting Framework for Education 4.0. In: IEEE Access, vol. 9, pp. 131286–131305(2021)
36. Jun, Q., Jing, X.: Innovation research on the emerging engineering talent cultivation mode in the era of industry 4.0. In: 2017 International Conference on Industrial Informatics - Computing Technology, Intelligent Technology, Industrial Information Integration (ICIICII pp. 333–336), Wuhan, China (2017)
37. Xu Lei, H., Bo, F.H., Weili, H.: On new engineering education in comprehensive universities. Res. High. Educ. Eng. 02, 6–12 (2017)

Multi-agent Reinforcement Learning Based User-Centric Demand Response with Non-intrusive Load Monitoring

Mohammad Ashraf[1], Sima Hamedifar[1], Shichao Liu[1(✉)], Chunsheng Yang[2], and Alanoud Alrasheedi[1]

[1] Carleton University, Ottawa, Canada
shichaoliu@cunet.carleton.ca
[2] Institute of Artificial Intelligence, Guangzhou University, Guangzhou, China

Abstract. This research proposes a multi-agent reinforcement learning framework as a home energy management algorithm that focuses on user needs and preferences as well. The proposed method aims to secure the smart grid from power outages due to overloading. The system predicts appliance-level load demand for the following day using non-intrusive load monitoring (NILM) and four neural network-based supervised learning methods to pick the more accurate forecasting method. The Python-based NILM toolkit is utilized to analyze disaggregation methods on the forecasted demand to obtain appliance-level energy consumption. The user feedback and time-based price values are employed to optimize appliance scheduling. The simulation results of each stage of the algorithm are presented. The results demonstrate a 15% reduction in the electricity cost.

Keywords: Demand Response · Home Energy Management · Non-Intrusive Load Monitoring · Reinforcement Learning

1 Introduction

The idea of Demand-Side Management (DSM) was first established in order to achieve the equilibrium of supply and demand for electricity [1]. The prior strategy of just expanding energy generation to keep up with demand growth was criticized due to various factors such as heavy financial investments, environmental concerns, and optimization theory. The DSM substantially transformed this paradigm. Demand Response (DR) program, as a class of DSM, allows for minimized electricity demand during peak periods in order to balance supply and demand and secure the grid from power outages due to overloading. A uniform and balanced energy profile also improves the robustness of the power system.

On the other hand, DR programs encourage users to save more energy, consume various renewable energy sources, save money on bills, and earn subsidies for selling surplus production back to the grid. Home Energy Management

J. Vaidya et al. (Eds.): AIS&P 2023, LNCS 14509, pp. 430–445, 2024.
https://doi.org/10.1007/978-981-99-9785-5_30

(HEM) as a DR program schedules the consumption pattern of residential users to off-peak hours to smooth the load profiles and increase energy efficiency [1,2]. Customers make smart decisions while operating appliances to modify load profiles and lower peak energy consumption to achieve more financial savings.

A DR technique can be designed according to incentive-based programs or price-based programs [1,3]. The incentive-based programs reward or penalize the consumers' consumption pattern during on-peak. In the price-based programs, consumption adapts according to the time-based tariffs. Despite all the benefits of DR, such as financial savings for both consumers and the utility and the increased stability and capacity of the grid, some challenges need to be addressed with the DR policies.

One of the most critical challenges is the user's dissatisfaction [4,5]. Many factors affect the comfort level and satisfaction of the users. The users have to adapt their electricity usage behavior to reduce some loads or shift loads to off-peak time intervals. The waiting time may lead to their dissatisfaction. Thermal comfort can be another criterion for the users' comfort level. A successful DR scheduling considers users' preferences to increase human consent. Despite the importance of user satisfaction, it is not considered in some designed DR algorithms [6] or is addressed with a few constraints in the problem formulation [7].

The designed DR solution needs to know about the interactions in the environment. The traditional model-based methods require extensive knowledge of the environment [5,8]. Many optimization-based solutions have been proposed in the literature, which uses the Hyper-Spherical Search algorithm [9], genetic algorithm [2], and ant colony algorithm [10]. However, uncertainties, unpredictable situations, missing information, etc., might make the designed model impractical. On the contrary, as a model-free approach, the Reinforcement Learning (RL) algorithm is a machine learning with highly intelligent decision-making capability. The RL process includes agents and their interaction with an environment. At each time step, the agents take actions according to the state of the environment and receive rewards if the chosen action is correct; otherwise, they are penalized. The RL algorithms have also been applied in various activities in the energy sector. For example, the authors of [1] use a single-agent Q-learning algorithm with a fuzzy reward function optimizing consumption and consumer satisfaction. In [8], price uncertainty is addressed by an NN-based extreme machine learning (EML), and DR is addressed by Q-learning.

A DR program design requires load forecasting [11,12]. Based on the user's consumption history, their required power in the future is predicted. The prior prediction of the demand level increases the effect and efficiency of the operator. A variety of methods have been used to solve the load forecasting problem accurately and efficiently, such as support vector machines (SVM) [13], fuzzy logic [11], cascaded Neural Networks (NN) [14], flower pollination algorithm (MOFPA) [12], clustering-based Seq2Seq LSTM [15], etc. In order to predict a home energy demand, load monitoring is required to measure the electricity consumption. The load monitoring methods are divided into two groups: intrusive and non-intrusive. The intrusive load monitoring (ILM) [16] uses separate smart sensors for each household appliance, which is costly and intrusive to the user's privacy and security [17,18]. To solve this issue, non-intrusive load monitoring (NILM), a single smart sensor method, is applied. NILM uses a single-point sensor system to measure the aggregated energy consumption.

The aggregated energy consumption in a home needs to be disaggregated into appliance-level power consumption, which is required for the HEM system. Many NILM algorithms have been designed [19] such as frequency-domain template filtering proposed in [20], and a technique based on support vector regression (SVR) and Elman Neural Network (ENN) in [21]. Besides all, the disaggregation problem is addressed by the non-intrusive load monitoring toolkit (NILMTK) based on Python [22,23]. It offers several disaggregation algorithms that make it possible to compare energy disaggregation strategies in a repeatable way. The toolkit consists of many reference benchmark disaggregation methods, a set of accuracy metrics, a collection of statistics for creating data sets, and parsers for various existent datasets. The toolkit made it possible to examine several publicly accessible data sets and compare the disaggregation methods used in these data sets. The development of NILMTK has simplified NILM research and diversified the study of multiple methods over diverse datasets. The NILMTK toolbox includes capabilities for dataset conversion and several operations for data modification, extraction, and displaying statistics and energy percentages.

This paper uses the data related to aggregated energy consumption for predicting the one-day-ahead load requirement. The Supervised Learning technique takes advantage of the Nonlinear Auto Regression with External Input (NARX) neural, Recurrent Neural Network (RNN), Regression Tree (RT), and Long-Short-Term Memory (LSTM) methods are evaluated, and the one with more accurate results is chosen. To obtain per-device energy usage, different NILM algorithms, including the Mean algorithm, Edge Detection (ED), Combinatorial Optimization (CO), and the Factorial Hidden Markov Model (FHMM), are evaluated. The algorithm with a higher precision is selected. Then, Multi-Agent Reinforcement Learning (MARL) is used to plan the energy consumption of the appliances. The appliances are divided into three groups: non-shiftable, power-shiftable, and time-shiftable, and for each group, a representative reinforcement learning agent is considered. The agents choose the actions for their group based on the Q-learning algorithm to get a higher reward. The agents make decisions regarding the electricity price and users' satisfaction at each time step. The REFIT dataset is applied to evaluate the proposed method. The results indicate that the LSTM-based forecasting method and the CO disaggregation have lower errors. The rescheduling of appliance usage using multi-agent Q-learning shows a significant decrease in home electricity consumption during on-peak, which results in more secure smart grids from overloading.

The contributions of this paper are as follows.

– A comparison between the accuracy of four electricity usage prediction methods based on supervised learning, including Nonlinear Auto Regression with

External Input (NARX) neural network, Recurrent Neural Network (RNN), Regression Tree (RT), and Long-Short-Term Memory (LSTM) methods.
- Using a Python-based NILMTK to disaggregate the energy consumption
- A comparison between four different disaggregation algorithms, including Mean algorithm, Edge Detection (ED), Combinatorial Optimization (CO), and Factorial Hidden Markov Model (FHMM)
- Designing a multi-agent Q-learning price-based HEM algorithm for lower consumption during on-peak and higher user satisfaction

The rest of the paper is organized as follows. Section 2 and Sect. 3 describe load forecasting and load disaggregation techniques, respectively. Section 4 gives information about the evaluation metrics used in this paper. Section 5 explains the MARL algorithm and the way it is used in this paper. Section 6 presents the simulation results. Finally, Sect. 7 concludes the paper.

2   Load Forecasting

The Nonlinear Auto Regression with External Input (NARX) neural network, Recurrent Neural Network (RNN), Regression Tree (RT), and Long-Short-Term Memory (LSTM) are the four widely utilized machine learning techniques for forecasting electrical load and are presented in this section.

2.1   Neural Network Model

Given a sufficient number of neurons, a Neural Network model can learn the input and output relationship, which can be utilized in the future. Dynamic feedback (recurrent) neural network called a nonlinear autoregression with external input (NARX) [8] employs a time series as the exogenous input and a delayed output value as the endogenous input to provide feedback. As a result, output is regressed on both the previous values of the independent variables and the prior values of the input signal. A NARX neural network is suitable for describing a nonlinear dynamic system and can be utilized in time series modeling.

2.2   Recurrent Neural Network

Recurrent neural networks (RNNs) use the output from a previous step as the input for the current step. The hidden layers, which retain some information about a sequence, are the primary and most crucial components of RNNs. Some of the points why RNN should be used are mentioned below [24]:

1. In time series predictor, the RNN model, with its ability to retain information over time, helps the model to perform better and give more accurate results
2. The model complexity does not grow even for a high input.
3. At each time step, the weights are updated.
4. RNNs have internal memory that they can use to process any set of inputs, which is not the case with feedforward neural networks

2.3 Regression Tree

Regression Tree (RT) is a nonlinear predictive model that employs a tree structure to describe the results of recursive partitioning [25]. A regression model is difficult to fit when there are several two-way interactions between the independent variables and a high number of parameters in a multiple regression. It is therefore necessary to develop a method to measure interactions' impact. A recursive partition is employed to handle complex non-linear interactions between all factors. It repeatedly partitions the covariate space into smaller areas to fit a straightforward model to a cell. A partitioned region's cell is represented by each node and leaf of the tree. A decision will be made in every node, known as a level. The bottom leaves are the fitted result, the mean estimation of the load utilization that applies to that specific location under the decisions made. We used the Regression Tree method available in MATLAB regression learner.

2.4 Long Short-Term Memory

In deep learning, the Long Short-Term Memory (LSTM) method is a synthetic Recurrent Neural Network (RNN) architecture that can learn order dependency [26]. The LSTM includes connected feedback neurons, which are utilized to reduce or remove severe instability difficulties generated by RNN effectively, referred to as vanishing gradient problems. This is different from ordinary feedforward neural networks. An LSTM is a good fit for classifying sequence and time-series data where the prediction or output of the network must be based on a recalled sequence of data points.

3 Non-intrusive Load Monitoring

The NILMTK is used to disaggregate the home energy into appliance-level energy consumption. The disaggregation methods of the toolkit, including the Factorial Hidden Markov Model (FHMM), Combinatorial Optimization (CO), Edge Detection (ED), and Mean algorithm, are evaluated in this research.

3.1 Mean Algorithm

The Mean algorithm is built to provide a clear benchmark against which more complex algorithms may be assessed. The trained mean model calculates and stores just the mean power state for each device. Despite its simplicity, the Mean method beats more complicated disaggregation algorithms on a number of metrics, making it a solid baseline.

3.2 Edge Detection

The Edge Detection approach divides the time series into steady and transient periods. A change in the status of an appliance (such as turning on or off) frequently correlates to an edge, which is the magnitude difference between two stable states. Even though the method is ostensibly unsupervised and does not require appliance-level data for training, the output must be mapped to appliance categories. In our solution, which uses the best case mapping, the algorithm outputs are assigned to the appliance categories that maximize the method's accuracy [23].

3.3 Combinatorial Optimization

Combinatorial optimization (CO) uses supervised learning in the context of real load disaggregation, which necessitates using signature libraries of appliances to train the model. The purpose of the CO approach is to reduce the inaccuracy between the total energy consumption of all considered individual appliances and the aggregated power demand of the home. Constructing the CO model to restrict the contributions of other household appliances that are not necessities or consume the same amount of power is necessary. The objective function is the minimization of the difference between the aggregate and sum of appliance power demand [22].

$$min|\hat{y}(t) - \sum_{i=1}^{N} y_i(t)| \tag{1}$$

where $\hat{y}(t)$ is the aggregated household power demand, $y_i(t)$ is the power demand of appliance i, and N is the number of appliances that are present in the house.

3.4 Factorial Hidden Markov Model

For training and testing, NILMTK separated the data into continuous sets and then used the hidden Markov models to simulate each appliance. This technique simulates the reliance between time slices to replicate each device's operating state. The total active or reactive power is the observation sequence for the NILM problem, and it is uncertain what state each piece of equipment is in or how much power it uses. For a Markov chain, all appliances are considered as HMMs. A FHMM made up of many HMMs is described as having a total power equal to the sum of the powers of all the appliances.

4 Model Performance Evaluation Metrics

Considering the advantages and disadvantages of statistical evaluation metrics, we employed various measures to evaluate the effectiveness of load prediction and disaggregation approaches. The metrics assessed how well the model can categorize and identify the appliance incident. The mean absolute error (MAE) indicates the relative error of the real and predicted energy. It is easy to interpret and provides information about the magnitude of the errors.

$$MAE = \frac{1}{N} \sum_{i=0}^{N} |Y_{real} - Y_{predicted}| \tag{2}$$

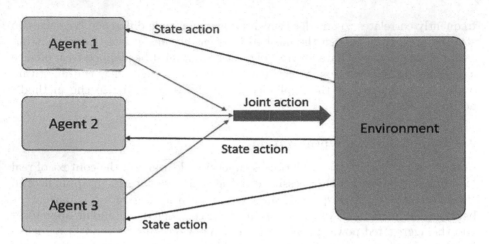

Fig. 1. Multi-Agent Reinforcement Learning Process

The root mean square error (RMSE) indicates is calculated as follows.

$$RMSE = \frac{1}{N}\sqrt{\sum_{i=0}^{N}(Y_{real} - Y_{predicted})^2} \tag{3}$$

It is a dominant metric in regression and forecasting algorithms. The RMSE gives more emphasis to larger relative errors. The mean absolute percentage error (MAPE) is expressed in percentage which makes it easily compared across different datasets and models. However, it does not work for zero values and places more emphasis on close to zero values.

$$MAPE = \frac{1}{N}\sum_{i=0}^{N}\frac{|Y_{real} - Y_{predicted}|}{Y_{real}} \times 100 \tag{4}$$

The standard deviation (SD) metric measures how spread out the numbers are. It is a measure of how much confidence can be placed in a model's performance on different datasets.

$$SD = \sqrt{\frac{1}{N}\sum_{i=0}^{N}(Y_{predicted} - Y_{mean})^2} \tag{5}$$

Fig. 2. Schematic Diagram of NILM-based MARL-based HEM System

5   Multi-agent Reinforcement Learning Demand Response

The designed HEM system includes multiple agents to enable decentralized control of different smart home items. This study takes into account three agents in a typical HEM system, characterizing appliances namely into non-shiftable appliances (Agent 1), power-shiftable appliances (Agent 2), and time-shiftable appliances (Agent 3), as shown in Fig. 1. All appliances must receive energy according to their priority. The non-shiftable appliances are of high priority because they always use fixed energy and are crucial to maintaining the comfort and safety of the living space. Then, the energy is provided to the other agents (appliance categories) in declining order of the dissatisfaction coefficients (Diss. Coef.)).

Each agent monitors the current state st at time slot t and then, according to the system, selects an action a_t. Then, it checks for the new state, s_{t+1}, and determines a new action, a_{t+1}, for the subsequent time period, $t+1$. This problem is formulated as a Factored Markov Decision Process (FMDP). The FMDP of this study has five tuples, i.e., $(\mathbf{S}, \mathbf{A}, \mathbf{R}, \gamma, \theta)$, where we denote $\mathbf{S}$ as the state space, $\mathbf{A}$ as the actions space, $\mathbf{R}$ as reward function, γ discount factor, and θ as learning rate.

5.1   State Space

The state s_t formulates the present situation in FMDP. In this paper, the state s_t in time slot t is defined as the price of electricity at time t.

$$s_t = \{\eta_1, \ \eta_2, \ \eta_t, \ ..., \ \eta_T\} \tag{6}$$

5.2   Action Space

The action space of each agent is related to the category of the appliances it represents and the features of those appliances. To provide everyday convenience and safety, non-shiftable appliances, including fridges, microwaves, kettles, toasters, and televisions, can only take one action: ON. Power-shiftable equipment such as overhead fans functions flexibly within a specified power range. Therefore, the power-shiftable agent can select various power levels represented as discrete actions. Time-shiftable loads, such as the dishwasher and washing machine, can be planned from peak to off-peak times. This group of the appliance has two operating points: ON and OFF.

5.3 Reward

Since the non-shiftable loads are immutable, their negative reward is their power costs.

$$r_{i,t}^{NS} = -\eta_t \left[E_{i,t}^{NS} \right]^+, \quad i \in \Phi^{NS}, \quad t - 1, 2, ..., T. \tag{7}$$

where $E^{NS_{j,t}]}$ is the energy consumption of non-shiftable appliance i at time t and Phi^{NS} is the set of non-shiftable appliances. The reward function of power-shiftable appliances' agent is

$$r_{j,t}^{PS} = -\lambda_t \left[E_{j,t}^{PS} \right]^+ - \alpha_j^{PS} (E_{j,max}^{PS} - E_{j,t}^{PS})^2, \quad j \in \Phi^{PS}, \quad t = 1, 2, ..., T. \tag{8}$$

where $E^{PS_{j,t}]}$ is the energy consumption of power-shiftable appliance j at time t with the upper boundary of $E_{j,max}^{PS}$ and Phi^{PS} is the set of power-shiftable appliances. The first term of the reward function is the power cost and the dissatisfaction cost due to variation in the power level of the appliance is represented by the second term. The dissatisfaction cost α_j^{PS}, can be changed to accomplish a trade-off between the price of electricity and the degree of satisfaction. The reward of time-shiftable appliances' agent is

$$r_{k,t}^{TS} = -\lambda_t \left[u_{k,t} E_{k,t}^{TS} \right]^+ - \alpha_k^{TS} (t_m^s - t_m^{init})^2, \quad j \in \Phi^{TS}, \quad t \in [t_k^{init}, t_k^{end}]. \tag{9}$$

where $u_{k,t}$ has only two states, i.e., the operating state of time-shiftable appliance k in time slot t, i.e., $u_{k,t} = 1$ (ON) or $u_{k,t} = 0$ (OFF). The first term of the reward function determines the price of electricity and the second term is the dissatisfaction cost caused by the waiting period. The waiting period dissatisfaction cost comes with the coefficient α_k^{TS}. The time-shiftable appliance k ought to begin functioning at its typical working hour $[t_k^{init}, t_k^{end}]$. The total reward at each time step is formed based on Eqs. 7 to 9.

$$R_t = r_{i,t}^{NS} + r_{j,t}^{PS} + r_{k,t}^{TS} \tag{10}$$

5.4 Action-Value Function

The selected action for each state is evaluated by the action-value function which is defined as the expected sum of rewards for all the future time steps. In this study Q-learning is used to train the agents.

$$Q_\pi(s_t, a_t) = E_\pi \left[\sum_{k=0}^{K} \gamma^k r_{t+1} | s_t = s, a_t = a \right] \tag{11}$$

where the action-value function is represented by $Q_\pi(s, a)$ and π is the policy mapping between a system state to an energy consumption schedule. The discount rate $\gamma \in [0, 1]$ indicates how important future benefits are in comparison to the current reward. Finding the best policy, π^*, is the goal of the energy consumption scheduling problem. In order to maximize the action-value function, the policy takes a series of optimized actions for all appliances. The algorithm builds a Q-table by updating each state-action pair's Q-value $Q(s_t, a_t)$ after each iteration until the convergence condition is met. The best action at each iteration has the maximum Q-value. The optimal Q-value $Q_\pi^*(s_t, a_t)$ can be obtained by using the Bellman equation, described as follows.

$$Q_\pi^*(s_t, a_t) = r(s_t, a_t) + \gamma * max(Q_\pi(s_{t+1}, a_{t+1})) \tag{12}$$

At each time step the Q-value is updated as follows.

$$Q_\pi(s_t, a_t) \leftarrow Q_\pi(s_t, a_t) + \theta[Q_\pi^*(s_t, a_t) - Q(s_t, a_t)] \tag{13}$$

where $\theta \in [0, 1]$ is the learning rate. If $\theta = 0$, the agent exclusively uses the previous information. If $\theta = 1$, the agent ignores the earlier information and just considers the present information. Therefore, a value of θ between 0 and 1 is selected, balancing the new Q-value and old Q-value.

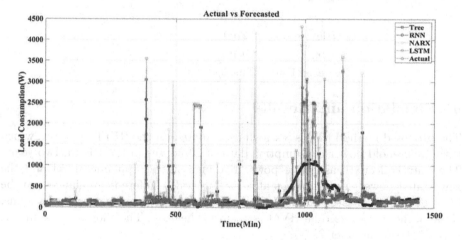

Fig. 3. The REFIT actual vs. forecasted load using NARX, LSTM, SVM, and RT methods

5.5   Structure of Proposed Multi-agent Q-Learning DR

The framework of the proposed HEM algorithm is demonstrated in Fig. 2. The components of the structure are (i) Receiving past 24-hour data of aggregated energy demand by NILM as the input, (ii) predicting the energy demand for the next day, (iii) load disaggregation and finding appliance level power demand, (iv) forming a multi-agent Q-learning algorithm for optimal load scheduling, and (v) obtaining the optimal power consumption schedule as the output.

Table 1. Errors of Load Forecasting

	MAPE	MAE	RMSE	SD
NARX	0.98	207.47	24.04	596.94
LSTM	**0.18**	**39.35**	**0.93**	**123.13**
RNN	0.77	149.02	9.50	346.86
DT	1.60	197.41	21.43	563.37

Table 2. List of Appliances and their Power Consumption in the Second Household

Appliance	Power Consumption (W)
Fridge-Freezer	95
Washing Machine	2000
Dishwasher	2250
Television	30
Microwave	1200
Kettle	2700
Toaster	950
Overhead Fan	400–800

6   Simulation and Results

The proposed method in this research is evaluated on the REFIT dataset, which is the household power consumption data of 20 houses in the UK for two years. The data collected has active power readings at the appliance level and the aggregated power consumptions at 8-sec intervals. Different appliances can be identified. This dataset has households that used gas central heating systems. None of the houses used an HVAC system for heating. The chosen house in this paper is house number 2.

Table 3. Disaggregation RMSE

Appliance	CO	FHMM	EDGE	Mean
Fridge-Freezer	159.06	31.27	51.98	41.29
Microwave	138.88	69.20	89.08	29.62
Kettle	302.92	436.81	86.47	209.27
Toaster	125.29	109.94	95.57	55.44
Television	27.34	41.30	81.43	19.73
Audio System	2.53	36.97	84.05	2.53
Washing Machine	258.38	199.92	304.09	53.43
Dishwasher	356.54	389.55	131.44	351.13
Overhead Fan	25.79	37.25	84.02	1.04

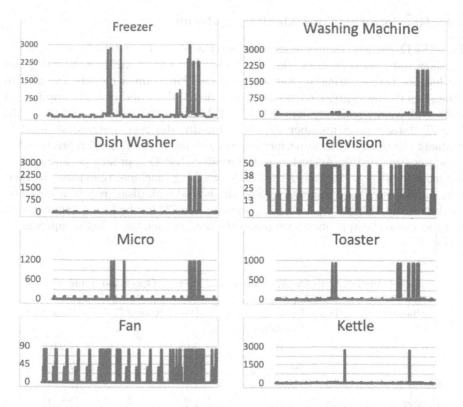

Fig. 4. Appliance-Level Power Consumption Obtained by CO Disaggregation Method

6.1 Load Forecasting Results

The power consumption of the second house in the REFIT dataset is forecasted using the NARX, LSTM, RNN, and RT algorithms trained using the data of September 16, 2013. Figure 3 the actual load versus forecasted load of four different prediction methods, NARX, LSTM, RNN, and RT, with a period of 1440 min, i.e., one day. The red line represents the actual aggregated data of house 2 in the REFIT dataset on September 27, 2013. As shown in Fig. 3, the blue line, which represents the LSTM method, follows the red line more accurately. The metrics in Eqs. 2–5 evaluate the prediction methods' effectiveness. The results are demonstrated in Table 1. The LSTM method has the lowest errors in comparison to other methods. It has a MAPE of 0.18, MAE of 39.35, RMSE of 0.93, and SD of 123.13. Therefore, the LSTM method is chosen to predict the next day's aggregated power. The experimental details of the LSTM method consist of a learning rate of 0.001, data size of 1440, 512 hidden units, Adam optimizer, and 1000 epochs.

6.2   Non-intrusive Load Monitoring Results

The REFIT dataset was first converted into the hierarchical data format (HDF5) to allow data analysis using the NILMTK. The CO, FHHM, ED, and Mean models are trained using appliance-level data. The trained models are used to disaggregate the aggregated house power demand to obtain each appliance's state and energy consumption, listed in Table 2, in the second building of the REFIT dataset on September 27, 2013. Finally, different metrics are used to evaluate the disaggregation performance of the methods. Table 3 represents the disaggregation RMSE. According to the results, the CO approach performed well for most appliances, including the fridge, washing machine, television, toaster, fan, audio system, and microwave. Still, it showed high disaggregation errors on the kettle and dishwasher. This paper chooses the CO method for disaggregation purposes, and the appliance-level power demand at each time step is represented in Fig. 4.

Table 4. Initialization for Appliances' Parameters, Operation Time Slots

Appliance	Type	Dissatisfaction Coefficient	Power Rating (kWh)	Time Slot
Fridge-Freezer	NS	100.0	0.08	[1,24]
Microwave	NS	100.0	1.2	[20,21]
Kettle	NS	100.0	2.5	[20,21]
Toaster	NS	100.0	1.2	[20,21]
Television	NS	100.0	0.05	[20,21]
Washing Machine	TS	0.10	2	[20,22]
Dishwasher	TS	0.06	2	[20,21]
Overhead Fan	PS	0.05	[0.04, 0.06, 0.08]	[1,24]

Table 5. Electricity Cost Comparison with and without DR

Appliance	Cost without DR (cents)	Cost with DR (cents)
Fridge-Freezer	40.248207	40.248207
Microwave	32.730943	32.730943
Kettle	68.189464	68.189464
Toaster	32.730943	32.730943
Television	1.363789	1.363789
Washing Machine	30.885735	14.988352
Dishwasher	30.885735	23.665837
Overhead Fan	40.248207	20.124103
Total	277.283023	234.041639

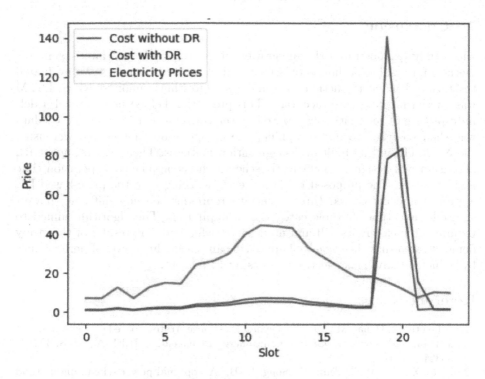

Fig. 5. Electricity Cost with and without DR

6.3   Multi-agent Reinforcement Learning Demand Response Results

At first, parameters are initialized. The discount factor γ is set to 0.9, so the obtained strategy is foresighted. To ensure that the agent can call all state-action pairs and learn new knowledge from the system, the learning rate θ is set to 0.1. Table 4 indicates the appliance list, including their type, the dissatisfaction coefficients, power ratings, and time slots. The three agents are trained using the Q-learning algorithm, the state space shown in Eq. 6. Agent 1 represents the non-shiftable appliances with an action set of $\{ON = 1\}$. Agent 2 represents the power-shiftable appliance, which is the overhead fan with the action set of $\{0.04, 0.06, 0.08\}$. Agent 3 is the representative of time-shiftable appliances with the action set of $\{ON = 1, OFF = 0\}$. The energy cost with and without DR is represented in Table 5 and in Fig. 5. The results demonstrate that the costs are decreased with the presence of the proposed energy scheduling method. There is a 15 percent drop in energy costs. Figure 5 illustrates that most of the energy is consumed during off-peak and the proposed DR algorithm has avoided energy consumption during on-peak period. The results prove that the proposed algorithm has a desirable performance.

7   Conclusion

This study demonstrated the integration of data analysis and multi-agent RL method for residential households to effective demand response. We evaluated the future demand prediction with multiple algorithms, amongst which LSTM was chosen for its greater accuracy. This prediction helped us to model a definite picture of residential aggregated power consumption. The predicted data was then analyzed to obtain appliance-level operation time and power using the NILMTK and its built-in disaggregation methods. Then, a multi-agent RL algorithm was designed to efficiently schedule home appliances' operation time and/or power. The proposed method used Q-learning with the price-based DR approach and considered three agents to represent the non-shiftable, power-shiftable, and time-shiftable categories of appliances. The algorithm aimed to minimize the electricity bill and users' dissatisfaction. The results of the study demonstrated that the suggested approach can reach a high level of performance by reducing daily home electricity costs by 15 percent.

References

1. Alfaverh, F., Denaï, M., Sun, Y.: Demand response strategy based on reinforcement learning and fuzzy reasoning for home energy management. IEEE Access **8**, 39310–39321 (2020)
2. Zhao, Z., Lee, W.C., Shin, Y., Song, K.-B.: An optimal power scheduling method for demand response in home energy management system. IEEE Trans. Smart Grid **4**, 1391–1400 (2013)
3. Zhou, K., Yang, S.: Understanding household energy consumption behavior: the contribution of energy big data analytics. Renew. Sustain. Energy Rev. **56**, 810–819 (2016)
4. Ogunjuyigbe, A.S.O., Ayodele, T.R., Akinola, O.A.: User satisfaction-induced demand side load management in residential buildings with user budget constraint. Appl. Energy **187**, 352–366 (2017)
5. Chen, S.-J., Chiu, W.-Y., Liu, W.-J.: User preference-based demand response for smart home energy management using multiobjective reinforcement learning. IEEE Access **9**, 161627–161637 (2021)
6. Zazo, J., Zazo, S., Macua, S.-V.: Robust worst-case analysis of demand-side management in smart grids. IEEE Trans. Smart Grid **8**, 662–673 (2017)
7. Pipattanasomporn, M., Kuzlu, M., Rahman, S.: An algorithm for intelligent home energy management and demand response analysis. IEEE Trans. Smart Grid **3**, 2166–2173 (2012)
8. Abbas, F., Feng, D., Habib, S., Rahman, U., Rasool, A., Yan, Z.: Short term residential load forecasting: an improved optimal nonlinear auto regressive (NARX) method with exponential weight decay function. Electronics **7**, 432 (2018)
9. Sanjari, M.-J., Karami, H., Yatim, A.-H., Gharehpetian, G.-B.: Application of Hyper-Spherical Search algorithm for optimal energy resources dispatch in residential microgrids. Appl. Soft Comput. **37**, 15–23 (2015)
10. Marzband, M., Yousefnejad, E., Sumper, A., Domínguez-García, J.-L.: Real time experimental implementation of optimum energy management system in standalone Microgrid by using multi-layer ant colony optimization. Int. J. Electr. Power Energy Syst. **75**, 265–274 (2016)

11. Chenthur Pandian, S., Duraiswamy, K., Christober Asir Rajan, C., Kanagaraj, N.: Fuzzy approach for short term load forecasting. Electr. Power Syst. Res. **76**(6–7), 541–548 (2006)
12. Xiao, L., Shao, W., Yu, M., Ma, J., Jin, C.: Research and application of a combined model based on multi-objective optimization for electrical load forecasting. Energy **119**, 1057–1074 (2017)
13. Chen, B.-J., Chang, M.-W., lin, C.-J.: Load forecasting using support vector Machines: a study on EUNITE competition 2001. IEEE Trans. Power Syst. **19**(4), 1821–1830 (2004)
14. Kouhi, S., Keynia, F.: A new cascade NN based method to short-term load forecast in deregulated electricity market. Energy Convers. Manage. **71**, 76–83 (2013)
15. Masood, Z., Gantassi, R., Ardiansyah, Choi, Y.: A multi-step time-series clustering-based Seq2Seq LSTM learning for a single household electricity load forecasting. Energies **15**(7), 2623 (2022)
16. Murray, D., et al.: A data management platform for personalised real-time energy feedback. In: Proceedings of the 8th International Conference on Energy Efficiency in Domestic Appliances and Lighting. IET (2015)
17. Yang, L., Chen, X., Zhang, J., Poor, H.-V.: Cost-effective and privacy-preserving energy management for smart meters. IEEE Trans. Smart Grid **6**(1), 486–495 (2015)
18. Beckel, C., Kleiminger, W., Cicchetti, R., Staake, T., Santini, S.: The ECO data set and the performance of non-intrusive load monitoring algorithms. In: Proceedings of the 1st ACM Conference on Embedded Systems for Energy-Efficient Buildings (BuildSys 2014), pp. 80–89. Association for Computing Machinery, New York (2014)
19. Zeifman, M., Roth, K.: Nonintrusive appliance load monitoring: review and outlook. IEEE Trans. Consum. Electron. **57**(1), 76–84 (2011)
20. Xin, W., Han, L., Wang, Z., Qi, B.: A nonintrusive fast residential load identification algorithm based on frequency-domain template filtering. IEEJ Trans. Electr. Electron. Eng. **12**, S125–S133 (2017)
21. Iliaee, N., Liu, S., and Shi, W.: Non-intrusive load monitoring based demand prediction for smart meter attack detection. In: International Conference on Control, Automation and Information Sciences (ICCAIS), pp. 370–374, Xi'an, China (2021)
22. Batra, N., et al.: NILMTK: an open source toolkit for non-intrusive load monitoring. In: Proceedings of the 5th International Conference on Future Energy Systems (e-Energy 2014), pp. 265–276. Association for Computing Machinery, New York (2014)
23. Batra, N., et al.: Towards reproducible state-of-the-art energy disaggregation. In: Proceedings of the 6th ACM International Conference on Systems for Energy-Efficient Buildings, Cities, and Transportation (BuildSys 2019), pp. 193–202. Association for Computing Machinery, New York (2019)
24. RNN. https://www.educba.com/recurrent-neural-networks-rnn/
25. Ding, Q.: Long-term load forecast using decision tree method. In: 2006 IEEE PES Power Systems Conference and Exposition, pp. 1541–1543, Atlanta, GA (2006)
26. LSTM. https://notesonai.com/LSTM

Decision Poisson: From Universal Gravitation to Offline Reinforcement Learning

Heqiu Cai[1], Zhanao Zhang[2], Zhicong Yao[1], Kanghua Mo[1], Dixuan Chen[1], and Hongyang Yan[1(✉)]

[1] Guangzhou University, Guangzhou 510006, China
chqstudy@gmail.com
[2] Nanjing University, Nanjing 210093, China

Abstract. Viewing offline reinforcement learning (RL) through the lens of conditional generative modeling has gradually become more accepted by researchers as a novel sequence modeling approach. Diffusion models have many advantages as state-of-the-art methods, but their repeated forward and reverse diffusion steps can be computationally demanding for large, high-dimensional data. Here we develop a new policy for offline RL based on Poisson flow generative modeling that does not rely on Gaussian assumptions. Our method achieves improved evaluation metrics, faster sample generation, and increased robustness to hyperparameters and model architectures. This also enables probing the significance of the underlying framework for offline sequence modeling. Ultimately, on D4rl and Minari benchmarks, our method matches state-of-the-art performance with fewer resources, further validating conditional generative modeling for decision tasks.

Keywords: Offline RL · Poisson flow generative modeling · Classifier-Guidance

1 Introduction

One consistent trend in deep reinforcement learning is that model performance improves with more parameters, effectively expanding training data requirements [1]. Rather than acquiring costly new human demonstrations or risking uncertain simulation-to-real transfers, leveraging readily available low-quality datasets is an appealing option [17]. Empirical evidence shows reinforcement learning algorithms' generalization ability tends to improve with larger model size [16]. However, for applications involving complex, real-world problems, simply scaling data quantity may not suffice due to data collection bottleneckscir [2]. To maximize information gained from limited high-quality data sources, offline reinforcement learning aims to augment training with vast amounts of unsupervised interactions [5].

J. Vaidya et al. (Eds.): AIS&P 2023, LNCS 14509, pp. 446–455, 2024.
https://doi.org/10.1007/978-981-99-9785-5_31

Offline reinforcement learning algorithms face significant stability challenges stemming from their reliance on temporal difference (TD) learning combined with function approximation [21], off-policy training, and imperfect behavior priors. In particular, the deadly triad of non-linear function approximation, updating policies with off-policy data, and incorporating prior bias can readily cause divergence during value learning [20]. Operating solely on static interaction logs further complicates algorithms, requiring careful regularization and heuristics to constrain policy updates to the support of the logged data distribution [7]. Addressing convergence and stability issues when incorporating non-linear function approximation has remained an active area of research [15]. The compounded effects of these factors pose substantial difficulties to reliably extending state-of-the-art offline RL methods [8].

Conditional generative modeling shows promise for circumventing issues with model-free value function learning. However, offline reinforcement learning (RL) planning poses unique challenges that have not been fully addressed by existing generative models. In this work, we develop a novel class of generative models directly suited for offline RL planning algorithms [23]. A key consideration for offline RL planning is that it requires optimal action sequences rather than just single-step predictions, as the model should represent long-horizon action distributions rather than solely state transitions [14]. Additionally, to generalize across different tasks, the model should remain reward-agnostic. By following offline planning principles over causal environment dynamics learned from logged data [24], we aim to develop a reward-agnostic generative model whose samples can be evaluated by a planner without distribution shift. Specifically, sampling from the model would produce trajectories that are directly evaluable by a planner, eliminating mismatches hindering policy optimization from logged data alone.

Inspired by planning diffuser [11], we instantiate this idea with a state-sequence based Poisson Flow Generative Model [22]called Decision Poissoner(DP). Compared to prior work, our main contributions are as follows:

1. We optimize trajectory fidelity over single timesteps: Our approach, referred to as DP, directly optimizes the fidelity of full generated trajectories as opposed to errors at each timestep. This circumvents compounding predictive mistakes seen in one-step models, improving scalability for long-term planning.

2. We produce globally coherent plans: By substituting the conventional U-Net with a retention mechanism for iterative local refinement, we are able to produce globally coherent trajectories by stitching together consistent subsequences. This endowed ability of DP allows it to synthesize novel long-term plans by recombining familiar motion patterns.

3. We conduct extensive experiments on benchmark tasks which demonstrate that our approach achieves markedly better performance than alternatives in terms of accuracy, rewards, and time efficiency - especially on long-range and complex scenarios. Our flexible integration of constraints and reward functions during sampling enables us to apply our approach to a broader range of tasks, including those unseen during our training.

2   Related Work

Conditional generative models have shown promising results for offline RL. Prominent examples include PETS [4] and MOReL [12], which learn compact latent dynamics models to plan online. SceneDiffuser [10] incorporates model uncertainty while planning. However, most methods focus on modeling state transitions rather than full action sequences [16].

Some recent works adopt sequential modeling. Decision Transformer [3]learns return-conditioned policies with transformers. Trajectory Transformer [8] models sequence probabilities via transformers. While powerful, transformers have high computational costs.

Diffusion models have shown remarkable image generation capabilities. Denoising diffusion probabilistic models (DDPM) [9] models noise-to-data transitions. D3P [18] improves sampling efficiency. GLIDE [25] generates images from text. Recent works apply diffusion models to RL [10]. Our work explores an alternative generation process based on Poisson flows.

3   Methodology

Leveraging offline data to solve RL problems is highly meaningful, but we do not auto regressively predict states and actions in temporal order. We also avoid reliance on TD learning or risking distribution shift. Instead, we formulate sequential decision making as a standard conditional generative modeling problem guided by classifiers: We choose a Poisson gravitational field model as a goal-conditioned trajectory generator, with states and actions at each timestep over a planning horizon H jointly generated for entire trajectories, so generated actions can directly inform planning. A function $R(\tau)$can guide the generator to find trajectories conforming to both the target distribution $p_\theta(\tau)$ and constraint conditions under $R(\tau)$using expected returns or cumulative discounted rewards. To prevent overly aggressive guidance by high-reward goals, conditionally generated trajectories may not always obey dynamics constraints, causing difficulties for a planner trying to follow the envisioned trajectory when interacting with the environment. First, in Sect. 4.1 we discuss the Poisson generative modeling choice. Next, in Sect. 4.2, we describe how goal-conditioned guidance is inc orporated to capture optimal trajectories. Then in Sect. 4.3, we discuss replacing the commonly-used U-Net backbone with a Retnet using a retention mechanism more suitable for sequence modeling. Finally, Sect. 4.4 covers the practical training details of our method (Fig. 1).

3.1   Constructing Trajectory Generation Model

The gravitational force between two point masses is directly proportional to the product of their masses and inversely proportional to the square of the distance between them. Due to the linear additivity of gravitational fields, if there are multiple sources of gravity, each representing a point where a real sample should

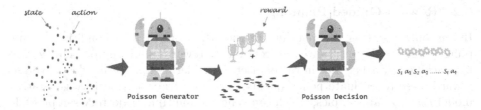

Fig. 1. Planning with Poisson Decision. Given the current state and action, Poisson Decision uses Classifier-Guidance generate a sequence of future states and action, It then extract the best trajectory by modifying the means μ of the reverse process according

be generated, then any distant point that moves along the field lines can evolve into a real sample point. We learn normalized Poisson fields from data by simulating a forward ODE, anchored by an equivalent backward ODE in the extra dimension. Generative modeling then proceeds via the reverse ODE, implying a certain "backwardness" in time during decision-making and dynamics prediction. Specifically, dynamics predictions follow causal relationships, treating the current state as determined by past states. But decision generation cannot be determined solely by the past and present; This approach considers multiple timesteps and future goal state constraints jointly during planning. Specifically, it generates the full sequence across the planning horizon in a reverse Markovian fashion. Additionally, it leverages Poisson fields to enable local consistency between neighboring states within the sequence. These locally consistent groups of states are then hierarchically stacked to gradually expand the scope of consistency approximation, aiming to finally achieve global consistency across the entire generated sequence over the full planning horizon.

In offline RL, predicting states alone may be easier, but to prevent generated state trajectories from departing physical realities and becoming unbridgeable by actions, we jointly predict actions as an extra state dimension. Concretely, with each time-step as a column within the planning horizon, the generator input and output are represented as 2D arrays. We introduce $s(x, t)$ to train the vector field function, with the training objective as follows.

$$\mathbb{E}_{\boldsymbol{x}_0 \sim \tilde{p}(\boldsymbol{x}_0)} \left[\left\| \boldsymbol{s}_\theta(\boldsymbol{x}, t) + \frac{(\boldsymbol{x} - \boldsymbol{x}_0, t)}{(\|\boldsymbol{x} - \boldsymbol{x}_0\|^2 + t^2)^{(d+1)/2}} \right\|^2 \right] \tag{1}$$

We construct (x, t) samples for each real trajectory by perturbation:

$$\tag{2}$$

$$\boldsymbol{x} = \boldsymbol{x}_0 + \|\boldsymbol{\varepsilon}_x\|(1 + \tau)^m \boldsymbol{u}, \quad t = |\varepsilon_t|(1 + \tau)^m$$

Among them: $(\varepsilon_x, \varepsilon_t) \sim N(0, \sigma^2 I_{(d+1)\times(d+1)})$, $m \sim \mathcal{U}[0, M]$, u is a unit vector uniformly distributed on the d-dimensional unit hypersphere, and τ, σ, M are constants

3.2 Reward-Guided Planning

In the offline setting, we reformulate the RL problem as a conditional sampling problem. We introduce the concept of rewards and approximate values by tracking reward signals in the dataset. Considering long-term effects, we use reward-to-go to evaluate policy quality. First, we train a trajectory generative model $p_\theta(\tau)$ on all samples, which can generate multiple trajectory samples following the data distribution as a sampling basis. Concurrently, we train a value prediction model J_ϕ that can assess the cumulative reward achievement of each sample trajectory τ.

Connecting the generator and classifier, we utilize the classifier's gradient signals to modify parameters inside the generator model. Based on value ratings from J_ϕ, we modify the mean parameters μ of the reverse process in p_θ. The gradient guidance steers p_θ to produce new samples biased toward high-value regions. We repeat sampling until acquiring a high-quality trajectory τ. Integrating the generator and classifier provides an effective approach to solving RL sampling and optimization problems in an offline manner.

Algorithm 1. Learning Normalized Poisson Field

Input: Training iteration T, initial model f_θ, dataset D, constant ϵ, learning rate η
1: Initialize f_θ with random weights θ
2: **for** $t = 1$ to T **do**
3:     Sample a large batch B_L from dataset D
4:     Simulate the ODE
5:     Calculate the normalized field
6:     Compute loss function: $\mathcal{L}(\theta) = \frac{1}{|B|} \sum_{i=1}^{|B|} |f_\theta(\tilde{y_i}) - v_{B_L}(\tilde{y_i})|_2^2$
7:     Update model parameters: $\theta \leftarrow \theta - \eta \nabla_\theta \mathcal{L}(\theta)$
8: **end for**
9: **return** f_θ

3.3 Strategies for More Robust Neural Network Architecture

To benefit from recent architectural trends through inheriting best practices and training methods from other domains, while retaining advantageous properties like scalability, robustness, and efficiency, we explored the impact of generator backbone networks on model performance. We found that the inductive bias of common U-Net architectures is not critical for generative model performance.

Considering our desire for better global influence, we adopted retentive networks (RetNets) [19] with multi-scale retention mechanisms in place of Transformer multi-head attention. Multi-scale retention enables efficient parallel computation, eliminating RNN dependence on sequence path lengths to fully utilize GPUs. Each position's token can directly interact and exchange information with the global context, allowing more thorough learning of global feature representations, superior to U-Nets' hierarchical aggregation through layers.

Secondly, the recursive representation achieves efficient O(1) inference in memory and computation. Deployment costs and latency can be significantly reduced. Moreover, no key-value caching techniques greatly simplify implementation. Chunk wise recursive representation enables efficient long sequence modeling. We encode local blocks in parallel to capture local neighborhood information, and more importantly can model longer-range dependencies, while recursively encoding global blocks to save GPU memory. This allows us to attain strong performance at lower cost, faster speed.

4   Experiments

4.1   Environments Settings

we first comprehensively evaluate how well each method performs when learning from different types of offline data which consist of three environments - half cheetah, hopper and walker2d. Each environment has three datasets collected using different data collection policies. Medium policy: A suboptimal policy whose performance is about one third that of the expert policy. Medium-expert mixture: A dataset containing a mix of transitions from both the medium and expert policies. Medium-replay: The replay buffer collected when training a policy to reach the performance level of the medium policy [6].

We also conducted experiments on more complex Minari Adroit tasks, which require controlling a 24-DoF robotic hand to perform tasks such as aligning a pen, hammering a nail, or opening doors. We use two types of datasets for each environment: human datasets, which contain 25 trajectories demonstrated by humans; and cloned datasets, which are a 50–50 mix of demonstration data and trajectories from a cloned policy of the demonstration (Fig. 2).

We evaluate long-horizon planning in the Maze2D environments. Maze2D is a 2D navigation task where the goal is for an agent to traverse from a randomly designated start location to a fixed goal location in the map. The reward scheme is: 1 if the agent succeeds in reaching the goal, 0 otherwise. Maze2D can evaluate the ability of RL algorithms to stitch together previously collected sub-trajectories to find the shortest path to evaluation goals. We use the agent's scores as the evaluation metric. So Maze2D tests both the planning and navigation abilities of the agent, as well as the agent's ability to leverage the offline dataset for online planning.

4.2   Baseline

We compare our method against existing offline RL algorithms, including model-free methods like CQL [14] and IQL [13], as well as model-based methods like trajectory transformers [8]and MoReL (Kidambi et al., 2020). We also compare against sequence models like decision transformers (DT) [3]. In order to ensure a fair evaluation across all methods, we tried to keep the general hyperparameters consistent as much as possible while tuning method-specific hyperparameters.

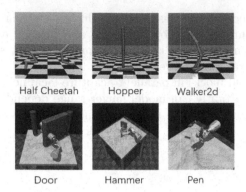

Half Cheetah   Hopper   Walker2d

Door   Hammer   Pen

Fig. 2. Performance of AsyDyn and HoneyBadger BFT

Table 1. Normalized average returns on D4RL Gym tasks, averaged over 3 random seeds.

Environment	Dataset	BC	CQL	IQL	DT	TT	MOPO	MBOP	DP(ours)
Halfcheetah	Med_expert	55.2	91.6	86.7	86.8	**95.0**	63.3	105.9	93.7 ± 0.6
Hopper	Med_expert	52.5	105.4	91.5	107.6	**110.0**	23.7	55.1	**109.2 ± 0.3**
Walker2d	Med_expert	107.5	108.8	**109.6**	108.1	101.9	44.6	70.2	108.4 ± 1.2
Halfcheetah	Medium	42.6	44.0	47.4	42.6	46.9	42.3	44.6	**48.2 ± 1.2**
Hopper	Medium	52.9	58.5	66.3	67.6	61.1	28.0	48.8	**76.2 ± 1.4**
Walker2d	Medium	75.3	72.5	78.3	74.0	79.0	17.8	41.0	**81.2 ± 0.7**
Halfcheetah	Med_Repaly	36.6	45.5	44.2	36.6	41.9	**53.1**	42.3	42.6 ± 0.2
Hopper	Med_Repaly	18.1	95.0	94.7	82.7	91.5	67.5	12.4	**97.3 ± 0..5**
Walker2d	Med_Repaly	26.0	77.2	73.9	66.6	82.6	39.0	9.7	**72.4 ± 0.3**

Specifically, when benchmarking our proposed approach against existing offline RL algorithms like CQL, IQL, trajectory transformers, MoReL [12], as well as sequence models like decision transformers and diffusion models like Diffuser, we used the same general experimental settings, dataset splits, and evaluation protocols. The hyperparameters we tuned in a method-specific manner were those intrinsically tied to each algorithm's internal workings, such as the Lagrangian multipliers in CQL. But aspects like the network architecture, batch size, dataset splits, and evaluation metrics were kept identical wherever possible for an apples-to-apples comparison. This consistent and fair experimental setup allows us to make meaningful comparisons between our proposed approach and the state-of-the-art in offline RL. The results thus accurately highlight the relative strengths and weaknesses of each method.

4.3   Results

We compared the performance of Poisson Decision Process (DP) against baseline methods on the Halfcheetah, Hopper, and Walker2d environments under the single-task setting. The results, summarized in Table 1, demonstrate that DP achieves competitive performance compared to the other baselines, outperforming model-based methods like MOReL and MBOP.

Table 2. Normalized average returns on Minari Adroit tasks, averaged over 3 random seeds.

Enviroment	Dataset	BC	SAC	CQL	DP(ours)
Adroit Pen	Expert	25.8 ± 8.8	4.3 ± 3.8	35.2 ± 6.6	$\mathbf{55.3 \pm 2.6}$
Adroit Hammer	Expert	$\mathbf{3.1 \pm 3.2}$	0.2 ± 0.0	0.6 ± 0.5	$\mathbf{2.6 \pm 1.5}$
Adroit Door	Expert	2.8 ± 0.7	-0.3 ± 0.0	1.2 ± 1.8	$\mathbf{13.8 \pm 1.7}$

Table 3. Normalized average returns on Maze2D task, averaged over 3 random seeds.

Enviroment	Dataset	MPPI	CQL	IQL	DP(ours)
Maze2D	U-Maze	33.2	5.7	47.4	$\mathbf{109.3 \pm 1.6}$
Maze2D	Medium	10.2	5.0	34.9	$\mathbf{117.3 \pm 3.3}$
Maze2D	Large	5.1	12.5	58.6	$\mathbf{121.3 \pm 2.7}$

Compared to other sequence modeling approaches such as Decision Transformer and Trajectory Transformer, which are both based on Transformer architectures, DP performs on par without any significant difference. The former models a mapping from past data and value estimates to actions (return-conditioned policy), while the latter models sequence probabilities autoregressively in a discrete space, searching for higher-value sequences. However, unlike their expensive computational complexity and decision times often in the order of seconds or tens of seconds, DP enjoys a huge advantage with average decision times less than one-tenth of a second In the more complex Adroit Hand tasks, the results are summarized in Table 2. The high-dimensional control problem leads to increased data complexity, the high-dimensional action space causes difficulties in exploration and policy optimization, and the sparse reward functions make effective learning signals hard to obtain. The complex, multi-stage behaviors pose even greater challenges for offline RL to grasp. Compared to the mediocre performance of previous algorithms, DP results are exciting. We hypothesize this stems from the parametric benefits of replacing the backbone with Retnet.

Finally, in the Maze2d task, the results are summarized in Table 3. The Maze2d dataset covers different planning paths, allowing the algorithm to effectively generalize using this data, finding feasible solutions for new start points as well. This demonstrates decent generalization capabilities. DP stitches together previously collected sub-trajectories to find the optimal path, and the longer horizon enables it to handle long-term planning and reasoning. This advantage in credit assignment allows DP to trace back and associate the final results with its earlier decisions. Ultimately leading to outstanding performance.

5   Conclusion

We have introduced DP, a new strategy representation for goal-conditioned behavior generation using a Poisson flow model. We leverage the expressiveness of PF and the outstanding efficiency of the Retnet backbone to learn task-agnostic behaviors from offline, reward-less game datasets, without the need for hierarchical structures or additional clustering. Additionally, we have validated the effectiveness of classifier guidance in simultaneously learning goal-dependent and goal-agnostic strategies in sequential settings. Experiments on multiple D4rl benchmarks demonstrate that DP significantly improves several state-of-the-art algorithms, as well as the foresight planning capability under long horizons. Our ablation studies confirm that the key components

Acknowledgment. The authors gratefully acknowledge the financial supports by the Guangzhou basic and applied basic research Project (2023A04J1725); Funded by National Natural Science Foundation of China (No.62102107).

References

1. Agarwal, R., Schuurmans, D., Norouzi, M.: An optimistic perspective on offline reinforcement learning. In: International Conference on Machine Learning, pp. 104–114. PMLR
2. Ajay, A., Du, Y., Gupta, A., Tenenbaum, J., Jaakkola, T., Agrawal, P.: Is conditional generative modeling all you need for decision-making?
3. Chen, L., et al.: Decision transformer: Reinforcement learning via sequence modeling. 34, pp. 15084–15097
4. Chua, K., Calandra, R., McAllister, R., Levine, S.: Deep reinforcement learning in a handful of trials using probabilistic dynamics models. Advances in neural information processing systems, 31 (2018)
5. Dhariwal, P., Nichol, A.: Diffusion models beat gans on image synthesis. 34, pp. 8780–8794
6. Fu, J., Kumar, A., Nachum, O., Tucker, G., Levine, S.: D4rl: datasets for deep data-driven reinforcement learning
7. Fujimoto, S., Meger, D., Precup, D.: Off-policy deep reinforcement learning without exploration. In: International Conference on Machine Learning, pp. 2052–2062. PMLR (2019)
8. Giuliari, F., Hasan, I., Cristani, M., Galasso, F.: Transformer networks for trajectory forecasting. In: 2020 25th International Conference on Pattern Recognition (ICPR), pp. 10335–10342. IEEE
9. Ho, J., Jain, A., Abbeel, P.: Denoising diffusion probabilistic models. 33, 6840–6851
10. Huang, S., et al.: Diffusion-based generation, optimization, and planning in 3d scenes. In: Proceedings of the IEEE/CVF Conference on Computer Vision and Pattern Recognition, pp. 16750–16761

11. Janner, M., Du, Y., Tenenbaum, J.B., Levine, S.: Planning with diffusion for flexible behavior synthesis
12. Kidambi, R., Rajeswaran, A., Netrapalli, P., Joachims, T.: Morel: model-based offline reinforcement learning. Adv. Neural. Inf. Process. Syst. **33**, 21810–21823 (2020)
13. Kostrikov, I., Nair, A., Levine, S.: Offline reinforcement learning with implicit q-learning
14. Kumar, A., Zhou, A., Tucker, G., Levine, S.: Conservative q-learning for offline reinforcement learning. 33, pp. 1179–1191
15. Laroche, R., Trichelair, P., Des Combes, R.T.: Safe policy improvement with baseline bootstrapping. In International conference on machine learning, pp. 3652–3661. PMLR (2019)
16. Peebles, W., Xie, S.: Scalable diffusion models with transformers
17. Prudencio, R.F., ROA Maximo, M., Colombini, E.L.: A survey on offline reinforcement learning: Taxonomy, review, and open problems. IEEE
18. Rombach, R., Blattmann, A., Lorenz, D., Esser, P., Ommer, B.: High-resolution image synthesis with latent diffusion models. In: Proceedings of the IEEE/CVF Conference on Computer Vision and Pattern Recognition, pp. 10684–10695
19. Sun, Y., et al.: Retentive network: a successor to transformer for large language models
20. Sutton, R.S., Barto, A.G.: Reinforcement learning: An introduction. MIT press (2018)
21. Tesauro, G., et al.: Temporal difference learning and td-gammon. Commun. ACM **38**(3), 58–68 (1995)
22. Xu, Y., Liu, Z., Tegmark, M., Jaakkola, T.: Poisson flow generative models. 35, pp. 16782–16795
23. Yang, L., et al.: Diffusion models: a comprehensive survey of methods and applications
24. Yang, S., Nachum, O., Du, Y., Wei, J., Abbeel, P., Schuurmans, D.: Foundation models for decision making: Problems, methods, and opportunities
25. Zhong, Z., et al.: Guided conditional diffusion for controllable traffic simulation. In: 2023 IEEE International Conference on Robotics and Automation (ICRA), pp. 3560–3566. IEEE

SSL-ABD: An Adversarial Defense Method Against Backdoor Attacks in Self-supervised Learning

Hui Yang[1], Ruilin Yang[1], Heqiu Cai[1], Xiao Zhang[2], Qingqi Pei[3],
Shaowei Wang[1], and Hongyang Yan[1(✉)]

[1] Institute of Artificial Intelligence and Blockchain, Guangzhou University,
Guangzhou, China
hyang.yan@foxmail.com
[2] Beihang University, Beijing, China
[3] Xidian University, Xi'an, China

Abstract. Recent research work has shown that self-supervised training encoders are susceptible to backdoor attacks. When the attacker is an untrusted service provider or a malicious third party, the attacker can manipulate the training process of the encoder at will. By adding specific patches or noise to the training dataset, the attacker successfully injects a backdoor into the image encoder and shares the backdoored encoder with downstream clients. While there have been many existing works on backdoor removal for supervised learning, most of them require labeled datasets and are not suitable for self-supervised training scenarios. Our work considers how to successfully remove the backdoor from the backdoored encoder when the defender has limited available training data. In this work, we propose SSL-ABD. The key idea behind our method is to formulate it as a min-max optimization problem: first, adversarially simulate the trigger pattern, and then remove the backdoor from the backdoored encoder through feature embedding distillation. We conducted experiments against various self-supervised attack algorithms such as CTRL [1] and SSL-Backdoor [2], and successfully removed the backdoor.

1 Introduction

Self-supervised learning (SSL) [3,4] has been widely applied in various fields and has made revolutionary advances, particularly in computer vision applications. SSL is an unsupervised learning method that trains models using the inherent structure and statistical information of the data, without the need for manually labeled annotations. SSL performs well in scenarios where labeled examples are scarce. Compared to supervised learning, SSL can avoid the expensive annotation costs by training on custom prediction tasks that can generalize to many downstream tasks. Several studies have shown that SSL can achieve comparable or even better performance than supervised learning in few-shot learning scenarios. This means that SSL can leverage unlabeled data to improve model performance when only a few labeled samples are available.

© The Author(s), under exclusive license to Springer Nature Singapore Pte Ltd. 2024
J. Vaidya et al. (Eds.): AIS&P 2023, LNCS 14509, pp. 456–467, 2024.
https://doi.org/10.1007/978-981-99-9785-5_32

Recent research has indicated that self-supervised learning is vulnerable to data poisoning attacks, where attackers inject carefully crafted samples, also known as "poisoned" samples, into the unlabeled training dataset. These samples often belong to a specific class and contain "triggers" chosen by the attacker, which can be specific patterns, colors, or hidden features. The inclusion of these poisoned samples during the training of image encoders introduces backdoor behavior, which means that when using the pretrained encoder for downstream tasks such as classification, incorrect predictions may occur. This is advantageous for attackers, as the backdoor behavior only manifests under specific trigger conditions, while the encoder behaves normally at other times, making the attack difficult to detect.

In recent years, significant progress has been made in defending against backdoors [5] in supervised models. These efforts can be broadly categorized into the following aspects: (1) Backdoor detection and removal [6,7]: This approach aims to detect and remove existing backdoors in models. It often relies on model auditing and analysis to identify potential backdoor trigger conditions and eliminate them. (2) Training data filtering [8,9]: This method involves filtering and cleaning the training data to remove potentially compromised samples containing backdoors. This helps reduce the available information that attackers can exploit. (3) Defense algorithm design: Some researchers are exploring the design of more robust and secure learning algorithms to defend against backdoor attacks. These algorithms can automatically identify and resist potential backdoor attacks during the training process. (4)Ensemble of multiple models [10,11]: Utilizing multiple independently trained models for inference and decision-making can enhance system robustness. This approach can identify inconsistent predictions and reduce erroneous outputs caused by a single model compromised by a backdoor. Due to differences in training data, training processes, and model structures between SSL and supervised models, most defense techniques developed for supervised models cannot be directly applied to SSL.

Recently, some research studies have started exploring backdoor defense in SSL. These works primarily focus on training data filtering and detecting backdoored models. For instance, [12] proposes a method to detect training samples containing triggers. It trains a backdoored encoder using poisoned data and then clusters the feature vectors of a subset of the training data based on the backdoored encoder. Poisoned samples are selected from each cluster. [13] introduces a method to detect if an encoder contains a backdoor. It utilizes trigger inversion to obtain an optimal trigger from the detected encoder and determines if the encoder contains a backdoor based on the trigger size. [14] also presents a trigger inversion-based detection method and designs a backdoor removal method using the detected backdoor trigger. However, this backdoor removal work requires the defender to have access to a significant amount of training data.

In this paper, we propose SSL-ABD, which aims to enable defenders to quickly remove backdoor patterns from backdoored encoders using a small amount of unlabeled data. Firstly, we attempt to mimic backdoor behavior by adding small perturbations to the inputs to make them as different as possible from the original features, creating adversarial examples. Then, we employ a minimax approach to decrease the weights of the encoder model's neurons that are associated with the backdoor trigger. Drawing inspiration from self-supervised model knowledge distillation methods [15] and model compression [16], we use a student model to extract benign knowledge from the suspicious teacher model while simultaneously eliminating the student model's own backdoor behavior.

2   Related Work

2.1   Self-supervised Learning

Self-supervised learning is a machine learning method that trains models by generating targets or tasks from the input data itself, without the need for manually labeled annotations. Unlike traditional supervised learning, self-supervised learning does not rely on externally provided human labels. Instead, it leverages the intrinsic information present in the input data for training. In self-supervised learning, algorithms perform some form of transformation or prediction task on unlabeled data. These tasks can include tasks such as filling in missing parts, image rotation recovery, colorizing black and white images, and more. By completing these tasks, the model can learn features such as the structure, semantics, and contextual relationships in the data, thereby obtaining useful representations. Once the model has learned good representations through self-supervised training, these representations can be transferred to other specific supervised tasks for fine-tuning, aiming to improve performance and generalization. Self-supervised learning has been widely applied in computer vision, natural language processing, speech recognition, and other fields. It provides an effective way to leverage unlabeled data for pretraining, overcoming the limitations of traditional supervised learning that requires a large amount of annotated data. SimCLR (Simple Contrastive Learning) and MoCo v2 (Momentum Contrastive Learning) are two common SSL methods. These SSL techniques rely on instance discrimination, where models learn competitive visual representations by training with contrastive losses.

2.2   Backdoor Attack on Self-supervised Learning

In neural networks, a backdoor attack refers to the manipulation of a trained model to trigger pre-defined malicious behavior under specific conditions. This attack method aims to introduce hidden vulnerabilities or malicious functionalities without affecting the normal performance of the model. One common approach for backdoor attacks is to embed a backdoor by intentionally modifying the training data or labels. Attackers can insert specific patterns, symbols, or noise into the training data or change the labels of training samples. Another method for backdoor attacks is to embed a backdoor by adjusting the weights or structure of the neural network. Attackers may utilize specific optimization algorithms or objective functions to incorporate the backdoor characteristics into the model during the training process. Backdoor attacks can pose a serious threat to the security and trustworthiness of neural network models, especially in scenarios involving sensitive data or critical tasks.

Recently, [2] proposed for the first time that self-supervised model training can be vulnerable to backdoor attacks. They suggested pasting triggers selected by the attacker at random positions in a small number of training samples. After data processing, some data still contain triggers, successfully injecting a backdoor into the encoder during training. However, triggers pasted at random positions may be partially filtered out after data processing, resulting in a low poisoning rate that affects the success rate of the backdoor attack. To address this issue, [17] proposed adding triggers after data augmentation to ensure a higher poisoning rate. Additionally, the authors suggested performing data augmentation three times for each image, with two augmentations used to train the model for high accuracy on clean samples, and the third augmentation used to train the correlation between the trigger and the target image. Furthermore, [1] introduced a covert backdoor attack scheme by adding the backdoor in the frequency domain of the image, making it invisible to the human eye. To ensure that triggers can bypass data augmentation operations, the authors used global triggers. The aforementioned scenarios consider attackers poisoning the data before or during encoder training. In addition to these scenarios, [18] assuming the attacker can obtain a pre-trained clean encoder, they can conduct a backdoor attack on it and then share the backdoored encoder with downstream clients. The success rate of this attack in their work can reach 99%.

2.3 Backdoor Defence on Self-supervised Learning

Recently, researchers have been investigating defense mechanisms against backdoor attacks in self-supervised model training. Among them, [12] focuses on detecting poisoned samples in the training dataset. They train a self-supervised learning (SSL) model on poisoned data and use it to identify poisoned samples. [19] accomplishes the detection of poisoned samples by computing the differential behavior of the model on poisoned samples and benign samples. [13] proposes a scheme for detecting whether an encoder contains a backdoor. The authors observe that the feature representations of poisoned samples exhibit higher similarity compared to clean samples. They leverage this observation by adding initialized triggers to a portion of the training data as poisoned samples and optimizing the triggers through the encoder to maximize the similarity between their feature representations until the triggers reach a minimum size. Backdoored encoders tend to have smaller reversed triggers compared to clean encoders, and the authors exploit this characteristic to detect backdoored encoders. Similarly, [14] proposes trigger reversal as a method to detect backdoored encoders. They first cluster a portion of the training data into k clusters using the encoder. For each cluster, they reverse multiple candidate triggers using data from other clusters. Finally, they check if there exists a trigger that is significantly smaller than the other triggers, indicating the presence of a backdoor in the encoder. Additionally, this work suggests using the reversed triggers and a subset of training samples to mitigate the impact of the backdoor.

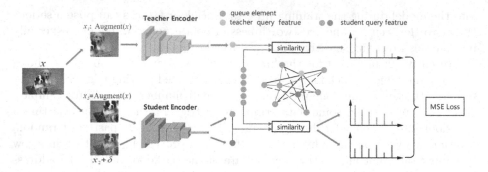

Fig. 1. Weight mask and feature embedding distillation

3   Methodology

In this section, we will introduce the method we propose. Inspired by adversarial work [20], we present Self-Supervised Adversarial Backdoor Defense (SSL-ABD) to remove backdoors from self-supervised training image encoders, even with limited available training data. Our approach consists of two steps: first, using adversarial perturbations, we find a "channel" from the feature space of clean samples to the feature space of backdoor samples, and then we use perturbations to break the boundary between these two feature spaces.

3.1   Simulating the Backdoor Trigger Pattern Using Adversarial Perturbations

[21] proposed a method for simulating the backdoor trigger pattern in a supervised learning scenario by adding small perturbations to inputs to mimic backdoor samples. The goal is to manipulate the predicted labels to deviate as much as possible from the true labels, thereby identifying the trigger pattern that classifies the backdoor samples into the target class. In contrast to the supervised learning scenario, self-supervised training data is unlabeled, and the injection of backdoors relies on establishing a relationship between the trigger pattern and the target class features. In Fig. 2, we can observe a significant difference in the features extracted by the backdoored encoder between clean samples and backdoor samples. Two views that originally had high similarity experience a reduction in feature similarity when one of them is augmented with a trigger pattern. Based on this observation, we attempt to simulate the backdoor trigger pattern using adversarial perturbations. By adding small perturbations to one of the views of a benign sample, we gradually reduce the feature vector similarity between the two views extracted by the backdoored encoder, thereby simulating the gap between the feature spaces of clean samples and backdoor samples in the presence of the backdoor model. Specifically, for an existing backdoored encoder f_θ, given a sample x_i that undergoes data augmentation to obtain different views

x_{i1} and x_{i2}, we can optimize the following objective to adversarially recover the trigger pattern:

$$\min_{||\delta||\leq\varepsilon} L_{similarity}(f_\theta(x_{i1}), f_\theta(x_{i2} + \delta)) \tag{1}$$

$|| \cdot ||$ is L_1 regularization, and δ represents the disturbance budget.The main objective of Equation (1) is to find an L_1 regularized perturbation that significantly deviates the features extracted by the encoder from the desired features.

Algorithm 1. Self-supervised Learning Adversarial Backdoor Defence(SSL-ABD)

Input: Infected Encoder f with θ; Clean dataset $D = \{x_i\}_{i=1}^n$; Batch size b; Learning rate η_1, η_2; Hyper-parameters α, Epochs E; Inner iteration loops T, L_1 norm bound τ; Queue Z;

Output: masks m for weights in encoder θ^S

 1: Initialize all elements in m as 1,Z
 2: **for** $i = 1$ to E **do**
 3:     Initialise δ as 0
 4:     **for** $t = 1$ to T **do**
 5:         $x_1 = \text{aug}(x)$, $x_2 = \text{aug}(x)$
 6:         $L_{inner} = \text{similarity}(f_\theta(x_1), f_\theta(x_2 + \delta))$
 7:         $\delta = \delta + \eta_1 \bigtriangledown_\delta L_{inner}$
 8:     Clip $\delta : \delta = \delta * \min(1, \frac{\tau}{||\delta||_1})$
 9:     Initialize $\theta^T = \theta^S = \theta$
10:     **for** $t = 1$ to T **do**
11:         $x_1 = \text{aug}(x)$, $x_2 = \text{aug}(x)$
12:         $S^T(x_1; \theta_T; Z) = similarity(f_\theta^T(x_1), Z)$
13:         $S^S(x_2; \theta_S; Z) = similarity(f_\theta^S(x_2), Z)$
14:         $S^{S*}(x_2 + \delta; \theta_S; Z) = similarity(f_\theta^S(x_2 + \delta), Z)$
15:         $L_{outer} = \alpha L_{MSE}(S^T, S^S) + (1 - \alpha) L_{MSE}(S^T, S^{S*})$
16:         $m = m + \eta_2 \bigtriangledown_m L_{outer}$
17:         Clip m to [0,1]
18:         Enqueue(Z, $f_\theta^T(x_1)$)

3.2　Weight Mask and Feature Embedding Distillation

Our work is similar to some research in the field of self-supervised knowledge distillation [15]. Our main objective is to remove the backdoor pattern from an existing backdoor model while preserving its feature extraction capability.We establish a queue to store the output features of a teacher encoder for the data. As shown in Fig. 1, given a data sample, we first perform data augmentation to generate two different views. For one of the views, we use the teacher encoder to calculate its similarity scores with all samples in the queue. For the other view, we create two copies and apply perturbations to one copy while leaving the other copy unchanged. Then, we use the student encoder to compute similarity scores between the two copies and the samples in the queue. We aim for the similarity scores computed by the student encoder to match those computed by the teacher encoder, thereby minimizing the discrepancy between the similarity scores of the student and teacher encoders.

Specifically, for an existing backdoor encoder f_θ, we set $f_\theta^T = f_\theta^S = f_\theta$, where f_θ^T acts as a teacher encoder and f_θ^S acts as a student encoder. Given a sample x_i, we perform data augmentation to obtain two different views, denoted as x_{i1} and x_{i2}. We input x_{i1} and x_{i2} separately to f_θ^T and f_θ^S , resulting in feature vectors $z_i^T = f_\theta^T(x_{i1})$ and $z_i^S = f_\theta^S(x_{i2})$. Additionally, we apply an adversarial perturbation δ to x_{i2} and input it to f_θ^S, resulting in $z_i^{S*} = f_\theta^S(x_{i2} + \delta)$. Let $Z = [z_1, z_2, ..., z_k]$ denote the instance queue, where k represents the queue length and z_i represents the feature vector obtained from f_θ^T. The existence of Z is similar to the queue in contrastive learning algorithms like Moco, where the queue stores the feature vectors obtained by inputting given samples x_{i1} to f_θ^T. During the distillation process, the content of Z is dynamic, with the earliest stored feature vectors being cleared as we enqueue the feature vectors of the current batch of samples. We calculate the similarity scores of z_i^{S*}, z_i^S, z_i^T and the samples in the queue Z respectively. Then, by minimizing the similarity scores between z_i^{S*} and z_i^T, as well as z_i^S and z_i^T, we update f_θ^S to ensure the successful removal of the backdoor from f_θ^S without causing a significant decrease in accuracy. Let $S^T(x_{i1}, \theta^T, Z)$ denote the similarity score between the feature vector z_i^T extracted by the teacher encoder and the instance feature queue $Z = [z_1, z_2, ..., z_k]$, defined as

$$S^T(x_{i1}, \theta^T, Z) = [s_1^T, s_2^T, ..., s_k^T], s_j^T = \frac{exp(z_i^T \cdot z_j/T)}{\sum_{d=1}^k exp(z_i^T \cdot z_d/T)} \qquad (2)$$

T represents the temperature parameter. Similarly, let $S^S(x_{i2}, \theta^S, Z)$ and $S^S(x_{i2} + \delta, \theta^S, Z)$ denote the similarity scores of the student-extracted clean sample features and perturbed sample features to the instance cohort, respectively, defined as

$$S^S(x_{i2}, \theta^S, Z) = [s_1^S, s_2^S, ..., s_k^S], s_j^S = \frac{exp(z_i^S \cdot z_j/T)}{\sum_{d=1}^k exp(z_i^S \cdot z_d/T)} \qquad (3)$$

$$S^{S*}(x_{i2} + \delta, \theta^S, Z) = [s_1^{S*}, s_2^{S*}, ..., s_k^{S*}], s_j^{S*} = \frac{exp(z_i^{S*} \cdot z_j/T)}{\sum_{d=1}^k exp(z_i^{S*} \cdot z_d/T)} \qquad (4)$$

Our backdoor removal formulation can be expressed as computing the sum of losses between $S^T(x_{i1}, \theta^T, Z)$ and $S^S(x_{i2}, \theta^S, Z)$, $S^T(x_{i1}, \theta^T, Z)$ and $S^{S*}(x_{i2} + \delta, \theta^S, Z)$ over all samples. We use the mean squared error to evaluate the loss value between the two. The model parameters of the teacher encoder f_θ^T are fixed, while the loss value is used to train the student encoder f_θ^S. For f_θ^S, we add an additional mask to all network weights. Instead of updating the weights of f_θ^S directly during training, an appropriate mask value is learned for each parameter weight. The formula is as follows:

$$\min_{m \in [0,1]} E_{x_i \sim D} \alpha L_{MSE}(S^T(x_{i1}, \theta^T, Z), S^S(x_{i2}, m \odot \theta^S, Z))$$
$$+(1-\alpha)L_{MSE}(S^T(x_{i1}, \theta^T, Z), S^{S*}(x_{i2}+\delta, m \odot \theta^S, Z)) \tag{5}$$

where α represents a hyperparameter. m represents the weight mask, the value range is [0,1], and the initial value is 1.

4   Experiments

In this section, we conduct experiments to verify the effectiveness of our proposed SSL-ABD method.

4.1   Experimental Settings

Datasets and Pre-training Image Encoders. We conduct experiments on three datasets: CIFAR10, CIFAR100 and Imagenet100. CIFAR10 includes 50,000 training images and 10,000 test images, the images are divided into 10 classes, and each image has a size of $32 \times 32 \times 3$. CIFAR100 includes 50,000 training images and 10,000 testing images of 100 classes, and each image has a size of $32 \times 32 \times 3$. Imagenet100 randomly selects 100 classes of data from the Imagenet dataset, including about 127,000 pieces of data. We pre-train the image encoder with all training images from the above dataset. By default, we use an encoder with ResNet18 [22] as the backbone and a two-layer MLP projector to map representations to a 128-dimensional latent space; moreover, we use a two-layer MLP with a hidden layer size of 128 as the downstream classifier. We use two representative contrastive learning methods: SimCLR [23] and MoCo [24].

Attacks. For the backdoor attack baseline, we considered 1) *SSL − Backdoor* with square trigger; 2) *BadEncoder*; 3) stealth-shaped attack *CTRL*. Specifically, SSL-Backdoor randomly selects 50% of the images from the specified category, applies square triggers at random positions to poison the training data set, and then uses the Moco v2 method to train for 500 rounds to obtain the poisoned encoder. Since there are only 500 images of each category in the CIFAR100 data set, in order to increase the success rate of the attack, we poison all images of the specified category. CTRL selects 1% of all training data images for poisoning in each attack, and then uses the SimCLR method to train for 500 rounds. BadEncoder first trains on clean samples for 1000 rounds to obtain a usable clean encoder, and then trains on the poisoned data set for 200 rounds to inject backdoors, also using the SimCLR method.

Evaluation Metrics. We employ two metrics: Accuracy Rate (ACC) and Attack Success Rate (ASR). ACC is the predictive accuracy of the model on a clean dataset, and ASR is defined as the ratio of the number of samples that divide samples containing triggers into the target class to the total evaluation images.

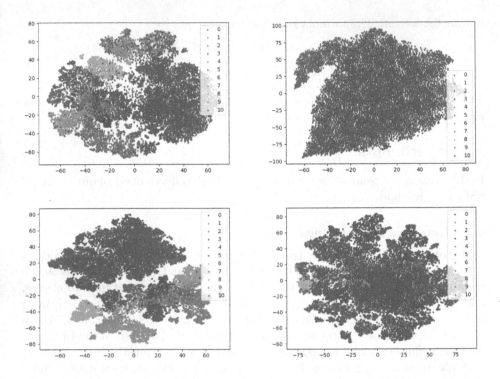

Fig. 2. The first row and second row respectively display the visualizations of features for clean and poisoned samples before(**left**) and after(**right**) the defense of poisoned encoders trained using SSL-Backdoor and CTRL attack methods on CIFAR10.

4.2   Defense Effectiveness

We studied the backdoor cleaning performance of SSL-ABD on various available data sizes. Table 1, Table 2 and Table 3 respectively list the defense effects on three different data sets. Specifically, we tested the defense effect on available data samples of different sizes in each defense, ranging from 500 to 100. For each dataset, we use all training data to train poisoning encoders. At the same time, we use 5000 training data to train downstream classifiers. For the CIFAR100 and Imagenet-100 data sets, we only select 10 categories of data for downstream classifier training and testing of various evaluation indicators.

Table 1. Backdoor removal results with various available data sizes on CIFAR10 dataset

Attack	Metric	Origin	Available Data Size			
			500	300	200	100
CTRL	ACC	85.18	84.44	84.30	85.22	83.60
	ASR	74.64	1.36	1.16	1.78	2.83
SSL-Backdoor	ACC	84.23	80.03	80.12	79.52	80.23
	ASR	93.74	3.48	5.46	7.38	7.45
BadEncoder	ACC	86.17	80.52	83.46	79	80
	ASR	99.14	9.24	7.76	8.56	6.43

Table 2. Backdoor removal results with various available data sizes on CIFAR100 dataset

Attack	Metric	Origin	Available Data Size			
			500	300	200	100
CTRL	ACC	87.69	86.6	86.2	86.3	86.8
	ASR	92.56	0.89	0.67	0.78	0.56
SSL-Backdoor	ACC	85.9	86.2	85.1	84.3	84.5
	ASR	25.67	0.89	1	1	2.11
BadEncoder	ACC	86.9	83.2	83.46	81.5	83.1
	ASR	99.3	12.67	9.5	9.67	12.11

Table 3. Backdoor removal results with various available data sizes on Imagenet-100 dataset

Attack	Metric	Origin	Available Data Size			
			500	300	200	100
CTRL	ACC	78.2	76.8	75.6	75	77.8
	ASR	17.11	1.78	1.58	1.78	2
SSL-Backdoor	ACC	85.6	85	83.6	83.2	84.4
	ASR	69.78	5.11	6.67	4	7.11

5 Conclusion

In this work, we proposed SSL-ABD, a novel approach for clearing backdoors in SSL encoders when there is limited available data. In this work, we used adversarial perturbations to recover the backdoor trigger's activation behavior. Then, we added weight masks to the encoder parameters and reduced the weights associated with the backdoor behavior using feature embedding distillation. Through experimentation, we demonstrated that our SSL-ABD method effectively clears the backdoor in the encoder in scenarios where only a few hundred clean unlabeled data samples are available. Currently, our method still requires a small amount of training data, which means that the defender needs to have some knowledge about the training dataset. However, in certain scenarios, the defender may not have any knowledge about the training set. In such cases, how to clear the backdoor in the encoder without any knowledge about the training set will be the focus of our next stage of work.

Acknowledgment. The authors gratefully acknowledge the financial supports by the Guangzhou basic and applied basic research Project (2023A04J1725); Funded by National Natural Science Foundation of China (No.62102107).

References

1. Li, C., et al.: Demystifying self-supervised trojan attacks. arXiv preprint arXiv:2210.07346 (2022)
2. Saha, A., Tejankar, A., Koohpayegani, S.A., Pirsiavash, H.: Backdoor attacks on self-supervised learning. In: Proceedings of the IEEE/CVF Conference on Computer Vision and Pattern Recognition, pp. 13337–13346 (2022)
3. Jaiswal, A., Babu, A.R., Zadeh, M.Z., Banerjee, D., Makedon, F.: A survey on contrastive self-supervised learning. Technologies **9**(1), 2 (2020)
4. Liu, X., et al.: Self-supervised learning: generative or contrastive. IEEE Trans. Knowl. Data Eng. **35**(1), 857–876 (2021)
5. Li, Y., Jiang, Y., Li, Z., Xia, S.T.: Backdoor learning: a survey. IEEE Transactions on Neural Networks and Learning Systems (2022)
6. Li, Y., Lyu, X., Koren, N., Lyu, L., Li, B., Ma, X.: Neural attention distillation: Erasing backdoor triggers from deep neural networks. arXiv preprint arXiv:2101.05930 (2021)
7. Wu, D., Wang, Y.: Adversarial neuron pruning purifies backdoored deep models. Adv. Neural. Inf. Process. Syst. **34**, 16913–16925 (2021)
8. Zeng, Y., Park, W., Mao, Z.M., Jia, R.: Rethinking the backdoor attacks' triggers: a frequency perspective. In: Proceedings of the IEEE/CVF International Conference on Computer Vision, pp. 16473–16481 (2021)
9. Chou, E., Tramer, F., Pellegrino, G.: Sentinet: Detecting localized universal attacks against deep learning systems. In: 2020 IEEE Security and Privacy Workshops (SPW), pp. 48–54, IEEE (2020)
10. Jia, J., Cao, X., Gong, N.Z.: Intrinsic certified robustness of bagging against data poisoning attacks. In: Proceedings of the AAAI Conference on Artificial Intelligence, vol.35, pp. 7961–7969 (2021)
11. Levine, A., Feizi, S.: Deep partition aggregation: Provable defense against general poisoning attacks. arXiv preprint arXiv:2006.14768 (2020)
12. Tejankar, A., et al.: Defending against patch-based backdoor attacks on self-supervised learning. In: Proceedings of the IEEE/CVF Conference on Computer Vision and Pattern Recognition, pp. 12239–12249 (2023)
13. Feng, S., et al.: Detecting backdoors in pre-trained encoders. In: Proceedings of the IEEE/CVF Conference on Computer Vision and Pattern Recognition, pp. 16352–16362 (2023)
14. Zheng, M., Xue, J., Chen, X., Jiang, L., Lou, Q.: Ssl-cleanse: Trojan detection and mitigation in self-supervised learning. arXiv preprint arXiv:2303.09079 (2023)

15. Fang, Z., Wang, J., Wang, L., Zhang, L., Yang, Y., Liu, Z.: Seed: Self-supervised distillation for visual representation. arXiv preprint arXiv:2101.04731 (2021)

16. Abbasi Koohpayegani, S., Tejankar, A., Pirsiavash, H.: Compress: Self-supervised learning by compressing representations. In: Advances in Neural Information Processing Systems, vol. 33, pp. 12980–12992 (2020)

17. Xue, J., Lou, Q.: Estas: Effective and stable trojan attacks in self-supervised encoders with one target unlabelled sample. arXiv preprint arXiv:2211.10908 (2022)

18. Jia, J., Liu, Y. and Gong, N.Z.: Badencoder: backdoor attacks to pre-trained encoders in self-supervised learning. In: 2022 IEEE Symposium on Security and Privacy (SP), pp. 2043–2059, IEEE, (2022)

19. Pan, M., Zeng, Y., Lyu, L., Lin, X. and Jia, R.: Asset: robust backdoor data detection across a multiplicity of deep learning paradigms. arXiv preprint arXiv:2302.11408 (2023)

20. Madry, A., Makelov, A., Schmidt, L., Tsipras, D., Vladu, A.: Towards deep learning models resistant to adversarial attacks. arXiv preprint arXiv:1706.06083 (2017)

21. Chai, S., Chen, J.: One-shot neural backdoor erasing via adversarial weight masking. Adv. Neural. Inf. Process. Syst. **35**, 22285–22299 (2022)

22. He, K., Zhang, X., Ren, S., Sun, J.: Deep residual learning for image recognition. In: Proceedings of the IEEE Conference On Computer Vision and Pattern Recognition, pp. 770–778 (2016)

23. Chen, T., Kornblith, S., Norouzi, M., Hinton, G.: A simple framework for contrastive learning of visual representations. In: International Conference on Machine Learning, pp. 1597–1607, PMLR (2020)

24. Chen, X., Fan, H., Girshick, R., He, K.: Improved baselines with momentum contrastive learning. arXiv preprint arXiv:2003.04297 (2020)

Personalized Differential Privacy in the Shuffle Model

Ruilin Yang, Hui Yang, Jiluan Fan, Changyu Dong, Yan Pang,
Duncan S. Wong, and Shaowei Wang[✉]

Institute of Artificial Intelligence and Blockchain, Guangzhou University, Guangzhou,
China
wangsw@e.gzhu.edu.cn

Abstract. Personalized local differential privacy is a privacy protection
mechanism that aims to safeguard the privacy of data by using personalized approaches, while also providing practical data analysis results. It
offers more flexible and precise privacy protection capabilities compared
to traditional local differential privacy. By employing distinct privacy
protection strategies for different users, it can better meet users' privacy
requirements while minimizing the impact on data. However, existing
mechanisms for personalized local differential privacy suffer from issues
such as low query accuracy and poor data utility. These issues need to be
addressed to improve the effectiveness and practicality of personalized
local differential privacy.

In this work, we have proposed a framework of personalized differential
privacy in the shuffle model. This framework introduces individualized
perturbation to the data locally and then reshuffles the records in the
dataset, disrupting the original order of the data and breaking the correlations between data points. This approach aims to achieve a higher level
of privacy protection. We have validated the practicality and superiority
of this framework on four different types of real-world datasets.

1 Introduction

With the rapid advancement of technology and the emergence of a data-driven
economy, a vast amount of personal information is being collected and stored for
various commercial purposes, such as personalized advertising, market research,
and user behavior analysis. However, this widespread collection and utilization
of data has also given rise to concerns regarding personal information leakage,
identity theft, and infringement upon individual rights. Furthermore, significant
events have further emphasized the need for privacy protection. For instance, the
disclosure of the surveillance program conducted by the US National Security
Agency, exposed by Edward Snowden, highlighted the government's capability
to extensively monitor citizens' communications and personal data, triggering
a debate on the delicate balance between privacy rights and national security.
Similar incidents, coupled with data breaches, cyber attacks, and other related
issues, have intensified people's focus on the security of personal information and
the necessity for safeguarding privacy.

© The Author(s), under exclusive license to Springer Nature Singapore Pte Ltd. 2024
J. Vaidya et al. (Eds.): AIS&P 2023, LNCS 14509, pp. 468–482, 2024.
https://doi.org/10.1007/978-981-99-9785-5_33

In recent years, local differential privacy(LDP) [1,2] has emerged as a privacy protection mechanism based on individual privacy. It requires each participant to locally perturb or process data to enhance privacy. This approach allows for the protection of individual data privacy while enabling the server to analyze and utilize the uploaded data. Currently, local differential privacy has been applied in many real-world scenarios and has shown excellent performance. For example, Chrome by Google [3] and Windows 10 by Microsoft [4] use it for user data collection, and there have been numerous research outputs and explorations related to it in the scientific community. [5] applied LDP to frequency and mean estimation.

However, Local Differential Privacy (LDP) sacrifices data utility while protecting privacy. Hence, the concept of Shuffle model emerged. This mechanism shuffles the perturbed data with the help of a third party before it is sent to the server. The purpose of this operation is to disrupt the original order between the data and break the correlation between them, thereby achieving a higher level of privacy protection. The exist works such as Amplification-by-shuffling lemmas quantify how well the privacy parameters are improved [6,7], the single-message protocol [8,9] and m-message Shuffle Privacy [10,11]. But, the characteristic of shuffle model and LDP, which requires the same level of privacy protection for each individual, is not suitable for the majority of real-world scenarios. In real-world situations, due to different backgrounds of individuals, the privacy preferences for the same type of information vary among different users. For example, celebrities have much stricter privacy protection requirements for their flight information compared to ordinary people.

Based on the different privacy needs of individuals, the concept of personalized differential privacy(PDP) has emerged [12,13]. Personalized local differential privacy(PLDP) [12] allows for the development of different data perturbation strategies based on the characteristics and requirements of individual users. Compared to traditional shuffle model and LDP, personalized local differential privacy offers more flexible and precise privacy protection capabilities. By adopting different privacy protection strategies for different users, it is possible to better meet their privacy needs while minimizing the impact on data. This approach can to some extent improve data availability and the accuracy of data analysis. Individual users can choose the appropriate level of privacy protection based on their privacy preferences, thereby customizing more suitable personalized privacy protection solutions. [14] proposed a histogram optimization strategy based on personalized differential privacy. [15] presented a perturbation algorithm (PDPM) that satisfies personalized local differential privacy (PLDP), addressing the problem of inadequate or excessive privacy protection for certain participants caused by using the same privacy budget for all clients.

Although the PLDP model avoids the risk of storing all data centrally on a single server and enhances the security of data privacy, it has a significant gap in terms of data availability compared to the central differential privacy model. The main challenges of personalized local differential privacy currently are: (1) How to accurately model users' characteristics and privacy preferences in order to generate appropriate personalized perturbations. (2) Balancing the level of privacy protection among different users, avoiding excessive protection or exposure of certain users' privacy. (3) Striking a balance between individual data privacy protection and data utility.

Based on the above challenges, in this paper, we propose several different privacy budget levels for users to choose from freely in order to meet the privacy preferences of different individuals and find the most suitable privacy protection strategy. We also separate the privacy budget parameters from the perturbed data during transmission to enhance the level of privacy protection. In addition, to further address the impact of local data perturbation on server-side data analysis, we have introduced the shuffle model. Finally, in the data aggregation phase on the server side, we assign fixed weights to each perturbed data for more accurate unbiased estimation. In this article, we investigated the frequency estimation problem in the context of personalized differential privacy in tne shuffle model. We improved the impact of noise perturbation on data availability by using a shuffling model and extended it to the multi-message shuffling model scenario. A large number of experimental results showed that our method significantly outperformed personalized local differential privacy.

In summary, we have made several contributions to this paper:

1. We have improved the basic framework of PLDP by separating the privacy parameters and perturbed data for enhanced privacy protection. Additionally, we have assigned fixed weights to each user to obtain more accurate unbiased estimates.
2. The first application of shuffle model in the PDP problem was introduced, and it was demonstrated that multi-message shuffling differential privacy method improve the utility of data statistics.
3. To demonstrate the effectiveness of our approach, we applied it to multiple real-world datasets of different types. The experimental results consistently showed that our method outperformed PLDP model, and we provided insights into the impact of certain parameters on utility.

2   Preliminaries

2.1   Differential Privacy

Differential Privacy(DP) is a privacy-preserving method that aims to minimize the risk of individual privacy disclosure during the analysis or processing of individual data. The core idea of differential privacy is to hide the specific information of individual data by introducing noise or randomness, allowing meaningful analysis of data while protecting privacy.

Definition 1. *Differential Privacy(DP) [16]: A random algorithm* $\mathcal{R} : R(D) \rightarrow \mathcal{S}$ *satisfies* (ϵ, δ)-*DP, where* $\epsilon \geq 0$ *and* $0 \leq \delta \leq 1$, *if and only if any Neighboring datasets* D, D', *for any possible of outputs* $z \subseteq \mathcal{S}$, *it has*

$$Pr[\mathcal{R}(D) \in z] \leq e^\epsilon \cdot Pr[\mathcal{R}(D') \in z] + \delta \tag{1}$$

where S represents the set of values output by R. If $\delta = 0$, random algorithm R satisfies pure DP(ϵ-DP), if $\delta > 0$, R satisfies approximate DP((ϵ, δ)-DP).

2.2 Shuffle Model

To address the issue of low utility in LDP models, some researchers have proposed Shuffle Differential Privacy(shuffle model) [6,8]. The shuffle model primarily provides privacy protection by introducing a semi-trusted third-party to reorganize or shuffle the perturbed data. The shuffle model serves as a compromise between central differential privacy models(CDP model) and LDP models. It neither relies on trusted servers nor requires excessive noise addition locally like the LDP model. Therefore, it can effectively enhance data utility in specific application scenarios.

Definition 2. *Shuffle model [17]: A shuffling model consists of three components: Randomizer, Shuffler and Analyzer.*

1. **the Randomizer** R: $R(a, \cdot) \to y^m$: *The randomizer R perturbs the input real data a and output y^m. When $m = 1$, it is referred to as a single-message protocol, and when $m > 1$, it is referred to as a multi-message protocol.*
2. **the Shuffler** S: $S(y^m) \to y^m$: *By recombining the attribute values of records or exchanging them with other records, the shuffler makes the correlation and individual information of the dataset more obscure and indistinguishable.*
3. **the Analyzer** A: $A(y^m) \to Z$: *The analyzer is responsible for collecting and processing the messages, as well as conducting subsequent analysis tasks.*

If we use protocol $P(R \circ S \circ A)$ to describe the above process, we define this process as:

$$P(a) = A(S(\cup_{i=1}^{n} R_i(a_i, \cdot))) \tag{2}$$

Based on the above content, we can derive the following definition 3.

Definition 3. *Differential Privacy in shuffle model(shuffle DP). the protocol P satisfies (ϵ, δ)-DP, where $\epsilon \geq 0$ and $0 \leq \delta \leq 1$, if and only if $S(\cup_{i=1}^{n} R_i(a_i, \cdot))$ satisfies central (ϵ, δ)-DP where $S(\cdot)$ is the shuffle operation.*

Definition 4. *Hockey-stick divergence. The hockey-stick divergence between two random variables P and Q is defined by:*

$$\mathbf{D}_{e^\epsilon}(P\|Q) = \int_{max} \{0, P(x) - e^\epsilon Q(x)\} dx \tag{3}$$

iff $max\{\mathbf{D}_{e^\epsilon}(P\|Q), \mathbf{D}_{e^\epsilon}(Q\|P)\} < \delta$ that P and Q are (ϵ, δ)-indistinguishable.

2.3 Personalized Differential Privacy

Personalized Differential Privacy(PDP) is an extended form of differential privacy that aims to provide more precise privacy protection while maintaining the availability and utility of individual data. Personalized differential privacy allows data processors to customize privacy protection mechanisms for each user based on their privacy requirements and data characteristics.

Definition 5. *Personalized Differential Privacy [12]. Given the local user sets* $\mathcal{U} = \{u_1, u_2, \ldots, u_n\}$ *and* $\varepsilon = \{\epsilon_1, \epsilon_2, \ldots, \epsilon_n\}$, *where each user* $u_i \in \mathcal{U}$ *has a privacy preference* $\epsilon_i \in \varepsilon$, *and a random mechanism* $R : R(D) \rightarrow \mathcal{S}$ *satisfying* (ε, δ)-*PDP where* $0 \leq \delta \leq 1, 0 \leq \epsilon_i \leq 1$. *if and only if any Neighboring datasets* D, D', *for any possible of outputs* $z \subseteq \mathcal{S}$, *it has*

$$Pr[\mathcal{R}(D) \in z] \leq e^{\epsilon_i} \cdot Pr[\mathcal{R}(D') \in z] + \delta \qquad (4)$$

where $\mathcal{S}$ *represents the set of values output by* $\mathcal{R}$ *and* ϵ_i *is the user* u_i*'s privacy preference.*

3   System Overview

3.1   Problem Definition

In a distributed scenario, data is typically stored across different nodes or participants, and it may involve multiple privacy-sensitive data owners. Frequency estimation can potentially leak information about individuals. Even without directly disclosing sensitive data, attackers may be able to infer individual attributes or behavioral patterns through analyzing frequency estimation results. Therefore, privacy concerns are an important consideration. In the case of an untrusted server, we typically employ the PLDP mechanism to provide personalized privacy protection for local users.

We use $\varepsilon = \cup_{i=1}^n \epsilon_i$ to define the set of privacy budget values, where ϵ_i represents the privacy preference of local user u_i. Smaller values of ϵ_i indicate stronger privacy. Suppose there is a local user population where each user holds an element $a_i \in \mathcal{T}$, where $\mathcal{T} = \{t_1, t_2, t_3, \ldots, t_k\}$. And let

$$f_t = \frac{1}{n} \sum_{i=1}^n 1_{(a_i = t)} \qquad (5)$$

f_t denotes the frequency of a specific element t in the entire population. n represents the total number of users. This method is commonly used in fields such as analyzing user behavior, conducting statistical surveys, and data mining.

Although PLDP improves certain limitations of the application scenarios, such as the single-purpose nature and limited privacy protection strength of LDP mechanisms, there still exists a significant gap in terms of accuracy compared to the CDP. Our goal is to maximize the statistical utility of server-side data while achieving personalized privacy protection.

3.2   Muti-messages Personalized Differential Privacy in Shuffle Model

We achieve this goal by introducing the shuffle model. Shuffle model is similar to LDP that perturbs individual data locally. However, unlike LDP, shuffle model enhances privacy protection by introducing a third-party shuffler that anonymizes and randomly reorders the perturbed data. Therefore, when both mechanisms provide the same level of privacy protection, the shuffle DP allows for the addition of less noise locally, resulting in a significant improvement in statistical utility at the server-side.

In this paper, we have introduced a scenario of Muti-messages personalized differential privacy in shuffle model(Muti-messages PSDP). We specify that in addition to the privacy budget ϵ_i, each user can send m messages, among which only one is a true message, and the rest are false messages sampled from a random uniform distribution. All messages are then sent to a random shuffler for shuffling before being sent to the server for data aggregation(see Fig. 1).

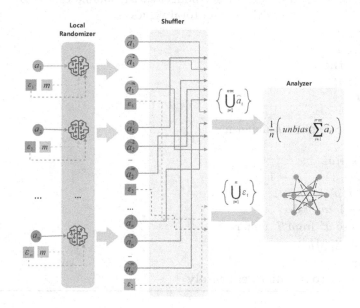

Fig. 1. The framework of multi-PSDP

The Local Randomization Perturbation(Randomizer). The local randomization perturbation (Randomizer) and server-side aggregation analysis(Analyzer). The local Randomizer perturbs the original data by modifying it with a certain probability or adding random noise to increase the indistinguishability between two pieces of data. The most commonly used mechanism is the random response mechanism [18] and its generalized versions [3]. Generalized Randomized Response (GRR) [19,20] is more suitable for situations where $k > 2$. Therefore, we have chosen GRR as the local randomizer to perturb the real data. The workflow is as follows:

$$\mathbf{P}\left[\mathcal{R}(a) = a^*\right] = \begin{cases} \frac{e^\epsilon}{e^\epsilon + k - 1}, & a^* = a \\ \frac{1}{e^\epsilon + k - 1}, & a^* \neq a \end{cases} \tag{6}$$

The local users u_i input their data a_i and the selected privacy budget ϵ_i into the local randomizer. The randomizer perturbs the real data and uniformly sample $m-1$ false messages from $\mathcal{T}$, then send the perturbed m messages and the privacy budget ϵ_i to the shuffler. The specific details are described in Algorithm 1.

Algorithm 1: the local random perturbation mechanism

Input: a_i: a ture value of local user
   $\mathcal{T}$: the dataset of total values
   m:The number of messages sent by local users.
Parameter: ϵ_i: the local privacy budget satisfy shuffle (ε, δ)-PDP
   k: the cardinality of total values
1 $\rho = \frac{e^{\epsilon_i} - 1}{e^{\epsilon_i} + k - 1}$
2 $y = Ber(\rho)$
3 **if** $y = 1$ **then**
4 | $\hat{a}_i = a_i$
5 **end if**
6 **else**
7 | choose $\hat{a}_i$ from $\mathcal{T}$ uniformly
8 **end if**
9 $\mathbb{S}_i \leftarrow \mathbb{S}_i \uplus \hat{a}_i$
10 $\rho = (m - 1) - \lfloor m - 1 \rfloor$
11 $y = Ber(\rho) + \lfloor m - 1 \rfloor$
12 **for** $j \leftarrow 1$ *to* y **do**
13 | choose $\hat{a}_i^j$ from $\mathcal{T}$ uniformly
14 | $\mathbb{S}_i \leftarrow \mathbb{S}_i \uplus \hat{a}_i^j$
15 **endfor**
16 **Send** $\mathbb{S}_i, \epsilon_i$ **to shuffler separately**

Third-Party Shuffler. At this stage, Shuffler shuffles the perturb values dataset $\mathbb{S}$ and privacy budget dataset ε to remove the correlation between data and send them to server.

Data Aggregation and Estimation(Analyzer). In the frequency estimation problem, the server needs to aggregate the received data and perform unbiased estimation to leverage the statistical utility of the data. he specific details are described in Algorithm 2. Since the correspondence between messages and privacy budgets is disrupted after being randomly shuffled, we simplify the data aggregation and unbiased estimation calculations by assigning a fixed weight w_i to each user.

The process of aggregation and unbiased estimation is as follows:

$$\hat{f}_t = \frac{1}{W}\left(\sum_{i=1}^{n}\left(\frac{\hat{a}_i - \frac{m-1}{k} - q_i}{p_i - q_i} \cdot w_i\right)\right)$$

$$= \frac{1}{W}\left(\hat{N}_t - n \cdot \frac{m-1}{k} - \sum_{i=1}^{n} q_i\right) \tag{7}$$

We set $w_i = p_i - q_i$, $W = \sum_{i=1}^{n} w_i$ and $\hat{N}_t$ be the total number of received value t after aggregation, the estimated frequency of value t is $\hat{f}_t$.

Algorithm 2: Server-side data aggregation and estimation

Input: $\mathbb{S}$: the perturb dataset
  ε: privacy budget dataset
Parameter: N: total numbers of user
  k: the cardinality of total values
Output: $\hat{f}_t$

1 **for** $\epsilon_i \in \varepsilon$ **do**
2 $\quad p_i = \frac{e^{\epsilon_i}}{e^{\epsilon_i}+k-1}$
3 $\quad q_i = \frac{1}{e^{\epsilon_i}+k-1}$
4 $\quad W = W + (p_i - q_i)$
5 **endfor**
6 get $\hat{N}_t = \sum_{i=1}^{n \cdot m} 1_{\hat{s}_i = t}$ from any data $\hat{s}_i \in \mathbb{S}$
7 $\hat{f}_t = \frac{1}{W}(\hat{N}_t - n \cdot \frac{m-1}{k} - \sum_{i=1}^{n} q_i)$
8 **Return** $\hat{f}_t$

3.3 Privacy Analysis

To analyze the privacy of the recommended method, we introduce the concept of personalized differential privacy in shuffle model and prove the privacy amplification effect of our method.

According to Definitions 3, 2 and 5, we have formalized the definition of personalized differential privacy in shffle model.

Definition 6. *Personalized differential privacy in shuffle model(shuffle-PDP): Given the sets* $\mathcal{U} = \{u_1, u_2, \ldots, u_n\}$, *and* $\varepsilon = \{\epsilon_1, \epsilon_2, \ldots, \epsilon_n\}$, *where each user* $u_i \in \mathcal{U}$ *has a privacy preference* $\epsilon_i \in \varepsilon$ *and a value* $a_i \in D$, *and a shuffle protocol* $\mathcal{P} : \mathcal{R} \circ \mathcal{S}(D) \to \mathcal{T}$ *satisfying* ε-PDP *,if and only if any Neighboring datasets* D, D'. *Then the algorithm :*

$$\mathcal{R} \circ \mathcal{S}(\cdot) = \mathcal{S}(R_1(a_1, \epsilon_1, m), R_2(a_2, \epsilon_2, m), \ldots, R_n(a_n, \epsilon_n, m)) \tag{8}$$

is satisfies (ε, δ)- *shuffle PDP, where the* $\mathcal{T}$ *is the output dataset, m represents the number of messages sent by each user.*

Here, to further enhance the privacy amplification effect, we employ the framework of the Variation-ratio [21], a unified, efficient, and comprehensive framework, to analyze privacy amplification in shuffling models. Therefore, we are given two neighboring datasets $\mathcal{A}$ and $\mathcal{A}'$. i.e. $\mathcal{A} = \{a_1 = a_1^0, a_2, a_3 ..., a_n\}$, $\mathcal{A}' = \{a_1 = a_1^1, a_2, a_3 ..., a_n\}$ where $\mathcal{A}, \mathcal{A}' \subset \mathbb{A}$.

These two datasets differ only in the first data entry, we defined some parameters $p > 1, \beta \in [0, \frac{p-1}{p+1}], q > 1$ on local randomzer $\{\mathcal{R}_i\}_{i \in [n]}$:

1. (p, β)-variation property: for all possible $a_1^0, a_1^1 \in \mathbb{A}$ that satifies the formula $\mathbf{D}_p(\mathcal{R}_1(a_1^0) \| \mathcal{R}_1(a_1^1)) = 0$ and $\mathbf{D}_{e^0}(\mathcal{R}_1(a_1^0) \| \mathcal{R}_1(a_1^1)) \leq \beta$.
2. q-ratio property:for all possible $a_1, a_2, a_3 ..., a_n \in \mathbb{A}$ and all $\{\mathcal{R}_i\}_{i \in [2:n]}$ that satifies $\mathbf{D}_q(\mathcal{R}_1(a_1) \| \mathcal{R}_i(a_i)) = 0$.

Lemma 1. *(Variation-ratio reduction [21]) For $p > 1, \beta \in [0, \frac{p-1}{p+1}], q > 1$, let $C \sim Bin(n - 1, \frac{2\beta p}{(p-1)q}), A \sim Bin(C, \frac{1}{2})$ and $\Delta_1 \sim Ber(\frac{\beta p}{p-1})$ and $\Delta_2 \sim Ber(1 - \Delta_1, \frac{\beta}{p-1-\beta p})$;let $P_{\beta,p}^q$ denote $(A + \delta_1, C - A + \Delta_2)$ and $Q_{\beta,p}^q$ denote $(A + \Delta_2, C - A + \Delta_1)$. Given any $a_1, a_2, a_3 ..., a_n \in \mathbb{A}$,if $\{\mathcal{R}_i\}_{i \in [n]}$ satisfy the (p, β)-variation property and the q-ratio property, then for any measurement D satisfying the data-processing inequality:*

$$D(S(\mathcal{R}_1(a_1^0), ..., \mathcal{R}_n(a_n)) \| S(\mathcal{R}_1(a_i^0), ..., \mathcal{R}_n(a_n)) \leq D(P_{\beta,p}^q \| Q_{\beta,p}^q) \qquad (9)$$

Now, we are attempting to prove that our method satisfies Definition 6 within the framework of the Variation-ratio.

According to the GRR mechanism used in our method, we convert the parameters p', β', q' in the framework of the Variation-ratio, where the privacy preference $\epsilon \in \varepsilon$ .The process is as follows:

First, according to the Lemma 1, given the randomizer $\mathcal{R}$ satisfy the (p', β')-property and q'-ratio property that:

$$\mathbf{D}_{p'}(\mathcal{R}(a_1^0) \| \mathcal{R}(a_1^1)) = 0$$

The application of Definition 4 can transform the above equation into:

$$\int_{max} \{0, \mathcal{R}(a_1^0) - p' \cdot \mathcal{R}(a_1^1)\} \, da = 0$$

therefore:

$$\mathcal{R}(a_1^0) - p' \cdot \mathcal{R}(a_1^1) = 0$$

so,we can get a p' satisfy the above conditions when the GRR mechanism is used in $\mathcal{R}$:

$$p' = \frac{\mathcal{R}(a_1^0)}{\mathcal{R}(a_1^1)} = \frac{\frac{e^\epsilon}{e^\epsilon + k - 1}}{\frac{1}{e^\epsilon + k - 1}} = e^\epsilon$$

Similarly, we can obtain a suitable β' according to the (p', β')-variation property:

$$\mathbf{D}(\mathcal{R}(a_1^0) \| \mathcal{R}(a_1^1)) \leq \beta'$$

The application of Definition 4 can transform the above equation into:

$$\int_{max} \{0, \mathcal{R}(a_1^0) - \mathcal{R}(a_1^1)\} \, da \leq \beta'$$

so,we can get the β' be bound by $[0, \frac{e^\epsilon - 1}{e^\epsilon + k - 1}]$.

And in multi-message scenario, we get false messages from $\mathcal{T}$ uniformly. Similarly, we get the parameter q' when $\mathcal{R}_{i \in [2:n]}$ satisfy q'-ratio property as follow:

$$\mathbf{D}_{q'}(\mathcal{R}(a_1) \| \mathcal{R}_i(a_i)) = 0$$

Similarity,wo can get the q' which satisfy the above conditions that: $q' = \frac{k \cdot e^\epsilon}{e^\epsilon + k - 1}$

if there are n' users and each user contributes $m - 1$ fake messages, the value of $n - 1$ in Defintion 1 becomes $n' \cdot (m - 1)$. Therefore, by applying Definition 1, our method satisfy:

$$D(\mathcal{R} \circ \mathcal{S}(R_1(a_1^0, \epsilon_1, m), \cup_{i=2}^n \mathcal{R}_i(a_i, \epsilon_i, m)) \| R \circ \mathcal{S}(R_1(a_1^1, \epsilon_1, m), \cup_{i=2}^n \mathcal{R}_i(a_i, \epsilon_i, m)) \leq D(P_{\beta', p'}^{q'} \| Q_{\beta', p'}^{q'})$$

are (ϵ_1, δ)-indistinguishable according to Definiton 4, where $\epsilon_1 \in \varepsilon$. So, our method satisfy (ε, δ)-shuffle PDP in framwork of the Variation-ratio.

4   Experiments

In this chapter, we evaluate and compare the performance of our proposed methods, single-message PSDP and multi-message PSDP, through extensive experiments. Since there is no existing work specifically addressing frequency estimation in the PSDP scenario, we compare our work with frequency estimation in the PLDP scenario. We conduct tests on various types of real data to demonstrate the impact of personalized perturbation and different hyperparameter settings on performance through experiments.

4.1   Datasets

Four commonly used datasets will be used in our experiments. These datasets are widely used for training and testing frequency estimation problems. The specific information about the datasets is provided in Table 1. The Adult dataset [22], which was extracted by Barry Becker from the 1994 Census database, contains 48,842 instances. For our dataset, we extracted a subset of 32561 samples from it. Bank Marketing [23] is related to a direct marketing campaign conducted by a Portuguese banking institution. We extracted a subset of 10,000 samples from this dataset to create a new dataset. Abalone [24] is derived from physical measurements of abalone (a type of sea snail). Car Evaluation [25] contains several basic parameters of cars and is derived from a simple hierarchical decision model.

Table 1. Dataset

Datasets	Feature	Sample numbers	dimension
Adult	education	32561	16
Bank marketing	job	10000	12
Abalone	sex	4177	3
Car evaluation	buy	1728	4

4.2 Performance Metrics

In this study, we use the Mean Squared Error (MSE) evaluation method to assess performance. The MSE is a commonly used evaluation metric to measure the degree of difference between predicted values and true values. It calculates the squared difference between predicted values and true values, and then takes the average. A smaller MSE indicates a smaller difference between the predicted results and the true values, indicating a better performance of the model. MSE is widely used in many machine learning and statistical fields.

$$MSE = \sum_{i=1}^{k}(f_t - \hat{f}_t)^2 \tag{10}$$

4.3 Experimental Performance

In this section, we compare our method with PLDP and single-PSDP on four real datasets under two different scenarios of privacy distributions (see Table 2). We further investigate the impact of message quantity and privacy distribution on experimental performance and present the results using line graphs for better visualization.

Table 2. Experimental parameter settings

	Scenario 1	Scenario 2
ϵ allocation	0.2,0.4,0.6,0.8,1.0	0.3,0.5,1.0
User ratio	0.25,0.25,0.25,0.25,0.25	0.33,0.33,0.34

PLDP allows users to freely choose the level of privacy protection according to their preferences, providing flexibility to adapt to various complex privacy-security scenarios. Although personalized local differential privacy partially addresses the limitations of LDP mechanisms, such as limited applicability and limited privacy protection strength, there is still a significant gap in accuracy compared to the CDP model. As shown in the Fig. 2, regardless of the scenario, the performance of the PLDP method is the worst. Both single-PSDP and local differential privacy perturb the local individual data items. However, unlike local differential privacy, single-PSDP further enhances privacy protection by introducing a third-party shuffler to anonymize and randomly reorder the perturbed data. When the privacy protection strengths of both mechanisms are the same, the single-PSDP mechanism allows for less noise to be added locally. Therefore, as shown in the graph, the results of single-PSDP have improved significantly compared to PLDP, but there is still a considerable gap compared to Our method.

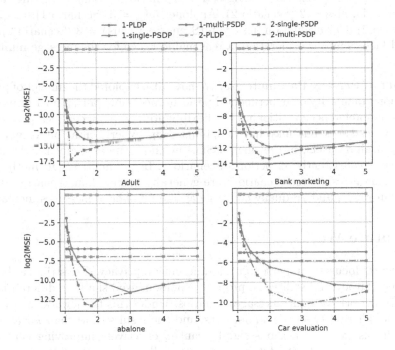

Fig. 2. Comparison of experimental results

Impact of the Dataset. We tested our method using four datasets of varying sizes and dimensions. Regardless of whether it was in scenario 1 or scenario 2, our method performed the best on the Adult dataset and the worst on the Car Evaluation dataset. The results indicate that with an increase in the number of samples, the experimental performance significantly improves. It can be observed that our method performs better and demonstrates its advantages more effectively on large-scale datasets.

Impact of Message Quantity. We explored the impact of the quantity of fake messages on the experimental performance. Although increasing the quantity of fake messages can amplify the privacy budget disturbance of the GRR mechanism on real data, excessive fake news still introduces a significant amount of noise to the aggregated results. Therefore, we set several reasonable message numbers(msg) to find the optimal parameter for the quantity of messages. In theory, as the quantity of messages increases, the MSE value initially decreases before increasing, with the rate of decrease gradually slowing down. From the Fig. 2, it can be observed that the optimal number of messages are as follows: Adult: msg = 2 (Scenario 1), msg = 1.4 (Scenario 2). Bankmarketing: msg=3 (Scenario 1), msg = 2 (Scenario 2). Analone: msg = 3 (Scenario 1), msg = 1.8 (Scenario 2). Car evaluation: msg = 5 (Scenario 1), msg = 3 (Scenario 2). Taking all factors into consideration, a reasonable range for the message number in $[1.4, 3]$.

Impact of Privacy Parameters ϵ. We have also explored the impact of privacy partition granularity on experimental performance. Generally, a finer granularity of privacy preference partitioning can flexibly address more complex privacy and security scenarios but may have an impact on experimental performance. In Fig. 2, in all datasets, in the single-PSDP method, the experimental results for Scenario 2 consistently outperform Scenario 1. In the multi-PSDP method, the experimental results for Scenario 2 are generally better than Scenario 1 within a certain range, and gradually converge as the number of messages increases.

5   Future Work

This work focuses on personalized differential privacy. To strike a balance between privacy and utility, we propose a personalized shuffling differential privacy framework for multi-message scenarios. Under this framework, we evaluate multiple different types of real datasets, and the results show that our approach outperforms existing methods signif- icantly, effectively improving centralized privacy protection. In the future, our work will focus on the following aspects of expansion: (1) Maintaining utility in more complex privacy protection scenarios for data analysis. (2) Exploring more efficient coding techniques or data sparsification tech- niques to balance communication costs and statistical utility.

References

1. Kasiviswanathan, S.P., Lee, H.K., Nissim, K., Raskhodnikova, S., Smith, A.: What can we learn privately? SIAM J. Comput. **40**(3), 793–826 (2011)
2. Duchi, J.C., Jordan, M.I., Wainwright, M.J.: Local privacy and statistical minimax rates. In: 2013 IEEE 54th Annual Symposium on Foundations of Computer Science, pp. 429–438. IEEE (2013)
3. Erlingsson, Ú., Pihur, V., Korolova, A.: Rappor: randomized aggregatable privacy-preserving ordinal response. In: Proceedings of the 2014 ACM SIGSAC Conference on Computer and Communications Security, pp. 1054–1067 (2014)

4. Ding, B., Kulkarni, J., Yekhanin, S.: Collecting telemetry data privately. In: Advances in Neural Information Processing Systems, vol. 30 (2017)
5. Bassily, R., Smith, A.: Local, private, efficient protocols for succinct histograms. In: Proceedings of the Forty-Seventh Annual ACM Symposium on Theory of Computing, pp. 127–135 (2015)
6. Erlingsson, Ú., Feldman, V., Mironov, I., Raghunathan, A., Talwar, K., Thakurta, A.: Amplification by shuffling: From local to central differential privacy via anonymity. In: Proceedings of the Thirtieth Annual ACM-SIAM Symposium on Discrete Algorithms, pp. 2468–2479. SIAM (2019)
7. Balle, B., Bell, J., Gascón, A., Nissim, K.: The privacy blanket of the shuffle model. In: Boldyreva, A., Micciancio, D. (eds.) CRYPTO 2019. LNCS, vol. 11693, pp. 638–667. Springer, Cham (2019). https://doi.org/10.1007/978-3-030-26951-7_22
8. Cheu, A., Smith, A., Ullman, J., Zeber, D., Zhilyaev, M.: Distributed differential privacy via shuffling. In: Ishai, Y., Rijmen, V. (eds.) EUROCRYPT 2019. LNCS, vol. 11476, pp. 375–403. Springer, Cham (2019). https://doi.org/10.1007/978-3-030-17653-2_13
9. Balcer. V. Cheu, A.: Separating local & shuffled differential privacy via histograms," arXiv preprint arXiv:1911.06879 (2019)
10. Beimel, A., Haitner, I., Nissim, K., Stemmer, U.: On the round complexity of the shuffle model. In: Pass, R., Pietrzak, K. (eds.) TCC 2020. LNCS, vol. 12551, pp. 683 712. Springer, Cham (2020). https://doi.org/10.1007/978-3-030-64378-2_24
11. Chen, L., Ghazi, B., Kumar, R., Manurangsi, P.: On distributed differential privacy and counting distinct elements, arXiv preprint arXiv:2009.09604 (2020)
12. Jorgensen, Z., Yu, T., Cormode, G.: Conservative or liberal? personalized differential privacy. In: 2015 IEEE 31St International Conference on Data Engineering, pp. 1023–1034. IEEE (2015)
13. Chen, R., Li, H., Qin, A.K., Kasiviswanathan, S.P., Jin, H.: Private spatial data aggregation in the local setting. In: 2016 IEEE 32nd International Conference on Data Engineering (ICDE), pp. 289–300. IEEE (2016)
14. Yiwen, N., Yang, W., Huang, L., Xie, X., Zhao, Z., Wang, S.: A utility-optimized framework for personalized private histogram estimation. IEEE Trans. Knowl. Data Eng. 31(4), 655–669 (2018)
15. Shen, X., Jiang, H., Chen, Y., Wang, B., Gao, L.: Pldp-fl: federated learning with personalized local differential privacy. Entropy 25(3), 485 (2023)
16. Dwork, C.: Differential privacy. In: Bugliesi, M., Preneel, B., Sassone, V., Wegener, I. (eds.) ICALP 2006. LNCS, vol. 4052, pp. 1–12. Springer, Heidelberg (2006). https://doi.org/10.1007/11787006_1
17. Bittau, A., et al.: Prochlo: strong privacy for analytics in the crowd. In: Proceedings of the 26th Symposium on Operating Systems Principles, pp. 441–459 (2017)
18. Warner, S.L.: Randomized response: a survey technique for eliminating evasive answer bias. J. Am. Stat. Assoc. 60(309), 63–69 (1965)
19. Kairouz, P., Oh, S., Viswanath, P.: Extremal mechanisms for local differential privacy. In: Advances in Neural Information Processing Systems, vol. 27 (2014)
20. Kairouz, P., Bonawitz, K., Ramage, D.: Discrete distribution estimation under local privacy. In: International Conference on Machine Learning, pp. 2436–2444. PMLR (2016)
21. Wang, S.: Privacy amplification via shuffling: Unified, simplified, and tightened, arXiv preprint arXiv:2304.05007 (2023)
22. Becker, B., Kohavi, R.: Adult, UCI Machine Learning Repository (1996). https://doi.org/10.24432/C5XW20

23. Moro, S., Cortez, P., Rita, P.: A data-driven approach to predict the success of bank telemarketing. Decis. Support Syst. **62**, 22–31 (2014)
24. Warwick, S. T. T. S. C. A. N., Ford, W.: Abalone, UCI Machine Learning Repository, (1995). https://doi.org/10.24432/C55C7W
25. Bohanec, M.: Car Evaluation, UCI Machine Learning Repository (1997). https://doi.org/10.24432/C5JP48

MKD: Mutual Knowledge Distillation for Membership Privacy Protection

Sihao Huang[1][iD], Zhongxiang Liu[1][iD], Jiafu Yu[1][iD], Yongde Tang[1][iD], Zidan Luo[1][iD], and Yuan Rao[1,2]([✉])[iD]

[1] Institute of Artificial Intelligence, Guangzhou University, 510006 Guangzhou, China
hsh7357@e.gzhu.edu.cn
[2] Guangdong Provincial Key Laboratory of Blockchain Security, 510006 Guangzhou, China
raoyuan@gzhu.edu.cn

Abstract. Machine learning models are susceptible to member inference attacks, which attempt to determine whether a given sample belongs to the training data set of the target model. The significant privacy concerns raised by member inference have led to the development of various defenses against Member Inference Attacks (MIAs). Existing techniques for knowledge distillation have been identified as a potential solution to mitigate the tradeoff between model performance and data privacy, demonstrating promising results. Nonetheless, the limitations in performance imposed by the teacher model in knowledge distillation, along with the scarcity of unlabeled reference data, present a challenge in achieving high-performance privacy-preserving training for the target model. To address these issues, we propose a novel knowledge distillation based defense method, i.e., Mutual Knowledge Distillation (MKD). Dividing the training set into subsets for the teacher and the student models, MKD trains them through mutual knowledge distillation for mitigating MIAs. Extensive experimental results demonstrate that MKD outperforms several existing defense methods in improving the trade-off between model utility and member privacy.

Keywords: Knowledge Distillation · Member Inference Attack · Data Partitioning

1 Introduction

Recently, The developing machine learning (ML) models, especially the deep learning based ones, have achieved considerable performance gains in areas such as image recognition [11], speech recognition [8,22], and natural language processing [5]. These substantial developments have successfully enhanced people's willingness to use ML in real-world scenarios [18,29]. Nowadays, even with the considerable progress of machine learning algorithms, a large amount of data is still needed for training to achieve the expected results. This source data may

J. Vaidya et al. (Eds.): AIS&P 2023, LNCS 14509, pp. 483–498, 2024.
https://doi.org/10.1007/978-981-99-9785-5_34

come from confidential industries such as finance and healthcare, causing concerns about the privacy and security of machine learning algorithms. However, recent research [24,27] indicates that the situation is not very promising. One critical flaw of machine learning models is that a large amount of training data could be unintentionally saved or leaked, making them susceptible to privacy attacks. These attacks include attribute inference attacks [12], dataset reconstruction attacks [23], and MIA [27]. Among them, MIAs have sparked intense discussions due to their ability to accurately infer whether a data sample is part of the target model's training dataset. MIA can accurately judge whether the current sample is part of the training sample through features such as output confidence [27,28], predictive entropy [24,28], predictive loss [32], and data robustness [4,31]. In a black-box ML situation, an attacker can easily obtain the previously mentioned features, making MIAs a significant threat to the privacy and security of data.

To prevent ML services from causing data leaks, current research on privacy and security is primarily focused on MIAs. Current countermeasures are based on changing the output of the ML model to interfere with the attacker's data classification judgment. This interference can be achieved in two dimensions: first, by affecting the training process of the model with techniques such as differential privacy [1], relaxed loss [2], adversarial regularization [20], and knowledge distillation [26]. Second, by affecting the model's inference process with methods like confidence score masking [13]. These techniques aim to affect the model's output or behavior, leading to security improvement. However, while enhancing security, these defensive techniques are also caught in the awkward dilemma of balancing privacy and utility. It is also undeniable that unpredictable attacks may still occur even with these measures in place.

In this paper, we introduce a novel framework, Mutual Knowledge Distillation (MKD), to defend against membership inference attacks. Most of the existing state-of-the-art knowledge improvement techniques use a teacher model to filter private training data and train a protected target model. These techniques reduce the risk of membership inference attacks by preventing the target model from remembering too much training data. However, these techniques often require additional training data [26] or fail to achieve promising task performance in environments without defense mechanisms [21]. The reason is that the teacher model creates a bottleneck in the distillation system and cannot provide additional representative information to the target model. Meanwhile, Fig. 1 clearly illustrates the significant distinction between MKD and the previous knowledge distillation technique known as Distillation for Membership Privacy (DMP) [26]. MKD overcomes the limitations of teacher model distillation and achieves improved privacy protection and data confidentiality. MKD differs from this series of techniques in that it innovatively references the training of mutual guidance between models. The teacher model and the student model are trained in different training sets, and the knowledge of the model is distilled from each other. This work helps to avoid the risk of privacy disclosure and improve the quality of model training. At the same time, MKD constantly iterates the interactive training process, which makes both the teacher model and the target model fully trained. The main contributions of this paper can be summarized as follows:

- We propose Mutual Knowledge Distillation (MKD) to defend against black-box membership inference attacks. MKD alternates the knowledge distillation learning process between the student model and the teacher model, leading to better privacy preservation during model training without additional data.
- Our innovation lies in the training stage of the student and teacher models. Instead of the teacher model directly acquiring knowledge from the training set, both the student and teacher models are trained using the same data, which effectively mitigates the need for excessive storage of member data.
- We extensively evaluate MKD to demonstrate the state-of-the-art tradeoffs between data privacy and model accuracy. For instance, when aiming at increasing member privacy, MKD exhibits a better testing performance compared to several state-of-the-art defense methods for various classification tasks.

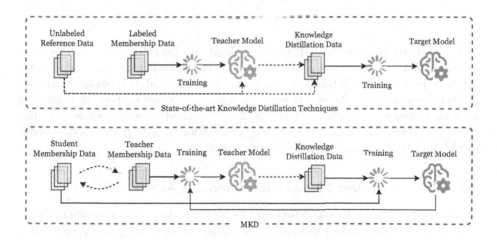

Fig. 1. The discernible dissimilarities between MKD and previous knowledge distillation techniques [26].

2 Related Work

2.1 Membership Inference Attack

The goal of MIA is to determine whether a piece of data belongs to the training set of a model. If it belongs, it is considered a member; otherwise, it is a non-member. Various types of MIA techniques have emerged, which can be roughly categorized into three types of attacks based on the information dimension of the model:

Label Information-Based Method. The label information-based method is an attack based on label information, which is more likely to occur in the real world. Previous research efforts [4,31] have designed a series of MIA attacks in this scenario. For example, Li et al. [19] designed a member inference attack that solely utilizes label information. In this attack, an adversary can make inferences and judgments about membership by analyzing partial data. This technique further demonstrates that machine learning models are more vulnerable to privacy attacks than previously expected.

Partial Output Information-Based Method. Shokri et al. [27] have designed effective membership inference attacks by utilizing partial output data to classify members and nonmembers. These attacks rely on the confidence of the predictions made by shadow models. A year later, Salem et al. [24]strengthened this attack by relaxing the constraints on architecture, training data, and the number of shadow models. This approach also provides a new method for inferring membership relationships by using the maximum output confidence. Song et al. [28] further enhanced this attack by designing thresholds based on different class labels.

Total Output Information-Based Method. In this attack, an adversary can perform MIA by accessing the statistical data of the target model. One popular attack is based on prediction loss, proposed by Yeom et al. [31] Shokri et al. [27] further demonstrated the effectiveness of using predictive entropy in launching attacks. In recent years, Song et al. [28] have designed an enhanced attack that considers predictive entropy, the training correlation of data points in the victim model, and their distances in the feature space.

2.2   Defense Against Membership Inference Attack

Noise Disturbance-Based Method. Noise Perturbation is a commonly used privacy protection method that aims to achieve differential privacy and confidence perturbation. Differential privacy ensures individual sample privacy and defends against MIA [1,3,6]. However, striking a balance between privacy and utility is challenging. To address performance issues, defense mechanisms introduce noise perturbation into the confidence score vector [13,14,30], making confidence score perturbation easier to implement. It has been demonstrated to be feasible in practical applications, even though it may not defend against certain attacks.

Suppressing Overfitting-Based Method. Overfitting in ML models poses a certain difficulty and increases the risk of membership inference attacks. Overfitting suppression is an exploratory approach for MIA defense, and its implementation is based on regularized training and relaxed learning. Adversarial regularization techniques [9,20]enhance generalization by incorporating adversarial examples. Relax loss [3] enhances generalization by using a new loss function. These methods improve the resilience of ML models against MIA, although striking a balance between privacy and utility remains challenging.

Knowledge Distillation-Based Method. Knowledge distillation is a model compression technique that aims to transfer knowledge from a higher-accuracy model (teacher model) to a smaller model (student model), reducing computational costs while maintaining accuracy. Zhang et al. [33] recognized that this technique can reduce the risk of MIA. Virat Shejwalkar et al. [26] proposed an innovative defense mechanism called DMP. DMP combines knowledge distillation strategies and uses partial member data to train the teacher model. The pre-trained teacher model is used to train the target model in an unlabeled manner. The practical performance of these techniques demonstrates the effectiveness of knowledge distillation-based defense.

3   Our New Method

In this paper, we propose Mutual Knowledge Distillation (MKD), a novel training framework designed to enhance privacy during training and mitigate the risk of member inference in the target class classifier model. The main steps of MKD, illustrated in Fig. 2, include an initial pre-training stage denoted as PTM, followed by two model knowledge distillation stages referred to as mutual distillation learning, i.e., KDS and ADT. It is worth noting that in order to meet the needs of knowledge distillation training, MKD also reserves the data partitioning step TSDP. All stages of MKD are introduced in Sects. 3.1, 3.2, 3.3, and 3.4.

This chapter introduces a simplified notation for clarity and analysis purposes. The teacher model will be denoted as θ_t, while the student model, will be denoted as θ_s. Next, the pre-training training set is represented by D, the dataset for training θ_s is represented by S_{tr}, and the dataset of the θ_t is represented by T_{tr}. Furthermore, θ_s serves as the target model in the MKD algorithm. The input data samples for all models are denoted as x, while their corresponding ground truths are labeled as y. In the implementation of MKD, the two distillation stages are interactive and iterative, and multiple exchanges of training sets improve the performance of θ_t and θ_s and reduce the data sensitivity of the direct learning from the dataset.

3.1   Pre-training of Teacher Model

The primary goal of Pre-training of Teacher Model (PTM) is to obtain high-quality initial parameters of θ_t so that it can better guide the learning of θ_s. To this end, PTM uses D to train θ_t before officially starting distillation training, so that θ_t can learn more discriminative feature representation capabilities in the initial stage. In this step, we minimize the loss objective, which can be represented by Eq. 1.

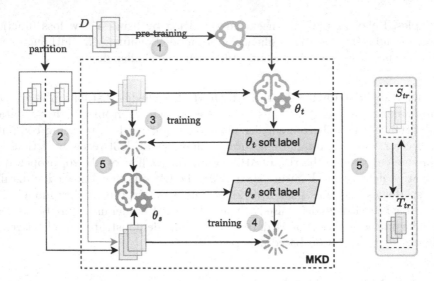

Fig. 2. Mutual Knowledge Distillation (MKD) defense. (1) In pre-distillation phase, MKD trains a θ_t on the full training set. (2)Then, the training set is divided into two parts for θ_t and θ_s in the distillation phase. (3) In distillation phase, MKD not only adopts the hard labels of S_{tr}, but also the soft labels generated by θ_t from S_{tr} to train θ_t. (4) In the same way as (3), the training of the θ_t comes from the soft labels of θ_s and the hard labels of T_{tr}. (5) T_{tr} and S_{tr} are exchanged, and then continue to execute the MKD process (3), (4).

$$l_{ptm}(\theta_t, D) = \sum_{x,y \in D} \theta_t(x) \cdot log(y) \tag{1}$$

3.2 Training Set Data Partitioning

Inspired by [31], directly training model parameters using all available training data increases the risk of overfitting and membership inference. To solve this problem, we propose the Training Set Data Partitioning (TSDP) step to reduce the sensitivity of the target model to the training set data by partitioning the training set. This method can achieve effective privacy training.

Before implementing the knowledge distillation training of MKD, we need to equally divide the original training set D to generate two dedicated training sets, called T_{tr} and S_{tr} respectively. These two training sets maintain the same size and content as the original training set D and will be used to train the θ_t and θ_s respectively.

The practical significance of TSDP is that it can change the data source of θ_s. Instead of directly utilizing the entire training data D to train model parameters, θ_s uses an alternative method of knowledge distillation for feature learning.

3.3 Knowledge Distillation of Student Model

The task of Knowledge Distillation of Student Model (KDS) is to train θ_s using a distillation learning approach with a pre-defined dataset S_{tr} and θ_t.

At each epoch, we use S_{tr} as the training set of θ_s. In this step, the real labels of S_{tr} are not used as the fitted labels of θ_s. Instead, the soft labels $\theta_t(x)$ obtained by accessing θ_t using inputs belonging to S_{tr} is used as the learning target. The purpose is to prevent θ_s from overmemorizing membership data, which is achieved by avoiding over-convergence on the real labels [26]. Finally, we minimize the loss function to the Eq. 2.

$$l_{tkd}(\theta_s, S_{tr}) = \sum_{x \in S_{tr}} \theta_s(x) \cdot log(\theta_t(x)) \tag{2}$$

Furthermore, KDS also incorporates real labels from S_{tr} during the training process, allowing it to provide enhanced guidance for θ_t. In this step, it is necessary to express the objective of minimizing the loss as Eq. 3.

$$l_{sh}(\theta_s, S_{tr}) = \sum_{x,y \in S_{tr}} \theta_s(x) \cdot log(y) \tag{3}$$

KDS uses the weighted sum of the loss functions l_{tkd} and l_{sh}, combined with the hyper-parameters α and β, to calculate the final loss of the target model. This weighted loss function design effectively balances the impact between the real labels and the soft labels in θ_t, thereby facilitating the target model to obtain knowledge from θ_t. In this step, it is necessary to express the objective of minimizing the loss as Eq. 4.

$$l_s(\theta_s, S_{tr}) = \alpha * l_{tkd} + \beta * l_{sh} \tag{4}$$

3.4 Anti-knowledge Distillation of Teacher Model

After performing knowledge distillation on θ_s, the Anti-knowledge Distillation of Teacher Model (ADT) is trained with the help of θ_s and Dataset on T_{tr} to enhance the guidance ability of θ_t. ADT contains two basic components: distillation learning of θ_s and label learning based on T_{tr}. This training method is similar to 3.3. These components work together to minimize losses by utilizing Eq. 5 and Eq. 6 respectively.

$$l_{skd}(\theta_t, T_{tr}) = \sum_{x \in T_{tr}} \theta_t(x) \cdot log(\theta_s(x)) \tag{5}$$

$$l_{th}(\theta_t, T_{tr}) = \sum_{x,y \in T_{tr}} \theta_t(x) \cdot log(y) \tag{6}$$

In the ADT, the loss Equation l_{skd} and l_{th} are weighted by hyper-parameter α and β to compute the final loss of θ_t. This weighted loss function design effectively balances the influence between the real labels and the soft labels from θ_s. Equation 7 represents the weight relationship between the two loss functions.

$$l_t(\theta_t, T_{tr}) = \alpha * l_{skd} + \beta * l_{th} \tag{7}$$

In the implementation of MKD, we introduce an important mechanism to enhance the model's adaptability and learning ability. After θ_t and θ_s complete a round of training, we exchange S_{tr} and T_{tr} and restart knowledge distillation learning. This mutually guided knowledge distillation process enables the model to learn different data distributions and tasks at different stages, thereby improving its generalization ability and adaptability.

4   Experimental Setup

4.1   Datasets

In our study, we perform classification tasks on three different datasets that are representative of previous classification efforts: the main benchmark dataset in the machine learning domain CIFAR100 [15] and two auxiliary validation datasets namely Texas100 and Purchase100. It is worth noting that these two auxiliary validation datasets are simplified versions provided by the author [27] and were used for the first membership inference attack on machine learning models. Table 1 shows the recorded results of the data partitioning process.

The CIFAR100 dataset is widely accepted and serves as a benchmark for evaluating the effectiveness of image recognition algorithms. This dataset comprises 60,000 images, each measuring 32×32 pixels. It is divided into 100 distinct categories, each representing a unique object class.

The Texas100 dataset consists of hospital discharge data. This dataset contains records containing inpatient information obtained from multiple healthcare providers, and this information is made publicly available by the Texas Department of State Health Services. Data records include information about the external cause of the injury, diagnosis, and procedure collected by the patient, as well as general information such as gender, age, race, hospital ID, and length of stay. This dataset consists of 67,330 records representing the 100 most common medical procedures and 6,170 binary features. These records are categorized into 100 classes, each representing a different type of patient.

The Purchase100 dataset is based on Kaggle's Identify Valuable Shoppers challenge. This dataset consists of the shopping records of thousands of individuals. The purpose of this challenge is to identify discounts to attract new shoppers to purchase new products. The dataset contains 197,324 data records. Each data record corresponds to a customer and consists of 600 binary features. Each feature corresponds to a product and reflects whether the customer purchased that product. The data is grouped into 100 categories and the task is to predict the category for each customer.

Table 1. The setting of data sample numbers used in the experiment. The privacy data consists of datasets D and D'. The shadow data consists of datasets D^A and $D^{A'}$. D is used to train the target classifier models, and D^A is used to train the shadow models. D' is considered non-membership data in the privacy data, and $D^{A'}$ is considered non-membership data.

Dataset	Private Data		Shadow Data	
	D	D'	D^A	$D^{A'}$
CIFAR100	25,000	5,000	25,000	5,000
Texas100	10,000	5,000	10,000	5,000
Purchase100	10,000	10,000	10,000	10,000

4.2 Models

To ensure a fair evaluation of the effectiveness of various defense mechanisms, we have separately created baseline classifier models for each dataset. Referring to Table 1 contains the data sample quantity settings used in the experiments. In these experiments, we use the privacy datasets D and D', as well as the shadow datasets D^A and $D^{A'}$. Specifically, dataset D is used to train S and T, while dataset D^A is used to train the shadow model. The dataset D' is considered non-member data within the privacy data, while dataset $D^{A'}$ is considered non-member data within the shadow data.

Taking this into consideration, we utilize different classifier models for various datasets. In specific, for the CIFAR100 dataset, we select AlexNet [16] and DenseNet [10] as classifier models. For the Texas100 dataset, we employ a 4-layer fully connected neural network (NN Net), while for the Purchase100 dataset, we use a 3-layer NN Net. Additionally, when dealing with image datasets (CIFAR100) and other datasets (Texas100, Purchase100), we also employ ResNet [7] and a 4-layer NN Net, which serve as adaptive noise generation models within MKD.

4.3 Attack Methods

To evaluate the effectiveness of multiple defense techniques, we employ four distinct membership inference attack methods. Below, we briefly introduce the four distinct membership inference attack methods we employ for validation:

Attack 1 (I_{corr}): Leino et al. [17] introduced a MIA method based on prediction accuracy. This occurs when the target model fails to generalize well, whereby it makes correct predictions on training data but performs poorly on test data. This method can infer whether a sample is a member or non-member based on whether it can be correctly predicted.

Attack 2 (I_{conf}): Song et al. [28] introduced a method that sets different thresholds for different categories to enhance the success rate of MIA. They assumed that the dataset might exhibit class imbalance, resulting in varying confidence levels among different categories. A sample is considered a member if its predicted confidence exceeds the previously set threshold. Otherwise, it is considered a non-member.

Attack 3 (I_{entr}): Song et al. [28] introduced a prediction-based entropy-driven MIA, similar to confidence-based membership inference attacks. Samples are considered members if their predicted entropy falls below a specific threshold, and non-members if it exceeds the threshold.

Attack 4 (I_{mentr}): Song et al. [28] discovered limitations in directly using predicted entropy. Therefore, they developed a method to compute prediction entropy correction by considering actual class labels.

4.4 Settings of MKD

Before proceeding to comparative experiments, we investigate different parameters for various data formats when aiming for excellent training outcomes (Table 2).

Table 2. Parameter settings for MKD. l_{rs} and l_{rt} are the learning rates of the student and teacher model, E_p denote the pre-training epochs of the teacher model, E_t and E_s are the training epochs of the student and teacher model.

Dataset	Model	l_{rs}	l_{rt}	E_p	E_s	E_t	α	β
CIFAR100	AlexNet	0.0001	0.0001	20	20	20	0.25	0.75
	DenseNet	0.0001	0.0001	15	30	30	0.3	0.7
Texas100	NN Net	0.0001	0.0001	10	10	10	0.3	0.7
Purchase100	NN Net	0.0001	0.0001	8	10	10	0.4	0.6

4.5 Methods of Comparison

To evaluate the superiority of MKD in privacy protection, we select various MIA defense methods from different defense perspectives as comparative methods. These methods encompass noise-based approaches such as Differential Privacy Stochastic Gradient Descent (DP-SGD) [1] and MemGuard [13], overfitting suppression methods like Adversarial Regularization (AdvReg) [20], as well as knowledge distillation-based methods like DMP [25].

Defense performance comparison with other competing methods on different datasets is shown in Table 4. All accuracy values in the table are expressed as percentages. In Table 4, we use our method's defense performance as the baseline, highlighted in gray. Additionally, $Test+$ indicates the percentage gain of MKD relative to other methods in terms of test accuracy, while $Avg.-$ represents the average percentage decrease of MKD relative to other methods in terms of the membership inference attack accuracy, with positive aspects marked in green and negative aspects marked in red.

5 Result and Discussion

5.1 Training Effectiveness of MKD

To validate the effectiveness of MKD, we conduct experiments using both MKD (represented as "w" in the Table 3) and without MKD (represented as "w/o" in the Table 3) on multiple datasets and classification models.

Table 3 displays the differences in classifier model performance and privacy inference risk before and after implementing MKD. The application of MKD significantly reduces the risk of privacy leakage, leading to a significant decrease in the accuracy of attackers in multiple membership inference attacks, dropping to a level as low as 50%, which is equivalent to random guessing. Additionally, while the application of MKD has some negative impact on model performance, it improves classification performance in CIFAR100 and Texas100 classification tasks. These findings demonstrate the superior capabilities of MKD as a defense mechanism in terms of both performance and privacy protection. Based on the overall experimental results, MKD achieves a favorable trade-off between member inference attack protection and model accuracy.

Table 3. The training and testing accuracy and member inference risk changes of models with or without MKD. All accuracies in this table are in %. The performance metrics of MKD are highlighted in gray . Moreover, we marked the optimal classification task accuracy in **blue**, as well as the change in member inference attack accuracy after defense in green.

Dataset	Model	Defense	Task Accuracy		Membership Inference Attack				
			$Train$	$Test$ ↑	$Avg.$ ↓	I_{corr}↓	I_{conf}↓	I_{entr}↓	I_{mentr}↓
CIFAR100	AlexNet	w/o	99.3	32.7	78.8	83.3	80.2	80.3	71.4
		w	50.4	**30.4**	54.7-24.1	60.0	53.2	53.2	52.4
	DenseNet	w/o	98.0	50.3	75.3	73.8	74.5	74.5	78.2
		w	68.0	**46.3**	55.6-19.7	61.9	52.3	52.3	55.8
Texas100	NN Net	w/o	92.4	56.3	65.4	68.0	62.5	61.7	69.2
		w	55.5	**52.1**	54.5-10.9	51.7	57.1	56.4	53.1
Purchase100	NN Net	w/o	96.3	70.1	65.6	66.1	63.8	63.7	68.9
		w	81.6	**74.8**	53.6-12.0	53.7	53.6	56.0	51.2

5.2 Comparative Analysis of the Noise-Based Methods

In Table 4, we compare MKD with noise-based defense methods (DP-SGD and MemGuard). The evaluation of DP-SGD is constrained by differential privacy theory and is limited to the application of AlexNet in the CIFAR100 classification task.

DP-SGD effectively reduces its test accuracy to approximately 50% in defending against membership inference attacks, significantly compromising the classification accuracy of the target model. On the other hand, MemGuard does not impair classification accuracy and performs well in preventing confidence-based membership inference attacks. However, it encounters difficulties when dealing with attacks that rely on model output labels.

Table 4. Comparing the defense performance with different data and under different attacks, all accuracies in this table are in %. In this table, the defensive effectiveness of our methods is set as a baseline, which is highlighted in gray . Meanwhile, the $Test^+$ shows the % increase in $Test$ of MKD over various methods, and $Avg.^-$ shows the % trimming in $Avg.$ of MKD over them, whose positive aspects are delineated in green, and negative aspects are marked in red.

Dataset	Model	Defense	Task Accuracy			Membership Inference Attack Accuracy					
			$Train$	$Test\uparrow$	$Test^+$	$Avg.\downarrow$	$Avg.^-$	$I_{corr}\downarrow$	$I_{conf}\downarrow$	$I_{entr}\downarrow$	$I_{mentr}\downarrow$
CIFAR100	AlexNet	DP-SGD	28.1	20.2	+50.5%	52.0	+5.1%	53.0	50.1	50.1	54,7
		MemGuard	99.3	32.7	-7.0%	58.4	-6.3%	83.3	50.0	50.0	50.1
		AdvReg	76.4	32.5	-6.4%	56.6	-3.3%	71.9	50.1	50.1	54.3
		DMP	50.4	30.4	+0.0%	55.2	-0.9%	60.0	54.2	54.2	52.4
		MKD(ours)	50.4	**30.4**		**54.7**		**60.0**	**53.2**	**53.2**	**52.4**
	DenseNet	MemGuard	98.9	50.3	-7.9%	55.9	-0.5%	72.6	50.0	50.0	51.1
		AdvReg	71.7	40.8	+13.5%	55.1	+0.9%	65.4	51.0	51.8	52.0
		DMP	89.5	49.5	-6.4%	58.9	-5.6%	65.1	51.2	59.6	59.8
		MKD(ours)	68.0	**46.3**		**55.6**		**61.9**	**52.3**	**52.3**	**55.8**
Texas100	NN Net	MemGuard	82.2	56.3	-7.4%	53.7	+1.4%	64.5	50.0	50.0	50.6
		AdvReg	76.0	50.1	+4.0%	57.0	-4.4%	61.6	54.4	53.8	58.3
		DMP	71.1	51.9	+0.4%	56.9	-4.2%	59.6	55.0	55.1	57.9
		MKD(ours)	55.5	**52.1**		**54.5**		**51.7**	**57.1**	**56.4**	**53.1**
Purchase100	NN Net	MemGuard	97.7	70.1	+6.7%	53.0	+1.1%	60.6	50.0	50.0	51.3
		AdvReg	95.3	70.3	+6.4%	56.6	-5.3%	60.9	55.1	55.2	55.2
		DMP	87.3	72.4	+3.3%	56.2	-4.6%	57.5	55.0	54.5	57.7
		MKD(ours)	81.6	**74.8**		**53.6**		**53.7**	**53.6**	**56.0**	**51.2**

In contrast, MKD significantly reduces the accuracy of all metrics for membership inference attacks while preserving or improving classification results. These results further confirm the superior effectiveness of MKD in balancing membership inference and classification task accuracy.

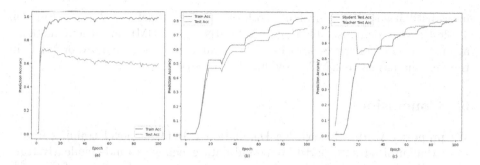

Fig. 3. The evolution of training and testing accuracy versus epochs using NN Net in the Purchase100 classification task. (a) MKD defense is not adopted in the target model. (b) The target model is incorporated with MKD defense. (c) The test accuracy of the student and teacher models evolves over epochs.

5.3   Comparative Analysis of the Suppressing Overfitting Methods

We process to compare MKD with the overfitting suppression-based defense method, i.e., AdvReg is based on suppressive overfitting, which effectively reduces the risk of member inference attacks by reducing overfitting, although the cost is a significant reduction in classification accuracy compared with the original model. It is ready to see that up to 18.8% of performance, degradation is achieved with AdvReg when adopting DenseNet on the CIFAR100 dataset. However, such defenses may overly depend on attack models, leading to an imbalanced trade-off between defense effectiveness and classification accuracy. This is demonstrated in the Test+ column of Table 4. Specifically, in DenseNet on CIFAR100, MKD achieves a classification accuracy that outperforms AdvReg by a clear margin (13.5%). In the case of relatively small datasets, such as Texas100 and Purchase100, MKD demonstrates superior performance compared to AdvReg in terms of model test accuracy and defense against MIA work.

Furthermore, MKD excels in combating overfitting, which is evident from the improvement in training and validation accuracy in Figs. 3a and 3b.

The observed result stems from the effective guidance provided by the teacher model, as illustrated in Figs. 3b and 3c. Mutual learning enables both the teacher and the target model to uphold remarkable anti-overfitting effects.

5.4   Comparative Analysis of the Knowledge Distillation Methods

Table 4 demonstrates a comparison between MKD and the defense method DMP, which utilizes knowledge distillation. DMP operates by utilizing the teacher model to transfer member data features, aiming to prevent the target model from retaining unnecessary details that may negatively impact classification accuracy. DMP ensures privacy effectively and maintains MIA accuracy below 60%, with a maximum classification accuracy degradation of 5% compared to undefended models. In contrast to DMP, MKD offers additional benefits through an inter-distillation process involving both the teacher model and the target model. Table 4 clearly demonstrates that MKD outperforms DMP in all four methods against member inference attacks. Additionally, the test accuracy of MKD in the four models rivals DMP, providing further evidence.

6   Conclusions

We proposed Mutual Knowledge Distillation (MKD), a novel training framework that utilized data partitioning and data exchange to independently train the student and teacher models, thereby improving the performance of the models through mutual distillation. By conducting empirical evaluations, we have verified the effectiveness of MKD in mitigating member inference attacks. Our method achieved an improved trade-off between model utility and member privacy in comparison with several existing defense methods.

Acknowledgements. This work was supported by the National Natural Science Foundation of China under Grants 62302117.

References

1. Abadi, M., et al.: Deep learning with differential privacy. In: Proceedings of the 2016 ACM SIGSAC Conference on Computer and Communications Security, pp. 308–318 (2016)
2. Chen, D., Yu, N., Fritz, M.: RelaxLoss: defending membership inference attacks without losing utility. arXiv preprint arXiv:2207.05801 (2022)
3. Chen, J., Wang, W.H., Shi, X.: Differential privacy protection against membership inference attack on machine learning for genomic data. In: BIOCOMPUTING 2021: Proceedings of the Pacific Symposium, pp. 26–37. World Scientific (2020)
4. Choquette-Choo, C.A., Tramer, F., Carlini, N., Papernot, N.: Label-only membership inference attacks. In: International Conference on Machine Learning, pp. 1964–1974. PMLR (2021)
5. Chowdhary, K., Chowdhary, K.: Natural language processing. In: Fundamentals of Artificial Intelligence, pp. 603–649 (2020)
6. Giraldo, J., Cardenas, A., Kantarcioglu, M., Katz, J.: Adversarial classification under differential privacy. In: Network and Distributed Systems Security (NDSS) Symposium 2020 (2020)
7. He, K., Zhang, X., Ren, S., Sun, J.: Deep residual learning for image recognition. In: Proceedings of the IEEE Conference on Computer Vision and Pattern Recognition, pp. 770–778 (2016)
8. Hong, Y., An, S., Im, S., Jo, J., Oh, I.: MONICA2: mobile neural voice command assistants towards smaller and smarter. In: Proceedings of the AAAI Conference on Artificial Intelligence, vol. 36, pp. 13176–13178 (2022)
9. Hu, H., Salcic, Z., Dobbie, G., Chen, Y., Zhang, X.: EAR: an enhanced adversarial regularization approach against membership inference attacks. In: 2021 International Joint Conference on Neural Networks (IJCNN), pp. 1–8. IEEE (2021)
10. Huang, G., Liu, Z., Van Der Maaten, L., Weinberger, K.Q.: Densely connected convolutional networks. In: Proceedings of the IEEE Conference on Computer Vision and Pattern Recognition, pp. 4700–4708 (2017)

11. Huang, T., Huang, J., Pang, Y., Yan, H.: Smart contract watermarking based on code obfuscation. Inf. Sci. **628**, 439–448 (2023)

12. Jayaraman, B., Evans, D.: Are attribute inference attacks just imputation? In: Proceedings of the 2022 ACM SIGSAC Conference on Computer and Communications Security, pp. 1569–1582 (2022)

13. Jia, J., Salem, A., Backes, M., Zhang, Y., Gong, N.Z.: MemGuard: defending against black-box membership inference attacks via adversarial examples. In: Proceedings of the 2019 ACM SIGSAC Conference on Computer and Communications Security, pp. 259–274 (2019)

14. Kaya, Y., Hong, S., Dumitras, T.: On the effectiveness of regularization against membership inference attacks. arXiv preprint arXiv:2006.05336 (2020)

15. Krizhevsky, A.: Learning multiple layers of features from tiny images. University of Toronto (2012). http://www.cs.toronto.edu/kriz/cifar.html. Accessed 13 May (2022)

16. Krizhevsky, A., Sutskever, I., Hinton, G.E.: ImageNet classification with deep convolutional neural networks. In: Advances in Neural Information Processing Systems 25 (2012)

17. Leino, K., Fredrikson, M.: Stolen memories: leveraging model memorization for calibrated {White-Box} membership inference. In: 29th USENIX Security Symposium (USENIX Security 20), pp. 1605–1622 (2020)

18. Li, Z., Liu, F., Yang, W., Peng, S., Zhou, J.: A survey of convolutional neural networks: analysis, applications, and prospects. IEEE Trans. Neural Netw. Learn. Syst. **33**, 6999–7019 (2021)

19. Li, Z., Zhang, Y.: Membership leakage in label-only exposures. In: Proceedings of the 2021 ACM SIGSAC Conference on Computer and Communications Security, pp. 880–895 (2021)

20. Nasr, M., Shokri, R., Houmansadr, A.: Machine learning with membership privacy using adversarial regularization. In: Proceedings of the 2018 ACM SIGSAC Conference on Computer and Communications Security, pp. 634–646 (2018)

21. Papernot, N., Song, S., Mironov, I., Raghunathan, A., Talwar, K., Erlingsson, Ú.: Scalable private learning with pate. arXiv preprint arXiv:1802.08908 (2018)

22. Qin, X., Tan, S., Tang, W., Li, B., Huang, J.: Image steganography based on iterative adversarial perturbations onto a synchronized-directions sub-image. In: ICASSP 2021–2021 IEEE International Conference on Acoustics, Speech and Signal Processing (ICASSP), pp. 2705–2709. IEEE (2021)

23. Salem, A., Bhattacharya, A., Backes, M., Fritz, M., Zhang, Y.: {Updates-Leak}: Data set inference and reconstruction attacks in online learning. In: 29th USENIX Security Symposium (USENIX Security 20), pp. 1291–1308 (2020)

24. Salem, A., Zhang, Y., Humbert, M., Berrang, P., Fritz, M., Backes, M.: ML-Leaks: model and data independent membership inference attacks and defenses on machine learning models. arXiv preprint arXiv:1806.01246 (2018)

25. Shejwalkar, V., Houmansadr, A.: Manipulating the byzantine: optimizing model poisoning attacks and defenses for federated learning. In: NDSS (2021)

26. Shejwalkar, V., Houmansadr, A.: Membership privacy for machine learning models through knowledge transfer. In: Proceedings of the AAAI Conference on Artificial Intelligence, vol. 35, pp. 9549–9557 (2021)

27. Shokri, R., Stronati, M., Song, C., Shmatikov, V.: Membership inference attacks against machine learning models. In: 2017 IEEE Symposium on Security and Privacy (SP), pp. 3–18. IEEE (2017)

28. Song, L., Mittal, P.: Systematic evaluation of privacy risks of machine learning models. In: 30th USENIX Security Symposium (USENIX Security 21), pp. 2615–2632 (2021)

29. Wang, M., Deng, W.: Deep face recognition: a survey. Neurocomputing **429**, 215–244 (2021)

30. Xue, M., et al.: Use the spear as a shield: an adversarial example based privacy-preserving technique against membership inference attacks. IEEE Trans. Emerg. Top. Comput. **11**(1), 153–169 (2022)

31. Yeom, S., Giacomelli, I., Fredrikson, M., Jha, S.: Privacy risk in machine learning: analyzing the connection to overfitting. In: 2018 IEEE 31st Computer Security Foundations Symposium (CSF), pp. 268–282. IEEE (2018)

32. Yeom, S., Giacomelli, I., Menaged, A., Fredrikson, M., Jha, S.: Overfitting, robustness, and malicious algorithms: a study of potential causes of privacy risk in machine learning. J. Comput. Secur. **28**(1), 35–70 (2020)

33. Zhang, Z., Lin, G., Ke, L., Peng, S., Hu, L., Yan, H.: KD-GAN: an effective membership inference attacks defence framework. Int. J. Intell. Syst. **37**(11), 9921–9935 (2022)

Fuzzing Drone Control System Configurations Based on Quality-Diversity Enhanced Genetic Algorithm

Zhiwei Chang[✉], Hanfeng Zhang, Yue Yang, Yan Jia, Sihan Xu, Tong Li, and Zheli Liu

College of Cyber Science, Tianjin Key Laboratory of Network and Data Security Technology, Nankai University, Tianjin, China
{2120210501,2120210508,yy99}@mail.nankai.edu.cn,
{jiay,xusihan,tongli,liuzheli}@nankai.edu.cn

Abstract. As drones are becoming widely used in various fields, drone security is a growing challenge nowadays. Drone control systems use various configuration parameters to control their positions and attitudes. If these parameters are misconfigured, drones will fall into abnormal flight states, such as trajectory deviation and crash to the ground. Existing works mainly focus on system memory errors which lead to obvious system failure but don't apply to drone flight state anomalies. This paper focuses on abnormal drone flight states caused by configuration parameter errors. We propose a novel state-guided fuzzing system called APFuzzer, which searches for incorrect configuration parameter values that would trigger abnormal flight states. To enhance the capability of searching for multiple optimal solutions, we design a quality-diversity enhanced genetic algorithm (QDGA) to mutate configurations to search for incorrect configuration parameter values and consider the effects of environmental factors and flight missions on the flight states. We evaluated APFuzzer on the drone control system ArduPilot and successfully searched 3389 incorrect configuration parameter values and triggered all predefined five abnormal flight states. In addition, APFuzzer automatically analyzed the fuzzing results and found five software bugs related to configurations.

Keywords: Drone security · Configuration parameters · Fuzzing · Quality diversity · Genetic algorithm

1 Introduction

Drones are unmanned aerial vehicles operated by radio remote control equipment and self-contained programmed control devices. The drone is a type of Autopilot and there are various airframes of drones, which can be divided into unmanned fixed-wing aircraft, unmanned multicopter, etc. Compared with manned aircraft, it has the advantages of small size and high maneuverability. Drones have been widely used in the world, such as cargo delivery, photography, etc.

In order to control drones for various missions, hundreds of configuration parameters have been designed to adjust the flight state of drones, such as yaw angle and flight

© The Author(s), under exclusive license to Springer Nature Singapore Pte Ltd. 2024
J. Vaidya et al. (Eds.): AIS&P 2023, LNCS 14509, pp. 499–512, 2024.
https://doi.org/10.1007/978-981-99-9785-5_35

speed. Although a large number of configurations bring flexibility to drone control, it also makes drone operation more complicated and error-prone. Incorrect configuration parameters can cause abnormal flight states, such as trajectory deviation and crash to the ground, which greatly threaten flight safety. In this paper, we focus on abnormal flight states caused by incorrect configuration parameters in the control system.

Existing techniques [11,25,30] are unable to detect errors in drone configuration parameters. Static program analysis methods [25] locate the root cause of a bug by analyzing the control dependencies of the data. However, the drone control system has over 100K+ lines of code and the time cost of program analysis is too huge. Techniques such as taint analysis [15,22] and symbolic execution [12,31] have been proposed to address the poor scalability of static analysis. However, searching the configuration space remains challenging due to the large space of values taken for each configuration parameter, and these techniques cannot handle this problem very well.

Fuzzing techniques [2,5,6,16] have proven to be effective methods for searching program vulnerabilities in many previous works. However, traditional coverage-guided fuzzing techniques are not applicable to drone control systems [18]. The drone control system is a stateful system and the coverage will quickly reach saturation and thus cannot continue to guide the mutation to find more errors. In addition, unlike traditional software, most drone flight state anomalies do not lead to obvious system failures like memory errors (e.g., buffer overflows). In this paper, we aim to design a fully automatic test framework to generate realistic inputs (including configuration parameters, flight missions, and environmental factors) to test drone control systems comprehensively.

Contributions. We propose APFuzzer, a state-guided fuzzing framework to search for incorrect configuration parameter values that could cause abnormal flight states. APFuzzer consists of three modules: (1) Flight State Detector. By analyzing official documents and source code, we define five abnormal flight states of drones and formulate accurate test oracles to detect them. (2) State-Guided Mutator. APFuzzer uses a quality-diversity enhanced genetic algorithm (QDGA) to mutate the configurations to drive the system into abnormal flight states, thereby finding incorrect configuration parameter values. In addition, APFuzzer considers the impact of flight missions and environmental factors on the correctness of configuration parameters. (3) Bug Analyzer. APFuzzer automatically analyzes the fuzzing results to locate the configuration parameter names and root causes of abnormal flight states.

We implemented the prototype of APFuzzer and applied it to ArduPilot [1], a mature and widely used autopilot control system. We fuzzed 46 configuration parameters related to flight state control, 16 flight mission commands, and 2 types of environmental factors. APFuzzer eventually found 3389 configuration parameter values that caused abnormal flight states within 6 h. Then, we systematically analyzed misconfiguration cases and found 5 software bugs.

To sum up, this paper makes the following novel contributions:

- We propose a fuzzing framework called APFuzzer which can automatically search for incorrect configuration parameter values of the drone control system.

- We design a quality-diversity enhanced genetic algorithm and a novel flight state metric to guide the mutation process to efficiently search for more configuration errors.
- We evaluated APFuzzer on ArduPilot and searched for 3389 incorrect configuration parameter values. We found 5 software bugs related to configuration parameters.

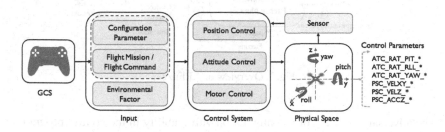

Fig. 1. Workflow of the drone control system.

2 Preliminaries

2.1 Drone Control System

Drones are typical cyber-physical systems that control the movement of drones in physical space through the interaction of multiple cyber and physical components. Drone control can be divided into three parts, shown in Fig. 1: (1) Ground Control System (GCS), where the user sets the drone configuration parameters, sends the flight mission and monitors the flight states of the drone in real-time. (2) Flight state control system, including many control functions, such as attitude control, position control and motor control, which accepts flight commands from GCS and data from sensors, and controls the movement and operation of the vehicle during the mission through a series of control algorithms. (3) Power system, which receives the regulation from the control system and drives the motors, propellers, and other components that make the drone move in the physical environment. In addition, configuration parameters are designed to control the flight state of the drone, such as flight speed, angular velocity, etc.

2.2 Attack Model

Similar to RVFuzzer [19], we assume that the adversary's goal is to stealthily drive the drone into an unstable flight state by manipulating configuration parameters, sending malicious missions or exploiting external environmental factors. We assume that the adversary has the following capabilities:

- Manipulating configuration parameters and missions before the drone takes off or is in flight by exploiting the vulnerabilities in Mavlink, a commonly used communication protocol [20].

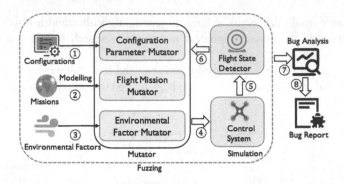

Fig. 2. Overview of APFuzzer

- The adversary is aware of a configuration vulnerability in the control program [19] and knows the exact value of the incorrect configuration parameters.
- The adversary is aware of the environmental conditions, such as wind speed, and can skillfully take advantage of them to attack the drone.
- The adversary's attack is stealthy and does not want to be detected by the real-time detection system or the post-incident investigation system [14].

In addition to the four capabilities above, the adversary does not need domain knowledge about the drone control system. The adversary would like to be able to stealthily launch an attack on the drone instead of directly sending malicious commands (e.g., closing the throttle) to directly drive the drone to crash, which can be easily detected by the log analysis tools [7, 10].

3   System Design

In this section, we present the design of APFuzzer. We first give an overview of APFuzzer's architecture and then present a detailed design of each component.

3.1   Overview

Figure 2 presents an overview of APFuzzer. APFuzzer first analyzes the official documentation and source code to select the test targets and then generates mutators based on the target characteristics (①②③). The output of the mutators is used as input to the drone control system, and then the simulations are further executed (④). The flight state detector detects the flight states of the drone in real-time during the simulation (⑤) and calculates the flight quality score to guide the mutator (⑥). APFuzzer uses a state-guided fuzzing approach to search for incorrect configuration parameter values, and the mutation method is based on QDGA (①②③④⑤⑥). Then, we analyze the results of fuzzing results. We first locate which configuration parameters caused the abnormal flight states and then search for the root cause to discover drone control system bugs (⑦⑧).

3.2 Flight State Detector

By analyzing the physical state of the drone and flight logs, we define five abnormal flight state types and design precise test oracles to detect each of them. We propose the flight state score as the fitness evaluation score of the QDGA to guide the mutation of configuration values. We define the flight state score as the sum of five abnormal state scores, represented as Eq. 1:

$$fitness\_score = \sum_{i=1}^{N_s} t_i * score_i \qquad (1)$$

Here, t_i denotes the measure of the importance assigned to each state, $score_i$ represents the score associated with each state and N_s represents the total number of abnormal flight states. Next, we will introduce the definitions of five abnormal flight states, their corresponding test oracles, and the method of calculating fitness scores.

(1) PreArm Failed. Before the drone takes off, the flight control system performs a range check on some of the configuration parameter values. However, many configuration values are not checked or are incorrectly checked in the source code, resulting in inconsistencies between the official documents and the source code.

Test Oracle. We identify the error by monitoring the log for error messages such as "PreArm: Bad Parameter" and "pre-arm fail". If such a flight abnormal state occurs, we define $score_{pf} = 1$.

(2) Thrust Loss. The drone relies on the throttle of the motor to control the thrust. If it is not correctly configured, even if the throttle is saturated, the drone will not be able to achieve the power required to maintain altitude and attitude, and a crash may occur.

Test Oracle. If the drone is under attitude control, the throttle is close to saturation, and the flight altitude is descending, we consider that Thrust Loss has occurred. If such an abnormal flight state occurs, we define $score_{tl} = 0.5$.

(3) Stuck. Incorrect configuration values can cause the drone to accidentally get stuck at a certain point, unable to continue driving or make a circular movement around a certain point.

Test Oracle. If the position of the drone in the flying state (non-hovering state) does not change for a long time or the change range is too small, it is determined that a stuck has occurred. If such a flight abnormal state occurs, we define $score_{st} = 1$.

(4) Crash. If the drone is unstable in flight and does not take emergency protections, the drone may crash on the ground.

Test Oracle. We check drone flight logs in real-time, and if we detect logs containing "Crash: Disarming", it means that this error has been caught. Also, we have observed that incorrect configuration can cause drones to land at excessive speed. With the community's help, we define that a crash is considered to have occurred if the landing speed is greater than 2 m/s when landing. According to our observation, the crash speed of drones generally does not exceed 20 m/s at most, so we define $score_{cr} = crash\_speed/20$.

(5) Deviation. The goal of the drone control system is to minimize the deviation between the observed state and the reference state. Incorrect setting of the configuration parameters will cause the control algorithm cannot drive the drone close to the reference state.

Test Oracle. The drone's position can be determined by longitude, latitude, and altitude. We calculate the distance of deviation of the drone from the reference position in real-time. The distance between any two waypoints of the flight mission we generated is less than 100 m. We define that if the deviation distance exceeds 10 m, it is considered that an abnormal flight state occurred. We define $score_{de} = distance/10$.

3.3   State-Guided Mutator

Fuzzing Targets. We mutate three input types to the drone control system: (1) Configuration parameter. The drone has six degrees of freedom (6DoF) in space, including the x-axis, y-axis, and z-axis for controlling linear movement, pitch, yaw, and roll for controlling rotation. We selected 46 configurations directly related to the drone's 6DoF as test targets. We use the official documents to get all the configuration parameter information, such as parameter type, value ranges, default values, etc. (2) Environmental factor. The drone system is a typical cyber-physical system, and environmental factors have a significant effect on the physical flight states. Especially, wind and obstacles are two important factors that affect the flight of drones. (3) Flight mission. Flight missions are used to plan the trajectory of the drone and what operations to perform during the flight. Each flight mission contains multiple flight commands and we chose 16 types of fight mission commands as test targets.

State-Guided Mutator Base on QDGA. Traditional fuzzing uses genetic algorithms [21,24] to heuristically mutate seeds to search for software bugs. However, the genetic algorithm is only suitable for searching one or a group of optimal solutions in the input space, and it is easy to fall into a local optimum. We aim to fully explore configuration space and find more optimal solutions that lead to abnormal flight states. Therefore, we use the quality-diversity [26,28] to enhance the searchability of the genetic algorithm to the solution space. Introducing the quality-diversity metric helps maintain the diversity of the population, makes the distribution of individuals within the population more even and prevents the algorithm from getting trapped in local optimum. This enhances the algorithm's exploration capabilities and contributes to discovering more software bugs.

We present our improved genetic algorithm in Algorithm 1. First, we use the function init_pop to randomly initialize the population pop_g (g=0) (line 2). Each individual in the population represents the combination of configuration parameters p_i, environmental factors e_i and flight missions m_i so that the individual can be expressed as $x_i = \{p_i, e_i, m_i\}$ and be encoded as a bit vector (e.g. 10111101001...). We use the function eval_fitness and eval_diversity to calculate each individual's fitness and quality-diversity scores in the pop_g, respectively. The function eval_fitness is obtained by Eq. 1 and the function eval_diversity is obtained by Eq. 2. Next, we enter the main loop: (1) Using the function select_parents to

---

Algorithm 1. State-guided fuzzing.

Input: P_C - Configuration parameters, P_M - Flight missions, P_E - Environmental factors, T - Maximum iterations, NP - Population size

Output: Incorrect configuration parameter values

1: $g \leftarrow 0$
2: $pop_g \leftarrow init\_pop(P_C, P_M, P_E, NP)$
3: $eval\_fitness(pop_g)$
4: $diversity_g \leftarrow eval\_diversity(pop_g)$
5: **while** $g < T$ **do**
6:     $parents \leftarrow select\_parents(pop_g)$
7:     $pop_{g+1} \leftarrow crossvoer(parents)$
8:     $pop_{g+1} \leftarrow mutate(pop_{g+1})$
9:     $eval\_fitness(pop_{g+1})$
10:    $diversity_{g+1} \leftarrow eval\_diversity(pop_{g+1})$
11:    **if** $diversity_{g+1} < diversity_g$ **then**
12:        $pop'_{g+1} \leftarrow best\_m(pop'_{g+1})$
13:        $pop''_{g+1} \leftarrow init\_pop(P_C, P_M, P_E, NP - P'_{g+1}.size)$
14:        $pop_{g+1} \leftarrow pop'_{g+1} \cup pop''_{g+1}$
15:    **end if**
16:    $eval\_fitness(pop_{g+1})$
17:    $g \leftarrow g + 1$
18: **end while**

---

select individuals with higher fitness scores from pop_g as the parent of the next generation population (line 6). The function `select_parents` is implemented by the `tournament selection operator` [27]; (2) Using the function `crossover` to generate new candidates (line 7). We use the `two-point operator` for the function `crossover`. We randomly select two points in the bit vectors of two parents and then swap the sub-vectors in the two points to generate new children; (3) Using the function `mutate` to mutate the candidate individuals (line 8). The function `mutate` is implemented by the `bit flip operator` which randomly flips a bit with a specific probability in the bit vectors; (4) Recalculating the fitness score of the individual of pop_{g+1} (line 9); (5) Recalculating the quality-diversity score of the population of pop_{g+1} (line 10). We use Euclidean distance [13] to evaluate the diversity of individuals. Finally, we calculate the population average distance as a diversity metric, as shown in Eq. 2.

$$diversity\_score = \sum_{i=1}^{NP}\sum_{j=i}^{NP} d(p_i, p_j)/(NP * (NP - 1)) \qquad (2)$$

Here, $d(p_i, p_j)$ represents the Euclidean distance between i and j, and NP represents the size of the population. If the diversity of the offspring is smaller than that of the parent, we select m individuals with the highest fitness scores, discard other individuals, and regenerate individuals to fill the population; (6) Updating the fitness score of the new population (line 16). When the population iterates T times, the algorithm is terminated.

3.4 Bug Analyzer

The target of the fuzzing includes 46 configuration parameters and each configuration parameter has hundreds of values. Exploring the root causes of configuration errors is highly time-consuming. First, we test each configuration parameter one by one to locate the configuration parameters that are most likely to cause abnormal flight states. After identifying the wrong configuration parameter name, we further analyze the wrong configuration parameter values and the root cause of the abnormal flight states.

4 Evaluation

We implemented a prototype of APFuzzer and verified our system on the open source control program ArduPilot. First, we verify the effectiveness and efficiency of APFuzzer to search for configuration errors and compare the performance of APFuzzer with the genetic algorithm (GA) and random algorithm. Then, we verify the impact of environmental factors and flight missions on flight states. Finally, we analyze the configuration errors and explore the root causes of the abnormal flight states caused by the configuration errors.

4.1 Experiment Setup

Flight Control System. We chose ArduPilot [1] flight control system as our test target, which is an open source and widely used system. We chose ArduCpoter as the target drone type.

Simulator. We chose mature and widely used Gazebo [4] to simulate the drone and physical environment. We validate it in one real drone with Pixhawk [23] (AMOVLab P450 [3]).

Experimental Machine Performance. All experiments are performed using a desktop PC with sixteen-core 3.2 GHz Intel Core i9 CPU and 32 GB RAM running Ubuntu 64-bit. We use QGroundControl [9] as the ground control station to manipulate the drone and MAVLink [8] as the communication protocol to interact with the drone. We implemented APFuzzer using 1472 lines of Python.

4.2 The Effectiveness and Efficiency of APFuzzer in Searching for Incorrect Configuration Parameter Values

Effectiveness. We select 46 configuration parameters directly related to the drone attitude control, 2 environmental factors and 16 flight mission commands as the test targets. In our QDGA implementation, we set the population size (NP) to 50, the crossover probability to 0.5, and the mutation probability to 0.01. We use average Euclidean distance to measure the quality diversity of the population and use the method of Sect. 3.2 to measure the fitness score of the population.

As shown in Fig. 3, we evaluate the performance of APFuzzer when m = 25, 35 and 45, and the test time is 6 h. When m = 25, APFuzzer completed 4788 simulations and found 2682 configuration errors, and when m = 35 and m = 45, APFuzzer searched for 3389 and 2633 configuration errors respectively. It can be seen that when m = 35, APFuzzer completes the most simulations, searches for the most configuration errors, and has the highest abnormal rate of search results. Additionally, Fig. 4(a) illustrates that when m = 35, APFuzzer searches for misconfigurations with the fastest growth rate and the best search performance. Figures 4(b) and 4(c) show the trends of average quality diversity and average fitness of populations with different values of m, respectively. Based on the comparison, APFuzzer performs best when m equals 35.

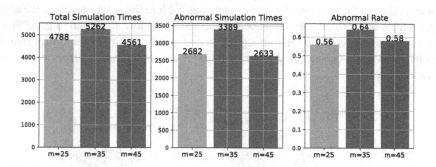

Fig. 3. Fuzzing results of APFuzzer with different parameter values of m.

Furthermore, we counted the number of each abnormal state caused by misconfigurations searched by APFuzzer when m = 35. APFuzzer completed 5262 simulations within 6 h, of which 3389 simulated flight results were abnormal. APFuzzer searched all five pre-defined abnormal states. Of all configuration errors, 3130 cases caused Crash, 215 cases caused Stuck, 37 cases caused Deviation, 6 cases caused PreArm Failed and 1 case caused Thrust Loss.

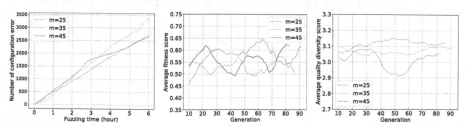

(a) Number of incorrect con-figuration parameter values.   (b) The average fitness score of each generation population.   (c) The average quality diversity score of each generation population.

Fig. 4. The performance of APFuzzer with different parameter values of m.

Efficiency. To evaluate the efficiency of the QDGA in the fuzzing of drone control system configuration parameters, we implemented comparative experiments and compared it with the traditional genetic algorithm (GA) and random algorithm. We set the value of m to 35 in QDGA and set the parameters of the genetic algorithm to be the same as QDGA, except for not using the quality diversity metric. The fuzzing time was set to 6 h. Figure 5 shows the fuzzing results of the three algorithms. APFuzzer completed 5262 simulations within 6 h, while the random algorithm and GA completed only 3677 and 958 simulations, respectively. GA only searched for 821 configuration errors and the abnormal rate was 86%, random algorithm found 939 configuration errors, but the abnormal rate was only 26%. Obviously, as the population evolves, GA falls into a local optimal solution, and the ability to explore the solution space becomes worse and worse. Although the random algorithm can better explore the solution space, the ability to search for the optimal solution is poor, reflected in the low abnormal rate. APFuzzer takes advantage of GA and random algorithm and has the ability to search for multiple optimal solutions. As a result, APFuzzer found 3389 configuration errors with a 64% abnormal rate.

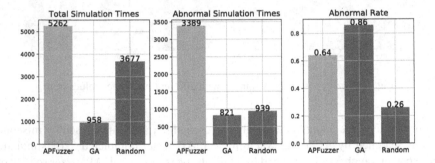

Fig. 5. Fuzzing results of APFuzzer, genetic algorithm and random algorithm.

Table 1. Fuzzing results with the impact of environmental factors and flight missions.

Factor	Crash	Deviation	Stuck
Environment	638	36	72
Flight Mission	994	2	69

4.3 The Influence of Environmental Factors and Flight Missions on Flight State

Environmental Factors. We assume that environmental factors have an important effect on the flight state of drones. Attackers can use specific environmental conditions to attack drones. In our experiments, we only consider the influence of wind speed and wind direction. We set the wind speed between [0,10] m/s and the wind direction between [0,360] degrees. We selected 600 correct configuration parameters from

APFuzzer's fuzzing results, and for each configuration parameter, we randomly introduced environmental variations ten times. In the end, we completed 6000 simulations in approximately 9 h. Table 1 shows the number of flight abnormal states caused by configuration errors, including 638 cases of Crash, 36 cases of Deviation and 72 cases of Stuck. Our experiment demonstrates that environmental factors can affect the correctness of configuration parameters, thereby influencing the flight state of drones.

Flight Mission. Unlike previous work LGDFuzzer, which only tested a single mission, we assume that the missions directly affect the correctness of the configuration values. We selected 600 correct configuration parameters from APFuzzer's fuzzing results and regenerated 1000 new missions as the mission seed pool. For each correct configuration, we randomly selected ten missions to run the simulation again. As shown in Table 1, after performing 6000 simulations, we found 994 cases of Crash, 2 cases of Deviation and 69 cases of Stuck. The test results show that the correctness of the configuration parameter values is related to the flight mission and prove that different mission scenarios change the range of incorrect configuration parameters.

4.4 Bug Analysis Result

APFuzzer searched for a total of 3389 configuration errors within six hours. To explore the root cause of configuration errors, we test each configuration parameter one by one to locate the specific configuration parameter name of each error, and then further analyze the specific reasons that lead to abnormal flight states. Table 2 shows our search results. After manual inspection, we confirmed 5 software bugs. Among them, 2 is related to PreArm Failed, 1 is related to Stuck and 1 is related to Crash.

Table 2. Bug analysis result of APFuzzer. The description presents the root cause of bugs caused by configuration errors.

No.	Flight State	Configuration	Description
01	PreArm Failed	PSC_ACCZ_I	$p \in [0, 3.0]$, $p = 0$ leads to floating-point exception
02	PreArm Failed	ATC_RAT_PIT_D	$p \in [0, 0.03]$, $p = 0$ leads to floating-point exception
03	PreArm Failed	ATC_ANG_PIT_P	$p \in [0, 12]$, $p < 1$ violates the constraint in program
04	Crash	ATC_RAT_PIT_D	$p \in [0, 0.05]$, $p > 0.42$ causes the drone to hit the ground at high speed
05	Stuck	PSC_ACCZ_D	$p \in [0, 0.4]$, $p > 0.13$ causes the drone to get stuck in climbing motion

Bug No. 01 and Bug No. 02 are floating point overflow vulnerabilities caused by a lack of range checking in the source code. Bug No. 03 is caused by the inconsistency between the source code and official documentation. The value range of the configuration parameter in the official documentation is [0,12], but configuring the value less than 1.0 will violate program constraints. Bug No.04 is caused by the value of the configuration parameter ATC_RAT_PIT_D being too large. When its value is between [0.042,0.05], the drone will crash on the ground at high speed due to losing control. Bug No.05 is similar to Bug No.04, in which configuration parameter values within a specific range will cause the drone to get stuck in an endless climbing motion.

5 Related Work

The traditional fuzzing methods [2,5,6,16,29] are used to find program memory vulnerabilities (e.g., buffer overflow) guided by code coverage. These methods improved the mutation algorithm and seed selection algorithm or used techniques such as taint analysis and symbolic execution to improve code coverage to search for more vulnerabilities. Still, they are unsuitable for searching for control system vulnerabilities caused by incorrect configuration parameters.

In recent years, a small amount of work has focused on vulnerabilities caused by configuration parameters. RVFuzzer [19] applies control-guided mutation to search for unsafe configuration parameter ranges called input validation bugs. However, this method has a high false positive rate and cannot accurately detect configuration errors. LGDFuzzer [17] applies a machine learning-guided fuzzing approach that uses a predictor and a genetic search algorithm to detect incorrect configurations. However, this predictor relies on specific flight logs, which is unsuitable for new flight missions, and the accuracy of the state predictor is not high. APFuzzer applies a QDGA and considers environmental factors and flight missions' impact on flight states. It efficiently searches for configuration errors and identifies the root causes of these errors.

6 Conclusion

In this paper, we design and implement a state-guided fuzzing system called APFuzzer, which is effective in searching for incorrect configuration parameter values that would trigger abnormal flight states on drones. We design a quality-diversity enhanced genetic algorithm to mutate configurations to detect incorrect configuration parameter values. We consider the effects of environmental factors and flight missions on the flight states and evaluate their effects on flight states. We implemented our system on ArduPilot, successfully detected 3389 incorrect configuration parameter values, and triggered all five predefined abnormal flight states within 6 h. Furthermore, APFuzzer can automatically analyze the fuzzing results and locate the root causes of configuration errors. In the end, we found five software bugs and fed them back to the developers.

References

1. Ardupilot (2022). http://ardupilot.org
2. American fuzzy lop (2023). http://lcamtuf.coredump.cx/afl/
3. Amovlab (2023). http://www.amovlab.com/
4. Gazobo (2023). http://gazebosim.org
5. Honggfuzz (2023). http://llvm.org/docs/LibFuzzer.html
6. Libfuzzer (2023). http://google.github.io/honggfuzz/
7. Mavexplorer (2023). http://github.com/ArduPilot/MAVProxy
8. Mavlink (2023). http://mavlink.io
9. Qgroundcontrol (2023). http://qgroundcontrol.com
10. Uav logviewer (2023). http://ardupilot.org/copter/docs/common-uavlogviewer.html

11. Attariyan, M., Flinn, J.: Automating configuration troubleshooting with dynamic information flow analysis. In: 9th USENIX Symposium on Operating Systems Design and Implementation (OSDI 10) (2010)

12. Baldoni, R., Coppa, E., D'elia, D.C., Demetrescu, C., Finocchi, I.: A survey of symbolic execution techniques. ACM Comput. Surv. (CSUR) **51**(3), 1–39 (2018)

13. Belkin, M., Niyogi, P.: Laplacian eigenmaps and spectral techniques for embedding and clustering. In: Advances in Neural Information Processing Systems 14 (2001)

14. Clark, D.R., Meffert, C., Baggili, I., Breitinger, F.: Drop (drone open source parser) your drone: forensic analysis of the DJI phantom iii. Digit. Investig. **22**, S3–S14 (2017)

15. Clause, J., Li, W., Orso, A.: Dytan: a generic dynamic taint analysis framework. In: Proceedings of the 2007 International Symposium on Software Testing and Analysis, pp. 196–206 (2007)

16. Haller, I., Slowinska, A., Neugschwandtner, M., Bos, H.: Dowsing for {Overflows}: a guided fuzzer to find buffer boundary violations. In: 22nd USENIX Security Symposium (USENIX Security 13), pp. 49–64 (2013)

17. Han, R., et al.: Control parameters considered harmful: Detecting range specification bugs in drone configuration modules via learning-guided search. arXiv preprint arXiv:2112.03511 (2021)

18. Kim, S., Liu, M., Rhee, J.J., Jeon, Y., Kwon, Y., Kim, C.H.: Drivefuzz: discovering autonomous driving bugs through driving quality-guided fuzzing. In: Proceedings of the 2022 ACM SIGSAC Conference on Computer and Communications Security, pp. 1753–1767 (2022)

19. Kim, T., et al.: {RVFuzzer}: Finding input validation bugs in robotic vehicles through {Control-Guided} testing. In: 28th USENIX Security Symposium (USENIX Security 19), pp. 425–442 (2019)

20. Kwon, Y.M., Yu, J., Cho, B.M., Eun, Y., Park, K.J.: Empirical analysis of mavlink protocol vulnerability for attacking unmanned aerial vehicles. IEEE Access **6**, 43203–43212 (2018)

21. Li, G., et al.: AV-FUZZER: finding safety violations in autonomous driving systems. In: 2020 IEEE 31st International Symposium on Software Reliability Engineering (ISSRE), pp. 25–36. IEEE (2020)

22. Maskur, A.F., Asnar, Y.D.W.: Static code analysis tools with the taint analysis method for detecting web application vulnerability. In: 2019 International Conference on Data and Software Engineering (ICoDSE), pp. 1–6. IEEE (2019)

23. Meier, L., Tanskanen, P., Fraundorfer, F., Pollefeys, M.: Pixhawk: a system for autonomous flight using onboard computer vision. In: 2011 IEEE International Conference on Robotics and Automation, pp. 2992–2997. IEEE (2011)

24. Mirjalili, S.: Genetic algorithm. In: Evolutionary Algorithms and Neural Networks. SCI, vol. 780, pp. 43–55. Springer, Cham (2019). https://doi.org/10.1007/978-3-319-93025-1_4

25. Møller, A., Schwartzbach, M.I.: Static program analysis. Notes. Feb (2012)

26. Mouret, J.B., Clune, J.: Illuminating search spaces by mapping elites. arXiv preprint arXiv:1504.04909 (2015)

27. Prayudani, S., Hizriadi, A., Nababan, E., Suwilo, S.: Analysis effect of tournament selection on genetic algorithm performance in traveling salesman problem (TSP). In: Journal of Physics: Conference Series. vol. 1566, p. 012131. IOP Publishing (2020)

28. Pugh, J.K., Soros, L.B., Stanley, K.O.: Quality diversity: A new frontier for evolutionary computation. Frontiers in Robotics and AI p. 40 (2016)

29. Stephens, N., et al.: Driller: augmenting fuzzing through selective symbolic execution. In: NDSS, vol. 16, pp. 1–16 (2016)

30. Xu, T., et al.: Do not blame users for misconfigurations. In: Proceedings of the Twenty-Fourth ACM Symposium on Operating Systems Principles, pp. 244–259 (2013)
31. Yao, Y., Zhou, W., Jia, Y., Zhu, L., Liu, P., Zhang, Y.: Identifying privilege separation vulnerabilities in IoT firmware with symbolic execution. In: Sako, K., Schneider, S., Ryan, P.Y.A. (eds.) ESORICS 2019. LNCS, vol. 11735, pp. 638–657. Springer, Cham (2019). https://doi.org/10.1007/978-3-030-29959-0_31

KEP: Keystroke Evoked Potential for EEG-Based User Authentication

Jiaxuan Wu, Wei-Yang Chiu, and Weizhi Meng[(⊠)]

SPTAGE Lab, DTU Compute, Technical University of Denmark,
Kongens Lyngby 2800, Denmark
{weich,weme}@dtu.dk

Abstract. In recent years, the rapid proliferation of Brain-Computer Interface (BCI) applications has made the issue of security increasingly important. User authentication serves as the cornerstone of any secure BCI systems, and among various methods, EEG-based authentication is particularly well-suited for BCIs. However, existing paradigms, such as visual evoked potentials and motor imagery, demand significant user efforts during both enrollment and authentication phases. To address these challenges, we introduce a novel paradigm–Keystroke Evoked Potentials (KEP) for EEG-based authentication, which is secure, user-friendly, and lightweight. Then, we design an authentication system based on our proposed KEP. The core concept involves generating a shared cryptographic session key derived from EEG data and keystroke dynamics captured during random button-pressing activities. This shared key is subsequently employed in a Diffie-Hellman Encrypted Key Exchange (DH-EKE) to facilitate device pairing and establish a secure communication channel. Based on a collected dataset, the results demonstrate that our system is secure against various attacks (e.g., mimicry attack, replay attack) and efficient in practice (e.g., taking only 0.07 s to generate 1 bit).

Keywords: User Authentication · EEG · BCI · Diffie-Hellman · Keystroke Evoked Potential · KEP

1 Introduction

The field of Brain-Computer Interface (BCI) is undergoing rapid development, with an increasing number of applications in areas such as neural engineering, robot control, gaming, entertainment, and security sectors [42]. These advancements significantly improve the quality of life for individuals, particularly those living with chronic disabilities and their caregivers. However, as the technology gains traction, it becomes increasingly vulnerable to newly emerging forms of attacks [43]. Several studies have explored the potential risks associated with BCI applications, identifying cybersecurity as a significant concern. For instance, Bhalerao et al. [3] has indicated that wireless BCI systems are susceptible to EEG signal tampering or attacks during transmission, potentially leading to system

© The Author(s), under exclusive license to Springer Nature Singapore Pte Ltd. 2024
J. Vaidya et al. (Eds.): AIS&P 2023, LNCS 14509, pp. 513–530, 2024.
https://doi.org/10.1007/978-981-99-9785-5_36

malfunctions. In the context of EEG-based authentication, tampered EEG data could deceive the authentication system, posing a critical security risk [11].

Surprisingly, this crucial issue has largely been overlooked in existing studies on EEG-based authentication systems [47]. These studies often replicate the workflow of traditional biometric authentication systems, which involves collecting EEG data from subjects and using machine learning algorithms for classification. However, such approaches fail to account for the unique cybersecurity challenges associated with BCIs. For example, BCI devices often lack a shared secret with users' laptops or smartphones, making them susceptible to Man-in-the-Middle (MITM) attacks during device pairing [46].

Another point worth noting is the limitations of current EEG-based authentication methods. The majority of these approaches rely on the detection and classification of motor imagery, which can be spoofed [7]. Also, utilizing intrinsic oscillatory brain signals and motor imagery for authentication necessitates several seconds to establish Event-Related Desynchronization (ERD) patterns. Moreover, users often require training to become accustomed to this method [34]. This contradicts one of the primary advantages of biometric systems over password-based systems: the elimination of the need to remember passwords [5]. For instance, current methods often require individuals to engage in imagined movements, increasing both cognitive workload and the time required for authentication, which can last up to 10 s. Therefore, there is a pressing need for alternative approaches that can enhance the Information Transfer Rate (ITR), ideally requiring less time and mental effort [34].

Another major type of BCI systems measures the P300 component of a Visual Evoked Potential (VEP), which occurs in response to flashed columns or rows of letters [9,23]. One major drawback of these 'visual' methods is that they require subjects to maintain gaze control for an extended period during the enrollment phase. Another significant limitation is the variability of EEG wave features, which can change over time due to many factors such as health, emotional state, and age. Consequently, these methods are inherently unreliable for high-security applications.

Contributions. In this work, we propose a novel paradigm for EEG-based authentication, termed **Keystroke Evoked Potential (KEP)**. To circumvent the downgraded user experience associated with prolonged and high concentration required in other methods, KEP only necessitates that users continuously keystroke a dedicated button randomly for 2 min using one fingertip, with any posture they prefer. During this process, a shared secret key is generated without communication and stored in both the device and the BCI system. Then, when users want to authenticate, they simply need to band the BCI device; no further action is required, making the process similar to token-based authentication. While this new approach can offer enhanced security compared to token-based methods. That is, even if the BCI device is stolen and used by someone else, the system can identify that the user is not the originally registered individual by analyzing the features of the EEG wave.

Research Scope. The primary goal of this work is to explore the viability of Keystroke Evoked Potential (KEP) as a foundation for EEG-based authentication. Specifically, we aim to address the following questions:

- Are the features of KEP both temporally consistent and sufficiently strong in amplitude?
- Can brain waves accurately reflect the dynamics of keystrokes?
- What is the time required to register a new user?

Paper Structure. The organization of this paper is shown as follows: Sect. 2 provides the background on KEP. Section 3 outlines our system design. Section 4 evaluates the security, effectiveness and practicability of our system, and Sect. 5 introduces related studies on biometric authentication and EEG. Finally, Sect. 6 offers a conclusion regarding our proposed system.

2   Background

In this section, we provide the essential background knowledge needed to understand the principles behind the Keystroke Evoked Potential and Autonomic Key Generation methods.

2.1   Keystroke Evoked Potential (KEP) Principle

Electroencephalography (EEG) is a medical imaging technique used to record the electrical activity of the brain. It is a non-invasive method in which electrodes are placed on the scalp to detect and measure the minute electrical voltages generated by brain cells (neurons) during communication. As illustrated in Fig. 1, when a charged button is touched by a person's finger, it induces changes in the body's electrical potential. This, in turn, leads to variations in the steady-state voltage observed in the EEG data.

This electrified button does not actually require specialized design. A common example encountered in daily life is the keyboard of a charging laptop. A Switching Mode Power Supply (SMPS) is an efficient method for power conversion that controls voltage output by switching the current at high frequencies. This switching operation can generate harmonics, which are integer multiples of the original switching frequency. While these dominant harmonics typically manifest in the high-frequency range, there are instances such as when an SMPS is designed to operate at 100 kHz where harmonics or intermodulation products may appear in lower frequency ranges, such as between 100 Hz and 120 Hz, as shown in Fig. 2.

2.2   Autonomic Device Pairing

Autonomic Device Pairing leverages a shared secret observation between communicating parties to independently generate matching keys, eliminating the need

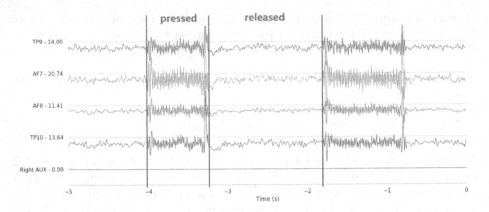

Fig. 1. Observed Potential Change in EEG

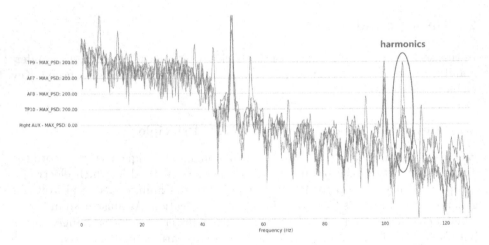

Fig. 2. Observed harmonics in EEG

for third-party intervention [44]. In recent years, researchers have explored various contexts for this secret observation, including vibrations from handshakes [8,12], user heartbeats [24,37], and user gait [39,45]. Although the specific algorithms for key extraction from these observed signals differ, all these techniques share a similar foundational model and signal processing pipeline, as illustrated in Fig. 3.

A workstation refers to a computing device such as a laptop or smartphone, while a BCI device can be a headset or headband capable of collecting and transmitting EEG data to the workstation. The device pairing process, as illustrated in Fig. 3, begins with the workstation sending a request to the BCI device to initiate pairing. Upon confirmation from the BCI device, a 30-second data collection phase commences. During this phase, users are instructed to randomly

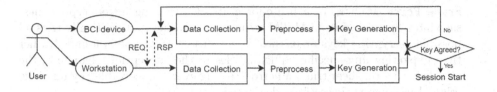

Fig. 3. General pipeline of device pairing

and continuously keystroke a designated button on the workstation. The BCI device records EEG data, while the workstation captures keyboard data.

In the preprocessing phase, both devices independently perform data alignment, scaling, and segmentation without inter-device communication. The EEG data is then fed into a transformer auto-encoder, followed by a hash function, to generate a session key. On the workstation side, key generation is more straightforward. It involves analyzing the segments of button states: '0' is output if the button is consistently released, '1' is output if the button state changes (i.e., from released to pressed or vice versa), and '2' is output if the button is consistently pressed. After analyzing all segments, the workstation hashes the resulting string to obtain its session key. If key generation is successful, fully matched symmetric session keys should be generated. Otherwise, the device pairing process restarts, looping through the pipeline until a symmetric key is successfully generated.

3   System Design

In this section, we provide a detailed overview of the key technological modules of KEP, ranging from identical sequence generation to identity verification.

3.1   Adversary Model

- **Man-In-The-Middle-Attack (MITM):** A Man-in-the-Middle (MITM) attack is a type of cyber attack where an unauthorized actor intercepts, relays, and may alter communications between two parties who believe they are directly communicating with each other. The attacker can capture and store any transmitted data, such as login credentials, personal information, and sensitive business data.
- **Mimic Attack:** In a mimic attack, adversaries observe the entire process of user enrollment and then attempt to imitate the user's actions. Their goal is to regenerate a key that closely resembles the original, thereby gaining unauthorized access.
- **Replay Attack:** In a replay attack, an unauthorized attacker intercepts and records a legitimate message exchanged between two devices that are in the process of pairing. The attacker then replays this captured message at a later time, either to gain unauthorized access to the system or to initiate actions that the original message was intended to trigger.

- **Brute Force Attack:** In a brute-force attack targeting an encrypted message intercepted during an MITM attack, the attacker systematically tries all possible decryption keys or passwords until the correct one is found. Rather than exploiting any specific vulnerabilities in the encryption algorithm, the attacker relies on sheer trial and error, testing every conceivable key until the original message is successfully decrypted.

3.2   Design Goals

To address the threats and challenges we have identified, we have set the following goals that our proposed KEP-based authentication system should aim to achieve in practice:

- **Security:** All transmitted data should be encrypted and designed in such a way that attackers cannot easily decrypt it. Additionally, the scheme should have the capability to detect any tampering with the transmitted data.
- **Usability:** The time required for both the enrollment and authentication stages should be minimal, and the process should demand a low effort from the user side.
- **Lightweight:** The scheme should be resource-efficient and lightweight, requiring minimal computational resources, as part of the system will be integrated into BCI devices.

3.3   Data Collection

After receiving an authentication request, the user types randomly on the keyboard with one fingertip, which serves as the source of a shared secret to establish the initial secure session. The workstation records the timestamps of button presses and releases, while the BCI device captures variations in voltage potential over time.

Temporal Alignment. Given that the devices independently sample EEG and keyboard data, temporal alignment becomes necessary. We employ an event-based approach where both devices detect the onset of the first pressing event and use this as a basis to segment the data. This approach eliminates the need for explicit time synchronization between the devices, under the assumption that they are equipped with sufficiently accurate real-time clocks to ensure alignment within the detected active segments.

Entropy of Sensor Data. During the enrollment phase, the workstation continuously calculates the resulting entropy of the collected keyboard data. If the entropy is found to be insufficiently high at the end of the enrollment process, users will be prompted to undergo the enrollment process again.

3.4   Data Processing

- **Scaling:** The EEG data is scaled within a specific range (e.g., 0 to 1) to reduce computational resource consumption during model training.

- **Segmentation:** The continuous EEG data is divided into segments, each with a specific window size (W). Each segment or window is then decoded into a number belonging to the set {0, 1, 2} by our transformer auto-encoder.

Unlike traditional EEG-based authentication approaches, we do not require the filtering of artifacts caused by eye blinks or facial muscle movements, nor require the noise filtering of irrelevant brainwave bands. According to our experimental results, the changes in brainwaves induced by KEP are sufficiently distinct to allow for accurate decoding.

3.5 Auto-encoder Training

We employ the standard transformer model, as described by Vaswani et al. [41], to serve as our auto-encoder. The structure of the model is depicted in Fig. 4.

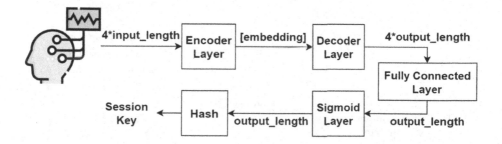

Fig. 4. The flowchart of the Transformer Auto-encoder.

In our configuration, we set d_model to 4, as the BCI device is limited to detecting data from four channels: AF7, AF8, TP9, and TP10. The architecture comprises four main layers: the Encoder Layer, the Decoder Layer, the Fully Connected Layer, and the Sigmoid Layer. The Encoder Layer processes the input sequence to create a 'memory' that encapsulates the contextual information of the input. This memory is subsequently utilized by the Decoder Layer. The Encoder Layer is composed of multi-head self-attention mechanisms and feedforward neural networks. We set the number of attention heads to 2, as it must be a factor of d_model, which in this case is 4.

The Decoder Layer receives the 'memory' from the Encoder Layer and generates the output sequence. During the training stage, it can also accept an additional 'target' sequence as input. The Fully Connected (FC) Layer serves to map the output of the Decoder Layer to the desired output shape. In this case, it maps each position in the sequence to a single value (scalar). The Sigmoid Layer then maps any real-valued number to the range of {0,1,2}. Finally, the hash module takes the tensor output from the transformer model and generates the session key.

3.6 Encrypted Key Exchange

If both devices require the key for secure communication, they can utilize a key agreement protocol to ensure that both possess the same key without having to transmit it directly. Specifically, we employ the Diffie-Hellman Encrypted Key Exchange [2], which is an example of a Balanced Password-Authenticated Key Exchange (PAKE) protocol (as illustrated in Fig. 5).

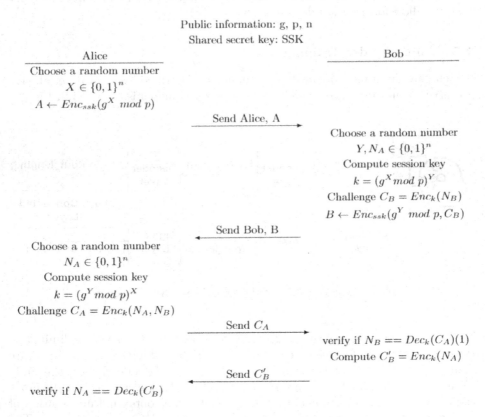

Fig. 5. Diffie-Hellman Encrypted Key Exchange

Following parameters are used for key generation:

- p: a big prime, called "modulus"
- q: a divisor of $p - 1$, called "subgroup order"
- g: "generator" an integer modulo p of order q (this means that the smallest integer $k > 0$ such that $g^k = 1 \bmod p$ is k = q)
- X and Y: random numbers chosen separately by the two parties

The basic idea is to use the shared secret key ssk to protect the public key swap of Diffie-Hellman Key Exchange. Only the one who possesses the key ssk

can encrypt and decrypt the ciphertexts. If A and B pass key verification successfully, then a secure channel can be established between A and B. Otherwise, the device pairing process restarts, looping through the pipeline until a shared secret key is successfully generated.

4   Evaluation

In this section, we introduce our goals and evaluation methodology. Then we discuss the security aspect of our system, and analyze the results of Bit Agreement Rate and Bit Rate.

4.1   Goals

In this section, we assess the performance of the proposed key generation scheme. The objectives of the evaluation are threefold: 1) assess the scheme's security against various types of adversarial attacks and evaluate the length of the generated keys; 2) determine the optimal key parameters, including the length of the input data and the window size (W); and 3) identify the optimal hyperparameters, including the number of attention heads, encoder layers, and decoder layers.

4.2   Methodology

Data Collection: The dataset used for evaluating the performance of the proposed system comprises 10 subjects (5 males and 5 females). As illustrated in Fig. 6, we collect EEG data from the following channels according to the EEG 10-10 system: AF7, AF8, TP9, TP10. The default sampling rate of the *MuseS* headband is 256 Hz.

During the data collection phase, participants were instructed to wear the BCI device and type randomly on the keyboard for approximately 30 s at a normal speed (fewer than 10 keystrokes per second). Data collection was conducted under both reasonable and slightly unreasonable wearing conditions to capture different scenarios in practical usage. Each subject was asked to repeat the random typing process five times.

4.3   Security Discussion

Man-In-The-Middle (MITM) Attack: The primary security concern in wireless communication over a public channel is the risk of an MITM attack during the session key exchange protocol. In our system, an MITM attacker cannot intercept the data from the communication channel because the secret travels through the user's body, from their fingertip to their brain. Attackers are unlikely to guess the 'secret,' as its length can be set to be sufficiently long (e.g., 256 bits) to ensure high entropy. Therefore, an MITM attack is highly improbable, given that the attacker would need access to the recorded data to derive the temporary key *ssk*, which the device uses to authenticate the BCI device.

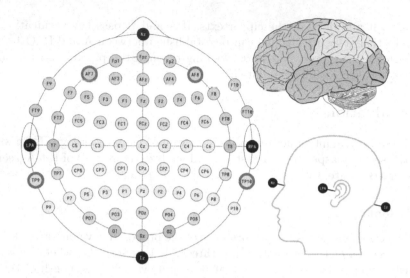

Fig. 6. EEG channels for data collection

Mimicry Attack: Our basic assumption is that the adversary would need to observe the entire process of user enrollment. However, most people have two hands. If users suspect that their environment is unsafe, they can register using one hand while using the other hand to completely cover the actions of the registering hand. As a result, an attacker attempting to derive the key solely from recorded videos would find it nearly impossible.

Replay Attack: In this type of attack, an intruder replays messages from a previous session key establishment procedure between two devices, A and B. By doing so, the intruder could potentially impersonate Device B to communicate with Device A, or vice versa. However, in our system, all transmitted messages are encrypted based on a shared secret, and timestamps are included in every message for data alignment. As a result, intruders are unable to execute a meaningful replay attack.

Entropy and Brute Force Attack: The success of a dictionary attack depends on the entropy of the shared secret, which measures the level of uncertainty or randomness associated with the generated bit strings. To validate the randomness of these bit strings, we can employ the NIST suite of statistical test algorithms [38]. If the entropy is found to be insufficient, users can be prompted to re-register.

Diffie-Hellman key exchange protocol has two key sizes, the discrete log group size, and the discrete log key size. More specifically, the discrete log group size is the length of the modulo p, while the discrete log key size is the length of the exponents, X or Y, chosen by Alice or Bob. The recommended value of the two sizes is 256-bit for key size and 2048-bit for group size. This ensures a 128-bit security level [1]. Note that to ensure this bit security level, it is required that the order of g is a prime q of at least 256 bits; which can be ensured by using a

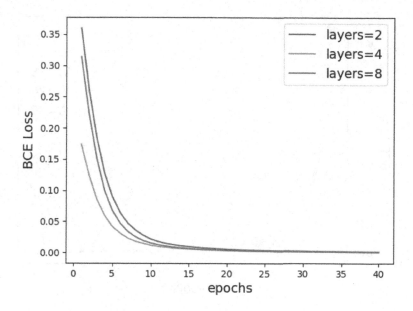

Fig. 7. Total loss values during training time with different depth

modulus p, which is a "safe prime" (prime integers p such that $(p-1)/2$ is also a prime) (Fig. 7).

4.4 Bit Agreement Rate

The bit agreement rate represents the percentage of bits that match in the secret keys generated by both parties. This metric is used to evaluate the likelihood of Alice and Bob agreeing on the same key. In our experiments, we set the key length to 256 bits and aim to train the transformer auto-encoder to identify the simplest model that can achieve a validation loss lower than 0.001.

Results show that even with just one layer of encoder and one layer of decoder, the loss can converge to zero. However, the model's performance converges very quickly when there are two layers of encoder and two layers of decoder.

4.5 Bit Generation Rate

The bit generation rate refers to the average number of bits generated from the acceleration samples per unit time, typically measured in bits per second (bps). This metric aims to assess the speed at which Alice and Bob can generate shared secret bits.

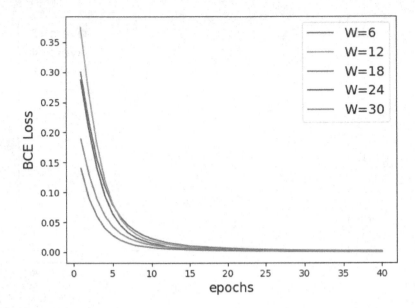

Fig. 8. Total loss values during training time with different window size

The Window Size (W) directly influences the bit generation rate. We investigate the impact of W on the generated key through a systematic exhaustive search. The objective of this search is to identify the optimal W that both maximizes the agreement rate and converges quickly. After selecting the appropriate hyper-parameters, we vary the window size within a specific range, i.e., W = 6, 12, 18, 24, 30, 36. This range is selected based on the sampling frequencies of EEG data (256 Hz) and keyboard data (21 Hz). The ratio is computed as: $256/21 \approx 12.19$. Therefore, when the window size is larger than 12, the model will learn the features of changes in keystrokes between two adjacent periods (i.e., from press to release or vice versa) (Fig. 8).

Results show that when $W = 18$, the performance of the model converges the most quickly. With $W = 18$, it takes approximately $\frac{18}{256} \approx 0.07$ seconds to generate 1 bit. Therefore, to generate a 256-bit key, it would take around 18 s, which falls within an acceptable range for the data collection phase.

Summary and Discussion. Based on our experimental results, we can answer our research questions proposed in the beginning. **(a)** First, the features of KEP are consistent across different subjects and are sufficiently distinct to be utilized in training transformer-based deep learning models. **(b)** Despite not performing a frequency domain transformation on the brainwave data, the transformer model was able to accurately decode the dynamics of keystrokes, namely the moments, when a button was pressed down and released. **(c)** In terms of generating a 256-bit secure session key, it takes approximately 18 s to register a new user, with a remarkably high success rate exceeding 99.99%. Since these results are based on a limited dataset consisting of 10 subjects, future research will investigate potential demographic biases and the generalizability of our approach for EEG-based user authentication.

5 Related Work

In this section, we introduce the related studies on biometric authentication and EEG-based authentication.

Biometric Authentication. Different from password-based or token-based authentication, biometric authentication takes use of a person's biological or behavioral features to verify the identity [31]. It includes various data sources such as face, eyes, palm, keystroke, mouse dynamics, touch dynamics, etc., and diverse platforms such as smartphones, tablets, laptops, etc. It is believed that biometric authentication can complement the existing authentication approaches, by deploying an additional layer of security [32].

Generally, there are two types of biometrics: physiological biometrics (based on features from a person's body) and behavioral biometrics (based on features from a person's action). To improve the authentication performance, one option is to select the high-quality features for authentication. Li et al. [15] explored the features of touch behavioral relating to a particular application–social networking applications on smartphones. They found that the authentication stability can be greatly enhanced. Meng et al. [27] proposed a cost-based intelligent mechanism that can be used to select a less costly learning algorithm for user authentication, thereby maintaining the authentication performance. More similar studies can refer to [25, 28–30].

In the literature, there are many hybrid (multimodal) system through integrating biometric characteristics with other biometrics. Casanova et al. [6] introduced a multimodal biometric recognition system by verifying touch dynamics and the characteristics extracted from the periocular area related to the pupils, blinks and fixations. Meng et al. [26, 33] introduced a method by checking multi-touch behaviours when a user creates a graphical password. They developed a multi-touch enabled authentication on smartphones and indicated multi-touch can make a positive impact on creating graphical passwords. Li et al. [18, 19] presented a double-cross-based unlock scheme (*Double-X*), which requires users to unlock the device by inputting two cross shapes on the selected dots and verifying the relevant touch movement actions. More similar studies can refer to DCUS (Double-Click) [20, 22], shape drawing [21], swiping action [16, 17], hand gesture [40] and ZoomPass [10, 14].

EEG-Based Authentication. Such kind of authentication can verify a person's identify by observing and analyzing the brainwaves under some given tasks or reactions. EEG still can be considered as a relatively new factor in user authentication, so most existing studies focus on hybrid schemes in the literature [36].

For instance, Klonovs et al. [13] introduced an EEG-based authentication system for mobile devices, in combination with facial detection and nearfield communication (NFC). Nakamura et al. [35] introduced an EEG-based system based on a "one-fits-all" viscoelastic generic in-ear EEG sensor. Bialas et al. [4] introduced a hybrid authentication system (NeuroSky MindWave device) by using the EEG signal combined with user image verification, which could provide a high accuracy and a low false rejection rate.

EEG-based authentication is developing very fast, but it may also suffer cyber-attacks. Chiu et al. [7] identified a kind of reaction spoofing attack, in which an attacker can imitate the mental reaction (either familiar or unfamiliar) of a legitimate user. This attack is particularly effective to an EEG-based computer-screen unlock mechanism. In this work, we follow the literature and develop a multimodal EEG-based authentication system by checking both EEG data and keystroke dynamics. Our results indicate that the combined system can defeat various attacks in practice.

6   Conclusion and Future Work

In this work, we proposed a concept of Keystroke Evoked Potentials and developed a KEP-based authentication system that can offer security, usability, and lightweightness. In particular, this paper presented the design and evaluation of a KEP-based authentication system using BCI devices. The core idea is to first generate a shared cryptographic session key from EEG and keyboard data collected during random button-pressing motions with a user's fingertip. This key is then employed in the Diffie-Hellman Encrypted Key Exchange (DH-EKE) to complete device pairing and establish a secure communication channel. Our results indicate that our approach successfully meets the requirements for security, usability, and lightweight.

Compared with existing authentication approaches, our system can provide many benefits. 1) Unlike traditional password-based authentication, our system eliminates the need for users to remember keys. 2) Compared to token-based authentication, our approach has the potential to prevent identity theft by analyzing brainwaves to ensure they match those stored signals during the enrollment stage. 3) Unlike traditional biometric systems (e.g., iris or face recognition), we do not need to worry about the security implications of leaked biometric features. 4) Further, our system requires less user effort in both the enrollment and authentication stages compared to traditional EEG-based authentication methods such as VEP and motor imagery.

One limitation of our work is that it currently serves as a proof of concept to validate the feasibility of KEP-based authentication. Further exploration is needed to assess the performance and resource costs with concrete hardware implementations. We also plan to extend our experiments to include more subjects with diverse background in our future work.

References

1. Barker, E.: NIST Special Publication 800–57 Part 1 Revision 5: Recommendation for Key Management. https://doi.org/10.6028/NIST.SP.800-57pt1r5
2. Bellovin, S.M., Merritt, M.: Encrypted key exchange: password-based protocols secure against dictionary attacks (1992)
3. Bhalerao, S., Ansari, I., Kumar, A.: Protection of BCI system via reversible watermarking of EEG signal. Electron. Lett. **56**(25), 1389–1392 (2020)
4. Bialas, K., Kedziora, M., Chalupnik, R., Song, H.H.: Multifactor authentication system using simplified EEG brain-computer interface. IEEE Trans. Hum. Mach. Syst. **52**(5), 867–876 (2022)
5. Buciu, I., Gacsadi, A.: Biometrics systems and technologies: a survey. Int. J. Comput. Commun. Control **11**(3), 315–330 (2016)
6. Casanova, A., Cascone, L., Castiglione, A., Meng, W., Pero, C.: User recognition based on periocular biometrics and touch dynamics. Pattern Recognit. Lett. **148**, 114–120 (2021)
7. Chiu, W.-Y., Meng, W., Li, W.: I can think like you! Towards reaction spoofing attack on brainwave-based authentication. In: Wang, G., Chen, B., Li, W., Di Pietro, R., Yan, X., Han, H. (eds.) SpaCCS 2020. LNCS, vol. 12382, pp. 251–265. Springer, Cham (2021). https://doi.org/10.1007/978-3-030-68851-6_18
8. Cornelius, C.T., Kotz, D.F.: Recognizing whether sensors are on the same body. Pervasive Mob. Comput. **8**(6), 822–836 (2012)
9. El-Fiqi, H., Wang, M., Salimi, N., Kasmarik, K., Barlow, M., Abbass, H.: Convolution neural networks for person identification and verification using steady state visual evoked potential. In: 2018 IEEE International Conference on Systems, Man, and Cybernetics (SMC), pp. 1062–1069. IEEE (2018)
10. Gleerup, T., Li, W., Tan, J., Wang, Y.: Zoompass: A zoom-based android unlock scheme on smart devices. In: Su, C., Sakurai, K., Liu, F. (eds.) Science of Cyber Security - 4th International Conference, SciSec 2022, Matsue, Japan, August 10–12, 2022, Revised Selected Papers. Lecture Notes in Computer Science, vol. 13580, pp. 245–259. Springer, Cham (2022)
11. King, B.J., Read, G.J., Salmon, P.M.: The risks associated with the use of brain-computer interfaces: a systematic review. Int. J. Hum. Comput. Interact. 1–18 (2022)
12. Kirovski, D., Sinclair, M., Wilson, D.: The martini synch. Microsoft Research, Cambridge, UK, Tech. Rep. MSR-TR-2007-123 (2007)
13. Klonovs, J., Petersen, C.K., Olesen, H., Hammershøj, A.: ID proof on the Go: Development of a mobile EEG-based biometric authentication system. IEEE Veh. Technol. Mag. **8**(1), 81–89 (2013)
14. Li, W., Gleerup, T., Tan, J., Wang, Y.: A security enhanced android unlock scheme based on pinch-to-zoom for smart devices. IEEE Trans. Consum. Electron. 1–9 (2023)
15. Li, W., Meng, W., Furnell, S.: Exploring touch-based behavioral authentication on smartphone email applications in IoT-enabled smart cities. Pattern Recognit. Lett. **144**, 35–41 (2021)
16. Li, W., Tan, J., Meng, W., Wang, Y.: A swipe-based unlocking mechanism with supervised learning on smartphones: design and evaluation. J. Netw. Comput. Appl. **165**, 102687 (2020)

17. Li, W., Tan, J., Meng, W., Wang, Yu., Li, J.: SwipeVLock: a supervised unlocking mechanism based on swipe behavior on smartphones. In: Chen, X., Huang, X., Zhang, J. (eds.) ML4CS 2019. LNCS, vol. 11806, pp. 140–153. Springer, Cham (2019). https://doi.org/10.1007/978-3-030-30619-9_11

18. Li, W., Tan, J., Zhu, N.: Double-x: Towards double-cross-based unlock mechanism on smartphones. In: Meng, W., Fischer-Hübner, S., Jensen, C.D. (eds.) ICT Systems Security and Privacy Protection - 37th IFIP TC 11 International Conference, SEC 2022, Copenhagen, Denmark, June 13–15, 2022, Proceedings. IFIP Advances in Information and Communication Technology, vol. 648, pp. 412–428. Springer, Cham (2022)

19. Li, W., Tan, J., Zhu, N.: Design of double-cross-based smartphone unlock mechanism. Comput. Secur. **129**, 103204 (2023)

20. Li, W., Tan, J., Zhu, N., Wang, Yu.: Designing double-click-based unlocking mechanism on smartphones. In: Wang, G., Chen, B., Li, W., Di Pietro, R., Yan, X., Han, H. (eds.) SpaCCS 2020. LNCS, vol. 12383, pp. 573–585. Springer, Cham (2021). https://doi.org/10.1007/978-3-030-68884-4_47

21. Li, W., Wang, Y., Li, J., Xiang, Y.: Toward supervised shape-based behavioral authentication on smartphones. J. Inf. Secur. Appl. **55**, 102591 (2020)

22. Li, W., Wang, Y., Tan, J., Zhu, N.: DCUS: evaluating double-click-based unlocking scheme on smartphones. Mob. Networks Appl. **27**(1), 382–391 (2022)

23. Liew, S.H., Choo, Y.H., Low, Y.F., Yusoh, Z.I.M.: Identifying visual evoked potential (VEP) electrodes setting for person authentication. Int. J. Adv. Soft Comput. Appl **7**(3), 85–99 (2015)

24. Lin, Q., et al.: H2B: heartbeat-based secret key generation using piezo vibration sensors. In: Proceedings of the 18th International Conference on Information Processing in Sensor Networks, pp. 265–276 (2019)

25. Meng, W., Li, W., Jiang, L., Zhou, J.: SocialAuth: designing touch behavioral smartphone user authentication based on social networking applications. In: Dhillon, G., Karlsson, F., Hedström, K., Zúquete, A. (eds.) SEC 2019. IAICT, vol. 562, pp. 180–193. Springer, Cham (2019). https://doi.org/10.1007/978-3-030-22312-0_13

26. Meng, W., Li, W., Kwok, L., Choo, K.R.: Towards enhancing click-draw based graphical passwords using multi-touch behaviours on smartphones. Comput. Secur. **65**, 213–229 (2017)

27. Meng, W., Li, W., Wong, D.S.: Enhancing touch behavioral authentication via cost-based intelligent mechanism on smartphones. Multim. Tools Appl. **77**(23), 30167–30185 (2018)

28. Meng, W., Li, W., Wong, D.S., Zhou, J.: TMGuard: a touch movement-based security mechanism for screen unlock patterns on smartphones. In: Manulis, M., Sadeghi, A.-R., Schneider, S. (eds.) ACNS 2016. LNCS, vol. 9696, pp. 629–647. Springer, Cham (2016). https://doi.org/10.1007/978-3-319-39555-5_34

29. Meng, W., Liu, Z.: TMGMap: designing touch movement-based geographical password authentication on smartphones. In: Su, C., Kikuchi, H. (eds.) ISPEC 2018. LNCS, vol. 11125, pp. 373–390. Springer, Cham (2018). https://doi.org/10.1007/978-3-319-99807-7_23

30. Meng, W., Wang, Y., Wong, D.S., Wen, S., Xiang, Y.: TouchWB: touch behavioral user authentication based on web browsing on smartphones. J. Netw. Comput. Appl. **117**, 1–9 (2018)

31. Meng, W., Wong, D.S., Furnell, S., Zhou, J.: Surveying the development of biometric user authentication on mobile phones. IEEE Commun. Surv. Tutorials **17**(3), 1268–1293 (2015)

32. Meng, W., Wong, D.S., Kwok, L.: The effect of adaptive mechanism on behavioural biometric based mobile phone authentication. Inf. Manag. Comput. Secur. **22**(2), 155–166 (2014)

33. Meng, Y., Li, W., Kwok, L.-F.: Enhancing click-draw based graphical passwords using multi-touch on mobile phones. In: Janczewski, L.J., Wolfe, H.B., Shenoi, S. (eds.) SEC 2013. IAICT, vol. 405, pp. 55–68. Springer, Heidelberg (2013). https:// doi.org/10.1007/978-3-642-39218-4_5

34. Muller-Putz, G.R., Scherer, R., Neuper, C., Pfurtscheller, G.: Steady-state somatosensory evoked potentials: suitable brain signals for brain-computer interfaces? IEEE Trans. Neural Syst. Rehabil. Eng. **14**(1), 30–37 (2006)

35. Nakamura, T., Goverdovsky, V., Mandic, D.P.: In-ear EEG biometrics for feasible and readily collectable real-world person authentication. IEEE Trans. Inf. Forensics Secur. **13**(3), 648–661 (2018)

36. Pham, T., Ma, W., Tran, D., Nguyen, P., Phung, D.Q.: Multi-factor EEG-based user authentication. In: 2014 International Joint Conference on Neural Networks, IJCNN 2014, Beijing, China, July 6–11, 2014, pp. 4029–4034. IEEE (2014)

37. Rostami, M., Juels, A., Koushanfar, F.: Heart-to-heart (H2H) authentication for implanted medical devices. In: Proceedings of the 2013 ACM SIGSAC Conference on Computer & Communications Security, pp. 1099–1112 (2013)

38. Rukhin, A., et al.: A statistical test suite for random and pseudorandom number generators for cryptographic applications, vol. 22. US Department of Commerce, Technology Administration, National Institute of ... (2001)

39. Schürmann, D., Brüsch, A., Sigg, S., Wolf, L.: Bandana-body area network device-to-device authentication using natural gait. In: 2017 IEEE International Conference on Pervasive Computing and Communications (PerCom), pp. 190–196. IEEE (2017)

40. Sun, Y., Meng, W., Li, W.: Designing in-air hand gesture-based user authentication system via convex hull. In: 19th Annual International Conference on Privacy, Security & Trust, PST 2022, Fredericton, NB, Canada, August 22–24, 2022, pp. 1–5. IEEE (2022)

41. Vaswani, A., et al.: Attention is all you need. In: Advances in Neural Information Processing Systems 30 (2017)

42. Wolpaw, J.R., del R. Millán, J., Ramsey, N.F.: Chapter 2 - brain-computer interfaces: Definitions and principles. In: Ramsey, N.F., del R. Millán, J. (eds.) Brain-Computer Interfaces, Handbook of Clinical Neurology, vol. 168, pp. 15–23. Elsevier (2020)

43. Wu, B., Meng, W., Chiu, W.: Towards enhanced EEG-based authentication with motor imagery brain-computer interface. In: Annual Computer Security Applications Conference, ACSAC 2022, Austin, TX, USA, December 5–9, 2022, pp. 799–812. ACM (2022)

44. Wu, Y., Lin, Q., Jia, H., Hassan, M., Hu, W.: Auto-key: using autoencoder to speed up gait-based key generation in body area networks. Proc. ACM Interact. Mob. Wearable Ubiquit. Technol. **4**(1), 1–23 (2020)

45. Xu, W., Revadigar, G., Luo, C., Bergmann, N., Hu, W.: Walkie-talkie: motion-assisted automatic key generation for secure on-body device communication. In: 2016 15th ACM/IEEE International Conference on Information Processing in Sensor Networks (IPSN), pp. 1–12. IEEE (2016)

46. Yadav, V.K., Yadav, R.K., Chaurasia, B.K., Verma, S., Venkatesan, S.: MITM attack on modification of Diffie-Hellman key exchange algorithm. In: Communication, Networks and Computing: Second International Conference, CNC 2020, Gwalior, India, pp. 144–155 (2021)
47. Zhang, S., Sun, L., Mao, X., Hu, C., Liu, P., et al.: Review on EEG-based authentication technology. Comput. Intell. Neurosci. **2021**, 20 (2021)

Verifiable Secure Aggregation Protocol Under Federated Learning

Peiming Xu[1], Meiling Zheng[2], and Lingling Xu[2](✉)

[1] Guangdong Provincial Key Laboratory of Power System Network Security, Electric Power Research Institute, CSG, Guangzhou, China
[2] School of Computer Science and Engineering, South China University of Technology, Guangzhou, China
csllxu@scut.edu.cn

Abstract. Federated learning is a new machine learning paradigm used for collaborative training models among multiple devices. In federated learning, multiple clients participate in model training locally and use decentralized learning methods to ensure the privacy of client data. However, although federated learning protects the privacy of client data, the update gradients uploaded by clients may still contain sensitive information. To solve this problem, this paper proposes a secure aggregation protocol which can verify the aggregation results under federated learning and protect gradient privacy. The core idea of this aggregation protocol is to use encryption technology to achieve secure computation between clients, ensuring the privacy of gradients during the aggregation process. At the same time, bilinear pairing technology is used to achieve the verifiability of aggregation results, ensuring the correctness and usability of the model after aggregation. In order to evaluate the security of the protocol, this paper conducts a detailed security analysis. The results show that this protocol has higher security properties compared to the existing related protocols. In addition, the computation and communication costs of the protocol are analyzed, which show that the protocol has good credibility and applicability in practical federated learning scenarios.

Keywords: Federated Learning · Secure Aggregation · Privacy Protection · Verifiability · Bilinear Pairing

1 Introduction

In recent years, artificial intelligence technology has shown great potential and advantages in various fields such as unmanned driving, smart finance, and healthcare. However, the development of these technologies typically relies on big data training models to drive [1]. In reality, the quantity and quality of data in many industries are limited, making it impossible to support model training alone. Therefore, is it possible to transmit data across domains and train together? In fact, in many industries, data often exists in the form of "islands". Due to the isolation and security requirements of data, breaking data barriers across domains has become very difficult. Meanwhile, due to the frequent occurrence of various data leakage issues [2], people's attention to data security is constantly

J. Vaidya et al. (Eds.): AIS&P 2023, LNCS 14509, pp. 531–547, 2024.
https://doi.org/10.1007/978-981-99-9785-5_37

increasing. For example, the European Union passed the General Data Protection Regulations (GDPR) in 2018, and a series of strict personal information protection regulations have also been issued domestically. These regulations have strengthened restrictions on the collection and use of personal privacy data. Therefore, the development and application of artificial intelligence face a dilemma: data is isolated and stored, and in most cases, data is prohibited from being collected, fused, and used in different places. In order to solve the problem of "data silos" caused by cross domain training of data, federated learning [3–5] was proposed and studied, aiming to reduce the risk of client data leakage.

In federated learning, in order to protect data privacy, data does not leave the local area, and each client uses local data to train local models. Then, they will periodically share the trained local model parameters, which are uploaded to the central server for aggregation to obtain a better global model. Clients can use the global model for the next round of training. This approach alleviates the privacy leakage issue that may arise from direct use of sensitive data. However, the model gradients transmitted in federated learning may also leak private information [6–8]. To achieve gradient privacy protection, secure aggregation can be used. The core idea of secure aggregation is that participants can process local gradients and transmit them to the server for secure aggregation. In this way, the central server will only learn the aggregated global model and will not know specific information about the local model. In order to achieve secure aggregation, Keith Bonawitz et al. [9] designed a secure aggregation protocol that tolerates client exit. They used double mask technology and secret sharing technology to encrypt the uploaded gradient and upload it to a central server. This method can both protect local model gradients and tolerate a certain number of clients exiting at any time. However, traditional double mask schemes require key negotiation between clients, which can incur significant communication overhead. In order to improve communication efficiency, Kalikinkar Mandal et al. [10, 11] entrusted the work of key negotiation to a trusted third party (TA), thereby avoiding communication between clients and greatly improving communication efficiency. The privacy of model gradients is protected through secure aggregation. The central server cannot directly or indirectly obtain specific information about any local gradient except for the aggregated gradient.

In addition to the privacy of the model gradient, the verifiability of the aggregation results from the central server is also an important factor in achieving privacy protection. Sometimes, due to network limitations or other reasons, aggregation servers may only aggregate the gradients of some clients. In addition, malicious servers may collude with clients, forge or tamper with aggregation results to obtain client privacy data. Therefore, researchers need to verify the correctness and completeness of server-side aggregation results to prevent server forgery or malicious tampering of data. When designing a validation plan, two issues need to be considered. The first issue is the complexity of verification, as verifiable federated learning requires encrypted computation and proof generation between various participants, which introduces additional computational and communication overhead. Secondly, the implementation of verifiable federated learning requires cooperation from all parties to ensure data privacy and the credibility of the model, so the validation scheme itself cannot directly or indirectly disclose the privacy of the data or model. In the current research process, Xu et al. [12] designed a verification scheme for secure aggregation results by combining the double mask scheme and homomorphic hash function. This scheme can safely verify the correctness of the aggregation results, but the verification process uses homomorphic hash functions and pseudo-random functions, which adds a lot of verification overhead. Guo et al. [13] improved the above verification scheme by combining the fuzzy commitment scheme with linear homomorphic hashing to propose a new protocol that reduces communication overhead. However, in reality, this protocol has security issues and cannot truly achieve verifiability. In addition, in verifiable federated learning, VERSA [14] adopts a dual aggregation protocol to achieve the verifiability of secure aggregation. Although it greatly reduces verification costs by using pseudo random generators, each client must hide the same secret vector from the server, which poses a serious security threat. In the verifiable secure aggregation model proposed by Wang et al. [15], a collector is used to calculate the auxiliary information required for clients authentication and server aggregation, respectively. And use the aggregation server to aggregate the encrypted gradient and validation labels, and then return the gradient aggregation value and validation information to the client. The client uses bilinear pairing method to verify the correctness of the aggregation results. However, during the verification process, malicious servers can use real aggregation values to disguise themselves as clients to obtain verification information, infer key information in the verification information, and forge gradient aggregation results and verification information.

This paper proposes a verifiable secure aggregation protocol based on bilinear pairing and aggregator oblivious encryption techniques. Specifically, the main contributions of this paper are as follows.

1) This protocol implements gradient privacy protection based on aggregator oblivious encryption [16, 17]. The client uses a key to encrypt the gradient and sends it to the aggregation server for unified aggregation. This method can complete the update of gradients without leaking anything about the gradients of each client to the collector and aggregation server.

2) This protocol achieves the verifiability of gradient aggregation results. Each client hides local gradients by using its secret key and a shared random array between all the clients to protect the gradient information in the verification tag. After the gradients are aggregated, each client can verify the results with the shared random array. Since the random arrays are kept confidential from the aggregation server, malicious servers cannot forge aggregation results to deceive clients.

3) This paper provides a detailed analysis on the security of the protocol and evaluates the computation and communication costs involved. By comparing with related existing works, this protocol can effectively ensure the confidentiality of client data and gradient information. It also minimizes computation and communication overhead as much as possible, enabling large-scale use in practical applications.

2 Preliminaries

2.1 Bilinear Pairing

Bilinear pairing [18] is the mapping of elements between two groups into a different group. It is defined as: for two multiplicative cyclic groups G_1 and G_2 of the same prime order q, satisfying the mapping $e : G_1 \times G_2 \to G_T$, where the generators of G_1 and G_2 are g_1 and g_2, respectively. Generally speaking, bilinear pairing e has the following properties:

(1) Bilinear: Given $\forall a, b \in Z_q^*$, for generators $\forall g_1 \in G_1$ and $\forall g_2 \in G_2$, satisfy $e\left(g_1^a, g_2^b\right) = e(g_1, g_2)^{ab}$;
(2) Non degenerate: $e(g_1, g_2) \neq 1$, where g_1 and g_2 are the generators of G_1 and G_2, respectively.
(3) Computability: for $\forall g_1 \in G_1$ and $\forall g_1 \in G_2$, $e(g_1, g_2)$ can be effectively calculates.

2.2 Bilinear Computational Diffie-Hellman Assumption (BCDH)

Let g_1 and g_2 be two generators for cyclic group G_1 and G_2 respectively, and $a, b, c \in Z_q^*$ be random numbers. The BCDH problem is to compute $e(g_1, g_2)^{abc} \in G_T$. The advantage of a probabilistic polynomial-time adversary $\mathcal{A}$ to computing the BCDH problem is:

$$Adv_{BCDH} = \Pr[\mathcal{A}(g_1, g_2, g_1^a, g_1^b, g_1^c, g_2^a, g_2^b) = e(g_1, g_2)^{abc}]$$

If the advantage $Adv_{BCDH} \leq \varepsilon$, where ε is a negligible function, it is assumed that the BCDH problem is a hard problem in polynomial time.

2.3 Aggregator Oblivious Encryption

Aggregator oblivious encryption (AO) [16] requires that aggregators should not learn any information other than aggregated values from honest clients' encrypted values within each time period. This scheme was first proposed by Shi [16]. Without compromising the privacy of each client, a set of encrypted values of participants are regularly uploaded to the aggregator, and then encryption technology is used to decrypt the sum of multiple ciphertexts encrypted by the aggregator under different client keys. However, this approach has certain limitations. That is, it is not conducive to large plaintext spaces and is only applicable to small pure text spaces. Therefore, Joye et al. [17] proposed a new aggregator oblivious encryption scheme which is divided into initialization stage, encryption stage, and aggregation decryption stage, to solve the above problem.

Initialization Stage:
On a certain input security parameter κ, a trusted dealer randomly generates a modulus $N = pq$, which is the product of two prime numbers of equal size. In addition, a hash

function is also defined as $H : \mathbb{Z} \rightarrow (\mathbb{Z}/N^2\mathbb{Z})^*$. Finally, set $s_0 = -\sum_{i=1}^{n} s_i$ and $sk_i = s_i (i \in [0, m])$.

Encryption Stage:
During time period t, client i generates:

$$c_{i,t} = (1 + x_{i,t}N)H(t)^{s_i} mod\ N^2 \tag{1}$$

Aggregation Decryption Stage:
The aggregator first calculates $V_t := H(t)^{s_0} \prod_{i=1}^{n} c_{i,t} mod\ N^2$, and then calculates X_t follows:

$$X_t = \frac{V_t - 1}{N} \tag{2}$$

Joye et al.'s scheme [17] can complete the aggregation of encryption and decryption with minimal interaction, and can be performed in both online and offline model.

3   System and Security Model

This section introduces our system model and security model of the proposed secure aggregation protocol under federated learning.

3.1   System Model

In the protocol, there are four types of entities, namely the key generation center KGC, clients, collector, and aggregation server. The entire process mainly consists of five stages as follows.

1) **Initialization**: KGC initializes the system and publishes the system parameters. Then all the other participants generate their secret keys respectively.
2) **Encrypt gradient and generate verification labels:** By interaction with the aggregation server and collector, each client encrypts the gradients and generates the verification labels. Then it sends the encrypted gradients to the aggregation server, and sends the verification labels to the collector.
3) **Collect and aggregate validation labels:** After the collector collects the verification labels from each online client, it aggregates the labels and then sends the result to the aggregation server.
4) **Calculate aggregation results:** The aggregation server calculates the aggregated gradients and proof information, and then sends them to each online client.
5) **Verify aggregation results:**
    Each client verifies the aggregated gradient results by using the proof information from the aggregation server.

3.2 Security Model

In our protocol, KGC is assumed to be honest. All clients and the collector are considered to be honest but curious. They may try to obtain some private information about other clients. The aggregation server can be malicious who may forge the aggregated gradients for some epoch except for inferring clients' privacy. We assume that the aggregation server will not collude with any client to get its secret keys.

Our protocol should achieve these security properties: aggregation unforgeability and verifiability, client gradient privacy. Client gradient privacy means that the gradients of each client should be private. Namely, for a client, the collector, the aggregation server and any other client cannot obtain any information about its gradients. For the aggregation unforgeability and verifiability, we define a game as follows. An aggregation protocol is aggregation unforgeable and verifiable if no probabilistic polynomial time adversary A can win the game below with non-negligible advantage. In the game, C is the game challenger and A is assumed to be an aggregation server.

Setup: C initializes the system and generates system parameters.

Phase 1: A can make the following queries.

(1) Hash queries: A makes queries for hash functions and C returns the answers.
(2) Encryption and label queries: A makes queries for the gradients in an epoch. C returns ciphertexts and the verification labels.

Challenge: A forges an aggregated gradient for some epoch. If it is verifiable, A succeeds. For the aggregation unforgeability and verifiability, we define a game as follows. An aggregation protocol is aggregation unforgeable and verifiable if no probabilistic polynomial time adversary A can win the game below with non-negligible advantage. In the game, C is the game challenger and A is assumed to be an aggregation server.

Setup: C initializes the system and generates system parameters.

Phase 1: A can make the following queries.

(1) Hash queries: A makes queries for hash functions and C returns the answers.
(2) Encryption and label queries: A makes queries for the gradients in an epoch. C returns ciphertexts and the verification labels.

Challenge: A forges an aggregated gradient for some epoch. If it is verifiable, A succeeds.

4 Verifiable Secure Aggregation Protocol

In the protocol [15], a malicious aggregation server can forge gradient aggregation results and deceive the clients. Namely, this protocol cannot achieve aggregation unforgeability and verifiability as the author claimed. In this section, we will present an improved verifiable secure aggregation protocol based on [15] and also give the security analysis for it. In this protocol, the participants consist of the key generation center KGC, clients, collector, and aggregation server. They conduct the protocol in the following five stages.

4.1 Concrete Protocol

In this protocol, the participants consist of the key generation center KGC, clients, a collector and an aggregation server. They conduct the protocol in the following five stages shown in Fig. 1.

1) Stage 1: Initialization
a) Assuming m clients $U = \{U_i, i \in [1, m]\}$ participate in the training of federated learning. Firstly, the key generation center KGC will initialize the model's parameter w_0. It selects two secure prime numbers p and q, and calculates $N = p \cdot q$. Let G_1 and G_2 be two multiplicative cyclic groups with order q, g_1 and g_2 be the generators of G_1 and G_2, respectively. KGC randomly selects a secret value $a \in Z_q^*$, and calculates $h_1 = g_1^a$ and $h_2 = g_2^a$. In addition, KGC also selects a computable bilinear pairing map $e : G_1 \times G_2 \rightarrow G_T$, two single term irreversible secure hash functions $H_0 : \{0, 1\}^* \rightarrow Z_{N^2}^*$ and $H_1 : \{0, 1\}^* \rightarrow G_1$. Then KGC publishes the parameters pm to all the participants:

$$pm = \{N, w_0, g_1, g_2, h_1, h_2, G_1, G_2, G_T, H_0, H_1\}.$$

b) KGC generates two random numbers $r_1, r_2 \in Z_{N^2}^*$, where r_1, r_2 are only disclosed to all clients and kept confidential from the server and collector. The client U_i chooses private keys$\{(sk_i, tk_i)|sk_i \in Z_{N^2}^*, tk_i \in Z_{N^2}^*\}$, where sk_i is used to protect the gradients of client U_i, tk_i and r_1, r_2 are used to generate verification labels.

c) The aggregation server chooses a private key $sk_A \in Z_{N^2}^*$.

Next, the participants conduct multiple rounds. In the following, we use the t-th round as an example.

2) Stage 2: Encrypt gradient and generate verification labels (Encryption)
The model trained by client U_i for neural network can be defined as $y' = f(x, \omega)$, where x is the input of the model, ω is the parameter of the model, and y' is the output of the model. The loss function on dataset $D_i = \{(x_j, y_j)|j \in [0, m]\}$ of client U_i can be defined as:

$$J(x, W) = \frac{1}{m} \sum_{j \in [0, m]} l(y_j, y_j') \tag{3}$$

where l is a specific loss function (sigmod, tanh, etc.). In order to minimize the loss-function, it is necessary to find the appropriate gradient parameter W, and the random gradient descent algorithm (SGD) is generally used to update the gradient of the local model as follows:

$$\omega_{i,t} \leftarrow \omega_{i,t} - \alpha \frac{\partial}{\partial \omega_{i,t-1}} J(x, \omega_{i,t-1}) \tag{4}$$

where α is the learning rate of the model.

Then the client U_i uses the key sk_i to encrypt the locally updated gradient parameter as follows:

$$C_{i,t} = (1 + \omega_{i,t}N) H_0(t)^{sk_i} \mod N^2 \tag{5}$$

At the same time, U_i use the verification key tk_i to calculate the verification label as follows:

$$T_{i,t} = H_1(t)^{tk_i} h_1^{r_1 \omega_{i,t} + r_2} \tag{6}$$

Subsequently, the client needs to send the encrypted gradient parameters and verification labels to the server for aggregation. At the same time, the current online client is recorded as U_1, and the online client list is sent to the aggregator.

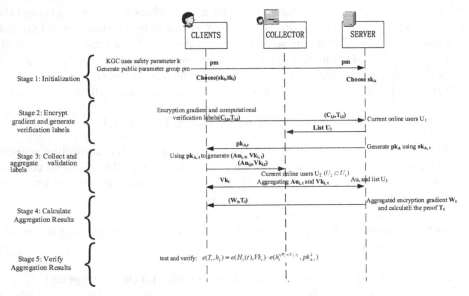

Fig. 1. Protocol Interaction Diagram

3) Stage 3: Collect and aggregate verification labels (Collection)

The server generates two public keys $pk_{A,t} = \left(pk_{A,t}^1, pk_{A,t}^2\right)$ using the private key sk_A, where $pk_{A,t}^1 = H_0(t)^{sk_A}$ and $pk_{A,t}^2 = h_2^{sk_A}$, and sends $pk_{A,t}$ to the client.

Clients use sk_i and tk_i to calculate auxiliary information $Au_{i,t}$ and verification information $Vk_{i,t}$ respectively as follows:

$$Au_{i,t} = (pk_{A,t}^1)^{sk_i} = H_0(t)^{sk_A sk_i} \tag{7}$$

$$Vk_{i,t} = (pk_{A,t}^2)^{sk_A} = h_2^{sk_A sk_i} \tag{8}$$

The client sends $\left(Au_{i,t}, Vk_{i,t}\right)$ to the collector and records the online client as U_2. Subsequently, the collector aggregates the collected information to obtain the necessary information for decryption Au_t and verification key Vk_t as follows:

$$Au_t = \prod_{i=1}^m Au_{i,t} = H_0(t)^{sk_A \sum_{i=1}^m sk_i} \tag{9}$$

$$Vk_t = \prod_{i=1}^{m} Vk_{i,t} = g_2^{sk_A a \sum_{i=1}^{m} tk_i} \tag{10}$$

Subsequently, the collector sends the verification key Vk_t to the client, and the decryption key Au_t and the list of current online clients $\mathbf{U_2}$ to the server.

4) Stage 4: Calculate Aggregation Results (CalAgg)

The server receives the gradient ciphertext $C_{i,t}$ and verification label $T_{i,t}$ sent by the client, as well as the decryption key Au_t sent by the collector. The server first calculates the aggregation value of ciphertext:

$$C_t = (\prod_{i=1}^{m} C_{i,t})^{sk_A} \bmod N^2$$

$$= (1 + sk_A \sum_{i=1}^{m} w_{i,t} N) H_0(t)^{sk_A \sum_{i=1}^{m} sk_i} \bmod N^2 \tag{11}$$

Subsequently, the aggregation value W_t of the plaintext gradient is calculated using the decryption key Au_t:

$$W_t = sk_A^{-1} \frac{\frac{C_t}{Au_t} - 1}{N} \bmod N$$

$$= sk_A^{-1} \frac{\frac{(1+sk_A \sum_{i=1}^{m} \omega_{i,t} N) H_0(t)^{sk_A \sum_{i=1}^{m} sk_i}}{H_0(t)^{sk_A \sum_{i=1}^{m} sk_i}} - 1}{N} \bmod N \tag{12}$$

$$= \sum_{i=1}^{m} \omega_{i,t} \bmod N$$

Then the server aggregates the validation label $T_{i,t}$ and generates proof T_t:

$$T_t = (\prod_{i=1}^{m} T_{i,t})^{sk_A} = H_1(t)^{sk_A \sum_{i=1}^{m} tk_i} g_1^{ask_A(r_1 \sum_{i=1}^{m} \omega_{i,t} + |U_2|r_2)} \tag{13}$$

Finally, it broadcasts proof T_t and plaintext aggregation result W_t to the client.

5) Stage 5: Verify Aggregation Results (Verification)

Online clients receive (T_t, W_t) and list U_2 broadcasted from the server, and the verification key Vk_t aggregated by the collector. The client determines whether the aggregation result is correct by verifying whether bilinear is true, that is, whether the equal sign in Eq. 14 is true. If the equal sign is true, the verification is passed. Otherwise, the verification fails and the aggregation value is discarded.

$$e(T_t, h_2) = e(H_1(t), Vk_t) \cdot e\left(h_1^{r_1 W_t + |U_2|r_2}, pk_{A,t}^2\right) \tag{14}$$

If the server honestly calculates and sends the aggregation results, the client verification process is as follows:

$$e(T_t, h_2) = e(H_1(t)^{sk_A \sum_{i=1}^{m} tk_i} g_1^{ask_A(r_1 \sum_{i=1}^{m} \omega_{i,t} + |U_2|r_2)}, g_2^a)$$

$$= e(H_1(t)^{sk_A \sum_{i=1}^{m} tk_i}, g_2^a) \cdot e(g_1^{ask_A(r_1 \sum_{i=1}^{m} \omega_{i,t} + |U_2|r_2)}, g_2^a)$$

$$= e(H_1(t), g_2^{ask_A \sum_{i=1}^{m} tk_i}) \cdot e(g_1^{a(r_1 \sum_{i=1}^{m} \omega_{i,t} + |U_2|r_2)}, g_2^{ask_A})$$

$$= e(H_1(t), Vk_t) \cdot e\left(h_1^{(r_1 W_t + |U_2|r_2)}, pk_{A,t}^2\right) \tag{15}$$

4.2 Security Analysis

Theorem 1. *The proposed protocol is client gradient private.*

Proof. In the protocol, the gradient $w_{i,t}$ of the client U_i in the epoch t is hidden in the stage **Encrypt-tag**. $w_{i,t}$ is first encrypted by using the client's secret key sk_i, to generate $C_{i,t} = \left(1 + \omega_{i,t} N\right) H_0(t)^{sk_i} \bmod N^2$, and then hidden in the tag $T_{i,t} = H_1(t)^{tk_i} h_1^{r_1 \omega_{i,t} + r_2}$ by using the secret key tk_i. The generation of the ciphertext $C_{i,t}$ is based on the encryption algorithm of Joye et al. [17] which is proved to be secure. Namely, other than the client, any participant cannot obtain any information about $w_{i,t}$. In the tag $T_{i,t}$, $w_{i,t}$ is hidden by using the secret key tk_i, r_1, r_2. From a tag $T_{i,t}$, an adversary cannot obtain $w_{i,t}$. For two gradients w and w', even if it can compute $h_1^{r_1(w-w')}$, since the privacy of r_1, it cannot obtain w.

Theorem 2. *The proposed protocol is aggregation unforgeable and verifiable if the hash function H_1 is collision resistant and the BCDH assumption holds.*

Proof. The malicious aggregation server is assumed to be the PPT adversary $\mathcal{A}$. The challenger is given a BCDH tuple $\left(g_1^a, g_1^b, g_2^a, g_2^b\right)$. $\mathcal{A}$ interacts with the game challenger $\mathcal{C}$ as follows.

Setup: the challenger $\mathcal{C}$ generates the system parameters $pm = \{N, w_0, g_1, g_2, h_1, h_2, G_1, G_2, G', H_0, H_1\}$, where $h_1 = g_1^a$, $h_2 = g_2^a$, and sends pm to $\mathcal{A}$. It also generates the secret key $sk_A \in Z_{N^2}^*$ for $\mathcal{A}$, computes $pk_{A,t} = \left(pk_{A,t}^1, pk_{A,t}^2\right) = (H_0(t)^{sk_A}, h_2^{sk_A})$ and $Vk_t = h_2^{sk_A \sum_{u_i \in U_2} tk_i}$.

Phase 1: $\mathcal{A}$ issues the queries as follows.

(1) Hash queries: $\mathcal{A}$ makes queries for an epoch t hash functions H_1 and $\mathcal{C}$ maintains a list H_1-list. $\mathcal{C}$ first selects a random number $r_t \in Z_{N^2}$. Then it flips a random coin $b_t \in \{0, 1\}$. If $b_t = 0$, $\mathcal{C}$ sets $H_1(t) = g_1^{r_t}$ and replied to $\mathcal{A}$ with it. If $b_t = 1$, $\mathcal{C}$ replied to $\mathcal{A}$ with $H_1(t) = g_1^{c \cdot r_t}$. $\mathcal{C}$ adds the tuple $\langle r_t, b_t, H_1(t) \rangle$ to H_1-list.

(2) Encryption and label queries: $\mathcal{C}$ select random values $r_1, r_2 \in Z_{N^2}^*$. $\mathcal{A}$ makes a query with $\langle t, U_i, w_{i,t} \rangle$. $\mathcal{C}$ responds as follows. $\mathcal{C}$ initializes an empty list ET-list and initializes W_t. $\mathcal{C}$ first checks $\langle r_t, b_t, H_1(t) \rangle$ from H_1-list. If $b_t = 1$, it aborts. If $b_t = 0$, $\mathcal{C}$ selects random $tk_i \in Z_{N^2}^*$ and computes $T_{i,t} = g_1^{b r_t tk_i} \cdot g_1^{a(r_1 w_{i,t} + r_2)} = H_1(t)^{b tk_i} \cdot g_1^{a(r_1 w_{i,t} + r_2)}$. It then selects $sk_i \in Z_{N^2}^*$ and computes the gradients ciphertext $C_{i,t} = \left(1 + w_{i,t} N\right) \cdot H_0(t)^{sk_i}$. $\mathcal{C}$ adds $< t, U_i, w_{i,t}, T_{i,t} >$ to the ET-list and sets $W_t \leftarrow W_t + W_{i,t}$. If $\mathcal{A}$ makes query for $< t, U_i, w_{i,t} >$ where $w_{i,t} \neq w_{i,t'}$, $\mathcal{C}$ aborts.

Challenge: $\mathcal{A}$ forges $\langle W_{t^*}^*, T_{t^*}^* \rangle$ for the epoch t^*, where $W_{t^*}^* \neq W_{t^*}$. $\mathcal{C}$ gets $\langle r_t^*, b_t^*, H_1(t^*) \rangle$ from H_1-list. Then it satisfies the equation $e\left(T_{t^*}^*, h_2\right) = e(H_1(t^*), Vk_{t^*}) \cdot e\left(h_1^{r_1 w_{t^*}^* + |U_2| \cdot r_2}, pk_{A,t^*}^2\right)$. $\mathcal{C}$ compute $Z = (\frac{T_{t^*}^*}{g_1^{a \cdot sk_A \cdot W_{t^*}^*}})^{\frac{1}{r_t \cdot sk_A \cdot \sum tk_i}}$, then there is $e\left(Z, g_2^a\right) = a(g_1, g_2)^{abc}$. If $\mathcal{A}$ can succeed with a non-negligible probability, then $\mathcal{C}$ can break BCDH assumption with a probability.

In our protocol, since each client hers secretly r_1 and r_2, the verification stage can only be conducted by the clients. The aggregation server cannot forge valid gradient as in the protocol [15].

5 Theoretical Analysis and Experiments

5.1 Accuracy Comparison

In our protocol, after the client encrypts the gradients and sends them to the aggregation server, the server calculates the aggregated results. The encryption part of gradients in our protocol is shown in Eq. (5) in the Sect. 4. It is the same in the protocol [15]. The experiment results in [15] has shown that their proposed realized similar model accuracy with plaintext federated learning. Since the model accuracy difference mainly depends on the encryption method of gradients, our protocol has the same model accuracy with [15] theoretically. So, we omit the simulation of this part.

5.2 Security Comparison

In the following, we will compare the security of our protocol with the related existing protocols [5, 9, 13, 15] from the four aspects: data privacy, gradient privacy, verification and unforgeability as shown in Table 1.

Table 1. Security Comparison

	data privacy	gradient privacy	Verification and unforgeability
FedAvg [5]	✓	×	×
Aggregation [9]	✓	✓	×
VeriFL [13]	✓	✓	×
VOSA [15]	✓	✓	×
Ours	✓	✓	✓

In order to protect data privacy, the classic method of federated learning FedAvg [5], utilizes local data from the client for training and uploads the trained local model parameters to the server for processing, thereby avoiding privacy leakage caused by data leaving the local area and protecting data security. However, later it was discovered that attackers could indirectly disclose the privacy of clients by analyzing sensitive information through uploaded local model parameters. In order to solve this problem, researchers such as Google [9] proposed a method of using double mask technology and Shamir secret sharing technology to protect local model gradients. In addition, in order to further protect security, clients hope to use a low-cost and secure way to verify the aggregation results. However, through theoretical analysis, it has been found that whether using homomorphic hashing verification methods [13] or bilinear pairing verification methods [15], there are security issues with the verification information. To solve this problem, in our protocol, a random array r_1, r_2 is introduced when the client generates validation labels, which is kept secret from the server. In short, it means using the privacy of the client to protect verification information. Due to the inability of the random array to be obtained by the server, malicious servers are unable to obtain critical information and therefore cannot forge verification information to deceive clients.

5.3   Theoretical Performance Analysis

This part analyzes the performance of our protocol by evaluating computation, communication.

a) Computation overhead

In the protocol, we mainly consider the computation overhead of the participants: clients, server and collector in each stage expect **Initialization**. Since **Initialization** is a one-time stage and the keys are all length constant, the computation overheads are very small compared with the other stages.

Assuming the number of clients participating in the protocol is m and the number of the total gradients is n. Let M_0 represent the time for multiplication in group $Z_{N^2}^*$, M_1 represent the time for multiplication in group G_1, M_T represent the time for multiplication in bilinear group G_T, e represent the time for bilinear pairing operation, E_0 represent the time for exponential operation in group $Z_{N^2}^*$, E_1 represent the time for exponential operation in group G_1, and E_2 represent the time for exponential operation in group G_2.

The computational costs of the client, server, and collector in each stage are shown in Table 2. From Table 2, it can be seen that the computation costs of the client increase linearly with the number of gradients, while the collector's and the server's computation cost both increase linearly with the number of clients and gradients.

Table 2. Computation overhead of the proposed protocol

	Encryption	Collection	CalAgg	Verification
Client	$n(2M_0 + E_0 + 2E_1 + M_1)$	$E_0 + E_2$	\	$3e + E_1$
Server	\	$E_0 + E_2$	$n((m+2)M_0 + E_0 + (m-1)M_1 + E_1)$	\
Collector	\	$(m-1)(M_0 + M_2)$	\	\

b) Communication overhead

For the convenience of analysis and representation, let m represent the number of clients, n represent the total gradients, G_0 represents an element in $Z_{N^2}^*$, and G_1, G_2, G_T respectively represent an element group G_1, G_2, G_T when analyzing communication between various parties. In order to clearly represent the communication cost of each stage, the communication status of participants is divided into two aspects: sent and received.

From the analysis in Table 3, it can be clearly seen that in the stage of **Encryption** stage, the communication costs between the client and server are proportional to the number of clients and the total gradients. During the stage of **Collection**, the communication costs between the three parties are proportional to the number of clients.

5.4 Experiments

This article mainly focuses on improving the privacy leakage issue in the verification process, improving the security of verifiable federated learning. In the experimental section, the costs are tested under different client and gradient numbers.

Table 3. Communication overhead of the proposed protocol

Participants	State	Encryption	Collection	CalAgg
Client	sent	$(nG_0 + G_1)$	$G_0 + G_2$	\
	received	\	$G_0 + 2G_2$	$n(G_0 + G_1)$
Server	sent	\	$G_0 + G_2$	$n(G_0 + G_1)$
	received	$mn(G_0 + G_1)$	G_0	\
Collector	sent	\	$G_0 + G_2$	\
	received	\	$m(G_0 + G_2)$	\

The hardware conditions for the experiment are 64-bits Windows 10 PC with Intel(R) Core (TM) i7-7700HQ CPU @ 2.80 GHz, 16 GB RAM, and NVIDIA GeForce GTX 1050 GPU. This article simulates the stages of **Initialization, Encryption, Collection and CalAgg** and **Verification** of the protocol, and uses JPBC library to test the running time of each stage. To test the impact of different gradients and client counts on computational overhead, we used the control variable method for testing. Firstly, with the number of clients set to 100, test the computational cost of each participant at each stage under different gradient numbers; Secondly, set the number of clients to 100, 150, 200, 250, 300, 350, 400, 450, and 500 respectively to test the computational overhead of the server and aggregator.

From Fig. 2(a), it can be seen that as the number of gradients continues to increase, the time for the client to encrypt the gradient in the **Encryption** stage increases linearly. In the **Collection** stage, and **Verifications** stage, from Fig. 2(b) and Fig. 2(c), it can be seen that the calculation cost of the above process is independent of the number of gradients, which can be considered fixed and unchanged. From experimental results, it can be concluded that the total computational cost of each client is linearly related to the number of gradients, as shown in Fig. 2(d). Combined with the above experimental process, the vast majority of the computational cost of the client comes from the computational cost of encrypting gradients.

The main stages of server participation in computing include the **Collection** stage and the **CalAgg** stage. During the **Collection** stage, the server calculates the public key and broadcasts it to the client. In this stage, the cost of calculating the public key is fixed and does not change with the gradient and the number of clients, as shown in Fig. 3(a)(b). In the **CalAgg** stage, the computational cost of the server is related to the number of clients and gradients. From Figs. 4(a) and Fig. 4(b), it is easy to see that the computational cost on the server side is linearly related to the number of gradients and users.

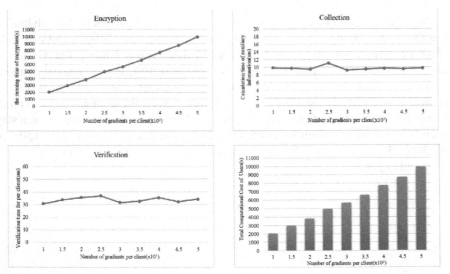

Fig. 2. (a): Time cost of per client in the **Encryption** stage; (b): Time cost of per client in the **Collection** stage; (c): Time cost of per client in the **Verification** stage; (d): Total computation cost of per client.

The aggregator mainly participates in the **Collection** stage. From Fig. 5(a), it can be seen that as the number of gradients changes, the computational overhead at the aggregator end remains within a certain range. From Fig. 5(b), it can be seen that the computational cost of the aggregator is positively correlated with the number of clients.

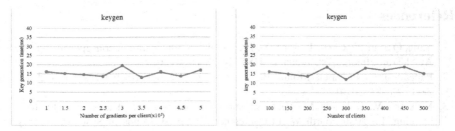

Fig. 3. (a): Public key generation time of per 100 clients; (b): Public key generation time for clients.

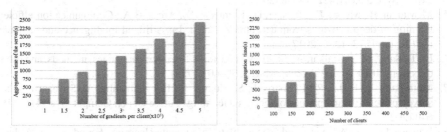

Fig. 4. (a): Aggregation time of server for per 100 clients; (b): Aggregation time for different number of clients.

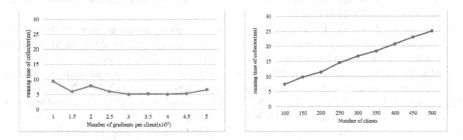

Fig. 5. (a): Time cost of Collector for per 100 clients; (b): Time cost of Collector for clients.

6 Conclusions

In this paper, we study verifiable secure aggregation under federated learning. Our protocol can achieve accurate model training while protecting data privacy. Meanwhile, the aggregated gradients can be verified by the clients. Namely, the malicious server cannot forge aggregation gradients and proof information to deceive clients. We analyze the security and performance of the proposed protocol. The experiment results show that our protocol is efficient and applicable for the real circumstances.

Acknowledgment. This paper is supported by Guangdong Provincial Key Laboratory of Power System Network Security.

References

1. Yang, Q., Liu, Y., Chen, T., Tong, Y.: Federated machine learning: concept and applications. ACM Trans. Intell. Syst. Technol. (TIST) **10**(2), 1–19 (2019)
2. Papadimitriou, P., Garcia-Molina, H.: Data leakage detection. IEEE Trans. Knowl. Data Eng. **23**(1), 51–63 (2010)
3. Kulkarni, V., Kulkarni, M., Pant, A.: Survey of personalization techniques for federated learning. In: 2020 Fourth World Conference on Smart Trends in Systems, Security and Sustainability (WorldS4), pp. 794–797. IEEE (2020)
4. Chen, Y., Su, X., Jin, Y.: Communication-efficient federated deep learning with layerwise asynchronous model update and temporally weighted aggregation. IEEE Trans. Neural Netw. Learn. Syst. **31**(10), 4229–4238 (2019)
5. McMahan, B., Moore, E., Ramage, D., Hampson, S., Arcas, B.A.: Communication-efficient learning of deep networks from decentralized data. In: Artificial intelligence and statistics, PMLR, pp. 1273–1282 (2017)
6. Li, T., Sahu, A., Talwalkar, A., Smith, V.: Federated learning: challenges, methods, and future directions. IEEE Signal Process. Mag. **3**, 50–60 (2020)
7. Mothukuri, V., Parizi, R.M., Pouriyeh, S., Huang, Y., Dehghantanha, A., Srivastava, G.: A survey on security and privacy of federated learning. Future Gener. Comput. Syst. **115**, 619–640 (2021)
8. Kairouz, P., et al.: Advances and open problems in federated learning. Found. Trends® Mach. Learn. **14**(1–2), 1–210 (2021)
9. Bonawitz, K., et al.: Practical secure aggregation for privacy-preserving machine learning. In: Proceedings of the 2017 ACM SIGSAC Conference on Computer and Communications Security, pp. 1175–1191 (2017)
10. Mandal, K., Gong, G.: PrivFL: practical privacy-preserving federated regressions on high-dimensional data over mobile networks. In: Proceedings of the 2019 ACM SIGSAC Conference on Cloud Computing Security Workshop, pp. 57–68 (2019)
11. Mandal, K., Gong, G., Liu, C.: Nike-based fast privacy-preserving high dimensional data aggregation for mobile devices. IEEE Trans. Dependable Secure, 142–149 (2018)
12. Xu, G., Li, H., Liu, S., Yang, K., Lin, X.: VerifyNet: secure and verifiable federated learning. IEEE Trans. Inf. Forensics Secur. **15**, 911–926 (2019)
13. Guo, X., et al.: VeriFL: communication-efficient and fast verifiable aggregation for federated learning. IEEE Trans. Inf. Forensics Secur. **16**, 1736–1751 (2020)
14. Hahn, C., Kim, H., Kim, M., Hur, J.: Versa: verifiable secure aggregation for cross-device federated learning. IEEE Trans. Dependable Secure Comput. (2021)
15. Wang, Y., Zhang, A., Shu, W., Shui, Y.: Vosa: verifiable and oblivious secure aggregation for privacy-preserving federated learning. IEEE Trans. Dependable Secure Comput. **20**(5), 3601–3616 (2023). https://doi.org/10.1109/TDSC.2022.3226508
16. Shi, E., Chan, H.T.H., Rieffel, E., Cho, R., Song, D.: Privacy-preserving aggregation of time-series data. In: Annual Network & Distributed System Security Symposium (NDSS). Internet Society (2011)
17. Joye, M., Libert, B.: A scalable scheme for privacy-preserving aggregation of time-series data. In: Sadeghi, A.-R. (ed.) FC 2013. LNCS, vol. 7859, pp. 111–125. Springer, Heidelberg (2013). https://doi.org/10.1007/978-3-642-39884-1_10
18. He, D., Chen, C., Chan, S., Bu, J.: Secure and efficient handover authentication based on bilinear pairing functions. IEEE Trans. Wireless Commun. **11**(1), 48–53 (2011)
19. Tsobdjou, L.D., Pierre, S., Quintero, A.: A new mutual authentication and key agreement protocol for mobile client—server environment. IEEE Trans. Netw. Serv. Manage. **18**(2), 1275–1286 (2021)

20. Xin, L., Yunyi, L., Miao, W.: A lightweight authentication protocol based on confidential computing for federated learning nodes. Netinfo Secur. **22**(7), 37–45 (2022)
21. Emura, K.: Privacy-preserving aggregation of time-series data with public verifiability from simple assumptions. In: Pieprzyk, J., Suriadi, S. (eds.) ACISP 2017, Part II. LNCS, vol. 10343, pp. 193–213. Springer, Cham (2017). https://doi.org/10.1007/978-3-319-59870-3_11

Electronic Voting Privacy Protection Scheme Based on Double Signature in Consortium Blockchain

Wei Xie, Wenmin Li[✉][iD], and Huimin Zhang

State key Laboratory of Networking and Switching Technology, Beijing University of Posts and Telecommunications, Beijing 100876, China
liwenmin@bupt.edu.cn

Abstract. Electronic voting can improve the efficiency, accuracy and fairness of voting, and save resources and costs. However, the identity privacy of voters may not be effectively protected during the counting process of third-party electronic voting. In order to solve the problems of transparency and identity privacy protection in electronic voting, this paper combines pseudonym mechanism and digital signature technology to propose a privacy protection scheme for double signature electronic voting based on Consortium blockchain. By introducing the Consortium Blockchain, the voting ciphertext is published on the Blockchain to realize the transparency of electronic voting and ensure the security of the voting; the introduction of the pseudonym mechanism can effectively protect the identity privacy of voters; the use of homomorphic encryption technology to realize ciphertext vote counting, it can effectively resist the internal attack of electronic voting; through double signature, that is, to sign the voting ciphertext and pseudonym to ensure the identity of the voter and the legitimacy of the vote. At the same time, a scoring mechanism is introduced to achieve more precise evaluation of candidates by voters, and the Bulletproof protocol is used to standardize the range of score. The electronic voting privacy protection scheme proposed in this paper can also satisfy multi-candidate voting and is applicable to various voting scenarios.

Keywords: Hyperledger Fabric · Paillier encryption · ECDSA · Electronic voting

1 Introduction

In recent years, blockchain technology has developed rapidly and has attracted the attention of many researchers. The blockchain-based voting scheme can well solve the problem of lack of security and reliability of third-party Internet electronic voting. Researchers have analyzed and studied the blockchain voting tech-

This work was supported by National Key R&D Program of China under Grant 2020YFB1005900, NSFC(Grant Nos. 62072051).

J. Vaidya et al. (Eds.): AIS&P 2023, LNCS 14509, pp. 548–562, 2024.
https://doi.org/10.1007/978-981-99-9785-5_38

nology [1], conducted in-depth research on its feasibility, and proposed a series of voting principles [2], voting rules and voting algorithms.

Homomorphic encryption is used in electronic voting systems to allow votes to be counted without decryption. Yang et al. [3] proposed a homomorphic weighted electronic voting system based on SEAL, employing the BFV to ensure resistance against quantum attacks, however, it still relies on proxy servers and trusted third parties.

Literature [4] optimizes the electronic voting scheme based on homomorphic encryption by using multi-threading technology, and verifies the validity of votes by using zero-knowledge proofs, but does not consider the weight of voters, so a more accurate voting mechanism cannot be realized.

At present, most blockchains are used for voting applications using public chains [5], such as Bitcoin [6] and Ethereum. Literature [7] proposes an electronic voting scheme based on blockchain technology, which satisfies the basic properties of electronic voting while providing a certain degree of decentralization.

Literature [8] proposes a blockchain system using fingerprint authentication, adding fingerprint authentication to the identity authentication link to ensure that the voter's identity is legal. Literature [9] proposes a secure online voting method based on blockchain and machine learning, which uses machine learning technology to automate the authentication process of legitimate users in the oracle platform and realize face recognition authentication.

A privacy protection scheme for protecting electronic voting through blockchain. However, few studies have combined consortium chain and voting, among which the document [10] proposes to build a scalable electronic voting system based on Hyperledger Fabric. For the weighted voting scheme, Yang et al. proposed a homomorphic weighted voting scheme based on the SEAL library [3] which lacks the protection of user identity information.

The aforementioned research employs blockchain technology to address the issue of traditional voting; however, there are still several challenges in integrating blockchain with electronic voting. Firstly, the identity privacy of voters during the process of submitting votes has not been effectively protected. Secondly, the current electronic voting system fails to achieve precise voting. For example, in an annual election for the most popular celebrity election, voters want to cast their votes for both Star A and Star B, but have a stronger preference for Star A. If they want to cast an extra vote for Star A or give Star A higher score, the current plan can only be achieved through two votes. Unfortunately, such flexibility is currently unattainable within existing schemes.

The blockchain is divided into permissioned blockchain and non-permissioned blockchain. Permissioned blockchain, also known as a public chain, allows any node to join or exit, granting access to data and transactions on the chain for all nodes. The non-permissioned blockchain can be further categorized into consortium blockchain and private blockchain. Consortium blockchain consists of multiple organizations joining the same blockchain network and sharing the blockchain ledger, while private blockchain is usually created and maintained by a single participant. Currently, most research combines electronic voting systems with public blockchain; however, traditional privacy protection methods without supervision in public blockchain is not well-suited for electronic voting systems. The endorsement in consortium blockchain provides greater convenience in achieving controllable anonymity of identity and rapid verification of hidden transactions.

Homomorphic encryption technology enables operations on ciphertexts, so this paper studies the privacy protection method of hidden votes in electronic voting based on homomorphic encryption technology. Paillier encryption has semi-homomorphic characteristics, which is suitable for electronic voting scenarios. Using Paillier homomorphic encryption technology, the scheme can support multiple candidates and multiple voters while satisfying the privacy protection of ballots. And in the final ticket stage, only one decryption is required. In order to protect the identity and privacy of voters, a pseudonym mechanism is used to hide the relationship between voters and ballots, and double digital signatures are used to ensure the identity of voters and the legitimacy of ballots. The scheme also includes a voting scoring mechanism, which uses Bulletproof protocol [11] to standardize the scoring range and achieve fine voting, which is more suitable for practical applications. Finally, an optional traceability mechanism is provided to prevent malicious voters from participating in voting.

2   Background

This section introduces the basic knowledge of ECDSA digital signature, Paillier homomorphic encryption, Consortium blockchain and so on.

2.1   ECDSA Digital Signature

ECDSA (Elliptic Curve Digital Signature Algorithm, Elliptic Curve Digital Signature Algorithm) is a secure digital signature algorithm, which uses the discrete logarithm problem in Elliptic Curve Cryptography to ensure the security of the signature [12].

ECDSA is a commonly used digital signature technology in blockchain. This paper will use ECDSA to implement a double signature mechanism. The following is a specific description of the ECDSA algorithm:

ECDSA consists of four sub-algorithms, defined as $ECDSA = (Setup, KeyGen, Sig, Ver)$, and its algorithm description is as follows:

1) **Initialization Algorithm** $Setup(1^\lambda) \rightarrow pp$: Given the elliptic curve E on the field:

$$E : y^3 = x^3 + ax + b(mod\ p) \tag{2.1}$$

and choose the base point as G, its order is q, H represents the selected hash function, and set the public parameter $pp = (E, G, H, q)$.

2) **Key generation algorithm** $KeyGen(pp) \rightarrow (pk, sk)$: Pick a random number $d \in [1, q-1]$ as private key sk, and calculate the public key:

$$pk = dG \tag{2.2}$$

3) **Signature algorithm** $Sig(pp, sk, M) \rightarrow (r, s)$: Input public parameters pp, private key sk and message M to be signed, the execution process of the signer is as follows:
   (1) Pick a random number $k \in [1, q-1]$;
   (2) Perform scalar multiplication:

$$R = kG = (x, y) \tag{2.3}$$

   (3) Compute:

$$r = x(mod\ q) \tag{2.4}$$

   x is the abscissa of R, if $r = 0$, return step 1;
   (4) Compute hash value $H(M)$ of message M;
   (5) Compute:

$$s = k^{-1}(H(M) + sk \cdot r)(mod\ q) \tag{2.5}$$

   if $s = 0$, return step 1;
   (6) Output signature result (r, s);
   (7) Check whether

$$r = x' mod\ q \tag{2.6}$$

are true. If established, the signature is legal, otherwise the signature is invalid.

2.2 Paillier Homomorphic Encryption

The security of Paillier homomorphic encryption [13] can be reduced to the decisional composite residuosity assumption (DCRA), and its algorithm consists of three parts $(KeyGen, Enc, Dec)$.

1) **Key generation algorithm** $KeyGen \rightarrow (pk, sk)$: Randomly choose two independent large prime numbers p, q, and

$$gcd(pq, (p-1)(q-1)) = 1 \tag{2.7}$$

; compute the product

$$n = p \cdot q \tag{2.8}$$

of p, q; choose an order to be a large prime number q_1 multiplicative group of $\mathbb{G}$, its generators are g and h, satisfy $g < n^2$; the least common multiple calculated by the LCM function is

$$\lambda = LCM(p-1, q-1) \tag{2.9}$$

; define the function L as

$$L(u) = \frac{(u-1)}{n} \tag{2.10}$$

, calculate the modular inverse element according to the least common multiple

$$\mu = L(g^\lambda mod\ n^2)^{-1} mod\ n \tag{2.11}$$

; set the public key to $pk = (n, g)$, private key is $sk = (\lambda, \mu)$.

2) **Encryption Algorithm $Enc(pk, M) \rightarrow c$**: Input the public key pk and the message to be encrypted $M \in \mathbb{Z}_n$, select a random number $r \in \mathbb{Z}_n^*$, and calculate the ciphertext:

$$E(M) = c = (g^M r^n) mod\ n^2 \tag{2.12}$$

3) **Decryption algorithm $Dec(c, sk) \rightarrow M$**: Enter the ciphertext c and the private key sk, and use the following formula to decrypt the corresponding plaintext M, the formula is:

$$D(c) = M = L(c^\lambda mod\ n^2) \cdot \mu\ mod\ n = \frac{L(c^\lambda mod\ n^2)}{L(g^\lambda mod\ n^2)} mod\ n \tag{2.13}$$

Let the ciphertexts c_1, c_2 corresponding to the plaintext m_1, m_2 product of two ciphertexts

$$c_1 \cdot c_2 = E(m_1) \cdot E(m_2)$$
$$= g^{(m_1+m_2)} r_1^n r_2^n\ mod\ n^2$$
$$= g^{(m_1+m_2)} (r_1 r_2)^n\ mod\ n^2$$
$$= E(m_1 + m_2) \tag{2.14}$$

It can be seen from the above formula, because $E(m_2) \cdot E(m_2) = E(m_1 + m_2)$, so the Paillier encryption algorithm has additive homomorphism.

Paillier homomorphic encryption is a very powerful and secure encryption technology, which has been widely used in many fields, such as digital signature, identity verification, key exchange [12] and so on.

2.3   Consortium Blockchain

Consortium blockchain is a special type of distributed ledger based on blockchain technology. Compared with public blockchain, it has the following characteristics:

1) **Efficiency.** The participants in the consortium blockchain are usually determined. Compared with the unlimited participants in the public blockchain, the confirmation speed of the transaction is faster, so it can support higher transaction throughput.
2) **Scalability.** The scalability of the alliance chain is higher than that of the public blockchain, because the participants of the alliance chain are determined, and the number of nodes and processing capacity can be expanded without affecting the entire network.

2.4   Bulletproof Protocol

The Bulletproof protocol is a zero-knowledge proof protocol utilized for verifying the veracity of a claim while preserving the confidentiality of its exact content. In 2018, Benedikt Bunz first proposed Bulletproof protocol, an efficient non-interactive zero-knowledge proof protocol. The basic idea of the Bulletproof protocol is to convert a declaration into a polynomial and then prove the correctness of that polynomial using the Pedersen commitment.

3   Scheme Design

3.1   System Composition

As shown in Fig. 1, the privacy protection scheme for double signature electronic voting based on the consortium chain is mainly composed of three entities: the chain code in the Fabric consortium blockchain network, the voter (the person who initiates the vote, the voter) and the regulatory agency. Its specific functions are described as follows:

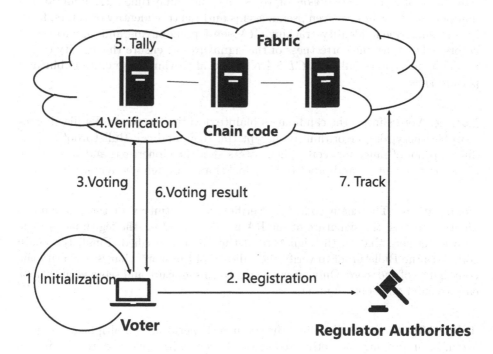

Fig. 1. System Model.

3.2   System Process Design

The double signature electronic voting privacy protection scheme based on Consortium blockchain mainly includes six stages: Initialization, Key Generation, Registration, Voting, Verification, and Vote Counting. This scheme also designs an optional malicious voter tracking. The detailed description is as follows:

Initialization. The initiator voter initializes the public parameters (Paillier homomorphic encryption parameters, ECDSA parameters and Pedersen commitment) through the initialization algorithm.

Key Generation. The voting initiator generates a homomorphic encryption public/private key. And publish voting information on the blockchain, including voting title, candidate list, public key, voting deadline, etc. RA calls the ECDSA.KeyGen algorithm to generate public and private keys.

Registration. Voters register with RA as pseudonyms with signatures for subsequent anonymous submission of votes. At the same time, RA maintains a mapping relationship between pseudon-yms and the true identity of voters, facilitating subsequent identity tracking. RA send pseudonyms and signatures to voters, who verify the correctness of the signatures to ensure the legality of the pseudonyms. Voters call the $ECDSA.KeyGen$ algorithm to generate public and private keys.

Voting. Voters input the candidate's plaintext voting score, use Paillier homomorphic encryption algorithm to encrypt the score and call Bulletproof to produce a proof of range for score. Then voters sign the ciphertext, and submit the ciphertext, signature and proof to the blockchain through anonymous identity.

Verification. The chain code first verifies the legitimacy of the anonymous identity, that is, the signature of the RA is verified. After, the legitimacy of the vote is verified, that is, the signature of the voter is verified. Final, the chain code uses the Bulletproof to verify the validity of the score, that is, to verify the compliance of the score. Only after passing both signature and score verifications can we enter the vote counting.

Tally. After the voting deadline, the chain code performs a multiplication operation, accumulating the voting scores of all voters in ciphertext to obtain the candidate's final score in ciphertext, and sending the multiplication result to the voting initiator; the initiator of the vote performs decryption operations to obtain plaintext of the final scores of each candidate, and publishes the scores of each candidate and the chain code of the winning candidate to complete the electronic voting.

Optional traceable mechanism. In the verification stage, the chain code detects malicious voters and publishes their pseudonyms on the blockchain. Afterwards, RA obtains pseudonym information from the blockchain and track the true identity of the voters based on the local mapping list, which is then published on the blockchain.

3.3   Specific Plan

A double signature electronic voting privacy protection scheme based on Consortium blockchain includes six stages: **Initialization, Key Generation, Registration, Voting, Verification,** and **Tally**.

Initialization. During the initialization phase,

(1) The system chooses two safe large primes p, q, calculates n.
(2) The system choose an order to be a large prime number e multiplicative group of $\mathbb{G}$, its generators are g and h, satisfy $g < n^2$,define the function $\mathbf{L}(x)$.
(3) The system calls $ECDSA.Setup(1^\lambda)$ algorithm, generates public parameters $pp = (E, G, H(x), w)$ for ECDSA, where G is the base point, its order is w, $H(x)$ is a hash function.
(4) The public parameter is $PP = (pp, n, g, h, \mathbf{L}(x), e, \mathbb{G})$.

3.4   Key Generation

(1) The RA calls the $ECDSA.KeyGen$ algorithm to generate public/private key (pk_{RA}, sk_{RA}).
(2) The voting initiator calls $Paillier.KeyGen$ to generate Paillier public/private key (pk_P, sk_P).
(3) The voting initiator publishes voting information on the blockchain and generates the current voting ID based on the timestamp and voting title. The voting information includes the voting title, m candidate lists $Candidate = \{cd_1, \cdots, cd_m\}$, public key pk_P, voting scoring range $[0, 2^l)$, voting deadline, etc.

3.5   Registration

During the registration phase, voter sends a request to register their pseudonym identity with RA and upload parameters such as the voter's true identity (UID) and the ID of the current vote. The RA verifies the qualifications of voters based on their UID and ID, after passing the verification, generates a pseudonym identity

$$AID = H(UID\|ID) \tag{3.1}$$

and calls $ECDSA.Sig(pp, sk_{RA}, AID)$ algorithm to generate ECDSA signature $S_{RA} = (r_{RA}, s_{RA})$ for AID, then add the pseudonym AID and signature S_{RA}

returns to the voter. Voters receive pseudonyms AID and signature S_{RA} from RA, call the algorithm $ECDSA.Ver(pk_{RA}, AID, S_{RA})$, to verify the legitimacy of pseudonyms. Voter calls the $ECDSA.KeyGen(pp)$ algorithm to generate a signed public private key pair (pk, sk). Finally, RA retains mapping relationships (AID, ID) to facilitate tracking of users' true identities in necessary cases, enabling malicious users to be monitored.

3.6 Voting

In the voting stage, voters rate all candidates and encrypt the score and generates proof of rang for the score, while signing the ciphertext of the score and submitting the ciphertext, proof and signature to the blockchain. The specific steps are as follows:

(1) The voter calls the $Paillier.Enc(pk_P, v_j)$ algorithm to generate voting score ciphertext

$$c_j = g^{v_j} r_j^n mod \ n^2 \tag{3.2}$$

, where v_j represents the voter's preference for the j-th candidate cd_j's score, $v_j \in [0, 2^l)$, $j \in [1, m]$, r_j is a random number.

(2) The voter calculates Pedersen commitment:

$$V_j = g^{v_j} h^{r_j} mod \ e, j \in [1, m] \tag{3.3}$$

(3) The voter invokes the Bulletproof, using the commitment V_j to produce a proof $proof_j$ of v_j, $j \in [1, m]$.

(4) Calculate auxiliary data:

$$K_{j,1} = r_j^n mod \ n^2, K_{j,2} = h^{r_j} mod \ e, j \in [1, m] \tag{3.4}$$

(5) The voter multiplies c_j to get

$$M = \prod_{j=1}^{m} c_j = \prod_{j=1}^{m} g^{v_j} r_j^n \tag{3.5}$$

, where r_j is the random number selected by voter when encrypting v_j.

(6) Voter calls $ECDSA.Sig(pp, sk, M)$ to sign their scores and get signature S.

(7) Let $\pi_j = (c_j, K_{j,1}, K_{j,2}, proof_j)$. The voter uploads $(\pi_1, \cdots, \pi_m, S, S_{RA})$ to the blockchain with the pseudonym AID.

3.7 Verification

The voter publishes the voting information on the blockchain, and the blockchain verifies the legitimacy of the voting information by calling the signature verification algorithm and Bulletproof.

(1) Call the $ECDSA.Ver(pk_{RA}, AID, S_{RA})$, verify the legitimacy of voters, if that fail, abort.

(2) All ciphertexts of voters are multiplied to get M.

(3) Call the $ECDSA.Ver(pk, M, S)$, verify the legitimacy of the ciphertext. If that fail, abort.

(4) Generate commitment V_j based on c_j and auxiliary data $K_{j,1}$, $K_{j,2}$, calculated commitment:

$$V_j = \frac{c_j K_{j,2}}{K_{j,1}} \tag{3.6}$$

(5) Call Bulletproof to verify the correctness of $proof_j$. If this fails, the voter's identity is sent to the RA for tracking.

After the two verifications and range proof are passed, the chain code will store the ciphertext c_j of the j-th candidate cd_j into the voting list CD_j, and then enter the vote counting stage.

3.8 Tally

After the voting time expires, it is assumed that there are N voters participating in the voting. In the vote counting phase, the chain code accumulates and counts the ciphertext ballot tickets obtained by the candidates, and sends the final result of the ciphertext to the voting initiator for decryption. Specific steps are as follows:

(1) The chain code performs ciphertext multiplication on the voting score c_j^i in the voting list CD_j to obtain the ciphertext of the final voting score of the candidate cd_j. The multiplication operation performed on the chain is as follows:

$$C_j = \prod_{i=1}^{N} c_j^i \, mod \ n^2 = (g^{v_j^1} r_1^n) \cdot \cdots \cdot (g^{v_j^N} r_N^n) mod \ n^2 \tag{3.7}$$

, where c_j^i represents the ciphertext of the i-th voter's voting score for the j-th candidate, and C_j represents the ciphertext of the sum of all voters' votes for the j-th candidate. The chain code sends the final result of the encrypted ballot to the voting initiator.

(2) After receiving C_j, he voting initiator calls $Paillier.Dec(sk_P, C_j)$ to decrypt and obtain the plaintext v_j^* of the final voting scores of each candidate.

(3) The scores obtained by the candidates are compared and the one with the highest score is the winner. The voting initiator publishes the voting scores of each candidate and the final winning candidate to the chain code.

4 Security and Performance Analysis

In this paper, the scheme introduces a scoring mechanism in order to achieve more accurate and secure electronic voting, and combines electronic voting with consortium blockchain by taking advantage of the fact that consortium blockchain is able to achieve identity-controllable anonymity. This section analyses the security of the scheme, the overhead of computation, communication and storage, and the experimental validation, and the results show that the scheme can be well applied to the electronic voting scenario.

4.1   Security Analysis

The double signature electronic voting privacy protection scheme based on the Consortium block-chain can provide the functions of recording votes and managing votes, using ECDSA signatures to ensure the legitimacy of voters and voting scores, and using pseudonym mechanisms to protect the privacy of voters. The scheme uses the Bulletproof protocol to ensure the compliance of voting scores. This section will analyze the security of the proposed scheme from the following aspects.

(1) **Correctness.** Paillier homomorphic encryption has additive homomorphism, so it can guarantee that the final score obtained by the candidate is the sum of the scores of N voters.

(2) **Safety.** The security of the Paillier homomorphic encryption algorithm is based on the difficulty of DCRA, so the attacker cannot obtain any information about the voting score from the ciphertext. In generating Paillier homomorphic encryption parameters, $n = p \cdot q$, where p, q are large prime numbers, and if the integer z is called the n-order residual class modulo n^2, then there is an integer $y \in \mathbb{Z}_{n^2}^*$ making $z = y^n \bmod n^2$. It is difficult to determine whether a given integer z is an n-order residual class modulo n^2. ECDSA digital signature is based on the intractability of elliptic curve discrete logarithm problem to realize the security of signature.

(3) **Anonymity.** In the scheme in this chapter, voters use pseudonym identity AID to submit ciphertext voting scores to ensure anonymity.

4.2   Performance Analysis

The communication, storage, and computation overheads of the proposed scheme are presented in the Table 1.

Table 1. Communication, Storage, Computation overheads

	Communication overheads	Storage overheads	Computation overheads
Voter	$O(1)$	$O(1)$	$O(1)$
Voter initiate	$O(N)$	$O(1)$	$O(mN + m)$

Communication Overhead and Storage Overhead. For the voter, the signature key pair is generated locally when each voter registers. During the voting process, the voter needs to submit their identity (UID) and participating vote ID to the RA. And, they only once to submit information to blockchain, which includes the ciphertext, signature, and zero knowledge proof. Thus, the communication overhead is a constant round, i.e. $O(1)$. The vote initiator collects the transaction information of all the ballots with a communication overhead of $O(N)$, where N represents the number of participating voters. Participating registered voters store the signed private key locally with a storage overhead of $O(1)$ and the vote initiator stores the homomorphic encrypted private key locally with a storage overhead of $O(1)$.

Computational Overhead. For the voter, each user's pseudonym is generated by RA. Users only need to verify the validity of the pseudonym once and generate the signature key pair once. Additionally, they perform ballot encryption and zero-knowledge proof generation only once. Therefore, the computational overhead for each voter is $O(1)$. For the vote initiator, the number of candidates is m. The vote initiator needs to perform mN times of product calculations to generate the ballot ciphertext for each candidate. Subsequently, they must carry out m times of decryption operations to obtain the ballot count for each candidate. Hence, the computational overhead of the vote initiator is $o(mN + m)$.

This section compares the voting time consumption of this scheme with the scheme given in literature [3,4], typical schemes of electronic voting. All three schemes use homomorphic encryption. The difference is that literature [3] uses fully homomorphic encryption. Algorithm BFV, while using the SEAL library to improve system performance, literature [4] is the same as this scheme, using the Paillier homomorphic encryption algorithm to calculate the ciphertext of voting scores.

Table 2. Comparison

Schemes	Homomorphic Encryption Algorithm	Counting Time	Difficult Problem	Safety	Anonymity	Traceable	Score
This plan	Paillier	3 ms	DCRA	Yes	Yes	Yes	Yes
Literature [3]	RFV	1 ms	RLWE	No	No	No	No
Literature [4]	Paillier	34 ms	DCRA	Yes	No	No	No

It can be concluded from Table 2 that this scheme has the following advantages:

1) Compared with literature [3,4], this scheme has the provable characteristics of voting score range, and has a slight disadvantage in terms of time performance.
2) In terms of identity anonymity, this scheme has the advantage of protecting the anonymity of the voter's identity, while literature [3,4] lacks identity protection measures.
3) This scheme has an openable mechanism, which can be tracked when malicious voting users are found, and the literature [3,4] is also lacking.

In order to test the performance and energy consumption of this solution, this article conducted an experimental test. The experimental environment configuration is as follows: Ubuntu 20.04TLS Intel(R) Core(TM) i5-10400 CPU @ 2.90 GHz, 24.0 GB RAM, the code used is Golang, the application Part of it is compiled by Vscode, and part of the chain code is executed by smart contracts. Assuming that there are 3 candidates and 5 voters, in this scheme, the time consumption of each stage is mainly evaluated, including six stages of initialization, key generation, registration, voting, verification, and vote counting. When shown in Fig. 2.

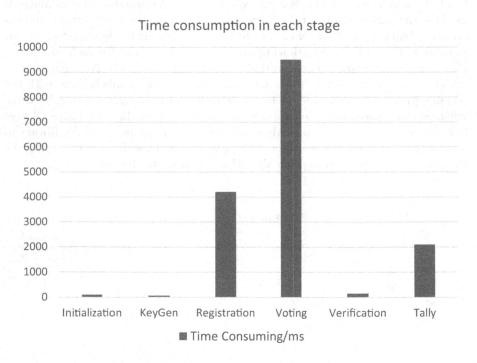

Fig. 2. Time Consumption.

At the same time, the scheme of literature [3,4] is compared with the scheme of this chapter in terms of system vote counting running time. The scheme of literature [3] has advantages in time performance, but there are problems in security. Compared with literature [4], this scheme is better than the comparison scheme in terms of system vote counting running time. The system running time comparison chart is shown in Fig. 3.

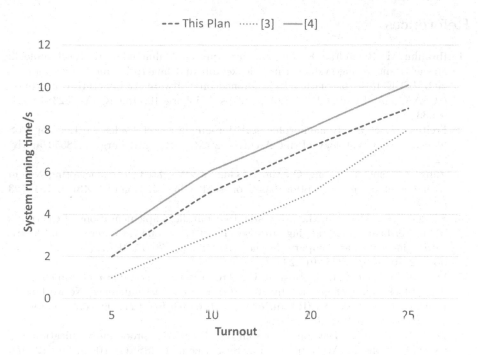

Fig. 3. System Runtime Comparison.

5   Conclusion

This paper proposes a privacy protection scheme for double signature electronic voting based on the Consortium blockchain, aiming at the problem that the identity information of the voter and the ballot information of the vote in the voting system may be leaked on the alliance chain. The main idea of the solution is to use Paillier homomorphic encryption technology to encrypt ballots during the voting process, and use ECDSA signature to ensure the legitimacy of voters and ballots. More accurate voting is achieved through the scoring mechanism. By adding regulatory agencies, the open mechanism of the voting privacy protection scheme can be realized, and malicious voting users can be verified and tracked.

In the future, the double signature electronic voting privacy protection scheme based on the Consortium blockchain will play a greater role, so as to ensure the voting security of participants, help improve voting efficiency, and realize safe voting in the true sense.

We will focus on the privacy protection scheme of electronic voting based on the Consortium blockchain, aiming to solve the problems of voting security and reliability, and how to prevent malicious attacks, and vigorously develop related applications of the voting scheme based on the Consortium blockchain to enhance voting security, reliability and controllability of the environment.

References

1. Ibrahim, M., Ravindran, K., Lee, H., Farooqui, O., Mahmoud, Q.H.: Electionblock: an electronic voting system using blockchain and fingerprint authentication. In: 2021 IEEE 18th International Conference on Software Architecture Companion (ICSA-C), pp. 123–129. IEEE (2021). https://doi.org/10.1109/ICSA-C52384.2021. 00033

2. Freitas, L., et al.: Homomorphic sortition-single secret leader election for PoS blockchains. Cryptology ePrint Archive (2023). https://doi.org/10.48550/arXiv. 2206.11519

3. Yang, Y., Zhao, Y., Zhang, Q., Ma, Y., Gao, Y.: Weighted electronic voting system with homomorphic encryption based on seal. Chin. J. Comput. **43**(4), 711–723 (2020)

4. Saadeh, I.A., Abandah, G.A.: Investigating parallel implementations of electronic voting verification and tallying processes. In: 2017 European Conference on Electrical Engineering and Computer Science (EECS), pp. 70–75. IEEE (2017). https:// doi.org/10.1109/EECS.2017.23

5. Stan, I.-M., Barac, I.-C., Rosner, D.: Architecting a scalable e-election system using blockchain technologies. In: 2021 20th RoEduNet Conference: Networking in Education and Research (RoEduNet), pp. 1–6. IEEE (2021). https://doi.org/10. 1109/RoEduNet54112.2021.9638303

6. Peng, K., Boyd, C., Dawson, E.: Batch zero-knowledge proof and verification and its applications. ACM Trans. Inform. Syst. Secur. (TISSEC) **10**(2), 6-es (2007). https://doi.org/10.1145/1237500.1237502

7. Specter, M.A., Koppel, J., Weitzner, D.: The ballot is busted before the blockchain: a security analysis of voatz, the first internet voting application used in {US}. federal elections. In: 29th USENIX Security Symposium (USENIX Security 20), pp. 1535–1553 (2020)

8. Tian, H., Fu, L., He, J.: A simpler bitcoin voting protocol. In: Chen, X., Lin, D., Yung, M. (eds.) Inscrypt 2017. LNCS, vol. 10726, pp. 81–98. Springer, Cham (2018). https://doi.org/10.1007/978-3-319-75160-3_7

9. Utah Green Party Hosts Dr. Stein. Elects new officers. using the following range voting system, the green party of utah elected a new slate of officers. SpinJ Corp, Boca Raton, FL, USA, Independent Political Rep (2017)

10. Rule twenty-two special rules for the visual effects award. five productions shall be selected using reweighted range voting to become the nominations for final voting for the visual effects award. In: 89th Annual Academy My Awards of Merit. Beverly Hills, CA, USA (2016)

11. Bünz, B., Bootle, J., Boneh, D., Poelstra, A., Wuille, P., Maxwell, G.: Bulletproofs: short proofs for confidential transactions and more. In: 2018 IEEE Symposium on Security and Privacy (SP), pp. 315–334. IEEE (2018). https://doi.org/10.1109/ SP.2018.00020

12. Meduza Moscow's online voting system has some major vulnerabilities, allowing votes to be decrypted before the official count. 01-Jul-2020

13. Bartolucci, S., Bernat, P., Joseph, D.: SHARVOT: secret share-based voting on the blockchain. In: Proceedings of the 1st International Workshop on Emerging Trends in Software Engineering for Blockchain, pp. 30–34 (2018)

Securing 5G Positioning via Zero Trust Architecture

Razy Youhana Adam and Weizhi Meng[✉]

SPTAGE Lab, Department of Applied Mathematics and Computer Science, Technical University of Denmark, Copenhagen, Denmark
weme@dtu.dk

Abstract. 5G is the future, regardless of what mobile network technology succeeds it, whatever comes will be built upon the 5G technology. The 5G positioning system in its part will also be the leading navigation system, once the 5G mobile network potentials are fully deployed and accessible everywhere and by every user and device. However, 5G technology needs firstly to be cyber secured, before its benefits and capabilities can be totally explored and reached. Almost the majority of cyber security experts agree that the 5G technology is vulnerable to various threats, i.e., the high risk that it brings regarding cyber hacking and data theft due to the absence of suitable access control and data encryption during the communication process between the user-end devices and the 5G mobile network. In order to securely benefit from the full potential and capabilities of 5G network, there is a huge demand to protect the 5G technologies through ensuring the trust factor regarding the access to the 5G mobile network, in order to achieve a secure 5G positioning system from start to end. This review paper introduces the background of 5G and how to secure the 5G positioning system by implementing zero-trust network access on top of the 5G mobile network combined with the network overlay virtualization.

Keywords: Cyber Security · 5G Positioning · Trust Management · Zero Trust · Overlay Network

1 Introduction

This paper provides a review to discuss how we are able to secure 5G positioning from many cyber threats by integrating with a zero-trust model solution. This is to prove that there is a method to make the 5G positioning a secure solution even though 5G technology is under various security threats.

1.1 Problem to Solve

The main problem to solve in this paper is how to make the 5G positioning a more secure solution by decreasing any trust effect regarding any access by a user or any app and making the whole 5G positioning from start to end a secure solution through implementing a zero-trust model/architecture [5,6].

© The Author(s), under exclusive license to Springer Nature Singapore Pte Ltd. 2024
J. Vaidya et al. (Eds.): AIS&P 2023, LNCS 14509, pp. 563–578, 2024.
https://doi.org/10.1007/978-981-99-9785-5_39

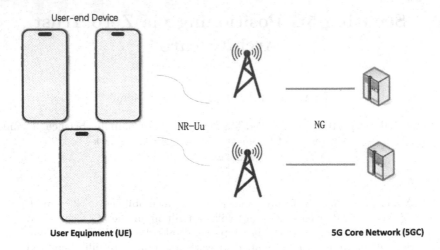

Fig. 1. Overview of 5G

1.2 Why Is This Problem Hard to Solve?

The short answer is because the 5G technology increases the risk of hacking and data theft [1,4]. One of the main weaknesses of the 5G mobile technology is its potential security vulnerabilities among components and processes [2]. Almost the majority of cyber security experts agree that 5G technology indeed raises the risk of hacking due to the absence of suitable encryption during the communication process between devices using the 5G technology network [3].

1.3 What Is 5G?

New technologies often come with new cyber security vulnerabilities and weaknesses. The security vulnerability in 5G positioning will be a huge issue and a headache to solve if not handled in a proper way [7].

The 5G wireless mobile network technology can promise a gigabyte speeds with low latency less than 1 millisecond and a huge capability to accommodate millions or even billions of devices. This capability will provide a huge step forward for 5G positioning because there would be almost no latency in response time, which is a huge advantage when considering navigation [8]. The 5G traffic capacity and the ability to accommodate a huge amount of devices could reach millions and even billions in the near future, which will help improve and widen its adoption in positioning systems. Thus, there is a huge demand and need to ensure the security of data exchanged during 5G positioning operations, and to ensure the security and safety of users and their equipment [9].

Figure 1 shows a high-level overview of 5G system, including User Equipment (UE)/User-end Device, the Radio Access Network (NG-RAN) and the Core Network (5GC). It is worth noting that gNB is the main entity of NG-RAN. Here "g" refers to "5G" and "NB" refers to "Node B", indicating radio transmitter (the same from 3G and afterwards). "NR-Uu" refers to radio interface, "NR" means "New Radio" and "Uu" is inherited from previous generations. The 5G core (5GC) can be represented by the AMF/UPF entity–the User Plane Function (UPF) to manage the user data and the Access and Mobility management Function (AMF).

When using 5G, we would have to reconstruct our whole telecommunication mobile network and this will have a huge impact on the way we develop the future mobile networks. 5G technology is almost a network built of software, resulting in many cyber-threats and vulnerabilities where software usually come with [10]. So, the main challenge regarding 5G technology including the 5G positioning is how to make it a cyber-secure mobile technology network [3,5].

As we already know the future is data and with that came the importance of 5G and this is due to that almost every single process that we perform now and, in the future, will have a device/app that needs data and a mobile network to perform the process it needs [11]. Therefore, we need to secure the network technology we are using. Actually 5G is more vulnerable to a cyber-attack than the older mobile technologies as in 4G/3G/2G, etc.

This is because its reliance depends more on software than hardware. For instance, the 5G technology network is totally relied on software, i.e., controlling its main process. One main issue regarding cyber security is that 5G technology including 5G positioning relies on software virtualization in its higher-level network functions and this is also a big cyber security threat [2,8].

Another problem with 5G is its great expansion in bandwidth, which also leads to a greater and wider possibilities of network attacks [12]. An additional cyber security issue or vulnerability regarding 5G is the countless number of IoT devices that will be connected to the network and in theory if attackers find any vulnerability in these millions or even billions of devices, then there is a huge risk that the network will collapse by only disturbing these billions of IoT devices [14].

1.4   Why Use Zero Trust Model with 5G Positioning?

A zero trust framework operates in a totally different way than many existing security models, as many of them define or declare that most parties inside an organization network can be trusted [18]. This is not the case in a zero trust model (aka. ZT model), which starts with defining trust as a cyber threat and a vulnerability that should be handled firmly with. It is actually an important type of trust management [45, 46].

Implementing a ZT model on a 5G positioning system means implementing the principles of a ZT model by making ZT principles more effective regarding performance execution and security throughout the whole positioning process from start to end [16]. The best element in a zero-trust architecture is that there are no assumptions made through software nor hardware regarding security [43]. There are no assumptions made regarding whether a certain device or user is trustworthy or not.

There is no trust in a zero-trust model and that is why it is also called a *perimeter less security model* [18]. The only way to gain access through a zero-trust model is: you are who your claim to be. The ZT model decreases the impact of trust and continuously validates every single step of any interaction within the network. Thus, the 5G positioning system will gain security through a zero-trust model by removing the trust risk on the network regardless of the process, user, location, application or data that 5G positioning is trying to access.

1.5   How Can Secure 5G Positioning System from Cyber Threats?

The use of a ZT model on the 5G positioning system is to protect and secure critical and sensitive data and applications [15]. The ZT model would actually change the way how 5G positioning system accesses users, apps and data by converting the existing solutions to a more secure, reasonable and sustainable solution, which would be more compatible with the current infrastructure, such as new cloud solutions, SDN environments and other 5G networks belonging to different service providers. This can be done by implementing the following activities on the 5G positioning system:

– Implementing or performing a segmentation on the network.
– Preventing any other outgoing data communication on the network that is not monitored or continuously validity-checked.
– Providing the highest level of layer threat prevention.
– Monitoring user access control in a more effective manner.

In summary, our review aims to discuss how to secure 5G positioning system by integrating different kinds of security strategies and solutions in order to authenticate and protect the access and communications between the user-end devices and the 5G positioning system (including the 5G core network).

The rest of this paper is structured as follows. Section 2 introduces the background on 5G and the positioning system. Section 3 introduces the background on zero trust and its main workflow. Section 4 describes how to secure 5G and the positioning system with a zero trust architecture. Section 5 provides a discussion on advantages and limitations of using an Overlay Network solution. Finally, Sect. 6 summarizes our work.

2   5G Background and Positioning System

There are still some questions when 5G is coming–why do we need 5G? What can 5G offer to us that 4G could not deliver? Is it all about streaming videos and games? As 4G with LTE is able to provide a high quality stream and online video games, what can 5G bring?

To answer these questions, 5G was announced clearly that it will bring a faster streaming speed, reduce the mobile network response time (also known as *latency*), enhance user experience with gaming and cloud applications. In addition, 5G mobile network promises the stability and speed that none of the existing mobile network predecessors 2G, 3G and even 4G could provide.

5G is a new kind of mobile network, which is mainly designed to connect everything to the Internet. The 5G technology is designed to deliver a network that can provide a higher Gbps speed than 4G LTE and the predecessors networks. It is a unified and more capable mobile network. 5G also promises that it will provide the ability to have all users including their LOT devices connected at once to the Internet and enable every one and each device to be online without the problem or the hassle of the network capacity and latency.

It is worth noting that 5G technology is still under development, i.e., 5G networks are expanding even further around the world. This next-generation wireless communication is partially enhanced by a new technology called mmWave. we summarize its pros and cons as below.

2.1 5G Pros

1. **High speed**. 5G brings faster speeds than the predecessors mobile networks, and can reach up to 10GBPS or higher.
2. **Lower Latency**. 5G has the lowest latency in all mobile network generations, e.g., about 1ms.
3. **Wide Bandwidth**. 5G can provide a wider bandwidth by expanding the usage of spectrum resources, from sub-3 GHz used in 4G to 100 GHz and beyond. This allows the connection and the support of a huge amount of devices at the same time.
4. **Smart support**. 5G mmWaves can support a narrow bandwidth (in the millimeter range), making it suitable for small cells.
5. **Compact antennas**. The 5G antennas are more compacts and smaller in size than the predecessor 4G LTE antennas.
6. **Coverage**. 5G coverage is limited not only to user-end devices that are in sight, but also to others that are not direct in sight of the 5G signal.
7. **Cloud support**. With a high speed, 5G can help business and states, even single users can easily transmit their data to the cloud.
8. **Green solution**. 5G can bring a green mobile network, with its high connection capacity and high speeds, users can move local data to the cloud, resulting in fewer laptops or desktops with less electricity towards a more green solution.

Overall, the 5G mmWave network signal can support multi-giga-bit (backhaul); that is, 5G is capable of transmitting data in high speeds because of its high bandwidth. Thus, 5G has a big future potential, due to its ability to connect a huge number of devices at the same time and the speed it provides. 5G has the ability not just to connect smart phones but also IoT devies, including AI devices, self-driving cars, AI technologies and many others.

2.2 5G Cons

1. `Slow replacement`. The pace of rolling out to all users is very slow, this means that 5G may take years and years to overtake the predecessors 4G LTE place in the mobile network communication.
2. `Availability coverage`. The 5G network coverage may be limited, e.g., within a city zone, a country-side area may have to wait some time until the 5G technology is capable to scale everywhere.
3. `Interference issue`. The 5G radio signal is weak and could be easily distracted in urban areas, because of the buildings and many obstacles like trees.
4. `Security issues`. 5G may suffer many cyber-security vulnerabilities, because some security features are still missing, e.g., the suitable encryption of transmitted data in the network, and the absence of access control system. Also it has many other security weaknesses, e.g., if there is a vulnerability found in one type of LOT devices within the network, then the whole network may be under threat.
5. `Environmental interference`. The 5G mmWave signal may become weak in bad weather, it cannot co-exists with rain.

Overall, 5G will cost more than previous predecessor to be available and spread to all users, because the amount of tower gNB needed in practical implementation. The distance coverage in 5G network when using the MMWave (high-band) is only 2 m indoors and up to 300 m outdoors. The 5G cell tower when transmitting via a low- and mid-band spectrum can only reach a coverage distance up to 600 m. By contrast, 4G LTE, depending on the area it covers, can provide a coverage from 3–6.5km in radius.

2.3 5G Positioning Security

5G positioning is a natural and necessary component in the expected coming era of 5G with various industrial use cases, e.g., logistics, smart factories, automotive [13,23]. Such system can have a big chance to handle users' data, especially private data. Due to its importance, security and privacy should be the priority [24]. Below we list a set of security and privacy concerns about the existing solutions to secure the 5G positioning system.

– What kind of user-end data can be stored? Does it include personal data like personally identifiable information (PII)? Is there any authority from the users to store that kind of data or information? Is the solution aligned with GDPR regulations according to relevant personal data and information?
– Is the solution aligned with the governmental or international regulation, regarding the collection and storage of personal data?
– Is the user privacy protected in the solution? Is the user consent regularly updated?
– Is the organization aligned with the governmental and the international regulation a) Regarding the transporting of users personal data and information in the cloud? b) Regarding the storage period or retention period? c) Who has access to the personal data that this organization stores?

3   Background on Zero Trust

Zero trust indicates a revolution regarding how an organization can protect data, connect with devices and ensure access control [19,20]. It should implement the following routine checks:

- It must be applied to every single user and each device including security and network administrators.
- The network should be continuously monitored and audited. The audit should include: a) Users' or devices' identities should continuously be checked. b) Access control should be continuously updated. c) Access polices should continuously be checked and updated.
- It should continuously update access protocols, by monitoring the network and gathering audit data in order to understand the users' and devices' behaviour.
- The zero-trust related solutions should have a management tool (database) that contains both users and devices, which have access to the network. Every user and device should be mentioned in this management tool (database) and there should also be mentioned to which application the user/device may have access and how long.

The management tool should also be able to update the access policy of the company and use this policy to enforce user access within the network [47], e.g., how the access is used, when and what for. To sum up, all users' and devices' roles in the database should continuously be checked and updated. Figure 2 depicts the workflow of a zero-trust strategy.

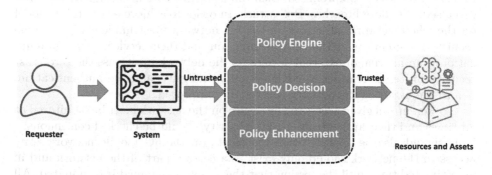

Fig. 2. Workflow of zero-trust solutions

In practice, a Zero Trust Security Model should consider the following factors:

- **User authentication.** To examine the identity of users, which is the core aspect of zero-trust mechanisms.

- Data integrity. To control the integrity of the device (without location consideration), as long as the device is authenticated (e.g., based on device identity, device health, and user authentication).
- Trustworthiness. The reputation about the device and data should be measured before any access is given. This can be achieved by monitoring and examining the health signal from a device or user, which at the end will help in gaining confidence about the user or device (e.g., this information can be saved in the policy engine in order to make access decision).

4   Securing 5G Positioning with Zero Trust Architecture

There are two main principles of zero trust architecture (ZTA): First, it has to protect all the services and resources at any time and at any cost. Second, ZTA always assumes that an attacker existing in the 5G network environment [22].

The 5G core network has a service based architecture (SBA), the provided services are a main part of 5G core network functions, where each of them supports a specific service included in the network repository function (NRF). Every network function service in the 5G core is an asset that needs to be protected. Hence, identifying these services (e.g., critical network resources) is the first step of implementing a ZTA on a 5G positioning system [21].

This means that if we want to achieve a zero-trust security model on 5G positioning system, then there is a need to protect the 5G core network functions and any services that it runs and meet the other 5G security requirements.

One of the best ways to protect the network function is to implement authentication and authorization processes on the network operators, which is a service provider that controls the infrastructure of the mobile network. If any network function is down or compromised, then the whole 5G core network may get compromised, e.g., data breaches, DoS attack. In order to achieve a zero-trust model on the whole network, all assets on the 5G network must implement the same security requirements as the network function and their services. All communication from internal or external devices on the network must pass the zero-trust security processes relating to confidentiality, integrity, and the authentication, before the access to the network is given and approved.

To sum up, all given access to any device on the network must be monitored at all times and their authentication and integrity should be checked continuously. This should also be applied to all the assets on the 5G mobile network. Any access to the network should only exist to a certain part of the network and in an estimated time until the session that the access was gained is terminated. All authentication and authorization processes and checks should be forced on all assets before the access is gained.

4.1   Implementation of Zero Trust

There are different ways to implement a zero-trust model on the 5G core network in order to secure the 5G positioning system, depending on different aspects, e.g., the organization security policy [18].

Nevertheless, an organization implementing a ZTA on its 5G mobile network should have the following systems in place:

- An ICAM system (Identity, Credential and Access Management system).
- An ASSETS management system.
- An Multifactor authentication (MFA) system for accessing the resources/ assets.
- A System for monitoring user authentication and authorization continuously on the mobile network and checking if the authentication or authorization steps are successful.

Below is an example of implementing ZTA on the 5G core network to secure the transport layer: that is, to secure the TLS protocol in the SBI (Service based Interface) protection, we can build an API-based communication software. This API software is used to establish communication between 5G Network Functions within the 5G SBA (Service Based Architecture) framework that lies inside the 5G Core. Here, the 5G architecture is built around an SBA. The services of an SBA are delivered to the 5G core network by connecting the network function through APIs and using the authentication security principle to allow network functions to connect with each other. By using this method, we have to protect the 5G core network by ensuring integrity and confidentiality at the same time. This solution works via an authentication method during the TLS session with the help of certificates. This can be done through a software authentication token.

4.2   Recommendations and Principles

We also need to address what it needs in order to secure the 5G positioning system, including the 5G core network, by considering the zero trust principles as below.

- To check the 5G security requirements of the network when implementing the zero-trust strategy.
- To continuously check if there are new technical enhancements for zero-trust strategy.
- To build a procedural security-enhanced program on the 5G network, in order to always be up-to-date with the latest threat information.
- A monitoring solution should be built in order to continuously check the trustworthiness, e.g., whether the user is authenticated and authorized to exchange data and gain access to the 5G network.
- To continuously check the network functions' activities to determine whether it has been compromised or not.
- To implement the ZTA on the SBA within the core network to ensure a zero trust security model.

4.3   An Overlay Network Solution

Securing 5G positioning with Zero Trust Network Access (ZTNA) could be depended on the vendors' network architecture and enterprise use cases. Technical details for implementing a ZTNA strategy on 5G positioning system may vary from one network to another, it depends on the specific use case and the network architecture. ZTNA is a security model based on the standard or strategy Zero trust. The ZTNA used in our case will verify each access request to the network, regardless whether the request comes from internal or external. Implementing ZTNA on 5G positioning involves integrating different security solutions. In order to achieve this, it requires securing an access point to the 5G network by using ZTNA to enforce security policies in order to limit the legitimate access to only identified, verified, authenticated and authorized users and devices.

One approach of implementing such a solution is to use an overlay network that can be built on top of the 5G infrastructure and help to provide an extra layer of security by granting access to only authenticated and authorized users and devices. An overlay network can be used here to provide an additional layer of security by creating a secure and isolated virtual network on top of the 5G mobile network. The Overlay Network will play a role as a military check, where it will control the users' and devices' identities in order to grant access. When implementing this kind of checks, the solution will ensure that only authenticated & authorized users and devices can be granted a secure access to the network. The goal of using an overlay network is to add a new layer without the need to redesign the whole network. These additional missing functionalities added to the overlay network are normally linked to the existing network through virtual or physical nodes.

An example for an overlay network can be the Internet, VPN, cloud computing and P2P networks. Administrators monitor and handle the network traffic in the overlay network without disturbing the traffic on the existing physical network. For instance, an overlay network can consist of an SDN (software-defined networking) technology [25,26] and virtualization. The SDN here can allow the creation of a virtual network segments that can be used to isolate traffic and provide security checks, authentication and authorization. Theses virtual network segments are normally created by using the tunnels. These tunnels are encrypted connections between the network and devices that allow the communication traffic to pass securely over the underlying network infrastructure.

To summarize, the 5G positioning system will totally reside on the network infrastructure to determine the exact positioning of the user equipment/device on the network. When a user-end device connects to the 5G network, it will then send multi-signals to nearby cell towers (gNB), which can then determine the approximate position or location of the user-end device in the 5G network, based on the signal strength and timing calculations. ZTNA will play a security-guard role in protecting the connections between the user-end devices and the 5G core network by verifying the identity of the devices and ensuring that only authorized users and devices can access the network. Once the user-end device is connected to the 5G network and its location has been determined, the network can then provide location-based services (LBS) such as navigation and tracking.

4.4   Implementation of Overlay Network Based 5G

Figure 3 shows the high-level architecture of Overlay Network based 5G system, including a trust controller and a policy engine. These two components can enhance the adopted zero-trust strategies with various supports from Overlay Network. Below are the key steps for implementing an Overlay Network.

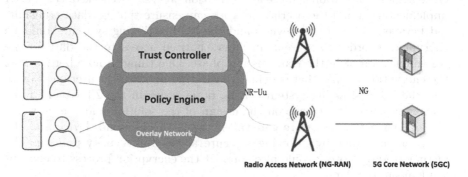

Fig. 3. 5G with Overlay Network

- First there is a need to build an Overlay Network by using the SDN and virtualization technologies, which will perform as a virtual network running on top of the existing physical networks [29,30]. It aims to secure communication and data transferring in the network. The virtual network segment will cover the entire network, not just at the boundary and access point to the network, e.g., perimeter and switches will manage the virtual LAN environment. SDN can use the applications (SDN controllers) [27,28] or APIs (application programming interfaces) to communicate with the underlying hardware network (the physical network under the overlay network) in order to direct and manage the traffic on the network.
- The tunnels are then used to create an encrypted communication connection between the user-end device and the virtual network segmentation. The next step is to implement the ZTNA model/strategy that the enterprise has selected on the virtual network segment in order to enforce access control policies on every network access. When implementing the zero-trust strategy, the enterprise has the chance to configure their access polices and authentication strategies. The ZTNA could be included in the 5G core network, in the radio access network (RAN) and finally in the user-equipment device.
- The ZTNA strategy that is implemented on the virtual network (and the physical) will enforce the authentication process to every user and device

before accessing the 5G positioning system (5G core network). The authentication could be done in different ways and methods such as using certificates or even by using biometric methods.

– Once the authentication process is preformed, the user and devices are authenticated to access the 5G positioning system, they then need to go through the next security step–authorization process. In the authorization process the users and devices should check their authorization in order to ensure only authorized users and devices have access to the 5G positioning system. For the authorization process, the enterprise needs to define and configure its authorization access polices.

– Once the authentication and the authorization process are checked, the ZTNA implementation and the overlay network can ensure that all data transmitted between the user-end devices and the 5G positioning system should be encrypted in order to prevent any unauthorized access to the data in the communication or in the transmission phase. One thing to note here is that the encryption of data that is being transmitted between the user-end device and the 5G positioning system can be done either in the ZTNA or in the overlay network. It depends on the design of the solution and the network. The ZTNA acts as a secure gateway between the user-end device and the 5G position system by enforcing a secure access and security policies. The virtual overlay network could also conduct the encryption process to encrypt the transmitted data.

– The last step needed, in order to have a secure 5G positioning system and a secure 5G mobile network from unauthorized users or devices, is to have a continuous monitoring system in place. This continuous monitoring should also include detecting and responding to any potential cyber-security threats or breaches that can occur on the whole network. There are different tools and systems that could be used to monitor and analyze the data traffic, such as intrusion detection systems (IDSs) [31,32], collaborative mechanisms [37,39], and security information and event management (SIEM) systems.

It is worth noting that the continuous monitoring and analyzing process on the network traffic can be performed by using different security solutions [33,34] such as firewalls, IPS, IDS and SIEM systems. All these tools can be implemented at a strategic point in the overlay network. By implementing this security solution or method, the virtual overlay network will then help implement the ZTNA security strategy in order to secure the 5G positioning system. In practice, the implementation can leverage various security-enhancement techniques such as cloud and edge computing devices [35,40], filtering mechanisms [36,39] and blockchain technique [38,41,42].

There is a need to clarify that the virtual overlay network is built on top of the ZTNA solution rather than the other way around. One important point that needs to be explained and clarified here is that the ZTNA and the overlay network have different roles and functions when being combined together to secure the 5G positioning system from unauthorized access and malicious attacks. The ZTNA is a security strategy that provides a secure access mechanism to the 5G positioning system by performing the authentication and authorization process on the user and devices before they gain access to the 5G positioning system. On the other hand, the overlay network is a virtual network that is build on top of the existing network to provide additional services and functionalities for the existing network without the need to redesign the whole network.

5 Discussion

Using the overlay network solution can be an effective method to secure the 5G positioning system, but it may also encounter some limitations. Below are some pros and cons when using the overlay network virtualization.

5.1 Pros

- One of the most important advantages of using an overlay network virtualization on the 5G network (including the 5G positioning system) is the capability of offering an additional (extra) layer of security on top of the existing network infrastructure.
- The overlay network is very easy to implement since it is a virtualization solution that does not need any hardware component.
- The overlay network can be implemented without requiring a redesign of the whole existing network infrastructure. This is because it adds additional services and functionalities to the existing network.
- The overlay network is a very flexible solution that can be easily customized and configured. This is an important point since 5G network is a continuously changing environment and its configuration in the network changes from time to time depending on what is running and implemented on the 5G network.
- The overlay network can provide improvements regarding the high data-transferring speeds and reduce latency of data exchange.

5.2 Cons

- The overlay network is an extendable solution, so it may face problems when the network scale gets larger. That means it has a scalability issue, which may become an unsuitable and unstable solution for large networks.
- The overlay network is built on top of the underlying infrastructure, so any performance and security issues from the underlying networks will make a huge impact on the overlay network.
- The overlay network in some scenarios can have a big impact on performance, i.e., it may increase data-transfer latency due to communication delay, as it is an additional layer that is built on top of the existing network infrastructure.
- The overlay network is very complex to implement and needs very experienced experts to perform implementation and configuration.

Therefore, a well-secured 5G mobile network should be established with the following three main factors:

– The first factor includes: a well-defined user awareness program,
– The second factor includes: a well-designed security program for user-equipment devices
– The third factor includes: a well-defined 5G core security policy and program.

6   Conclusion

5G is the 5th generation mobile network, which is designed to connect virtually everyone and everything together including machines, objects, and devices. 5G also leverages time of flight and angular resolution to bring multiple positioning techniques for different deployment scenarios and use-cases. However, 5G positioning system is still under threat, and there is a need to protect its practical implementation. In this work, we provided a review on 5G technology and 5G positioning system, especially discussing how zero-trust concept can help secure 5G network and the positioning system. We also discussed the advantages and limitations of one effective solution of using Overlay Networks.

References

1. Oruma, S.O., Petrovic, S.: Security threats to 5G networks for social robots in public spaces: a survey. IEEE Access **11**, 63205–63237 (2023)
2. Boualouache, A., Engel, T.: A survey on machine learning-based misbehavior detection systems for 5G and beyond vehicular networks. IEEE Commun. Surv. Tutorials **25**(2), 1128–1172 (2023)
3. Banda, L., Mzyece, M., Mekuria, F.: 5G business models for mobile network operators - a survey. IEEE Access **10**, 94851–94886 (2022)
4. The impact of 5G on location technology: whats real and whats hype? https://www.pointr.tech/blog/5G-indoor-positioning
5. 5G security - enabling a trustworthy 5G system. https://www.ericsson.com/en/reports-and-papers/white-papers/5G-security--enabling-a-trustworthy-5G-system
6. NSA - CSI : Embracing a Zero Trust Security Model: https://media.defense.gov/2021/Feb/25/2002588479/-1/-1/0/CSI_EMBRACING_ZT_SECURITY_MODEL_UOO115131-21.PDF
7. Sullivan, S., Brighente, A., Kumar, S., Conti, M.: 5G security challenges and solutions: a review by OSI layers. IEEE Access **9**, 116294–116314 (2021)
8. Tezergil, B., Onur, E.: Wireless backhaul in 5G and beyond: issues, challenges and opportunities. IEEE Commun. Surv. Tutorials **24**(4), 2579–2632 (2022)
9. Shahzadi, R., Ali, M., Khan, H.Z., Naeem, M.: UAV assisted 5G and beyond wireless networks: a survey. J. Netw. Comput. Appl. **189**, 103114 (2021)
10. Chettri, L., Bera, R.: A comprehensive survey on Internet of Things (IoT) toward 5G wireless systems. IEEE Internet Things J. **7**(1), 16–32 (2020)
11. Sharma, S.K., Bogale, T.E., Le, L.B., Chatzinotas, S., Wang, X., Ottersten, B.E.: Dynamic spectrum sharing in 5G wireless networks with full-duplex technology: recent advances and research challenges. IEEE Commun. Surv. Tutorials **20**(1), 674–707 (2018)

12. Agiwal, M., Roy, A., Saxena, N.: Next generation 5G wireless networks: a comprehensive survey. IEEE Commun. Surv. Tutorials **18**(3), 1617–1655 (2016)
13. Bodi, B., Chiu, W.Y., Meng, W.: Towards blockchain-enabled intrusion detection for vehicular navigation map system. ISPEC **2022**, 3–20 (2022)
14. Wu, Y., Khisti, A., Xiao, C., Caire, G., Wong, K.K., Gao, X.: A survey of physical layer security techniques for 5G wireless networks and challenges ahead. IEEE J. Sel. Areas Commun. **36**(4), 679–695 (2018)
15. Syed, N.F., Shah, S.W., Shaghaghi, A., Anwar, A., Baig, Z.A., Doss, R.: Zero trust architecture (ZTA): a comprehensive survey. IEEE Access **10**, 57143–57179 (2022)
16. Chen, B., et al.: A security awareness and protection system for 5G smart healthcare based on zero-trust architecture. IEEE Internet Things J. **8**(13), 10248–10263 (2021)
17. Ge, Y., Zhu, Q.: MUFAZA: Multi-Source Fast and Autonomous Zero-Trust Authentication for 5G Networks. MILCOM 2022: 571–576
18. NIST Special Publication 800–207- Zero Trust Architecture https://nvlpubs.nist.gov/nistpubs/SpecialPublications/NIST.SP.800-207.pdf
19. Bello, Y., Hussein, A.R., Ulema, M., Koilpillai, J.: On sustained zero trust conceptualization security for mobile core networks in 5G and beyond. IEEE Trans. Netw. Serv. Manag. **19**(2), 1876–1889 (2022)
20. Hireche, O., Benzaïd, C., Taleb, T.: Deep data plane programming and AI for zero-trust self-driven networking in beyond 5G. Comput. Netw. **203**, 108668 (2022)
21. Li, Y., Liu, S., Yan, Z., Deng, R.H.: Secure 5G positioning with truth discovery, attack detection, and tracing. IEEE Internet Things J. **9**(22), 22220–22229 (2022)
22. Liu, S., Yan, Z.: Efficient privacy protection protocols for 5G-enabled positioning in industrial IoT. IEEE Internet Things J. **9**(19), 18527–18538 (2022)
23. Fan, S., Ni, W., Tian, H., Huang, Z., Zeng, R.: Carrier phase-based synchronization and high-accuracy positioning in 5G new radio cellular networks. IEEE Trans. Commun. **70**(1), 564–577 (2022)
24. Bai, L., Sun, C., Dempster, A.G., Zhao, H., Cheong, J.W., Feng, W.: GNSS-5G hybrid positioning based on multi-rate measurements fusion and proactive measurement uncertainty prediction. IEEE Trans. Instrum. Meas. **71**, 1–15 (2022)
25. Li, W., Wang, Y., Meng, W., Li, J., Su, C.: BlockCSDN: towards blockchain-based collaborative intrusion detection in software defined networking. IEICE Trans. Inf. Syst. **105-D**(2), 272–279 (2022)
26. Meng, W., Li, W., Zhou, J.: Enhancing the security of blockchain-based software defined networking through trust-based traffic fusion and filtration. Inf. Fusion **70**, 60–71 (2021)
27. Li, W., Meng, W., Liu, Z.G., Au, M.H.: Towards blockchain-based software-defined networking: security challenges and solutions. IEICE Trans. Inf. Syst. **103-D**(2), 196–203 (2020)
28. Sahay, R., Meng, W., Jensen, C.D.: The application of software defined networking on securing computer networks: a survey. J. Netw. Comput. Appl. **131**, 89–108 (2019)
29. Meng, W., Choo, K.K.R., Furnell, S., Vasilakos, A.V., Probst, C.W.: Towards bayesian-based trust management for insider attacks in healthcare software-defined networks. IEEE Trans. Netw. Serv. Manag. **15**(2), 761–773 (2018)
30. Li, W., Meng, W., Kwok, L.F.: A survey on OpenFlow-based software defined networks: security challenges and countermeasures. J. Netw. Comput. Appl. **68**, 126–139 (2016)

31. Meng, W., Li, W., Kwok, L.F.: EFM: enhancing the performance of signature-based network intrusion detection systems using enhanced filter mechanism. Comput. Secur. **43**, 189–204 (2014)

32. Meng, W., Luo, X., Li, W., Li, Y.: Design and evaluation of advanced collusion attacks on collaborative intrusion detection networks in practice. In: Trustcom/BigDataSE/ISPA, pp. 1061–1068 (2016)

33. Li, W., Meng, W.: Enhancing collaborative intrusion detection networks using intrusion sensitivity in detecting pollution attacks. Inf. Comput. Secur. **24**(3), 265–276 (2016)

34. Li, W., Meng, W., Kwok, L.F.: Horace Ho-Shing Ip: enhancing collaborative intrusion detection networks against insider attacks using supervised intrusion sensitivity-based trust management model. J. Netw. Comput. Appl. **77**, 135–145 (2017)

35. Meng, W., Wang, Y., Li, W., Liu, Z., Li, J., Probst, C.W.: Enhancing intelligent alarm reduction for distributed intrusion detection systems via edge computing. In: ACISP, pp. 759–767 (2018)

36. Meng, W., Li, W., Kwok, L.F.: Towards effective trust-based packet filtering in collaborative network environments. IEEE Trans. Netw. Serv. Manag. **14**(1), 233–245 (2017)

37. Meng, W., Li, W., Jiang, L., Choo, K.-K.R., Su, C.: Practical bayesian poisoning attacks on challenge-based collaborative intrusion detection networks. In: Sako, K., Schneider, S., Ryan, P.Y.A. (eds.) ESORICS 2019. LNCS, vol. 11735, pp. 493–511. Springer, Cham (2019). https://doi.org/10.1007/978-3-030-29959-0_24

38. Li, W., Tug, S., Meng, W., Wang, Y.: Designing collaborative blockchained signature-based intrusion detection in IoT environments. Future Gener. Comput. Syst. **96**, 481–489 (2019)

39. Li, W., Meng, W., Kwok, L.F., Ip, H.H.S.: Developing advanced fingerprint attacks on challenge-based collaborative intrusion detection networks. Clust. Comput. **21**(1), 299–310 (2018)

40. Wang, Y., Meng, W., Li, W., Li, J., Liu, W.X., Xiang, Y.: A fog-based privacy-preserving approach for distributed signature-based intrusion detection. J. Parallel Distributed Comput. **122**, 26–35 (2018)

41. Chiu, W.Y., Meng, W., Li, W., Fang, L.: FolketID: a decentralized blockchain-based NemID alternative against DDoS attacks. In: ProvSec, pp. 210–227 (2022)

42. Chiu, W.Y., Meng, W., Jensen, C.D.: My data, my control: a secure data sharing and access scheme over blockchain. J. Inf. Secur. Appl. **63**, 103020 (2021)

43. Li, W.W., Meng, W., Yeh, K.H., Cha, S.C.: Trusting computing as a service for blockchain applications. IEEE Internet Things J. **10**(13), 11326–11342 (2023)

44. Li, W., Meng, W., Kwok, L.F.: Surveying trust-based collaborative intrusion detection: state-of-the-art, challenges and future directions. IEEE Commun. Surv. Tutorials **24**(1), 280–305 (2022)

45. Li, W., Meng, W.: BCTrustFrame: enhancing trust management via blockchain and IPFS in 6G Era. IEEE Netw. **36**(4), 120–125 (2022)

46. Li, W., Meng, W., Yang, L.T.: Enhancing trust-based medical smartphone networks via blockchain-based traffic sampling. In: TrustCom, pp. 122–129 (2021)

47. Meng, W., Li, W., Zhu, L.: Enhancing medical smartphone networks via blockchain-based trust management against insider attacks. IEEE Trans. Eng. Manag. **67**(4), 1377–1386 (2020)

Email Reading Behavior-Informed Machine Learning Model to Predict Phishing Susceptibility

Ning Xu[1,2], Jiluan Fan[1,2], and Zikai Wen[3(✉)]

[1] Instituite of Artificial Intelligence, Guangzhou University, Guangzhou, China
{xuning,fanjiluan}@e.gzhu.edu.cn
[2] Guangdong Provincial Key Laboratory of Blockchain Security, Guangzhou, China
[3] Computational Media and Arts Thrust, The Hong Kong University of Science and Technology (Guangzhou), Guangzhou, China
zikaiwen@ust.hk

Abstract. As phishing threats intensify, incidents like the "COVID-19 vaccination form" phishing website underscore the limitations of relying solely on traditional firewall-based defenses. Consequently, there is a growing inclination towards user-centered anti-phishing solutions, exemplified by training games such as *What.Hack*. But could we proactively notify users in real time when they are on the brink of a scam or when their attention wanes? Our research explores machine learning and eye-tracking to identify email-reading weak spots and gauge a user's risk of succumbing to phishing lures. We put forth innovative hybrid models, *TransMLP Link* and *TransMLP Hybrid*, melding the strengths of both Transformer and MLP. Our method also facilitates consistent interpretation of eye-tracking data across varied email interfaces and displays. Our *TransMLP Hybrid* model boasts an 88.75% accuracy rate, outperforming the standard Transformer model. Our research points to the future of anti-phishing tools that elegantly combine technological advancements with insights into human behavior.

Keywords: Anti-Phishing · User Modeling · Machine Learning

1 Introduction

Phishing is a growing concern in the digital age. It involves seemingly genuine emails, messages, and links that trick users into revealing personal data or downloading harmful software. The rise of such attacks, especially those leveraging pandemic themes, has been alarming [2,9]. An infamous example is the fake "COVID-19 Vaccination Form" site that falsely posed as an official NHS platform, leading users into fraudulent vaccine registrations [19].

Traditionally, firewalls have been used to combat phishing by maintaining updated blocklists and allowlists [7,13,14]. However, they struggled to

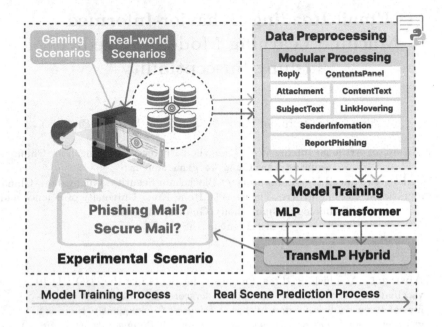

Fig. 1. Overview of the Model Training and Real-World Prediction Process for Phishing Email Detection.

counter new domains that are not yet listed [21]. As a solution, recent methods stressed the importance of educating users [3,24,30]. Training platforms like *What.Hack* [30] have sprung up to strengthen this first line of defense. Nonetheless, an unsettling 95% of phishing breaches result from human oversights [1]. This brings forth a question: Can we alert users in real time if they are about to fall for a scam or if their attention drifts?

To tackle this problem, we employed machine learning and eye-tracking techniques to analyze how users engage with emails, aiming to predict their vulnerability to phishing. Our research delved into the Transformer model, assessing its potential to gauge user focus; the Multilayer Perceptron (MLP) model, fine-tuned for eye-tracking data; and innovative hybrid models, *TransMLP Link* and *TransMLP Hybrid*, blending the best of both Transformer and MLP.

Moreover, we developed a technique to consistently interpret eye-tracking data across various devices and email applications, associating specific gaze points with their meaning in the email's layout. This approach translated the raw eye-tracking data into eight key areas reflecting the main regions of an email interface. A detailed overview of our approach, from data collection to applying our hybrid model in real scenarios, can be found in Fig. 1.

In our experiment with 25 participants, we gathered eye-gazing patterns and user interactions while they interacted with genuine and phishing emails and played the *What.Hack* anti-phishing game. We utilized the in-game data to train our phishing prediction models and the real-world data for testing. While the Transformer model delivered an 80.63% accuracy, the *TransMLP Hybrid* model stood out by achieving an impressive 88.75% accuracy rate.

In essence, our contributions are threefold:

1. The new design of hybrid models, *TransMLP Link* and *TransMLP Hybrid*, synergizing Transformer and MLP.
2. The new approach to uniformly interpret eye-tracking data across diverse email reading environments.
3. The experiment showcased the outstanding performance of *TransMLP Hybrid* with an 88.75% accuracy.

In the evolving landscape of anti-phishing, the dual challenges of innovative phishing tactics and human vulnerabilities necessitate more comprehensive defense strategies. This paper studies the intricate relationship between email reading behaviors, eye-tracking data, and their potential to inform machine learning models that predict phishing susceptibility.

We begin by examining the historical context of phishing attacks and the defense mechanisms in place, laying the groundwork for our innovative approach. Subsequently, we elucidate our machine learning models, emphasizing the novel integration of Transformer and MLP architectures. Following this, we detail our designed experiment, setting the stage for a thorough analysis of our results and their broader implications. By evaluating the effectiveness of our models and examining the underlying factors, we present a feasible strategy that combines advanced technological methods with deep insights into human behavior, paving the way for a significantly enhanced anti-phishing defense.

2   Related Work

The related work section explores phishing tactics, human vulnerabilities, and defense strategies designed to counteract these threats. The limitations of existing anti-phishing strategies led us to study the ability to leverage email reading eye-tracking data to train machine learning models to predict phishing susceptibility more effectively.

2.1   Phishing Email Attacks and Defense

Phishing is a cyber-attack where attackers pose as trustworthy entities to steal credentials or introduce malware. Research has identified three primary human vulnerabilities in defending against phishing attacks: a lack of system and security knowledge [4], challenges in detecting visual deception [10], and inattention [20]. For example, phishing emails often employ deceptive hyperlinks and subtle cues, such as spelling mistakes, to mislead users [12].

To address these vulnerabilities, a range of strategies has been developed to counteract phishing due to user negligence. These include anti-phishing training, active warning systems, and detection techniques using machine learning [8, 16, 24, 25, 27, 30]. Role-playing phishing simulation games [24, 30] aim to increase users' security knowledge and awareness, and alert mechanisms were designed to notify users of potential threats [16]. Therefore, modeling how humans recognize phishing emails and implementing protective measures are crucial in preventing successful breaches.

Machine learning models are a mainstay in the detection of phishing emails [23]. For instance, Shie et al. [25] utilized deep learning and feature extraction to identify phishing emails. Additionally, Subasi et al. [27] assessed Adaboost and other boosting algorithms for detecting phishing websites, leveraging a dataset from the UCI repository to improve classifier accuracy. Despite these advances, even the most sophisticated machine learning model occasionally misses phishing threats. Thus, creating automated detection methods for phishing risks when users access their emails could provide an added layer of protection, significantly reducing the chances of successful attacks.

2.2 Eye-Tracking for User Intention Prediction

The Eye-Mind Hypothesis (EMH) suggests that during a task, an individual's focal point and cognitive thought are intrinsically linked — what they see often mirrors what they think [18]. In this context, eye-tracking data becomes pivotal in decoding visual attention and cognitive operations. With this premise, we postulate that specific eye-tracking patterns might be indicative of an individual's vulnerability to phishing emails.

Recent research in intent recognition through eye-tracking [5,15,17,29] predominantly revolves around predicting the location or object of a user's attention. A research direction in this area aims to forecast subsequent attentional shifts of users [17,29]. For instance, leveraging eye movement patterns from VR goggles, Nicolas et al. [26] developed a model to predict users' upcoming focal points. Deng et al. [11] utilized logistic regression to project user menu selections. Bhattacharya et al. [6] took a step further to investigate if readers' eye movements alone could gauge the authenticity of news headlines. Despite these advancements, such models remain unable to assess user susceptibility to phishing endeavors.

In a parallel development, Huang et al. [16] designed an array of visual cues to deter phishing, aiding users in distinguishing malicious emails from legitimate ones. However, the trigger for these alerts rests upon conclusive firewall detections. If a firewall deems an email safe, no alert is generated. This underscores an opportunity: if we can determine a user's lack of attentiveness while reading a phishing email, a timely alert could also be triggered.

3   Prediction Models Design

The section explores machine learning models for analyzing email reading behaviors using eye-tracking data. We start with the Transformer model, detailing its mechanics and applications in understanding user attentiveness. We then discuss the Multilayer Perceptron (MLP) model and how to make it process eye-tracking statistics. Finally, we introduce two new variant models, combining the best of both Transformer and MLP, to better predict phishing email susceptibility.

3.1 Transformer Model

Model Background: The Transformer model, proposed by Vaswani et al. [28], addresses the performance bottlenecks of recurrent neural networks in processing long data sequences. It comprises an encoder and a decoder, both stackable with multiple layers that comprise the self-attention layer and the feed-forward layer.

While the self-attention mechanism of the Transformer model processes data, it does not inherently consider the order of the input sequence. To enable sequence processing, positional encoding (PE) is necessary. The formula for positional encoding is:

$$PE_{(pos,2i)} = \sin(pos/10000^{2i/d}), \ PE_{(pos,2i+1)} = \cos(pos/10000^{2i/d}), \quad (1)$$

where d denotes the embedding vector's dimension, pos signifies the position in the data processing sequence, and $i \in [0, d]$ represents the dimensions of the positional encoding vector. $2i$ and $2i + 1$ designate the even and odd dimensions of the positional embedding vector respectively.

The Transformer model may employ an h multi-head attention mechanism to capture richer feature information, which is essential for our application's purpose. Within the multi-head self-attention layer, the input vector undergoes three linear transformations to obtain the query vector Q, key vector K, and value vector V. The formula for multi-head attention computation is:

$$\text{MultiHead}(Q, K, V) = \text{Concatenation}(\text{head}_1, \ldots, \text{head}_h)W^O, \quad (2)$$

where each head_i represents the output vector of the i-th attention head, and W^O is a linear transformation matrix. The formula for each head is:

$$\text{head}_i = \text{Attention}(QW_i^Q, KW_i^K, VW_i^V), \quad (3)$$

with the matrices W_i^Q, W_i^K, and W_i^V being linear transformations. The dimension of each head helps define the scaled dot-product attention:

$$\text{Attention}(Q, K, V) = \text{softmax}\left(\frac{QK^T}{\sqrt{d/h}}\right) V. \quad (4)$$

The feed-forward network within the Transformer model is a two-layer neural network, which employs residual connections [17] or layer normalization [5] to facilitate model convergence and prevent gradient disappearance or explosion.

Model Implementation: We trained a series of Transformer-based models using temporal features to perform binary classification on email reading behavior. The aim is to determine whether users are careless about verifying the authenticity of emails.

The eye-tracking data for each user and email serves as a sequence input to the Transformer model. The eye-tracking data are chronologically organized into sequences according to the user's history of processing emails.

To provide consistency in interpreting eye-tracking data, regardless of the screen size or email application in use, we developed a method to map location points from the eye-tracking data to their respective semantic meanings. The transformed data comprises eight spatial attributes, specifically: *Sender-Information*, *SubjectText*, *Reply*, *ReportPhishing*, *ContentText*, *ContentsPanel*, *Attachment*, and *LinkHovering*. These attributes align with the core email functions' UI regions, as depicted in Fig. 2. Furthermore, our model incorporates a temporal feature. Each feature vector captures the needed eye-tracking information during each time step.

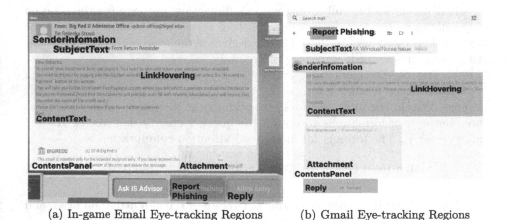

(a) In-game Email Eye-tracking Regions   (b) Gmail Eye-tracking Regions

Fig. 2. Eye-tracking Mapping for Email Interaction Zones across Two Different Email Application Interfaces.

Our Transformer encoder consists of two blocks, each containing one multi-head self-attention layer and one feed-forward layer. Within the self-attention layer, the input vector is divided into three segments, each of which undergoes a linear transformation. Subsequently, these transformed segments are subjected to scaled dot-product attention calculations. The resulting output vectors from each head are combined and processed through a linear transformation matrix to produce the final output of the self-attention layer as the input of the feed-forward layer. The feed-forward layer includes two linear layers with a *ReLU* activation function in between them. After the input undergoes transformation by a fully connected layer, the activation function provides a nonlinear transformation. A subsequent fully connected layer further modifies the output, producing a tensor that maintains the input's dimensions. Within each encoder block, the input is processed by both the self-attention mechanism and the feed-forward network, with the output being reintegrated with the original input through a residual connection.

3.2   Multilayer Perceptron Model

Model Background: The Multilayer Perceptron (MLP) model [22] employs multiple layers of neurons to enact nonlinear transformations, facilitating the extraction of higher-level features from input data. An MLP is composed of input, hidden, and output layers. Each layer houses multiple neurons, and each neuron processes the output from the preceding layer. The calculations performed by a neuron involve both a linear transformation that weights the output of the previous layer by the neuron's own weights and a subsequent nonlinear transformation via an activation function. This combination generates the neuron's final output. The computational formulation for MLP is given by:

$$r = f(W^{(L)}f(W^{(L-1)}f(W^{(L-2)}...f(W^{(1)}x+B^{(1)})...+B^{(L-2)})+B^{(L-1)})+B^{(L)}), \quad (5)$$

in this equation, f denotes the activation function, x is the input data, and $W^{(i)}$ and $B^{(i)}$ symbolize the weights and biases for the i-th layer, respectively. Generally, the terminal layer of the MLP model uses the *sigmoid* function to transform the previous network's output into two probability values, and the model picks the higher probability value as the final output.

Model Implementation: We employed an MLP model comprising six fully connected layers, using *ReLU* as the activation function and incorporating a dropout method to combat overfitting. The input to this model is derived from eye-tracking data, which we processed into 16 statistical features. These features come from eight previously identified spatial features related to the UI areas of core email functions. For each spatial feature, we calculated two values: the count and the total duration of user fixations. This data was then flattened into a one-dimensional vector. The model produces an output in the form of a probability value, representing the likelihood of a sample being a phishing email that successfully deceives the recipient.

For the training phase, we opted for the Adam optimizer over the stochastic gradient descent algorithm, enabling faster convergence and allowing distinct learning rates for individual parameters. Furthermore, we set the learning rate of each parameter group using a cosine annealing schedule to dynamically modify the learning rate, progressively decreasing it throughout training for better control and stability.

For the loss function, we used the binary cross-entropy for better training stability. This function is mathematically represented as:

$$Loss(y,p) = -(ylog(p) + (1-y)log(1-p)), \quad (6)$$

in this equation, y represents the ground truth, indicating if the user failed to recognize the deceptive phishing email. Meanwhile, $p \in [0,1]$ denotes the model's predicted probability that we aim to align with the y value.

3.3   TransMLP Model Variants Design

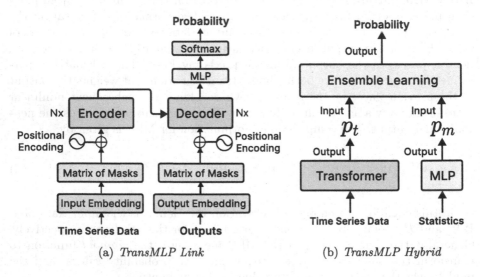

Fig. 3. Architectures of Two Proposed TransMLP Variants.

TransMLP Link: The *TransMLP Link* model is a new variant that diverges from the traditional Transformer model. The model integrates a multi-layer transformer encoder to enhance input data feature extraction, as shown in Fig. 3(a). In addition to this integration, the model employs an MLP model in lieu of the standard single fully connected layer to facilitate nonlinear transformations on the Transformer output. This design choice enables the *TransMLP Link* to achieve better nonlinear modeling capabilities relative to ML models that rely solely on a single fully connected layer.

TransMLP Hybrid: Rather than merely linking the output of the Transformer directly to the MLP's input, we designed the *TransMLP Hybrid* model to harmoniously integrate the strengths of both Transformer and MLP paradigms, as shown in Fig. 3(b). This model harnesses eye-tracking statistical features derived from time series data to train the MLP component. To produce the final output, an ensemble learning strategy is employed, judiciously weighing the predictions from both the Transformer and MLP models to optimize performance.

4   Experiment Design

Our experiment centered on leveraging eye-tracking data to enhance the ability of machine learning models to assess the phishing risk of an email as read by a user. We also aimed to evaluate the two Transformer model variants that we proposed, *TransMLP Link* and *TransMLP Hybrid*, comparing their performance to the basic Transformer model.

4.1 Participant Recruitment

We collected eye-tracking and user interaction data from 25 participants (10 females and 15 males) for model training and testing. Participants were recruited via social media and snowball sampling. The participants need to be over 18 years old, have no prior training in anti-phishing, and be affiliated with the first author's institution, which was targeted by the collected real phishing emails.

4.2 Experimental Method

First, we gathered data from participants in real-world application settings to assess their reactions to phishing and legitimate emails while using their everyday email applications. We designed simulated interfaces mimicking Gmail and NetEase Mail, which contained 6 phishing emails and 5 safe emails. These email addresses and contents were sourced from actual reported phishing cases. During the exercise, participants chose to reply or report the emails while we recorded their gaze data using the 7invensun A3 eye tracking device.

Then, we captured their gaze behavior while they engaged with the anti-phishing training game, *What.Hack*. This game comprises 5 levels, each emphasizing different email attributes to identify phishing attempts. In Level 1, volunteers inspected the sender's email address. By Level 3, they were also evaluating potential phishing links, and by Level 5, they assessed attachments alongside previous checks. We observed that all participants had completed the game.

We recorded gaze positions, mouse movements, and link-hovering events throughout these two activities.

4.3 Data Post-processing

For the dataset obtained from participants playing *What.Hack*, we preserved the time series data for the Transformer model and computed the accumulated statistical data for the MLP model. We collected a total of 18,720 eye-tracking fixation events. We processed them into 1,019 events of user reaction to emails. We also computed the overall fixation duration and the number of event occurrences. All data has been anonymized. We will release the database[1] after implementing differential privacy measures to enhance user data protection.

5   Findings and Discussions

In our comprehensive analysis of phishing email susceptibility prediction models, the *TransMLP Hybrid* model distinctly stood out for its accuracy and adaptability to various phishing email challenges. Furthermore, our findings highlighted that eye movement patterns offer valuable insights into factors that influence prediction accuracy.

---

[1] https://github.com/zikaiwen/EmailEye-PhishPredict.

In the subsequent discussions, we introduced our novel technique for organizing eye-gazing data in email interfaces, ensuring consistent data collection across diverse devices and email applications. Additionally, we discussed the potential of merging malicious link detection with behavior-driven alerts, providing a robust defense against phishing attacks.

5.1   Findings

Our three primary findings are the results from the model accuracy comparison, the relationship between phishing email complexities and model error rates, and the correlation between saccade counts and model error rates.

Model Accuracy Comparison: We evaluated the performance of three models: Transformer, *TransMLP Link*, and *TransMLP Hybrid*. The Transformer achieved an accuracy of 80.63% in predicting the phishing email's susceptibility. This accuracy saw a slight increase to 80.75% when augmented with MLP using *TransMLP Link*. However, the *TransMLP Hybrid*, which was trained on both game statistical and time series data, outperformed the others, achieving an accuracy of 88.75%. This suggests the *TransMLP Hybrid* is the most effective model when considering both statistical and sequential data. Detailed outcomes are provided in Table 1.

Table 1. Accuracy Rates of Transformer, *TransMLP Link*, and *TransMLP Hybrid* Models in Real-World and In-game Scenarios

Model Name	Real-World Accuracy(%)	In-game Accuracy	
		Testing(%)	Training(%)
Transformer	80.63	80.88	82.58
TransMLP Link	80.75	79.90	82.21
TransMLP Hybrid	**88.75**	**89.82**	**90.16**

Phishing Email Difficulty and Prediction Error Rates: We delved into the performance of the models as they predicted user intent during the game *What.Hack*, which is designed with escalating complexities across its 5 levels to simulate varying phishing email attributes. Starting at Level 1, participants primarily focused on scrutinizing the sender's email address. By Level 3, their evaluation expanded to include potential phishing links. By the time they reached Level 5, they were also assessing email attachments in addition to their previous tasks. Of all the models, *TransMLP Hybrid* stood out by consistently registering the lowest error rate across every level of difficulty. Conversely, the other two models struggled more with discerning user intent in the face of complex phishing emails, as depicted in Fig. 4(a).

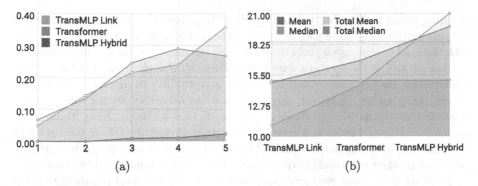

Fig. 4. Model Performance Analysis. (a) represents the prediction error rate of different models at different difficulty levels, and the vertical ordinate represents the prediction error rate. (b) represents the statistical data of different saccades when different models predict incorrectly, and the vertical ordinate represents the number of saccades.

Saccade Counts and Prediction Errors Rates: The average number of saccades (rapid eye movements) observed when models made inaccurate predictions was 18.37, with a median of 15.00. Notably, *TransMLP Link* exhibited more errors when there were fewer saccades. Conversely, as the number of saccades increased, the accuracy of *TransMLP Hybrid* predictions appeared to decline. These trends are illustrated in Fig. 4(b).

5.2   Discussions

Modularizing Eye-Gazing Points in Email UI for Enhanced Feasibility and Effectiveness: To ensure broader applicability and improved feature learning for classifying phishing email susceptibility, we developed a modularization technique for eye-gazing location data. This method divides email application interfaces into eight specific modules: *SenderInformation, SubjectText, Reply, ReportPhishing, ContentText, ContentsPanel, Attachment,* and *LinkHovering*. This structure enables the formation of a consistent dataset that is not tied to absolute coordinate positions. It thus overcomes the challenges posed by differing screen resolutions and email applications, ensuring the collected data from mouse and eye-gazing events remains relevant and usable.

Integrating Malicious Link Detection and Behavior Intervention for Comprehensive Anti-Phishing: Building on our research, there is potential to merge malicious link detection and behavioral intervention alerts. This holistic approach, fusing user intent recognition, machine learning classification, and effective UI warnings, can substantially lower the risk of phishing incidents.

6   Conclusion and Future Work

In our explorative research into machine learning's capabilities, we honed in on the Transformer model and its variants, particularly in the context of predicting phishing email susceptibility using eye-tracking data. Our ambition was to discern and understand the nuances of user attentiveness during email interactions, with the goal of leveraging this information to optimize phishing risk evaluations.

Among the models we evaluated, the *TransMLP Hybrid* emerged as a clear frontrunner. Its precision, coupled with its adaptability to diverse phishing scenarios, set it apart. Moreover, our study underscored the pivotal role that eye movement patterns play in determining prediction accuracy. Even though the *TransMLP Hybrid* model was exemplary in its performance. There lies an exciting challenge in enhancing this model further by augmenting its model architecture that marries eye-tracking data with other related behavioral indicators.

Looking ahead, our research has paved the way for several promising trajectories. The innovative technique we introduced for standardizing eye-gazing data in email interfaces marks a substantial advancement in ensuring consistent and reliable data collection across varying platforms. Furthermore, our discussions around merging malicious link detection with behaviorally-driven alerts have underscored a pressing need and significant opportunity for creating comprehensive defense mechanisms against phishing attacks. This multi-faceted approach, blending technology with human behavioral insights, could form the cornerstone of next-generation anti-phishing solutions.

Acknowledgments. The authors gratefully acknowledge support from the China Postdoctoral Science Foundation under grant number 2022M720889. The authors would like to thank the anonymous reviewers for their valuable comments and helpful suggestions.

References

1. Alkhalil, Z., Hewage, C., Nawaf, L., Khan, I.: Phishing attacks: a recent comprehensive study and a new anatomy. Front. Comput. Sci. **3**, 563060 (2021)
2. Aonzo, S., Merlo, A., Tavella, G., Fratantonio, Y.: Phishing attacks on modern android. In Proceedings of the 2018 ACM SIGSAC Conference on Computer and Communications Security, pp. 1788–1801, 2018
3. Arachchilage, N.A.G., Love, S.: A game design framework for avoiding phishing attacks. Comput. Hum. Behav. **29**(3), 706–714 (2013)
4. Arachchilage, N.A.G., Love, S.: Security awareness of computer users: a phishing threat avoidance perspective. Comput. Hum. Behav. **38**, 304–312 (2014)
5. Bednarik, R., Eivazi, S., Vrzakova, H.: A computational approach for prediction of problem-solving behavior using support vector machines and eye-tracking data. In: Nakano, Y.I., Conati, C., Bader, T. (eds.) Eye Gaze in Intelligent User Interfaces: Gaze-based Analyses, Models and Applications, pp. 111–134. Springer London, London (2013). https://doi.org/10.1007/978-1-4471-4784-8_7
6. Bhattacharya, N., Rakshit, S., Gwizdka, J., Kogut, P.: Relevance prediction from eye-movements using semi-interpretable convolutional neural networks. In: Proceedings of the 2020 Conference on Human Information Interaction and Retrieval, pp. 223–233, 2020
7. Caputo, D.D., Pfleeger, S.L., Freeman, J D., Johnson, M.E.: Going spear phishing: Exploring embedded training and awareness. IEEE Secur. Priv. **12**(1), 28–38, 2014

8. Chanti, S., Chithralekha, T.: Classification of anti-phishing solutions. SN Comput. Sci. **1**(1), 11 (2020)
9. Cui, Q., Jourdan, G-V., Bochmann, G V., Couturier, R., Onut, I-V.: Tracking phishing attacks over time. In: Proceedings of the 26th International Conference on World Wide Web, pp. 667–676, 2017
10. Das, S., Christena, N-E., Camp, L.J.: Evaluating user susceptibility to phishing attacks. Inf. Comput. Secur. **30**(1), 1–18, 2022
11. John, B.D., Peacock, C., Zhang, T., Murdison, T.S., Benko, H., Jonker, T.R.: Towards gaze-based prediction of the intent to interact in virtual reality. In: ACM Symposium on Eye Tracking Research and Applications, pp. 1–7, 2021
12. Dhamija, R., Tygar, J.D., Hearst, M. :Why phishing works. In: Proceedings of the SIGCHI Conference on Human Factors in Computing Systems, pp. 581–590, 2006
13. Jr, R.C.D., Carver, C., Ferguson, A.J.:Phishing for user security awareness. Comput. Secur. **26**(1):73–80, 2007
14. Han, X., Kheir, N., Balzarotti, D. Phisheye: live monitoring of sandboxed phishing kits. In: Proceedings of the 2016 ACM SIGSAC Conference on Computer and Communications Security, pp. 1402–1413, 2016
15. Huang, C.-M., Andrist, S., Sauppé, A., Mutlu, B.: Using gaze patterns to predict task intent in collaboration. Front. Psychol. **6**, 1049 (2015)
16. Huang, L., Jia, S., Balcetis, E., Zhu, Q.: Advert: an adaptive and data-driven attention enhancement mechanism for phishing prevention. IEEE Trans. Inf. Forensics Secur. **17**, 2585–2597 (2022)
17. Ishii, R., Ooko, R., Nakano, Y.I., Nishida, T. Effectiveness of gaze-based engagement estimation in conversational agents. In: Eye Gaze in Intelligent User Interfaces: Gaze-Based Analyses, Models and Applications, pp. 85–110, 2013
18. Just, M.A., Carpenter, P.A.: A theory of reading: from eye fixations to comprehension. Psychol. Rev. **87**(4):329, 1980
19. Kay, R., phish, F.: Fake mandatory Covid-19 vaccine form, 2023. https://www.inky.com/en/blog/fake-mandatory-Covid-19-vaccine-form
20. Koggalahewa, D., Yue, X., Foo, E.: An unsupervised method for social network spammer detection based on user information interests. J. Big Data **9**(1), 1–35 (2022)
21. Miyamoto, Daisuke, Hazeyama, Hiroaki, Kadobayashi, Youki: An Evaluation of Machine Learning-Based Methods for Detection of Phishing Sites. In: Köppen, Mario, Kasabov, Nikola, Coghill, George (eds.) ICONIP 2008. LNCS, vol. 5506, pp. 539–546. Springer, Heidelberg (2009). https://doi.org/10.1007/978-3-642-02490-0_66
22. Murtagh, F.: Multilayer perceptrons for classification and regression. Neurocomputing **2**(5–6), 183–197 (1991)
23. Sharma, P., Dash, B., Ansari, M F.: Anti-phishing techniques-a review of cyber defense mechanisms. Int. J. Adv. Res. Comput. Commun. Eng. ISO, 3297:2007, 2022
24. Sheng, S., et al.: Anti-Phishing Phil: the design and evaluation of a game that teaches people not to fall for phish. In: Proceedings of the 3rd Symposium on Usable Privacy and Security, pp 88–99, 2007
25. Shie, E.W.S.: Critical analysis of current research aimed at improving detection of phishing attacks. Sel. Comput. Res. pap. **45**, 2020
26. Stein, N., Bremer, G., Lappe, M.: Eye tracking-based LSTM for locomotion prediction in VR. In: 2022 IEEE Conference on Virtual Reality and 3D User Interfaces (VR), pp. 493–503. IEEE, 2022

27. Subasi, A., Molah, E., Almkallawi, F., Chaudhery, T.J.: Intelligent phishing website detection using random forest classifier. In: 2017 International Conference on Electrical and Computing Technologies and Applications (ICECTA), pp. 1–5. IEEE, 2017
28. Vaswani, A., et al.: Attention is all you need. Advances in neural information processing systems, **30**, 2017
29. Wei, P., Liu, Y., Shu, T., Zheng, N., Zhu, S-C.: Where and why are they looking? jointly inferring human attention and intentions in complex tasks. In: Proceedings of the IEEE Conference on Computer Vision and Pattern Recognition, pp. 6801–6809, 2018
30. Wen, Z.A., Lin, Z., Chen, R., Andersen, E.: What. hack: engaging anti-phishing training through a role-playing phishisng simulation game. In: Proceedings of the 2019 CHI Conference on Human Factors in Computing Systems, pp. 1–12, 2019

Author Index

© The Editor(s) (if applicable) and The Author(s), under exclusive license
to Springer Nature Singapore Pte Ltd. 2024
J. Vaidya et al. (Eds.): AIS&P 2023, LNCS 14509, pp. 593–595, 2024.
https://doi.org/10.1007/978-981-99-9785-5

Printed in the United States
by Baker & Taylor Publisher Services

Printed in the United States
by Baker & Taylor Publisher Services